AF540287

BIOLOGICAL CHEMISTRY

BIOLOGICAL CHEMISTRY

By

Dr. Lata Bhattacharya

Professor

School of Studies in Zoology & Biotechnology

Vikram University

Ujjain (M.P.)

(India)

DISCOVERY PUBLISHING HOUSE PVT. LTD.

NEW DELHI-110 002

First Published – 2009
Reprinted – 2025

ISBN: 978-81-8356-486-1

Biological Chemistry

Published by:
DISCOVERY PUBLISHING HOUSE
4383/4B, Ansari Road, Darya Ganj
New Delhi-110 002 (India)
Phone: +91-11-23279245; 23253475; 43596065
Mobile: +91 9811179893 / +91 9871656464
E-mail: discoverybooksindia@gmail.com
orderdphbooks@gmail.com
namitwasan9@gmail.com
web: www.discoverypublishinggroup.com

Printed at:
Infinity Imaging Systems
Delhi

Preface

Biochemistry today has made spectacular progress in unraveling the mysteries of animate nature. This progress has allowed us to gain deeper insight into the principles of vital activity and has to a very significant extent stimulated the development of applied disciplines, especially medicine. The present title *"Biological Chemistry"* is intended for those who wish to understand living organisms, especially man. Biochemistry is essential for this purpose, but it would be almost impossible for a student to survey on his own the massive body of existing knowledge, constantly augmented by a remarkable torrent of brilliat discoveries. The purpose of the book, then is to organize our knowledge into something that can be comprehended in a relatively short time and still convey a reasonable complete picture of the chemical structure and function of man. Readability without sacrifice of coverage has been a prime goal. An important device in gaining that goal is to keep attention constantly focused on function, with repeated use of rationalization to show that the chemical facts are not isolated, but part of a whole.

Biochemistry has two major goals as a fundamental science, it treats the vital functions from the standpoint of physical chemistry, and as an applied discipline, it points out practical applications for the wealth of scientific knowledge it has acquired. The dual purpose has been, as far as possible, take into account in the writing of this book. We have also tried to summarize our pedagogical experience in teaching biochemistry to students specializing in medicine and pharmacy. The material of this book has been organized according to the principle of functionally in order to trace the close relationship between the functions of the living organism and the structures of its constituents molecules as well as the chemical and physico-chemical process in which they are involved.

The aim of this book is to present a core of biochemical knowledge that is desirable for undergraduate and postgraduate students and also those involved in the field of medical, microbiology, biotechnology and pharmaceutical. Every attempt has been made to keep abrest of the advances in the subject and at the same time to include the fundamentals.

To make the work more comprehensive and informative, the author has consulted many authoritative books, research journals, abstracts, monographs etc. He is grateful to all those great scholars whose work are cited or substantially reproduced.

There can be no claim to originality except in the manner of treatment and much of the information has been obtained from the books and scientific journals available in the different libraries.

The author expresses his thanks to his friends and colleagues whose continue inspirations have initiated him to bring out this book.

The author expresses his gratitude to Mr. Wasan and Staff of M/s Discovery Publishing House Pvt. Ltd., for their whole hearted co-operation in the publication of this book.

In the mean time, the author will remain sincerely responsible for any shortcomings of the book and be grateful to the readers for their suggestions and constructive criticism for the continuous betterment of the book. He takes this opportunity to appeal to the readers to send their suggestions straightway to his publisher.

Author

CONTENTS

1. **Analyzing Macromolecules** 1

Describing Binding Equilibria—Analyzing Data—Multiple Binding Sites on A Single Molecule—Cooperative Processes—Complementarily and the Packing of Macromolecules—Rings and Helices—Oligomers With Twofold (Dyad) Axes—Cooperative Changes in Conformation—Unequal Binding of Substrate and "Induced Fit"—Binding Equilibria for a Dimerizing Protein—Higher Oligomers—D. The Oxygen-Carrying Proteins 353D. The Oxygen-Carrying Proteins—1. Myoglobin and Hemoglobin—Abnormal Human Hemoglobins—3. Comparative Biochemistry of Hemoglobin—E. Self-Assembly of Macrontolecular Structures—1. Bacteriophages 34—2. "Kringles"and Other Recognition Domains and Motifs—F. The Cytoskeleton 38—1. Intermediate Filaments—2. Microfilaments—3. Microtubules—Life and Death for Protens: Chaperonins and Protesomes—Sickle Cell Disease, Malaria and Blood Substitutes.

2. **Intragastric Intubation** 58

Materials—Alcohol Administration During the Prenatal Period—Alcohol Administration during the Postnatal Period—Measurement of Blood Alcohol Concentrations (BACs)—Methods—Alcohol Administration During the Prenatal Period—Alcohol Administration During the Postnatal Period—BACs—Notes.

3. **Gene Expression in A Complex Tissue** 65

Primate Endometrium—Spatiotemporal Regulation—Animal Model—Laser Capture Microdissection—Lcm Analysis of Endometrial Tissue—Procedure—Synthesis and Amplification of Cdna Populations—Cdna Amplification Strategy—Differential Display—Cloning and Sequencing—Window of Receptivity—Serum P Level—Gene Microarray Analysis—Inadequate Secretory Phases—Lcm and Differential—Lcm and Slpi, Wfdc2, Bat2, and Syncytin—Microarray Analysis of Endometrial Cells—Conclusion.

4. **Analyzing Gene Expression** 80

Materials—Dna Microarrays—Reagents—Methods—Experimental Design—tocolProtocol—Microarray Quality Control—Affymetrix-specific Quality Control Metrics—Generalized Quality Control Measures— Primary Analysis—Statistical Filtering—Muitivariant Analysis—Bioinformatics and Network Analysis—Validation of Candidate Genes—Note.

5. **Sensory Systems** 91

Organic Compounds J&J& Detected By Olfaction—Seven-transmembrane-helix Receptors—Combinatorial Mechanism—Magnetic Resonance Imaging—Combination of Senses—Bitter Receptors—Sweet Compounds—Salty Tastes—Ambodde—Effects of Hydrogen Ions—Glutamate Receptor—Photoreceptor Molecules—Rhodopsin—Light Absorption—Light-induced the Calcium Level—Colour Vision—Green And Red Pigments—Detection of Mechanical Stimuli—Stereocilia—Mechanosensory Channels—Touch Sensing—Calo Receptors Other Painful Stimuli—Subtle Sensory Systems —Summary—Signal–Transuction Pathways—Olfaction—Taste—Photoreceptor—Mechanical Stimuli—Touch.

6. **Excretory System** 116

A. Fixation of N_2 and the Nitrogen Cycle—2. Nitrogenases—3. Intercpnversion of Nitrate, Nitrite, and Ammonium Ions —B. Incorporation of NH, Into Amino Acids and Proteins—1. Uptake of Amino Acids by Cells—2. Glutamate Dehydrogenase and Glutamate Synthase—3. Glutamine Synthetase—4. Catabolism of Glutamine, Glutamate, and Other Amino Acids—C. Synthesis and Catabolism of Proline, Ornithine, Arginine, and Polyamines—1. Synthesis And Catabolism of Proline—2. Synthesis of Arginine and Ornithine and the Urea Cycle—3. Amidino Transfer and Creatine Synthesis—4. The Polyamines—D. Compounds Derived from Aspartate—1. Control of Biosynthetic Reactions of Aspartate—2. Lysine Diaminopimelate, Dipicolinic Acid, and Car Ni Tine—3. The Catabolism Fo Lysine—4. Metabolism of Homocysteine And Methionine—Metabolism of Threonim—E. Alanine And The Branched-chain Amino Acids—1. Catabolism—2. Ketogenic and Glucogenic Amino Acids—F. Serine and Glycine—1. Biosynthetic Pathways From Serine—2. Metabolism of Glycine—3. Porphobilinogen, Porphyrins, and Related Substances—G. Cysteine and Sulfur Metabolism—1. Synthesis and Catabolism of Cysteine—2. Cysteine Sulfinate and Taurine—3. Mercaptopyruvate, Thiosulfate, and Assembly of Iron - Sulfur Centers—Mhc Proteins—Foreign Peptides on Antigen-presenting Cells—Diverse Mhc Proteinslife and Death for Protens: Chaperonins and Protesomes—Human Immunodeficiency Viruses—Responses Against Self-antigens A—Positive and Negative Selection in the Thymus—utoAutoimmune Diseases— The Immune Cancer.

7. **Immunological Responses** 183

Materials—Isolation of T Cells from Spleen, Thymus, or Lymph Node—Column Fractionation of Spleen Cell Suspension—Flow Cytometry Staining—Stimulation And Cytokine Response of Cells—Antigen-specific T Cells, Using Attenuated, L. Monocytogenes as A Model—Adoptive Transfer—Methods—Isolation of T Cells From Spleen, Thymus, or Lymph Node—Column Fractionation of Spleen Cell Suspension—Column Purification of Cd^{4+} T Cells by Negative Selection—Column Purification of Cd^{8+} T Cells by Negative Selection—Column Purification of B Cells With Miltenyi B-cell Kit; Overnight Activation—B-cell Depletion of Splenocytes—Flow Cytometry Staining—Stimulation and Cytokine Response of

Cells After Polyclonal Stimulation—(Antigen-specific T Cells, Using Attenuated, L. Monocytogenes as a Model Organism—Inoculation—Cell Preparation—Culture—Antigen Presenting Cells (Apcs))—Intracellular Staining for Cytokines After Brefeldin a Treatment—Time Course of Response—Adoptive Transfer—Adoptive Transfer Utilizing Congenic Strains of Mice—Fluorescent-labeled Mhc Class I Pentamers snd Mh(Class II Tetramers—Notes.

8. The Defensive System **197**

Principles of Evolution—Antigen-binding Units—Consists of A Rets-sandwich Frarriework with Hypervariable Loops—Specific Molecules—X-ray Analyses—Antigens and Antibodies with Numerous Interactions—Gene Rearrangements—Antibody Diversity—Combinatorial Association and Somatic Mutation—Oligomerization of Antibodies—Different Classes of Antibodies—Major-histocompatibility-complex Proteins—Peptides Presented by Mhc Proteins—Antibody-like Proteins—Cytotoxic T Cells.

9. Biochemistry of Nucleic Acid **221**

A, The Topology and Environment of DNA—1. DNA in Viruses —2. Bacterial Chromosomes and Plasmids—3. Protamines, Histones, and Nucleosomes—The Cell Nucleus—B. Organization of DNA—Repetitive DNA—2. Genes For Ribosomal Rna and Small Rna Molecules—3. Other Gene Clusters and Fseudogenes—4. Introns, Exons, and Overlapping Genes—5. DNA of Organelles—6. Methylation of DNA—C. Replication—1. Early Studies—2. DNA Polymerases—3. Other Replication Proteins—4. Replication of Bacterial DNA—5. The Replication of Viral DNA—6. Packaging or Viral Cenomes—7. Plasmids--8. Chromosome Ends—9. Mitochondrial and Chloroplast DNA—10. Replication of Bukaryotic Nuclear and; Viral DNA—D. Integration, Excision, and Recombination of DNA—1. Recombination Mechanisms—2. Nonreciprocal Recombination and Unequal Crossing-over—3. Site-specific Recombination and the Integration and Excision—4. Transposons and Insertion Sequences—5. Other Causes of Genetic Recombination—E. Damage and Repair of DNA—1. Causes of Mutations—2. Fidelity of Replication—3. Repair of Damaged DNA—F. Mutagens in the Environment.

10. Artificial Rearing **304**

Materials—Milk Diet—Stock Milk Diet—Treatment Diet—Gastrostomy Surgery—Artifical Rearing—Artificial Rearing Components—Methods—DMilk Diet— Stock Milk Diet—Treatment Diet—Gastrostomy Surgery —Gastrostomy Surgery Tools—Guide Wire—Intragastric Cannual——Washers—Gastrostomy Surgery—ReArtificial Rearing—Preparation—Determination of Diet Infusion Rate— DietDelivery of Milk Diet— Care of Artificially Reared Pups— Fostering of Pups to Lactating Dam—Notes.

11. Ethanol in Culture **313**

Two Culture Models—Dissociated Cell Cultures—Organotypic Slice Cultures—Materials—Animals—Primary Cultures of Cortical Neurons—Surgical

Equipment—Reagents—Culturing Methods—Preparation of the Culture Dishes—Dissection of the Rat Cortices—Yield and Plating of the Cells—In Alternative Method for Obtaining Primary Cultures of Cortical Neurons—Reagents—Methods—Organotypic Slice Cultures—pment—Reagents—Methods—Dissection—Preparation and Culture of Mouse Cortical Slices—Preparation and Culturing of Rat Cortical Slices—Controlled Ethanol Exposure—Equipment—Reagents—Methods—Application of Exogenous Substances and Harvesting -the Cells/Tissue—Fres Samples—Fixed Samples—Notes.

12. Apoptosis Analysis **322**

Materials—Cell Line Maintenance and Storage—CYP2E1 Oxidation Activity—Cell Toxicity—Measurement of Malondialdehyde and Ros—Determination of Gsh—Apoptosis Analysis—Caspase Activity—Mitogen- Activated Protein Kinase (Mapk) Phosphorylation—Methods—Cell Culture and Storage—CYP2E1 Activity—Mtt Assay to Determine Cytotoxicity—Assay to Determine Cytotoxicity—Trypan Blue Exclusion Assay to Determine Cytotoxicity—Lipid Peroxidation Analysis—Measurement Of Ros Productioi—Measurement of Gsh Levels—Mitochondrial Membrane Potential Analysis—Apoptosis Determination By Flow Cytomet—Apoptosis Analysis By Determination of DNA Fragmentation (Dna Ladder)—Immunohistochemistry Analysis to Determine Nuclear DNA Fragmentation—ActCaspase Activity—of Activation of p38 Mapk—Notes.

Index **333**

ANALYZING MACROMOLECULES

The complicated shapes and internal structures of cells are determined to a large extent by the way in which proteins and other macromolecules are bonded one to another. In addition, intimate association of macromolecules is essential to such biological processes as the motion of flagella, the contraction of muscle, the action of antibodies, the transmission of nerve impulses, the replication of DNA, and the synthesis of proteins. Equally important is the binding of small molecules to large ones. In this chapter we will first examine methods of measuring binding with an emphasis on protons and small molecules. Then we will consider the ways in which macromolecules stick together as well *as* the role of conformational changes within macromolecules.

DESCRIBING BINDING EQUILIBRIA

In discussions of pH we have dealt with dissociation constants, but in this section we will use formation constants K_f, where $K_f = 1/K_d$. Measurement of the strength of association of molecules is an everyday aspect of modern biochemical research. It may be important to know how strongly a hormone binds to a receptor in a cell membrane or how well a feedback inhibitor binds to an enzyme to determine whether the interaction is significant physiologically. The binding of O_2 to hemoglobin and other oxygen carriers is vitally important, but the description of these oxygenation reactions is mathematically complex. This is especially so because we must consider effects of pH changes and of changing concentrations of allosteric effectors on the binding equilibria. In considering such equilibria we must first examine the individual interactions of different domains of a protein, one with another. These can be described by association constants or, alternatively, by the Gibbs energy changes for the association reaction.

The average kinetic energy of motion of a molecule in solution is about $3 / 2k_BT$, where k_BT is Boltzmann's constant. For one mole the kinetic energy is $3 / 2RT$ or 3.7 kJ (0.89 kcal) mol^{-1} at 25°C Thus, if $K_f = 10$ M^{-1} ($\Delta G° = -5.7$ kJ mol^{-1} or -1.36 kcal mol^{-1}) the binding energy is only slightly in excess of the thermal energy of the molecules and the complex is weakly bound. In this instance, if X and P are both present in 10^{-4} molar concentrations (typical enough for biochemical systems), only 0.1 % of the molecules will exist as the complex ([complex] = K_f [X][P]). If the formation constant is higher by a factor of 1000, *i.e.*, $K_f = 10^4$ M^{-1} ($\Delta G° = -22.8$ kJ mol^{-1}), 38% of the molecules will exist as the complex; while if $K_f = 10^7$ M^{-1} (extremely strong binding, $\Delta G° = -40$ kJ or -9.55 kcal mol^{-1}), 97% of the molecules will be complexed.

Analyzing Data

The extent of binding of a molecule X to another molecule P Eqs. 1.1, 1.2 is measured by varying the concentrations of X and P and observing changes in the concentration of the complex [PX]. The first

$$X + P \rightleftarrows PX \tag{1.1}$$

$$K_f = [PX]/[P][X]) \tag{1.2}$$

prerequisite is to find a measurable property that is different for the complex than for either of the free components. For example, the complex may be colored and the components colorless. More commonly, the complex simply has a different light absorbance (A) at a certain wavelength than do the components. Likewise, the circular dichroism or the chemical shift of a peak in the NMR spectrum may change. If P is an enzyme, only the complex PX will undergo decomposition to products. Sometimes (but not always) the rate of breakdown of PX (the enzyme-substrate complex) to form products will be relatively slow compared to the rate at which the equilibrium between X, P, and PX is established.

In this case the concentration of complex PX will be proportional to the observed rate of formation of product. Whatever change of property is measured, its value will increase with increasing concentrations of X if the total concentration of the macromolecule P is kept constant. In the usual experimental design, the molar concentration of P is small and it is possible to increase the concentration of X to quite large values.

When this is done, it is usually observed that at high enough values of [X] almost all of the P is converted to PX, and the change being measured (*e.g.*, ΔA for increased light absorption) no longer increases. This effect is known as *saturation* and is observed in most binding studies and also in many physiological phenomena. The property being measured (ΔA) reaches a maximum value ΔA_{max} at saturation and when all of compound P has been converted to PX.

The ratio of [PX] to the total concentration of all forms of P present $[P]_t$ is known as the *saturation fraction* and is often given the symbol Y. If P has more than one binding site for X, Y is defined as the fraction of the total binding sites occupied. If n is the number of sites per molecule, the total number of sites is $n[P]$. The value of Y is often taken as $\Delta A/\Delta A_{max}$, an equality that holds for multisite macromolecules only if the change in A is the same for each successive molecule of X added. This is not always true, but when it is Eq. 1.3 is followed.

$$\sum_i i\frac{[PX_i]}{n[P]_t} = Y = \frac{\Delta A}{\Delta A_{max}} \qquad \text{...(1.3)}$$

Here i represents the number of ligands X bound to P and may vary from 0 to n. When $n = 1$ the saturation fraction Y and ΔA are related to the concentration of free unbound X and the formation constant as follows:

$$Y = \frac{K_f[X]}{1+K_f[X]} \qquad \Delta A = \frac{\Delta A_{max} K_f[X]}{1+K_f[X]} \qquad \text{...(1.4)}$$

A plot of Y or ΔA against [X] is shown in Fig. 1.1. This kind of plot is sometimes called an *absorption isotherm* because it describes binding only at a constant temperature. Notice, from both Fig. 1.1 and Eq. 1.4, that Y reaches a value of 0.5 when [X] is just equal to $1/K_f$ (or to K_d). Note also that as [X] increases saturation is reached slowly and that even at the point representing the highest concentration of X ($8\ /K_f$ in Fig. 1.1) saturation is less than 90%.

Since in the usual experimental situation, we do not know Y but only ΔA, it is difficult to estimate the limiting value ΔA_{max} from a plot of this type unless K_f is very high. However, we need to know ΔA_{max} to evaluate K_f. For this reason, plots like that of Fig. 1.1 are seldom used, this one being included mainly to illustrate a point of nomenclature. The curve shown in Fig. is a rectangular hyperbola, and the type of saturation curve shown is frequently referred to as *hyperbolic*. This is in contrast to certain other binding curves which, when plotted in this way, are *sigmoidal*, (S–shaped). A better type of plot is often that of Y against log [X].

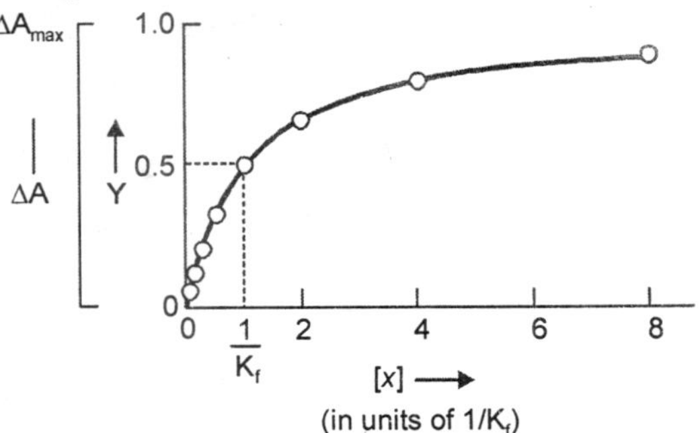

Fig. 1.1. An adsorption isotherm, a plot of the saturation fractionY or of some change in a measured property ΔA vs [X], the concentration of *a* substance that binds reversibly to a macromolecule. The curve is hyperbolic and [X] = $1/K_f$ when Y = 0.5.

It has the following features. (1) The "curve is symmetric about the midpoint at log [X] = log K_f. (2) No matter how high or low the concentration range used in the experiments, it is easy to choose a scale that puts all the points on the same sheet of paper. (3) Spacing between points tends to be more uniform than in a plot against [X]; *e.g.*, compare Figs. 1.1, and 1.2 for which the experimental points represent the same data and for which values of [X] for successive points are each twofold greater than the preceding one. (4) The same logarithmic scale can be used for all compounds, no matter how strong or weak the binding, and the same shape curve is obtained for all 1:1 complexes.

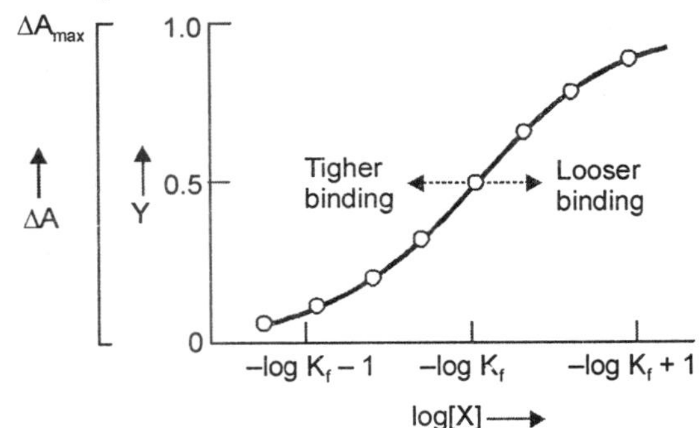

Fig. 1.2. A saturation curve plotted on *a* logarithmic scale for [X]. The data points are the same as those used in Fig. 1.1.

The midpoint slope, dY/$d$log [X], is 0.576; the change in log [X] in going from 10 to 90% saturation is 1.81. The curve is familiar to most chemists because it is frequently used for pH titration curves in which pH substitutes for -log [X]. To represent a complex with tighter binding, the curve is simply moved to the left, and for weaker binding, it is moved to the right.

Saturation data are often plotted in yet another form known as the *Scatchard plot* Fig. 1.3. The value of ΔA/[X] (or of Y/[X]) is plotted against ΔA (or Y) and a straight line is fitted to the points, preferably using the "method of least squares." The intercept on the x axis and the slope of the fitted line give values of $\Delta A_{max}/K_f$ and K_f, respectively, as indicated by Eq. 1.5, which follows directly from Eq. 1.4.

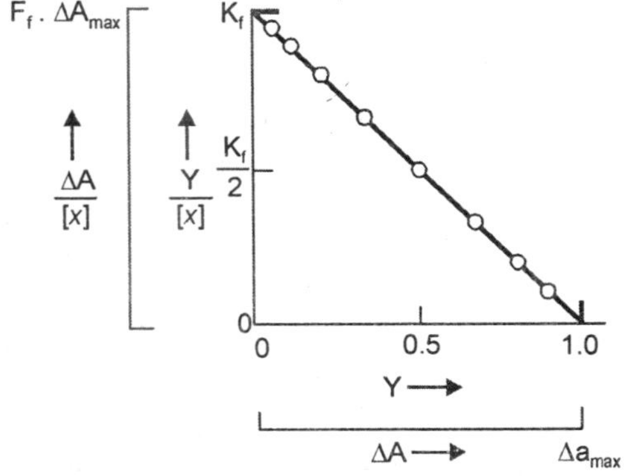

Fig. 1.3. A Scatchard plot of the same data shown in Figs. 1.1 and 1.2. This is the best of the linear plots for studying binding.

$$Y/[X] = K_f - YK_f$$
$$\Delta A/[X] = \Delta A_{max} K_f - \Delta AK_f \quad ...(1.5)$$

The Scatchard plot is the best of the various linear transformations of the saturation equation and is preferred to "double reciprocal plots".

Scatchard's original equation was formulated to deal with the binding of two or more ligands to a single macromolecule. If we let $[X]_b$ represent the concentration of bound X and $[P]_t$ the total molar concentration of the protein or other macromolecule and if there is only one binding site on the protein, $[X]_b/[P]_t$ will equal Y. However, if there are n independent binding sites that have the same binding constant K_f Eq. 1.5a will hold.

$$\frac{[X]_b}{[P]_t} \cdot \frac{1}{[X]} = n K_f - \frac{[X]_b}{[P]_t} K_f \quad ...(1.5a)$$

If $[X]_b/[P]_t[X]$ is plotted against $[X]_b/[P]_t$ the resulting linear plot will have an intercept of K_f on the y axis and n on the X axis. Thus, n is directly apparent, whereas in Eq. 1.5 it is incorporated into Y. A problem arises if, as discussed in the next section, the multiple binding sites are not independent but interact. Curved Scatchard plots result and attempts to extract more than one binding constant can lead to very large errors. Before measuring saturation curves, the student should read additional articles or books on the subject.

Multiple Binding Sites on a Single Molecule

A macromolecule may often be able to bind several molecules of a second compound X. Consider the case in which the macromolecule P binds successively one molecule of X, then a second, and a third, up to a total of n. We define the stepwise formation constants, K_1, K_2,..., *Kn, as* follows:

$$P + X \xrightarrow{K_1} PX$$
$$PX + X \xrightarrow{K_2} PX_2 ...$$
$$PX_{n-1} + X \xrightarrow{K_n} PXn \quad ...(1.6)$$

Remember that these are reversible reactions even though unidirectional arrows are used. The general expression for the ith stepwise formation constant is given by Eq. 1.7.

$$K_i = \frac{[PX_i]}{[PX_i - 1][X]} \quad ...(1.7)$$

Remember that Y is the fraction of total binding sites saturated. The number of moles of X bound per mole of P is nY and is obtained by summing the concentrations $[PX] + 2[PX_2] + ...$ and dividing the sum of all the forms of P:

For two binding sites($n = 2$)

$$2Y = \frac{[PX] + 2[PX_2]}{[P] + [PX] + [PX_2]} \quad ...(1.8)$$

For the general case

$$nY = \left(\sum_{i=1}^{n} i[PX_i]\right) \Big/ \left([P] + \sum_{i=1}^{n} [PX_2]\right) \quad ...(1.9)$$

The summations are over all of the integral values of i from 1 to n. Now, *by* expressing each concentration, $[PX_i]$, in terms of the concentrations [X] and [P] *of free* X and P, together with the stepwise formation constants, we obtain Eq. 1.10.

For $n = 2$

$$2Y = \frac{K_1[X]+2K_1K_2[X]^2}{1+K_1[X]+K_1K_2[X]^2} \qquad ...(1.10)$$

A similar equation can be written for the general case. Note that the concentration of P does not appear in Eq. 1.10 and that Y is a function only of [X] and the stepwise formation constants. Such equations define the isotherms for binding of two or more molecules of X to P. From an experimental plot of Y (or of ΔA) vs [X] or log [X], it is possible in favourable cases to determine the stepwise constants K_1, K_2, ..., K_n. However, this becomes quite complicated. To simplify Eq. 1.10 and the corresponding equation for the general case, we can group the constants together and designate the products of constants (K_1, K_1K_2, $K_1K_2K_3$, etc.) as ψ_1, ψ_2,..., ψ_n. Our equations are now as follows:

For $n = 2$

$$2Y = \frac{\psi_1[X]+2_{\psi_2}[X]^2}{1+\psi_1[X]+\psi_2[X]^2} \qquad ...(1.11)$$

For the general case

$$nY = \sum_{i=1}^{n} i\psi_i[X]^i \Big/ \left(1+\sum_{i=1}^{n}\psi_i[X]^i\right) \qquad ...(1.12)$$

From experimental data, it is usually easiest to first determine the ψ's (there are n of them), and then to calculate from the ψ's the stepwise constants. For example:

$$K_1 = \psi_1 \quad K_2 = \psi_2/K_1, \text{ etc.} \qquad ...(1.13)$$

While Eq. 1.12. Known as the *Adair equation,*[11] might seem to provide a complete description of the binding process, it usually does not. In many cases, there is more than one kind of binding site on a macromolecule and Eq. 1.12 tells us nothing about the distribution of the ligand X among different sites in complex PX. To consider this problem we must examine the *microscopic binding constants.*

Microscopic binding constants and statistical effects. As discussed already microscopic binding constants represent the constants for binding to specific individual sites. Now, consider a straight-chain dicarboxylic acid which has two *identical* binding sites for protons. If the chain connecting the two carboxylate anions is long enough, the carboxylate groups will be far enough apart that they do not influence each other through electrostatic interaction.

^-OOC ～～～～ COO^-

↑ $K_f^* = 5 \times 10^4$ ↑ $K_f^* = 5 \times 10^4$

Each group will have a microscopic binding constant (K_f^*) of 5×10^4 M^{-1}. The constant K_f^* can also be called an *intrinsic binding constant* because it is characteristic of a carboxylate group that is free of interactions with other groups. Intuition tells us (correctly) that, in its binding of protons, a solution of this dicarboxylic acid dianion will behave exactly like a solution of the monovalent anion $R - COO^-$ at twice the concentration. A single intrinsic binding constant suffices to describe both binding sites. It may seem surprising then that the *stepwise formation constants* (also called *stoichiometric* or *macroscopic* formations constants) K_1 and K_2 differ: $K_1 = 10 \times 10^4$ M^{-1} and $K_2 = 2.5 \times 10^4$ M^{-1}. This fact reflects the so-called statistical effect. Either of the two carboxylate groups in the molecule can bind a proton in the first step to give two indistinguishable molecules, PH:

P(COOH)(COO⁻) (A)

K* ↗ ↘ K*

P(COO⁻)(COO⁻) P(COOH)(COOH) ...(1.14)

K* ↘ ↗ K*

P(COO⁻)(COOH) (B)

P PH PH_2

If we label the two forms of PH as A and B (Eq. 1.14) and consider that each one of them is independently in equilibrium with P through formation constant K_f^*, we obtain Eq. 1.15 (which may be compared with Eqs. 6.75 and 6.76 which are written for *dissociation* constants).

$$K_1 = \frac{[\mathrm{PH}]\mathrm{A} + [\mathrm{PH}]_\mathrm{B}}{[\mathrm{P}][\mathrm{H}^+]} = 2K^* \text{ and } K_2 = K^*/2 \qquad \text{...(1.15)}$$

This result is related to probability and arises for the same reason that if you reach into a barrel containing 50% white balls and 50% black balls, you will pull out one of each just twice as often as you will pull out a pair of white or a pair of black. In the general case of n equivalent binding sites, the microscopic formation constants K_i^* are related to the stepwise constants K_i as follows:

$$K_i = \frac{(n+1-i)}{i} K_i^* \qquad \text{...(1.16)}$$

It is also easy to show,[14] using Eqs. 1.12 and 1.16 that for *n completely equivalent and independent* binding sites Eqs 1.17 and 1.18 hold:

$$Y = \frac{K^*[\mathrm{X}](1 + K^*[\mathrm{X}])^{n-1}}{(1 + K^*[\mathrm{X}])^n} \qquad \text{...(1.17)}$$

or

$$Y = \frac{K^*[\mathrm{X}]}{1 + K^*[\mathrm{X}]} \qquad \text{...(1.18)}$$

In this case the microscopic association constants are all identical and represent a single *intrinsic* constant applicable to all of the sites. In fact, Eq. 1.18 is identical to that for association of a single proton (or other ligand) with a single binding site, satisfying our intuitive notion that a set of n completely independent binding sites should behave just like a solution of an n-fold more concentrated compound with a single binding site.

Thus, our arithmetic has led us to a conclusion that was already obvious. However, it is rarely true that binding sites on a single macromolecule are completely independent; there is almost always *interaction* between them, and the equations that we have derived for evaluation of stepwise and intrinsic constants cannot be applied without modification.

Electrostatic repulsion: anticooperativity. As we have seen, a hypothetical acid with an infinite distance between the carboxylate groups and log $K_f^* = 4.8$ would have two macroscopic binding constants separated by the statistical distance (log 4 = 0.6). Compare these values with the observed binding constants for protons with the dianions of acids taining 7, 4, 2, and 1 CH_2 groups given in Table 1.1

Table 1.1. Binding Constants of Protons to Dianions of dicarboxylic Acids

Acid dianion	*No. of CH_2 groups*	*log K_1 (pK_2)*	*log K_2 (association) (pK_1) (dissociation)*
Hypothetical dianion with log $K^* = 4.8$	∞	5.1	4.5
Azelaic	7	5.41	4.55
Adipic	4	5.41	4.42
Succinic	2	5.48	4.19
Malonic	1	5.69	2.83

For the longest chain (that of azelaic acid) the log K_f values are not very different from those of the hypothetical long-chain acid. However, as the groups come closer together, the first binding constant is increased markedly because of the additional electrostatic attraction and the second is decreased. The spread between the two log K_f values increases from 0.6 to as much as 2.9 as a result of interaction between the binding sites. In malonic and succinic acids the first proton bound can be shared by both carboxyl groups through formation of a hydrogen bond.

Additional factors operate in oxalic acid where the carboxyl groups are connected directly and for which pK_a values ($pK_a = \log K_f$) are 4.19 and 1.23. In all of these examples the binding of the first proton makes it harder to bind a second proton. Such negative interaction or *anticooperativity* between binding sites is very common and always leads to a spread of the formation constants and a broadening of the curve of Y vs log [X]. This is shown graphically on the right side of Fig. 1.4 where the binding curves for protons with acetate ion and with succinic acid dianion are compared.

Notice that binding of protons *increases* as log $[H^+]$ increases, giving the curves an unfamiliar appearance when compared with the more familiar curve of *dissociation vs* pH. Can we predict the pK_a values in Table 1.1? With an appropriate dielectric constant chosen Eq. 1.8 can be applied. The difference between the two successive log K_f values reduced by 0.6 (the statistical factor) is a measure of the electrostatic effect. For malonic acid this ΔpK_a is 2.25 and for succinic acid it is 0.69. In 1923, N. Bjerrum proposed that the value of ΔpK_a could be equated directly with the work needed to bring the two negative charges together to a distance representing the charge separation in the malonate dianion.

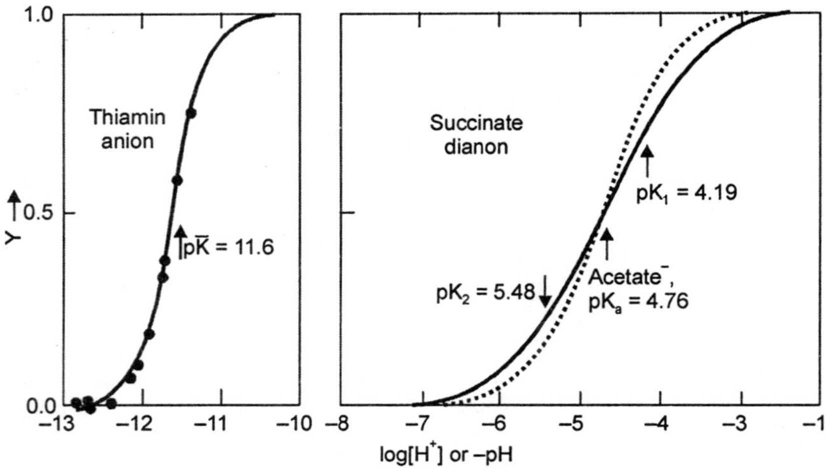

Fig. 1.4. Binding of protons to the thiamin anion, the succinate dianion, and the acetate anion. Acetate (dashed line) binds a single proton with a normal width binding curve. Succinate dianion binds two protons with anticooperativity, hence a broadening of the curve. The thiamin anion binds two protons with complete cooperativity and a steep binding curve.

$$^{-}O-C(=O)-CH_2-C(=O)-O^{-}$$

$r = 0.4$ nm

Thus, applying Eq.1.31, $\Delta G = 5.708\ \Delta pK_a$ kJ mol^{-1} = 12.84 kJ mol^{-1} for malonate. Equating this with W in Eq. 1.8 and assuming a dielectric constant of 78.5 (that of water), the distance of charge separation r is calculated to be 0.138 nm. This is much too small.

The computation was improved by Westheimer and Kirkwood, who assumed a dielectric constant of 2.0 *within* the molecule. By approximating the molecule as an ellipsoid of revolution, they were able to make reasonably accurate calculations of electrostatic effects on pK_a values. Thus, for malonic acid Westheimer and Shookhoff predicted r = 0.41 nm for malonic acid dianion. Recently more sophisticated calculations have been used to predict pK_a values for the compounds in Table 1.1 and others. Electrostatic theory has also been used successfully to interpret titration curves of proteins in which the net negative or positive charge distributed over the surface of the protein varies continuously from high pH to low as more protons are added. Electrostatic effects can be transmitted extremely effectively through aromatic ring systems, a fact that explains some of the significance of heterocyclic aromatic systems in biochemical molecules.

Consider the microscopic binding constant of the phenolate anion of pyridoxine as influenced by the state of protonation of the ring nitrogen. These are shown in Eq. 6.75, where pK_a^* =

4.94 and $pK_d^* = 8.20$ define the binding constants for protonation of a phenolate ion when the ring nitrogen is protonated or unprotonated, respectively. We see that $\Delta\ pK_a = 3.26$, even greater than that of malonic acid.

Cooperative Processes

Can it ever happen' that interaction between groups leads to a *decrease* from the statistical separation between values of the stepwise constants instead of to an increase? At first glance, the answer seems to be no. A decreased separation would imply that the intrinsic binding constant for the second proton bound is higher than that for the first, but common sense tells us that the first proton ought to bind at the site with the highest binding constant. However, look at the experimental binding curve of protons with the anion of thiamin shown in Fig. 1.4 Instead of being broadened from the curve of acetate, it is just half as wide.

The explanation depends upon some rather amusing chemistry of thiamine. Under suitable conditions, this vitamin can be crystallized as a yellow sodium salt, the structure of whose anion is shown in Eq. 1.19. Weak binding of a proton to one of the nitrogens as shown in Eq. 1.19 creates an electron deficiency at the adjacent carbon and the $-S^-$ anion adds to the C=N group, closing the ring to an unstable tricyclic form of thiamin. This tricyclic form can be observed in methanol and can be crystallized. It is unstable in water because the central ring can open, with the electrons flowing as indicated by the small arrows to create a strongly basic site on the same nitrogen.

A second proton combines at this basic nitrogen with a high binding constant to form a cation. *The key to the reversed order of strength of the binding constants lies in the molecular rearrangements intervening between the two binding* steps. In this particular case, we cannot measure the successive binding constants K_1 and K_2 directly from the titration curve because K_2 is almost two orders of magnitude larger than K_1.

Consequently, the binding curve shown in Fig 1.4 is (within the experimental error) twice as steep at the center (the slope is 2 × 0.576) as that for acetate ion and is accurately represented in Eq 1.20. Comparison of Eq 1.20 with Eq. 1.10 shows how he latter has been simplified because no significant concentration of the form *PX* is present in the cooperative case.

$$Y = \frac{\overline{K}2[X]^2}{1+\overline{K}^2[X]^2} \qquad ...(1.20)$$

This is only part of the story about the acid-base chemistry of thiamin.

The binding of protons by the thiamin anion.is an example of a *cooperative process*, so named because binding of the first proton makes binding of the second easier. Although relatively rare among small molecules, cooperative processes are very common and important in biochemistry. A cooperative binding curve is sometimes referred to as *sigmoidal* because the plot of Y against *[X]* (the binding isotherm) is S shaped. The maximum possible cooperativity is observed when the binding of the first ligand enhances the affinity of all other sites so much that no species other than P and PX_n are present in significant concentration.

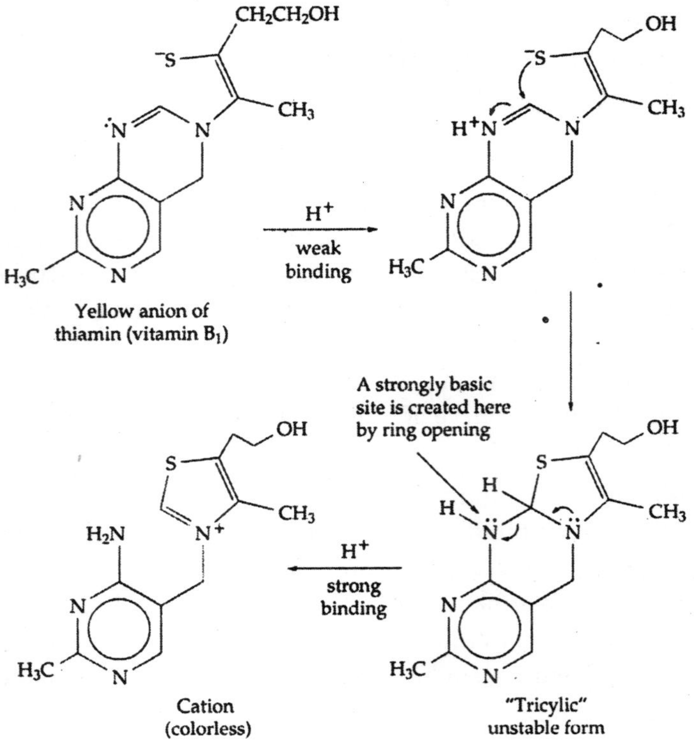

H^+ weak binding

Yellow anion of thiamin (vitamin B_1)

A strongly basic site is created here by ring opening

H^+ strong binding

Cation (colorless)

"Tricylic" unstable form

It is easy to show that, for *n* binding sites with such *completely cooperative binding,* the saturation fraction is:

$$Y = \frac{\overline{K}^n [X]^n}{1 + \overline{K}^n [X]^n} \quad ...(1.21)$$

where $\overline{K} = (K_1 \ldots K_n)^{1/n}$. The midpoint slope in the binding curve (Y vs log [X]) is 0.576n and the change, $\Delta \log [X]$, between Y = 0.1 and Y = 0.9 is $1.81/n$.

Equation 1.21 can be rewritten as

$$Y/(1-Y) = \overline{K}^n [X]^n \quad ...(1.22)$$

Taking logarithms (Eq. 1.22)

$$\log [Y/(1-Y)] = n \log \overline{K} + n \log IX] \quad ...(1.23)$$

A plot of log [Y/ (1 – Y)] vs log [X] is known as a *Hill plot.* According to Eq. 1.22. it is linear with a slope of *n*. Remember that this equation was derived for an ideal case of completely cooperative binding at *n* sites. However, Hill plots are often used to plot experimental data for systems in which cooperativity is incomplete. Thus, the experimentally measured slope of a Hill plot (n_{Hill}) an integer and is usually less than *n,* the number of binding sites.

A comparison n_{Hill} with *n* is often used as a measure of the degree ot cooperativity: $n_{Hill}/n = 1.00$ for complete cooperativity but is less than one if cooperativity, is incomplete. An example of a very high degree of cooperativity is provided by the hexameric enzyme glutamate dehydrogenase, whose saturation curve for substrate displays n_{Hill} approaching six. It is not necessary to make a Hill plot to get n_{Hill}. From the usual binding curve of Y (or ΔA) vs log [X] the midpoint slope can be measured with satisfactory precision. Alternatively, the difference, Δlog [X], between 0.1 and 0.9 saturation can be evaluated and n_{Hill} calculated from Eq. 1.24.

$$n_{Hill} = \frac{\text{midpoint slope}}{0.576} = \frac{1.81}{\Delta \log[X]} \qquad ...(1.24)$$

Binding curves sometimes show more than one step; in such cases Hill plots are not linear and no simple measure of cooperativity can be defined.

A second example of cooperativity is provided by the reversible denaturation of coiled peptide chains. Some proteins can be brought to a pH of 4 by addition of acid but without protonation of buried groups with intrinsic pK_a values greater than four. When a little more acid is added, some less basic group is protonated, permitting the protein to unfold and to expose the more basic hidden groups. Thus, cooperative proton binding is observed.

As in the case of thiamin the cooperativity depends upon the occurrence of a conformational change in the molecule linked to protonation of a particular group. The reversible transformation between an α helix and a random coil conformation is also cooperative. In this case, once a helix is started, additional turns form rapidly and the molecule is completely converted into the helix.

Likewise, once it unfolds it tends to unfold completely. Melting of DNA or, indeed, of any crystal is cooperative. The stacking of nucleotides alongside a template polynucleotide can also be cooperative. For example, the binding of an adenylate residue to two strands of polyuridylic acid leads to cooperative formation of a triple-helical complex. Here the stacking interactions make helix growth energetically easier than initiation of new helical regions.

COMPLEMENTARITY AND THE PACKING OF MACROMOLECULES

Because the forces acting between them are weak, two molecules will cling together tightly only if there is a close fit between their surfaces. For a firm bond to be formed many atoms must be in contact and the two molecular surfaces must be *complementary* one to the other. If a *"knob,"* such as a $-CH_3$ group, is present on one surface, there must be an appropriate hollow in the complementary surface. A positive charge in one surface is likely to be opposite a negative charge in the other.

A proton donor group can form a hydrogen bond only if it is opposite a group with unshared electrons; nonpolar (hydrophobic) groups must be opposite each other if hydrophobic interaction is to occur. An important principle is that *two molecules with complementary surfaces tend to join together and interact, whereas molecules without complementary surfaces do not interact.* Watson called this "selective stickiness." Selective stickiness permits the *self-assembly* of biological macromolecules having surfaces of complementary shape into fibers, tubes, membranes, and polyhedra.

It also provides the means for specific pairing of purine and pyrimidine bases during the replication of DNA and during the synthesis of RNA and of proteins. Complementarity of surfaces is equally important to the chemical reactions of cells. Each of these reactions is catalyzed by an enzyme, which contains reactive chemical groupings in the right places and in the right orientations to interact with and promote a chemical change in another molecule, the *substrate.* Specific catalysis is one of the most basic characteristics of living things. Enzymatic catalysis provides the basis not only for the reactions of metabolism but also for the movement of muscle fibers, the flowing of the cytoplasm in the ameba, and virtually all other biological responses.

To understand these phenomena requires an examination of the structures of the macromolecules involved and of the ways in which they can fit together. Just as the amino acids, sugars, and nucleotides are the building blocks for formation of proteins, polysaccharides, and nucleic acids, these three kinds of macromolecule are the units from which larger subcellular structures are assembled. Fibers, microtubules, virus *"coats,"* and small symmetric groups of *subunits* in *oligomeric proteins* all result from the packing of macromolecules in well-defined ways, something that is often called *quaternary structure*.

Rings and Helices

Consider first the aggregation of identical protein subunits. While many protein molecules are nearly spherical, they are nevertheless asymmetric. In the drawings that follow the asymmetry is exaggerated, but the principles illustrated are valid. One easily observed lesson from nature is that even though living things are made up of asymmetric materials, a great deal of symmetry is evident. At the molecular level the symmetry of crystalline arrays of atoms or molecules is described mathematically by the elements of symmetry present in *space groups*. There are 230 of these but only 65 accommodate asymmetric objects. Two of the natural ways for identical asymmetric subunits to interact lead to rings and helices, respectively.

Molecules with cyclic symmetry. Consider a subunit (*protomer*) of the shape shown in Eq. 1.25 and containing a region *a* that is complementary to the surface *j* on another part of the same molecule. Two such protomers will tend to stick together to form a dimer, region a of one protomer sticking to region *j* of the other. The dimer will still contain a free region a at one end and a region *j* at the other which are not involved in bonding. Other protomers can stick to these free ends. In some instances long chains can be formed. However, if the geometry is just right, a third subunit can fit in to form a closed ring (a trimer).

Depending on the geometry of the subunits, the ring can be even smaller (a dimer) or it can be larger (a tetramer, pentamer, etc.). The bonding involved is between two different regions (*a* and *j*) of a subunit and is sometimes described as ***heterologous.*** To obtain a closed ring of subunits, the angle between the bonding groups *a* and *j* must be correct or the ring cannot be completed. A ring formed using exclusively heterologous interactions possesses *cyclic symmetry*. The trimer in Eq. 1.25 has a *threefold axis:* Each subunit can be superimposed on the next by rotation through 360°/ 3. The oligomer is said to have C_3 symmetry. Many real proteins, including all of those with 3, 5, or another uneven number of identical protomers, appear to be formed of subunits arranged with cyclic symmetry.

An example is the cholera toxin from *Vibrio cholerae,* which forms a pentamer with an outer ring of subunits with C_5 symmetry. Now consider the quantitative aspects of heterologous interactions with ring formation. Let K_f be the formation constant and $\Delta G°$ the Gibbs energy change for the reaction of the *j* end of protomer P with the *a* end of a second protomer to form the dimer P_2. In the second step (Eq. 1.25) a third protomer combines. It forms *two* new *aj* interactions. If we assume for this step that $\Delta G°_f$ is 2 $\Delta G°$, K_f will be equal to K^2.

The overall association constant for formation of a trimer from three protomers will be given by Eq. 1.26.

$$3\,P \rightarrow P_3$$
$$K_f = K^3$$
$$\Delta G_f^\circ = 3\,\Delta G^\circ \qquad ...(1.26)$$

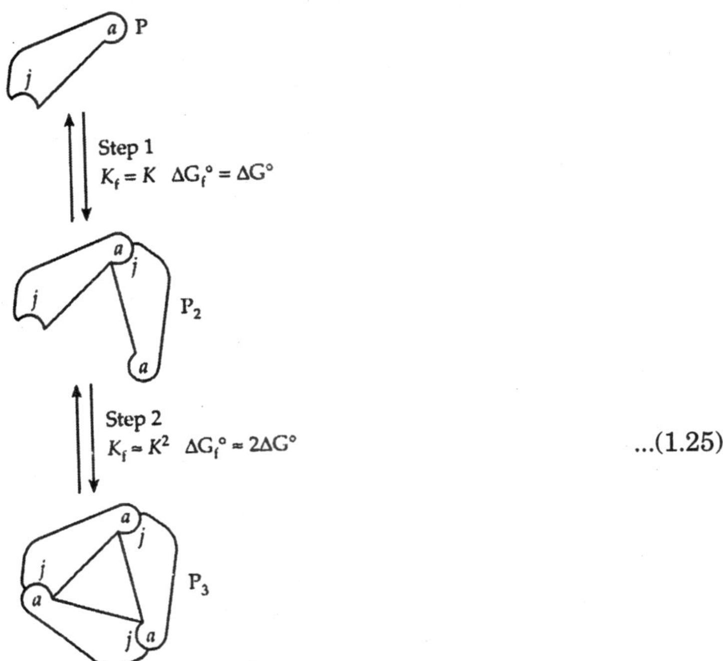

...(1.25)

This will be true only if ΔG° for formation of both new *aj* bonds in the trimers is exactly the same as that for formation of the *aj* bond in the dimer. The reader may wish to criticize this assumption and to suggest conditions that might lead to overestimation or underestimation of K_f for the trimer as calculated previously. Now consider a hypothetical example: Protomer P is continuously synthesized by a cell. At the same time some subunits are degraded to a nonaggregating form via a second metabolic reaction. The two reactions are balanced so that [P] is always present at a steady state value of 10^{-5} M. Suppose that a value for a single *aj* interaction of K = 10^4 (and ΔG° = – 22.8 kJ mol^{-1}) governs aggregation to form dimers and trimeric rings. What concentration of dimers and of trimer rings will be present in the cell in equilibrium with the 10^{-5} concentration of P. Using Eq. 1.25 we see that the concentration of dimers [P_2] is $10^4 \times (10^{-5})^2 = 10^{-6}$ M. (Note that the amount of material in this concentration of dimer is equivalent to 2×10^{-6} M of the monomer units.) The concentration of rings [P_3] is $(10^4)^3 \times (10^{-5})^3 = 10^{-3}$ M, equivalent to 3×10^{-3} M, of the monomer units.

Thus, of the *total* P present in the cell ($10^{-5} + 0.2 \times 10^{-5} + 300 \times 10^{-5}$ M), 99.6% is associated with trimers, 0.33% is still monomers, and only 0.07% exists as dimers. Thus, the formation of two heterologous bonds simultaneously to complete a ring imparts a high digree of cooperativity to the association reaction of Eq. 1.25. We will find in a cell mostly either rings or monomer but little dimer.

Now consider what will happen to the little rings within the cell if the process that removes P to a non-associating form suddenly becomes more active so that [P] falls to 10^{-6} M. If *K* is still 10^4, what will be the percentages of P, P_2, and P_3 at equilibrium? Here we note a characteristic of cooperative processes: A higher than first power dependence on a concentration.

Helical structures. If the angle at the interface *aj* is slightly different, instead of a closed ring, we obtain a helix as shown in Fig. 1.6a. The helix may have an integral number of subunits per turn or it may have a non-integral number, as in the figure. The same type of heterologous interaction *aj* is involved in joining each subunit to the preceding one, but in addition other interactions occur. If the surfaces involved in these additional interactions are complementary and the M geometry is correct, groups from two different parts of the molecule (*e.g., b* and k) may fit together to form another heterologous bond.

Still a third heterologous interaction cl may be formed between two other parts of the subunit surfaces. If interactions *aj, bk,* and *cl* are strong (*i.e.,* if the surfaces are highly complementary over large areas), extremely strong microtubular structures may be formed, such as those in the flagella of eukaryotic organisms. If the interactions are weaker, labile microfilaments and microtubules, such as are often observed to form and dissociate within cells, may arise. The geometry of subunits within a helix is often advantageously displayed by imagining that the surface of the structure can be unfolded to give a radial projection Fig. 1.6B.

Here subunits corresponding to those in the helix in Fig. 1.6a are laid out on a plane obtained by slitting the cylinder representing the surface of the helix and laying it out flat. In the example shown, the number of subunits per turn is about 4.8 but it can be an integral number. The interactions *bk* between subunits along the direction of the fiber axis may sometimes be stronger than those *(aj)* between adjacent subunits around the spiral.

In such cases the microtubule becomes frayed at the ends through breaking of the *aj* interactions. This phenomenon can be observed under the electron microscope for the microtubules from flagella of eukaryotic organisms. Fig. 1.7 to 1.10 show four helical structures from the molecular domain. They are a filamentous bacteriophage, a plant virus, a bacterial pilus, and an actin microfibril. Each is composed largely of a single kind of protomer. A larger and more complex helical structure, the microtubule, is shown in Fig. 1.34.

Filamentous bacteriophages. Bacteriophages of the Ff family include the fd, f1, and M13 strains. Phage M13 is widely used in cloning genes and for many other purposes. The genome is a circular, single-stranded DNA of ~ 6400 nucleotides which is held in an elongated double-stranded form by a helical sheath of about 2700 subunits of a 50–residue protein.

The rod is about 6 nm in diameter and 880 nm long and it is capped by two specialized proteins at one end and a different pair of proteins at the other end. The five coat proteins are encoded by five of the 11 genes present in these little viruses. Each coat subunit in the Ff viruses is coiled into an ∝-helical rod of 7 nm length. These are arranged in the virus in a right-handed helical pattern with a pitch of 1.5 nm and. with 4.4 subunits per turn. The protein rods are inclined to the helix axis and extend inward. This arrangement permits a "knobs-in-holes" hydrophobic bonding between subunits.

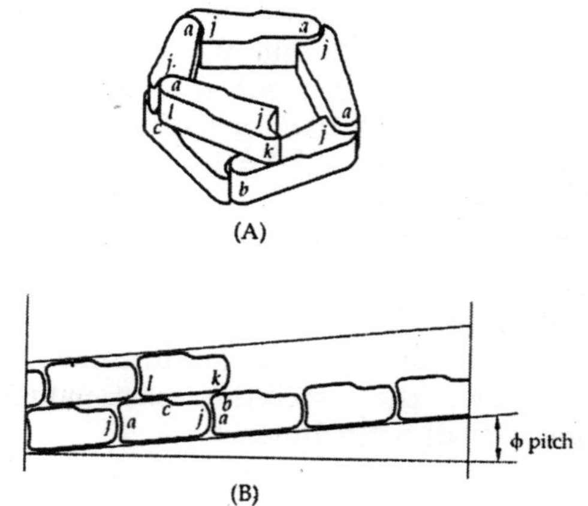

Fig. 1.5. (A) Heterologous bonding of subunits to form a helix. (B)T?adial projection of subunits arranged as in helix A. Different bonding regions of the subunit are designated *a, b, c, j, k,* and *l.*

The helix of pitch 1.5 nm is *the primary* or *one-start helix*. However, in every regular helical structure we can also trace a two-start helix, a three-start helix, etc. In this instance the five-start helix is easiest to see. The protein coat of these viruses provides an elongated cylindrical cavity to protect the circular, single-stranded DNA molecule that is the genome. Although there are two antiparallel strands of DNA, a regular base-paired structure is impossible and the DNA is probably not present in a highly ordered form.

There are about 2.4 nucleotides in the DNA per protein subunit. However, there are related viruses with ratios' as low as one nucleotide per protein subunit and containing more highly extended DNA.

A rod-shaped plant virus. The tobacco mosaic virus is a 300-nm-long rod constructed from 2140 identical wedge-shaped subunits whose detailed molecular structure is known. Each 158–residue subunit contains five helices and a small β sheet. A single strand of RNA containing 6395 nucleotides (~ 3 per protein subunit) lies coiled in a groove where it interacts with side chains from two of the helices. The virus is assembled by the binding of a region of the RNA 800 – 1000 nucleotides from the 3' end to a two-turn helix of subunits that appears to form spontaneously. Additional subunits then add at each end, binding to the RNA as well as the adjacent protein subunits. A relative with a very similar structure is cucumber green mottle mosaic virus.

Bacterial pili : The adhesion pili, or fimbriae, of bacteria are also helical arrays of subunits. The P pili of *E. coli* are encoded by a cluster of 11 genes in the *pap* (pilus associated with pyelonephritis) cluster. They are needed to allow the bacteria to colonize the human urinary tract. The bulk of the ~1–μm–long pilus is made up of about 1000 subunits of a 185–residue protein encoded by gene *Pap A*. They form a right-handed helix of ~7 nm diameter with 3.28 subunits per turn and a pitch of -2.5 nm. The rod is anchored to the bacterial outer membrane by a protein encoded by gene *PapH*, while subunits encoded by *PapE* and *PapF*

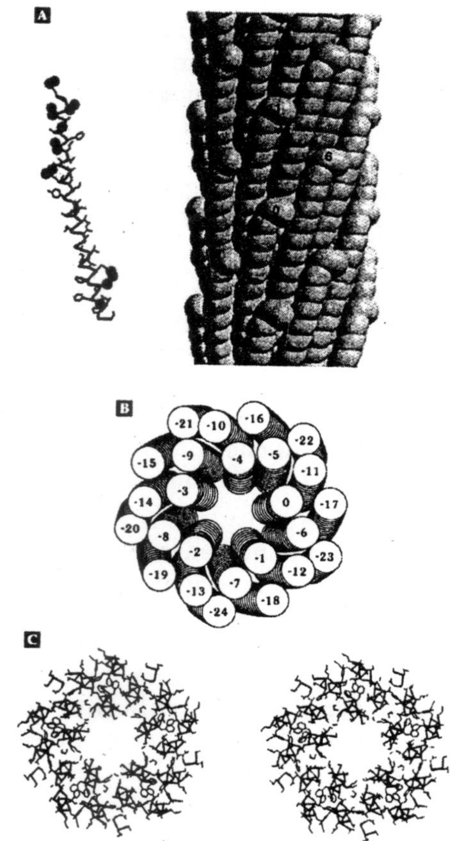

Fig. 1.6. Structure of the virus fd protein sheath. (A) Left. A single coat subunit, with its N terminus towards the top, as if moved from the left side of the sheath. The dark circles represent charged atoms of Asp, Glu, and Lys side chains. The backbone of the protein is a C^{α} diagram. The positively charged atoms near the C terminus line the inner surface of the sheath neutralizing the negative charge of the DNA core. Right. Each subunit is represented by a helical tube through successive C^{α} atoms. Three nearest neighbours, indexed as 0, 6, and 11, are indicated. The axial slab shown represents ~1% of the total length of the virion. (B) A 2.0 nm section through the virus coat with the helices shown as curved cylinders. The view is down the axis from the N-terminal ends of the rods. The rods extend upward and outward. The rods with indices 0 to –4 start at the same level, forming a five-start helical array. The rods with more negative indices start at lower levels and are therefore further out when they are cut in this section. (C) The same view but with "wire models" of the atomic structure of the rods.

fasten the adheson protein (PapG) to the tips of the pili. The adheson binds to the Galα1 → 4Gal ends of glycolipids in the kidney. The PapE, F, and G subunits form a thin ~2.5–nm–thick by 15 nm "fibrillum" which is attached by an adapter protein encoded by gene *PapK.* A special chaperonin *(PapD* gene) is also required for pilus assembly in *E. coll* as well as other bacteria.

Another *E. coli* pilus adheres to mannose oligosaccharides. Similar pili of *Neisseria gonnorheae* are used by that bacterium. The three-dimensional structure of the 158–residue pilin subunit is that of a globular subunit with an 8.5 nm ∝-helical spine at one end. A proposed model of the intact pilus shaft is shown in Fig. 1.9B,C.

Notice the similarity of the packing of the ∝-helical spines in the center to the packing arrangement in the bacterio-phage coat in Fig. 1.7 Similar features may be present in the P pilus rod shown in Fig. 1.9A. However, there is uncertainty about the packing arrangement. The *E. coli* type 1 pilus subunits contain immunoglobulin folds that are completed by donation of an N–terminal strand from a neighbouring subunit.

In thin fimbriae of *Salmonella* extended, parallel β helices may be formed. Other types of pili are also well known. F pili or conjugative pili are essential for sexual transfer of DNA between bacterial cells F^+ strains of E. *coli* form hollow pili of 8.5 nm diameter with a 2.0-nm central hole. Their 90-residue subunits apparently form rotationally symmetric pentamers which stack to form the pili. These pili are essential to establishing the initial contact between conjugating bacterial cells.

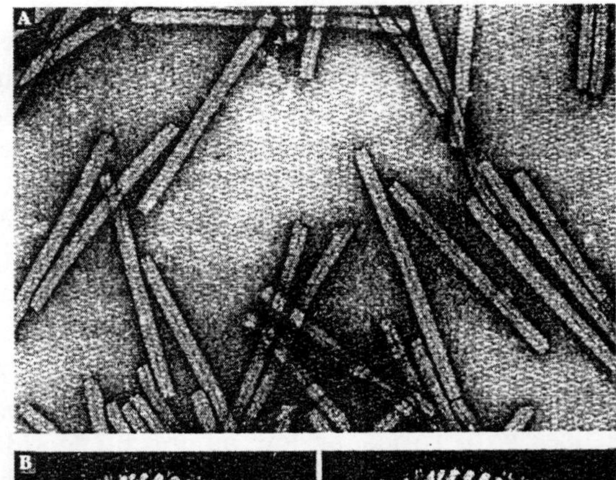

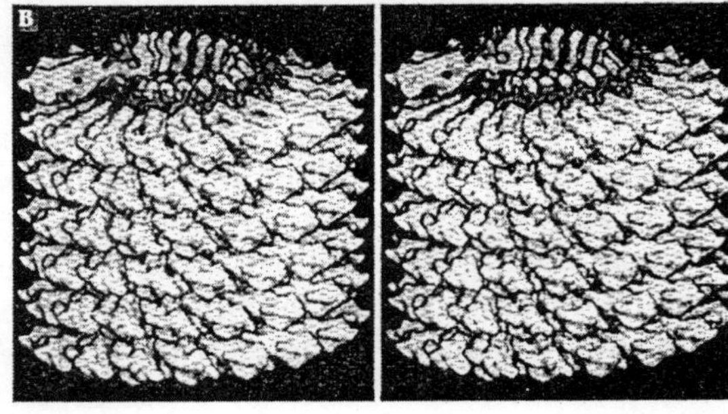

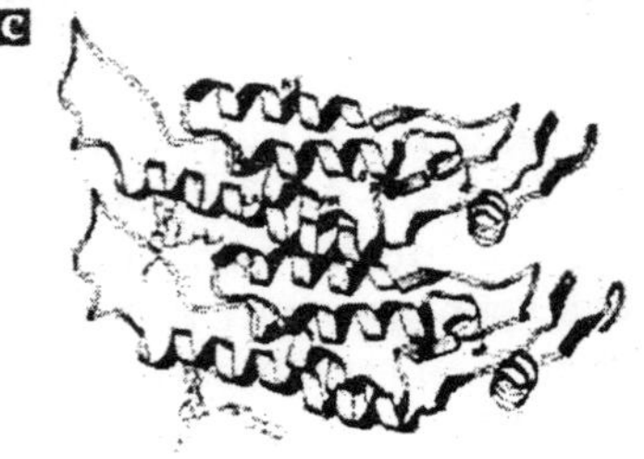

Fig. 1.7. (A) Electron micrograph of the rod-shaped particles of tobacco mosaic virus. (B) A stereoscopic computer graphics image of a segment of the 300 nm long tobacco mosaic virus. The diameter of the rod is 18 nm, the pitch of the helix is 2.3 nm, and there are 1613 subunits per turn. The coat is formed from 2140 identical 17.5–kDa subunits. The 6395-nucleotide genomic RNA is represented by the dark chain exposed at the top of the segment. The resolution is 0.4 nm. (C) A MolScript ribbon drawing of two stacked subunits.

The thin filaments of muscle. An essential component of skeletal muscle is filamentous *actin* (F-actin). It is composed of 375–residue globular subunits of a single type and with a highly conserved sequence. It is found not only in muscle but also in other cells where it is a component of the cytoskeleton. The actin microfilament has the geometry of a left handed one–start or *primary helix* with a pitch of only 0.54 nm and with approximately two subunits per turn. It can also be described as a right-handed two-start helix in which two chains of subunits coil around one another with a long pitch.

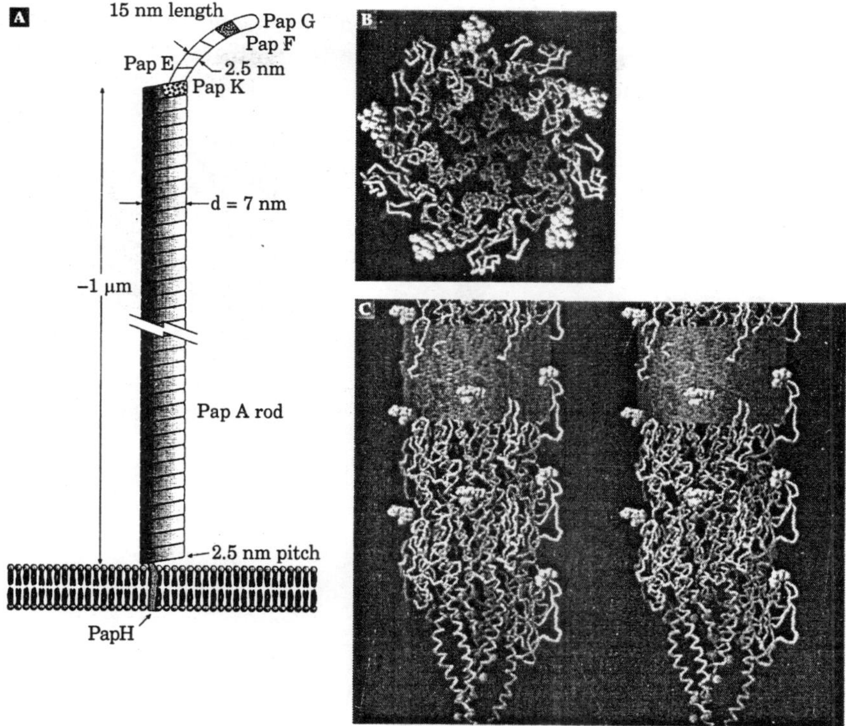

Fig. 1.8. (A) Schematic diagram of a bacterial P pilus. The ~1–μm-long helical rod is anchored to the outer cell membrane by protein Pap H. The adhesion Pap G binds to galactosyl glycolipids of the host. (B, C) The structure of pilus fiber from *Neisseria gonnorrheae* modeled from the atomic structure of the 54–residue pilin subunit. The exact structure of the fiber is uncertain, but the model generated here by trying various possible helical packings matches the dimensions obtained from fiber diffraction patterns and electron microscopic images. (B) Cross section. (C) Stereoscopic view. The experimental dimensions of 4.1–nm pitch and 6.0–nm diameters are shown by the "transparent" ring in (B).

Oligomers with Twofold (Dyad) Axes

Paired interactions. If two subunits are held together with interactions aj and are related by a twofold axis of rotation as shown in Fig. 1.11, we obtain an isologous dimer. Each point such as a in one subunit is related to the same point in the other subunit by reflection through the axis of rotation. In the center, along the twofold axis, points c and C' are directly opposite the same points in the other subunit. Fig. 1.11 is drawn with a hole in the center so that groups *c* and *c'* do not actually touch, and it is the paired interactions such as *aj* of groups not adjacent to the axis that contribute most to the bonding. However, a real protein dimer may or may not have such a hole. The pair of identical interactions in an isologous dimer may be referred to as a single *isologous bond.* Such a bond always contains the paired interactions between complementary groups *(aj)* and has pairs of identical groups along the axis.

However, because they are identical those groups usually cannot interact in a specific complementary manner. Isologous bonding is very Important in oligomeric enzymes, and it has been suggested that isologous interactions evolved early. Initially there may not have been much complementarity in the bonding but two *"hydrophobic spots"* on the surface of the subunits came together in a nonspecific association. Later in evolution the more specific paired interactions could have been added.

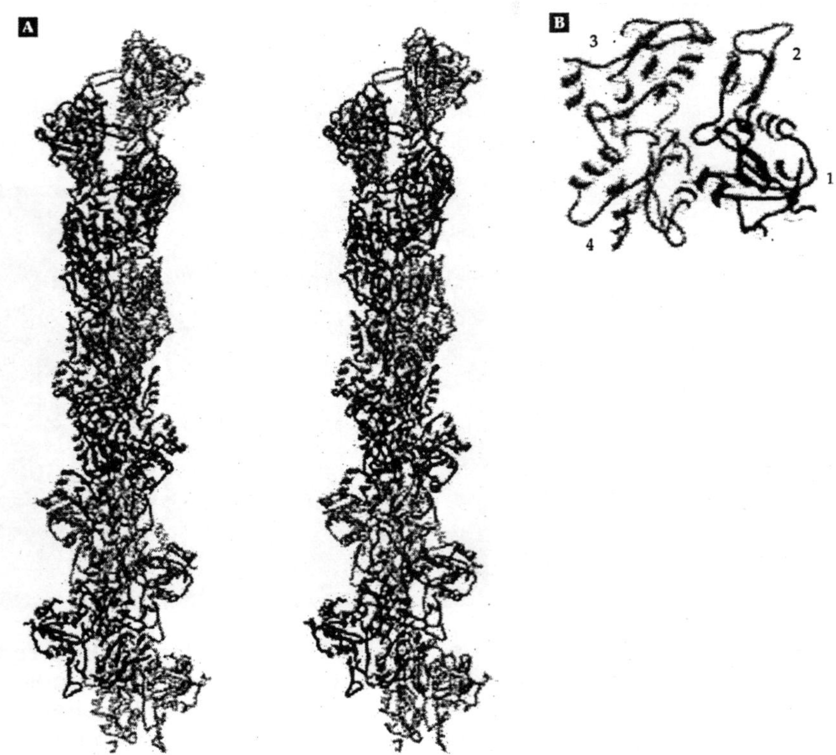

Fig. 1.9. (A) Model of the F-actin helix composed of eight monomeric subunits. The model was constructed from the known structure of the actin monomer with bound ADP using X-ray data from oriented gels of fibrous actin to deduce the helical arrangement of subunits. The main interactions appear to be along the two-start helix. (B) Ribbon drawing of an actin monomer with the four domains labeled.

Dihedral symmetry. Isologous dimers can serve as subunits in the formation of larger closed oligomers and helices; for example, an isologous pair of the sort shown in Fig. 1.11A can be flipped over into the top of another similar pair as shown in Figs. 1.11B and 1.llC. Again, if the proper complementary surfaces exist bonds can form as shown (*bk* in Fig 1.11B and *bk* and *cl* in Fig. 1.11C. Both structures in Figs. 1.11B and 1.11C possess dihedral (D_2) symmetry. In addition to the twofold axis of rotation lying perpendicular to the two rings, there are two other twofold axes of rotation as indicated in the drawings. Again, the interactions are paired and isologous; of many possible contacts two *bk* interactions and two *cl* interactions are marked for each pair of subunits in Fig. 1.11. There are a total of six pairs of these interactions, one between each combination of two subunits. This may be a little more difficult see in Fig. 1.11B than in Fig.1.11C because in the former the subunits are arranged in a more or less square configuration.

Nevertheless, a pair of interactions between the left-hand subunit in the top ring and the subunit in the lower ring at the right does exist, even if it is only electrostatic and at a distance. An example of a tetrameric enzyme with perfect dihedral symmetry of the type shown in Fig. 1.11B is *lactate dehydrogenase.* The plant agglutinin *concanavalin* A has a quateruary structure resembling that in Fig. 1.11C.

Square arrays of four subunits can be formed using either heterologous or isologous interactions. Both types of bonding can occur in larger aggregates. For exarnple, two trimers such as that shown in Eq. 1.25 can associate to a hexamer having dihedral (D_3) symmetry; a heterologous "square tetramer" can dimerize to give a dihedral (D_4) octamer. The enzymes *ornithine decarboxylase* and *glutamine synthetase* each consist of double rings of six subunits each. The upper ring is flipped over onto the lower giving dihedral symmetry (D_6) with one 6-fold axis and six 2-fold axes at right angles to it.

Oligomers with cubic symmetry (polyhedra). Symmetrical arrangements containing more than one axis of rotation of order higher than 2-fold are said to have cubic symmetry. The *tetrahedron* is the simplest example. It contains four 3-fold axes which pass through the vertices and the centers of the faces and three 2-fold axes which pass through the midpoints of the six edges. *Since protein subunits are always asymmetric, a tetrameric protein cannot possess cubic symmetry.* As we have already seen, tetrameric enzymes have dihedral symmetry.

However, a heterologous trimer with 3-fold symmetry can form a face of a tetrahedron containing a total of 12 asymmetric subunits. Twenty-four subunits can interact to form a *cube.* Three 4-fold axes pass through the centers of the faces, four 3-fold axes pass through the vertices, and six 2-fold axes pass through the edges. The largest structure of cubic symmetry that can be made is the *icosahedron,* a regular solid with 20 triangular faces. Sixty subunits,

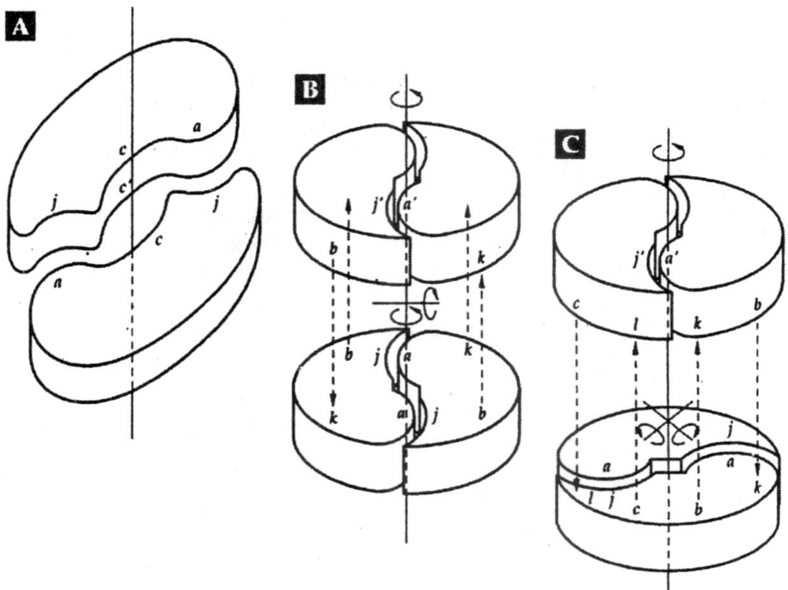

Fig. 1.10. (A) Isologous bonding between pairs of subunits; (B) an "isologous square" arrangement of subunits; (C) an apparently "tetrahedral" arrangement of subunits. Note the three twofold axes.

or some multiple of 60, are required and at each vertex they form a heterologous pentamer. As with the tetrahedron, each face contains a heterologous trimer, while isologous bonds across the edges form dimers.

Many viruses consist of roughly spherical protein shells (coats) containing DNA or RNA inside. As with the filamentous viruses, the protein coats consist of many identical subunits, a fact that can be rationalized in terms of economy from the genetic viewpoint. *Only one gene is needed to specify the structure of a large number of subunits.* Under the electron microscope the viruses often have an icosahedral appearance and chemical studies show that the number of the most abundant subunits is usually a multiple of 60.

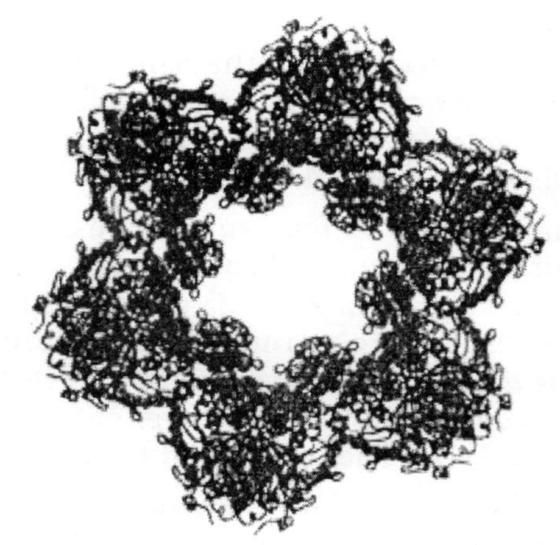

Fig. 1.11. A ribbon representation of the ornithine decarboxylase dodecamer. Six dimers of the 730–residue subunits are related by C_6 crystallographic symmetry.

An example is the tiny *satellite tobacco necrosis virus,* diameter ~18 nm, whose coat contains just 60 subunits of a 195-residue protein. Its genome is a 1239–nucleotide molecule of RNA. The structure of the coat has been determined to 0.25–nm resolution. Many virus coats have 180 subunits or a number that is some other multiple of 60. However, in these coats the subunits cannot all be in identical environments. Two cases may be distinguished. If all of the subunits have identical amino acid sequences they probably exist in more than one distinct conformation that permit them to pack efficiently. Alternatively, two or more subunits of differing sequence and structure may associate to form 60 larger subunits that do pack with icosahedral symmetry.

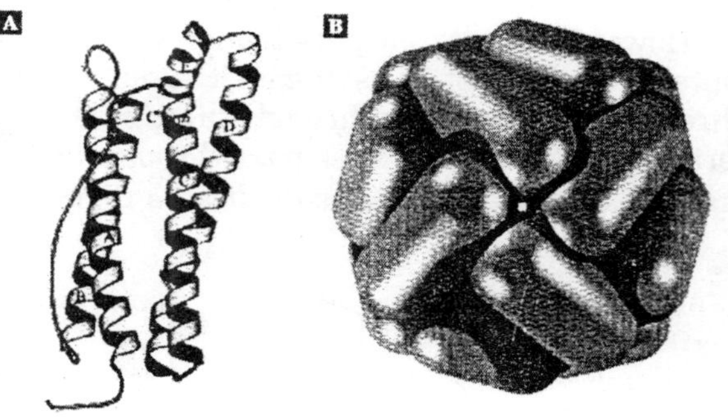

Fig. 1.12. (A) MolScript ribbon drawing of a subunit of the iron oxide storage proteins L–ferritin from amphibian red cells. This 4-helix bundle is represented by cylinders of 1.3 nm diameter in the oligomer. (B) Helices A and C of the monomer are on the outer surfaces of the oligomer and helices B and D are on the inner surface. The oligomer consists of a shell of 24 subunits and is viewed down a 4–fold axis illustrating its 423 (cubic) symmetry.

For example, the polioviruses (diameter 25 nm) contain three major coat proteins (α, β, and γ or VP1, VP2, and VP3). These are formed by cleavage of a large precursor protein into at least four pieces. The three largest pieces of ~33–, 30–, and 25–kDa mass (306, 272, and 238 residues, respectively) aggregate as (αβγ). Sixty copies of a foruth subunits of 60 residues are found within the shell. Related picorna viruses such as human rhinoviruses foot–and–mouth disease virus, parvovirus, and Mengo virus have similar architectures. The small (diameter 25 nm) single–stranded DNA bacteriophages such as ϕX174 also have three different coat proteins, one of which forms small hollow spikes at the vertices of the icosahedral shell.

Asymmetry and quasi-equivalence in oligomers. It is natural to think about association of subunits in symmetric ways. Consequently the observation of square, pentagonal, and hexagonal arrangements of subunits directly with the electron microscope led to a ready acceptance of the idea that protomers tend to associate symmetrically. However, consider the predicament of the two molecules shown in Fig. 1.16.

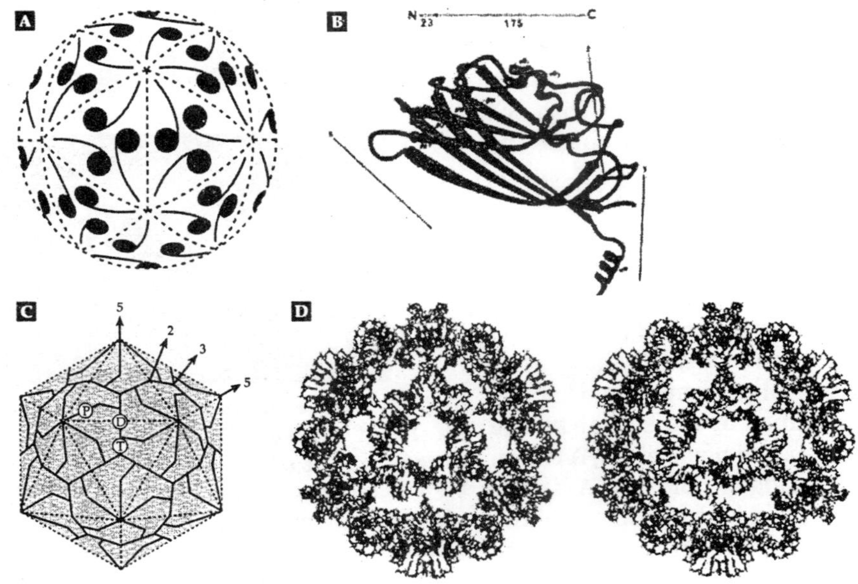

Fig. 1.13. (A) Schematic drawing illustrating an icosahedrally symmetric structure with sixty identical asymmetric subunits all in equivalent positions. The 5-fold axes are located at the vertices of the icosohedron and the 2-fold and 3-fold axes can readily be seen. (B) Ribbon drawing of the 195-residue polypeptide chain of the coat subunit of satellite tobacco necrosis virus. The protein folds into an inwardly projecting N-terminal segment and a "(b-jellyroll" domain. The packing of this subunit in the virus particle is shown schematically in (C). The symmetry axes drawn next to the subunit diagram (B) can be used to position it in the structure. Contacts between subunits are labeled D, T, and P ('dimer', 'trimer', 'pentamer'). Diagrams courtesy of Drs. Strandberg, Liljas, and Harrison. (D) The distribution of RNA helical segments in a hemisphere of a virion of a similar small virus, the satellite tobacco mosaic virus. The virion viewed down a 3-fold axis from the virus exterior. The helical axes of the RNA segments are along icosahedral edges.

They might get together to form an isologous dimer if it were not for the fact that their "noses" are in the way. Despite the obvious steric hindrance, an isologous dimer can be formed in this case if one subunit is able to undergo a small change in conformation Fig. 1.16. In the resulting dimer the two subunits are only *quasi–equivalent*. Unsymmetrical dimerization of proteins appears to be a common phenomenon that is often observed in protein crystals. For example, the enzymes malic dehydrogenase and glyceraldehyde phosphate dehydrogenase are both tetramers of approximate dihedral symmetry but X-ray crystallography revealed distinct asymmetries which include a weaker binding of the coenzyme NAD^+ in one subunit. This may simply reflect differences in environment within the crystal lattice.

However, negative cooperativity in coenzyme binds has also been revealed by kinetic experiments. The polypeptide hormone *insulin* is a small protein made up of two chains (designated A and B) which are held together by disulfide bridges. Fig. 1.17B a sketch of the structure as revealed by X-ray crystallography, with only the backbone of the peptide chains and a few side chains shown.

In the drawing, the B chain lies behind the A chain. Beginning with the N–terminal Phe 1 of the B chain the peptide backbone makes a broad curve, and then falls into an helix of three turns lying more or less in the center of the molecule. After a sharp turn, it continues upward on the left side of the drawing in a nearly completely extended β structure. The A chain has an overall U shape with two roughly helical portions. The U shape is partly maintained by a disulfide bridge running between two parts of the A chain. Two disulfide bridges hold the A and B chains together, and hydrophobic bonding of internal side chain groups helps to stabilize the molecule. Insulin in solution dimerizes readily, the subunits occupying quasi–equivalent positions.

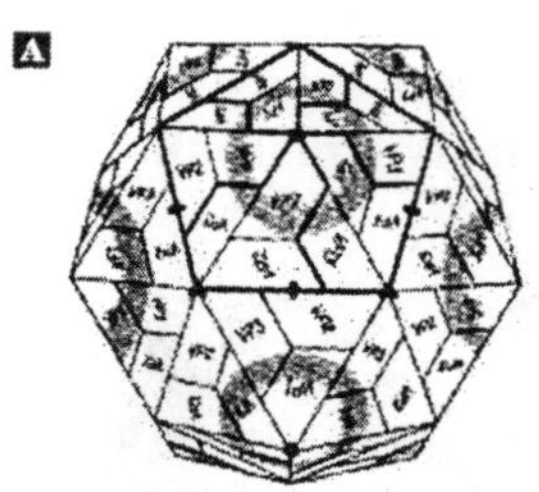

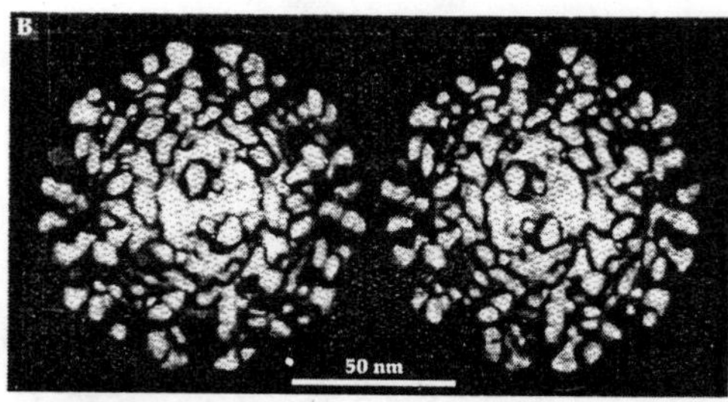

Fig. 1.14. (A) Schematic diagram of the icosahedral shell of a human rhinovirus showing the arrangement of the three subunits VP1, VP2, and VP3, each present as 60 copies. (B) Stereoscopic view of an image of the virus "decorated" by the binding of two immunoglobulinlike domains of the intercellular adhesion molecule ICAM-1, a natural receptor for the virus. Part of this receptor binds into a groove or "canyon," which in marked in (A) by the dark bands.

Fig. 1.17C shows some details of the bonding between the subunits in the insulin dimer with a view from the outside of the molecule down the 2–fold axis (marked by the X in the center of the Phe 25 ring in the right-hand chain) through the dimer. The C–terminal ends of the B chains are seen in an extended conformation. The two anti-parallel chains form a β structure with two pairs of hydrogen bonds. If there were perfect isologous bonding, the two pairs would be entirely equivalent and symmetrically related one to the other. A straight line drawn from a position in one chain and passing through the twofold axis (X) would also pass approximately through the corresponding position in the other chain.

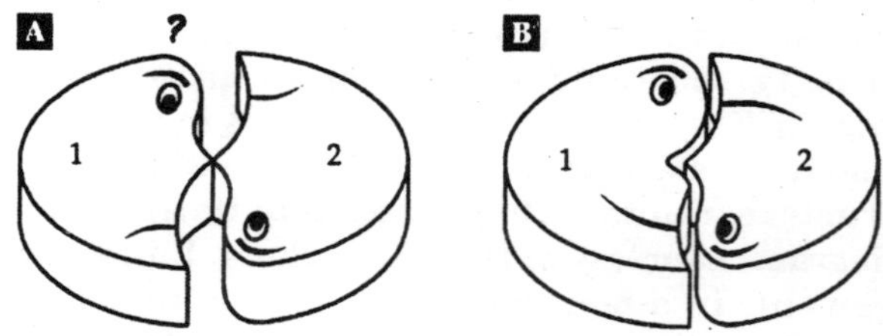

Fig. 1.15. Nonsymmetric bonding in a dimer. (A) Two molecules which cannot dimerize because of a bad fit at the center. (B) A solution: Molecule 1 has refolded its peptide chain a little, changing shape enough to fit to molecule 2.

However, there are many deviations from perfect symmetry, the most striking of which is at the center where the Phe 25 from the right–hand chain projects upward and to the left. If the symmetry were perfect the corresponding side chain from the left–hand chain would project upward and to the right and the two phenylalanines would collide, exactly as do the "noses" in Fig. 1.16.

In insulin one phenylalanine side chain has been flipped back out of the way. Under proper conditions, three insulin dimers associate to form a hexamer of approximate dihedral (D_3) symmetry that is stabilized by the presence of two zinc ions. Fig. 1.17D is a crude sketch of the hexamer showing the three dimers, the 3–fold axis of symmetry, and the two pseudo 2–fold axes, one passing between the two subunits of the dimer and the other between two adjacent dimers. Fig. 1.18) is a stereoscopic ribbon diagram of the atomic structure, with the A chains omitted, as obtained by X–ray diffraction.

The structure has also been obtained by NMR spectroscopy. Note that each of the two zinc atoms lies on the threefold axis and is bound by three imidazole rings from histidines B–10. The significance of the zinc binding is uncertain but these hexamers readily form rhombohedral crystals, even within the pancreatic cells that synthesize insulin. The structure illustrates a feature that is common to many oligomers of circular or dihedral symmetry.

A central *"channel"* is often quite open and protruding side chain groups, such as the imidazole groups in insulin, form handy nests into which ions or molecules regulating activity of proteins can fit. Conformational differences in insulin are induced by the binding of phenol. In Fig. 1.18A the C–terminal ends of the chains are extended but in the phenol complex (B) they have coiled to extend the $\propto$ helices.

Quasi-equivalence in virus coats. A large number of icosahedral viruses have coats consisting of 180 identical subunits. For example, the small RNA-containing bacteriophage MS 2 consists of an eicosahedral shell of 180 copies of a 129-residue protein that encloses one molecule of a 3569-residue RNA. The virus also contains a single molecule of a 44–kDa protein, the A protein, which binds the virus to a bacterial pilus to initiate infection. Related bacteriophages GA, fr, f2, and Qβ have a similar architecture.

Many RNA-containing viruses of plants also have 180 subunits in their coats. Much studied are the *tomato bushy stunt virus* (diameter ~33 nm, 40–kDa subunits), and the related *southern bean mosaic virus*. The human *wart virus* (diameter ~56 nm) contains 420 subunits, seven times the number in a regular icosahedron. *Adenoviruses* (diameter ~100 nm) have 1500 subunits, 25 times more than the 60 in a regular icosahedron. Caspar and Klug proposed a theory of quasi-equivalence of subunits according to which the distances between the centers of subunits are preserved in a family of *icosadeltahedra* containing subunits in multiples of 20.

However, the angles must vary somewhat from those in a regular icosahedron (compare with geodesic shells in which the angles are constant but the distances are not all the same). The resulting polyhedra contain hexamers as well as pentamers at vertices; for example, the shells of the 180–subunit viruses contain clusters of subunits forming 12 pentamers and 20 hexamers. There are also 60 trimers (on the faces) and 90 dimers (across the edges). Such structures can be formed only for certain values of *T* where the number of subunits is 60*T* and there are 12 pentamers (pentons) and 10 (T-l) hexamers hexons. *T* can assume values of $h^2 + hk + k^2$ where h and k may be positive integers or zero. Some allowed *T* values are 1,3,4, 7,9,13, 25. The subunits in virus coats with *T* greater than one are not all in equivalent positions.

For example, the three subunits labeled A, B, and C in Fig. (1.19) are each slightly differently positioned with respect to neighbouring subunits. Since virus coats are usually tightly packed the subunits must assume more than one conformation. One kind of conformational change that allows quasi–equivalence of subunits is observed in the tomato bushy stunt virus.

Two structural domains are connected by a hinge which allows an outer protruding domain to move slightly to preserve good isologous interactions with a corresponding domain in another subunit. The southern bean mosaic virus has an eight–stranded antiparallel β–barrel structure closely similar to that of the major domain of the bushy stunt viruses but lacking the second hinged domain.

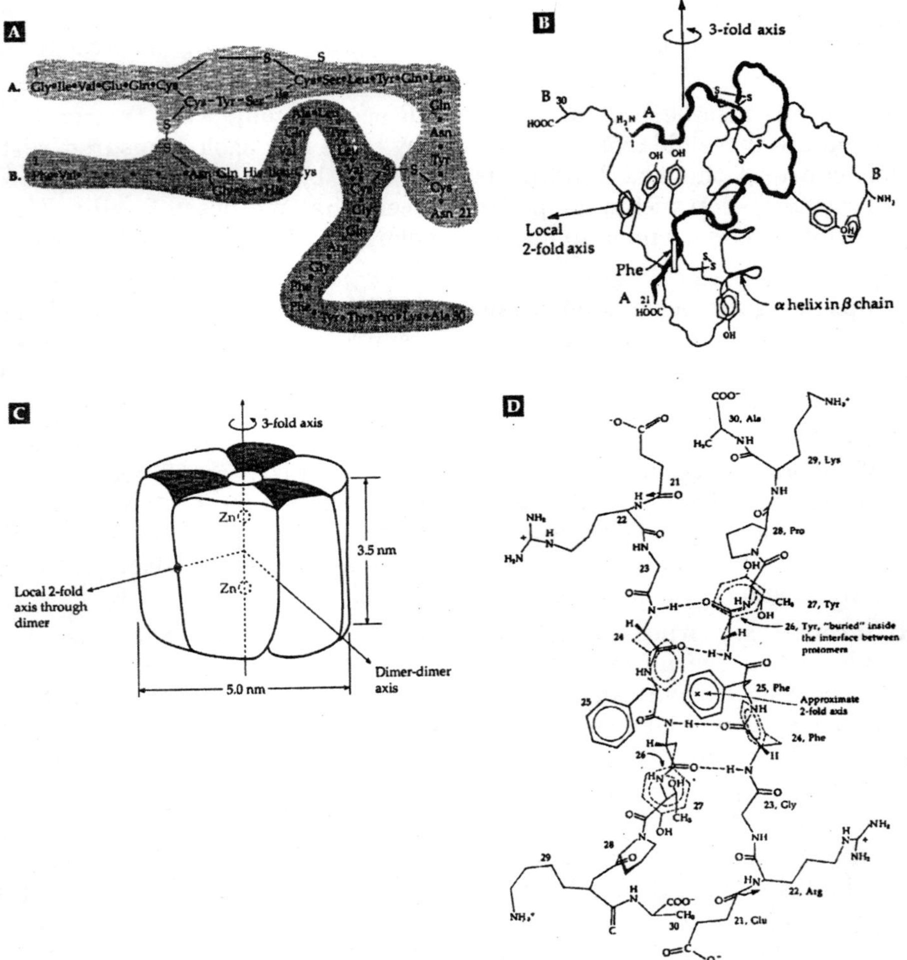

Fig. 1.16. The structure of insulin. (A) The amino acid sequence of the A and B chains linked by disulfide bridges. (B) Sketch showing the backbone structure of the insulin molecule as revealed by X–ray analysis. The A and B chains have been labeled. Positions and orientations of aromatic side chains are also shown. (C) View of the paired N–terminal ends of the B chains in the insulin dimer. View is approximately down the pseudo-twofold axis toward the center of the hexamer. (D) Schematic drawing showing packing of six insulin molecules in the zinc-stabilized hexamer.

The problem of quasi equivalence is resolved by the presence of an N–terminal extension that binds onto a subunit across the quasi-six-fold axis to give a set of three subunits that associate with true three–fold symmetry and another set (B) with a slightly different conformation fitting between them. The subunits A, which have a third conformation, fit together around the five-fold axis in true cyclic symmetry. A surprising finding is that the polyoma virus coat, which was expected to contain 420 (7 × 60) subunits, apparently contains only 360. The result is that the hexavalent morphological unit is a pentamer and that quasi-equivalence appears to be violated. Flexible arms tie the pentamers together. Quasi–equivalence of subunits also provides the supercoil in bacterial flagella and accounts for some interesting aspects of the structure of tobacco unosaic virus.

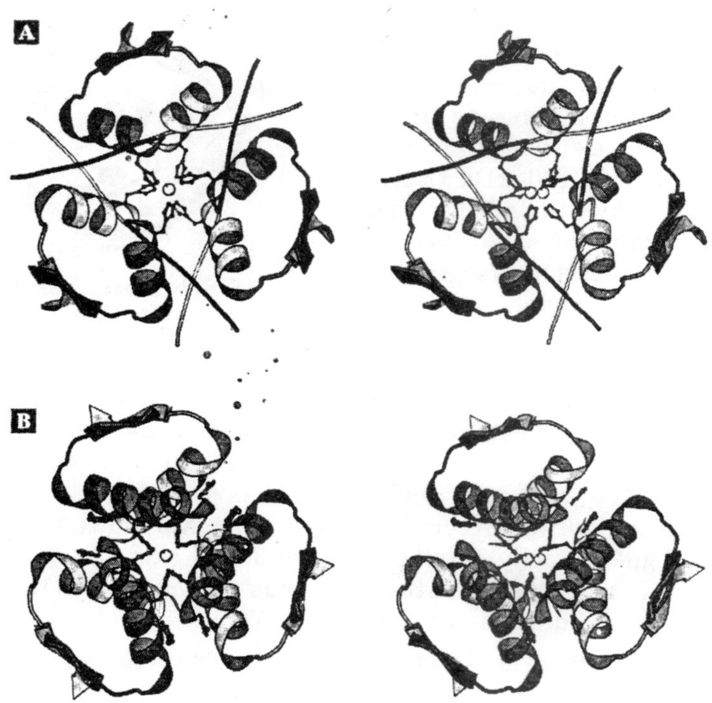

Fig. 1.17. Stereoscopic MolScript ribbon drawings of the B chains (A chains omitted) of (A) hexameric 2-zinc pig insulin. (B) A phenol complex of the same protein. Within each dimer the B chains are shaded differently. The Zn^{2+} ions are represented by white spheres and the coordinating histidine side chains are shown. Six noncovalently bound phenol molecules can be seen, as can several conformational differences.

The protein subunits of the virus can exist either as a helix with 16.3 subunits per turn or as a flat ring of 17 subunits. A very small conformational difference is involved. These rings dimerize but do not form larger aggregates. What is surprising is that the dimeric rings do not have dihedral symmetry, all of the subunits in the dimeric disk being oriented in one direction but with two different conformations.

The disk may serve as an intermediate in virus assembly. The inner portions of the quasi–equivalent disk subunits have a jawlike appearance as if awaiting the incorporation of RNA. As the RNA becomes bound, the disks could dislocate to a *"lock washer"* conformation to initiate and to propagate growth of the helical virus particle. However, there is uncertainty about this interpretation. Some enzymes, such as yeast hexokinase and creatine kinase associate in extremely asymmetric ways.

A dimer is formed by means of heterologous interactions but steric hindrance prevents the unsatisfied sets of interacting groups from joining with additional monomers to form higher polymers. As Galloway pointed out, many biological structures are not completely ordered but nevertheless possess well–defined and functionally important local relationships.

Regulatory subunits and multienzyme complexes. Proteins are often organized into large complexes, sometimes for the purpose of regulating metabolism. An example is *aspartate carbamoyltransferase* which catalyzes the first step in the synthesis of the pyrimidine rings of DNA and RNA. The 310-kDa enzyme from *E. coli* can be dissociated into two 100–kDa trimers, referred to as *catalytic subunits*. and three $34X_2$ – kDa dimers, the *regulatory subunits* which alter their conformations in response to changes in the ATP, UTP, and CTP concentrations.

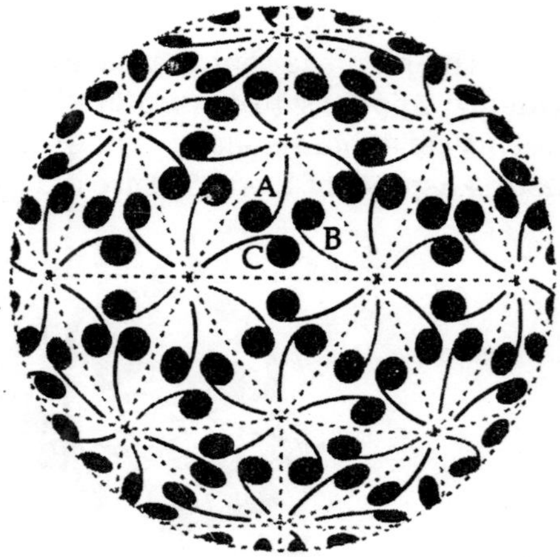

Fig. 1.18. Schematic icosahedrally symmetric structure with 180 subunits. The quasi-equivalent units A, B, and C are necessarily somewhat differently positioned with respect to their neighbours and must therefore assume different conformations in order to fit together tightly.

The molecule is roughly triangular in shape with a thickness of 9.2 nm and a length of the triangular side of 10.5 nm. The symmetry is 3:2, *i.e.,* it is dihedral with one 3–fold axis of rotation and three 2–fold axes. The two trimers of catalytic subunits lie face–to–face with the dimeric regulatory subunits fitting between them into the grooves around the edges of the trimers. The dimers are not aligned exactly parallel with the 3–fold axis, but to avoid eclipsing, the upper half of the array is rotated around the 3-fold axis with respect to the lower half. In the center is an aqueous cavity of dimensions ~2.5 × 5.0 × 5.0 nm.

The active sites of the enzyme are inside this cavity which is reached through six ~1.5 – nm opening around the sides. Many other oligomeric enzymes and other complex assemblies of more than one kind of protein subunit are known. For example, the *2–oxoacid dehydrogenases* are huge 2000–to 4000–kDa complexes containing three different proteins with different enzymatic activities in a cubic array. The filaments of striated muscle antibodies and complement of blood and the tailed bacteriophages all have complex molecular architectures.

COOPERATIVE CHANGES IN CONFORMATION

A substrate will bind better to some conformations of a protein than it will to others. This simple fact, together with the tendency for protein monomers to associate into clusters, allows for cooperative changes in conformation within oligomeric proteins. These changes provide the basis for important aspects of the regulation of enzymes and of metabolism. They impart cooperativity to the binding of small molecules such as that of oxygen to hemoglobin and of substrates and regulating molecules to enzymes.

Many of the most fundamental and seemingly mysterious properties of living tilings are linked directly to cooperative changes within the fibrils, membranes, and other structures of the cell. In 1965 a simple, appealing mathematical description of cooperative phenomena was suggested by Monod, Wyman, and Changeux and focused new attention on the phenomenon.

They suggested that conformational changes in protein subunits, which could be associated with altered binding characteristics, occur cooperatively within an oligomer. For example, binding of phenol to hexameric 2–zinc insulin could induce all six individual subunits to change their conformation together, preserving the D_6 symmetry of the complex. The four subunits of hemoglobin could likewise change their conformation and affinity for O_2 Synchronously.

This is very nearly true and is of major physiological significance. Consider an equilibrium between protein molecules in two different conformations A and B (T and R in the MWC terminology) and containing a single binding site for molecule X. In the Monod–Wyman–Changeux (MWC) model the conformations are designated T (tense) and R (relaxed) but in the interest of providing a more general treatment the terminology used in this book is that of Koshland *et al.*

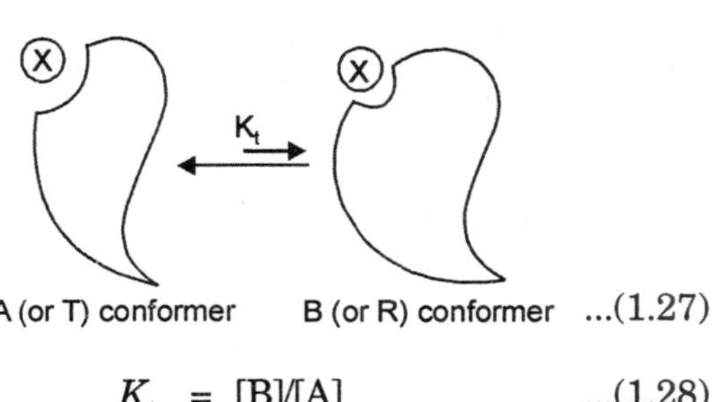

...(1.27)

$$K_t = [B]/[A] \qquad ...(1.28)$$

If the equilibrium constant K_t is approximately 1, the two *conformers* have equal energies, but if $K_t < 1$, A is more stable than B.

Unequal Binding of Substrate and "Induced Fit"

Assume that conformer B binds X more strongly than does conformer A (as is suggested by the shapes of the binding sites in Eq.1.27. The intrinsic binding constants to the A and B conformers K_{AX} and K_{BX} (or K_T and K_R) are defined by Eq. 1.29.

$$K_{AX} = [AX]/[A][X]$$
$$K_{BX} = [BX]/[B][X] \qquad ...(1.29)$$

The entire set of equilibria for this system are shown in Eq. 1.30. Note that the constant relating BX to AX is not independent of the other three constants but is given by the expression

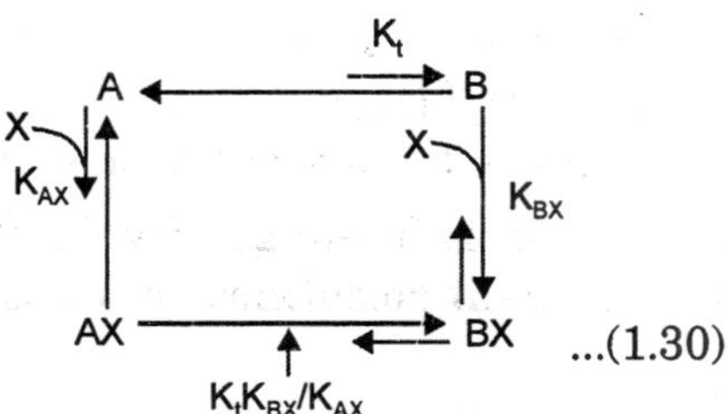

...(1.30)

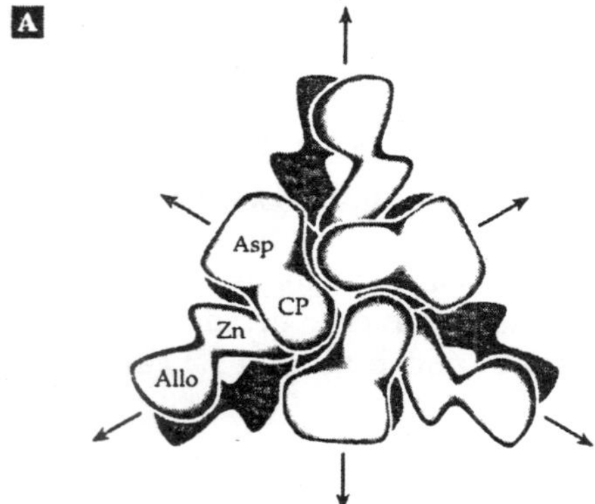

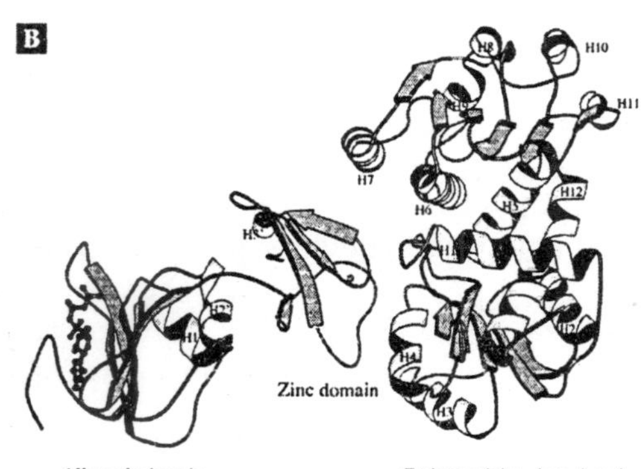

Fig. 1.19. (A) Subunit assembly of two C_3 catalytic trimers (green) and three R_2 regulatory dimers around the periphery in aspartate carbamoyltransferase. The aspartate and carbamoylphosphate–binding domains of the catalytic subunits are labeled Asp and CP, respectivley, while the zinc and allosteric domains of the regulatory subunits are labeled Allo and Zn, respectively. (B) Ribbon drawing of a single pair of regulatory (left) and catalytic (right) subunits with the structural domains labeled.

K_tK_{BX} / K_{AX}. Now consider the following situation. Suppose that A predominates in the absence of X but that X binds more tightly to B than to A. There will be largely either free A or BX in the equilibrium mixture with smaller amounts of AX and B.

An interesting kinetic question arises. By which of the two possible pathways from A to BX will the reaction take place? The first possibility, assumed in the MWC model, is that X binds only to performed B, which is present in a small amount in equilibrium with A. The second possibility is that X can bind to A but that AX is then rapidly converted to BX. We could say that X *induces a conformational change* that leads to a better fit. This is the basis for the *induced fit* theory of Koshland. Bear in mind that the equilibrium constants can give us the equilibrium concentrations of all four forms in Eq. 1.30. However, rates of reaction are often important in metabolism and we cannot say *a priori* which of the two pathways will be followed.

If K_{BX} / K_{AX} is very large, an insignificant amount of AX will be present at equilibrium. In such a case there is no way experimentally to determine K_{AX}. The two constants K_t and K_{BX} are sufficient to describe the *equilibria* but an induced fit mechanism may still hold.

Now consider the association of A and B to form oligomers in which the intrinsic binding constants K_{AX} and K_{BX} have the same values as in the monomers. Since more enzymes apparently exist as isologous dimers than as any other oligomeric form, it is appropriate to consider the behaviour of such dimers in some detail. Monod *et al.* emphasized that both conformers A and B (T and R) can associate to form isologous dimers in which symmetry is preserved Eq. 1.31.

K = 1/L (in MWC terminology)

A A ⇄ B B ...(1.31)

On the other hand, association of B and A would lead to an unsymmetric dimer in which bonding between subunits might be poor:

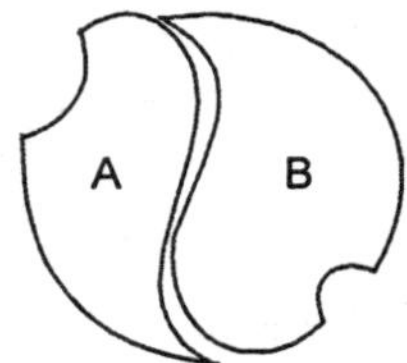

Mixed AB dimer
which associates weakly

In the MWC treatment, the assumption is made that the mixed dimer AB can be neglected entirely. However, a more general treatment requires that we consider all dimeric forms. The formation constants K_{AA}, K_{BB}, and K_{AB} are defined as follows.

$$2 \rightleftarrows A_2 K_{AA} = \frac{[A_2]}{[A]^2} \quad ...(1.32)$$

$$2B \rightleftarrows B_2 \; K_{BB} = \frac{[B_2]}{[B]^2} = \frac{[B_2]}{K_t^{\,2}[A]^2} \quad ...1.33$$

$$A + B \rightleftarrows AB \qquad K_f = 2K_{AB} = \frac{[AB]}{[A][B]} = \frac{[AB]}{K_t[A]^2} \qquad ...(1.34)$$

Binding Equilibria for a Dimerizing Protein

All of the equilibria of Eqs. 1.28 through 1.34 involved in formation of dimers A_2 AB, and B_2 and in the binding of one or two molecules of X per dimer are depicted in Fig. 1.21. Above each arrow the microscopic constant associated with that step is shown multiplied by an appropriate statistical factor. The fractional saturation Y is given by Eq. 1.35. Each of the nine terms in the numerator gives the concentration of bound X represented by one of the nine

$$2Y \text{ (based on dimer)} = \frac{[AX] + [BX] + [A_2X] + 2[A_2X_2] + [ABX] + [AXB] + 2[ABX_2] + [B_2X] + 2[B_2X_2]}{\frac{1}{2}([A] + [AX] + [B] + [BX]) + [A_2] + [A_2X] + [A_2X_2] + [AB] + [ABX] + [AXB] + [ABX_2] + [B_2] + [B_2X] + [B_2X_2]} \qquad ...(1.35)$$

forms containing X in Fig. 1.21. The 14 terms in the denominator represent the concentration of protein in each form including those containing no bound X. Protein concentrations are given in terms of the molecular mass of the dimer; hence, some of the terms in the denominator are multiplied by 1/2.

All of the terms in both the numerator and the denominator of Eq. 1.35 can be related back to [X], using the microscopic constants from Fig.1.21 to give an equation which presents Y in terms of [X], K_{AX} and K_{BX}, K_t, and the interaction constants K_{AA}, K_{AB}, and K_{BB}. Since the equation is too complex to grasp immediately, let us consider several specific cases in which it can be simplified.

The Monod–Wyman–Changeux (MWC) model. If both K_{AA} and K_{BB} are large enough, there will be no dissociation into monomers. The transition between conformation A and conformation B can occur cooperatively within the dimer or higher oligomer, and the mathematical relationships shown in Fig. 1.21 are still appropriate. One further restriction is needed to describe the MWC model. Only symmetric dimers are allowed. That is, K_{AA} and $K_{BB} \gg K_{AB}$ and only those equilibria indicated with green arrows in Fig. 1.21 need be considered. In the absence of ligand X, the ratio $[B_2]/[A_2]$ is a constant, 1/L in the MWC terminology

$$\frac{[B_2]}{[A_2]} = \frac{1}{L} = \frac{K_{BB}}{K_{AA}} K_t^2 \qquad ...(1.36)$$

Both of the association constants K_{AA} and K_{BB} and the transformation constant K_t affect the position of the equilibrium. Thus, a low ratio of $[B_2]$ to $[A_2]$ could result if K_{BB} and K_{AA} were similar but K_t was small. If K_t were ~1 a low ratio could still arise because $K_{AA} > K_{BB}$, *i.e.*, because the subunits are associated more tightly in A_2 than in B_2. For this case Eq. 1.35 simplifies to Eq. 1.37.

$$2Y = \frac{[A_2X] + 2[A_2X_2] + [B_2X] + [B_2X_2]}{[A_2] + [A_2X] + [A_2X_2] + [B_2] + [B_2X] + [B_2X_2]}$$

$$2K_{AA}K_{AX}[X] + 2K_{AA}K_{AX}^2[X]^2$$

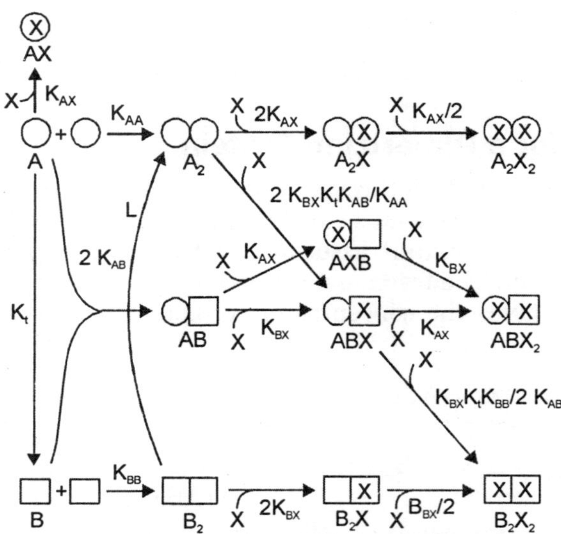

Fig. 1.20. Possible forms of a dimerizing protein existing in two conformations with a single binding site per protomer for X. Green arrows indicate equilibria considered *by* MWC. Solid arrows indicate equilibria considered *by* Koshland *et al.* Heavy gray arrows are for the simplest induced fit model with no dissociation of the dimer. Note that all equilibria are regarded as reversible (despite the unidirectional arrows). K_{AX} and K_{BX} are assumed the same for subunits in monomeric and dimeric forms.

$$= \frac{+2K_{BB}K_{BX}K_t^2[X] + 2K_{BB}K_{BX}K_t^2[X]^2}{KAA + 2K_{AA}K_{AX}[X] + K_{AA}K_{AX}^2[X]^2 + K_{BB}K_t^2}$$

$$+ 2K_{BB}K_{BX}K_t^2[X] + K_{BB}K_{BX}^2K_t^2[X]^2 \qquad ..(1.37)$$

Substituting from Eq. 1.36 into Eq. 1.37 we obtain Eq. 1.38:

Y (for dimer)

$$= \frac{L \cdot K_{AX}[X](1 + K_{AX}[X]) + K_{BX}[X](1 + K_{BX}[X])}{L(1 + K_{AX}[X])^2 + (1 + K_{BX}[X])^2} \qquad ...(1.38)$$

For an oligomer with n subunits Monod *et al.* assumed that all sites in either conformer are independent and equivalent. The equation for Y (based on Eq. 1.17 is

$$Y = \frac{L \cdot K_{AX}[X](1 + K_{AX}[X])^{n-1} + K_{BX}[X](1 + K_{BX}[X])^{n-1}}{L(1 + K_{AX}[X])^n + (1 + K_{BX}[X])^n} \qquad ...(1.39)$$

We assume initially that B_2 binds X more strongly than does A_2. Hence, if the equilibrium in Eq. 1.36 favours B_2 strongly (L is small), the addition of X to the system will not shift the equilibrium between the two conformations and binding will be noncooperativ. However, if the equilibrium favours A_2 (L is large), addition of X will shift the equilibrium in favour of B_2 (which binds X more tightly). Furthermore, since the expression for Y Eq. 1.39 contains a term in $K^2_{BX}[X]^2$ in the numerator, binding will tend to be cooperative. In the extreme case that L is large and K_{AX} ~0, most of the terms in Eq. 1.39 drop out and it approaches the equation previously given for completely cooperative binding Eq. 1.21 with $K = K_{BX}^2L$. With other values of K_{AX}, K_{BX}, and L incomplete cooperativity is observed.

The induced fit model. In this model, only A_2, ABX, and B_2X_2 are considered (heavy arrows in Fig. 1.15. The expression for 2Y is:

$$2Y = \frac{[ABX]+2[B_2X_2]}{[A_2]+[ABX]+[B_2X_2]}$$

$$= \frac{2K_{BX}K_t\frac{K_{AB}}{K_{AA}}[X]+2(K_{BX}K_t)^2\frac{K_{BB}}{K_{AA}}[X]^2}{1+2K_{BX}K_t\frac{K_{AB}}{K_{AA}}[X]+(K_{BX}K_t)^2\frac{K_{BB}}{K_{AA}}[X]^2} \quad \text{...(1.40)}$$

The constants used here are defined by Eqs. 1.8 through 1.10 and differ from those of Koshland, who sometimes arbitrarily set $K_{AA} = 1$ and redefined K_{BB} as an *interaction constant* equal to K_{BB} / K_{AA}. Although this simplifies the algebra it is appropriate only for completely associated systems and might prove confusing.

When K_{AB} is small Eq. 1.16 (no "mixed" dimer) also simplifies to Eq. 1.45 for completely cooperative binding with the value K given by Eq. 1.17. On the other hand, if K_{AB} is large compared to K_{AA} and K_{BB}, anticooperativity (negative cooperativity) will be observed. The saturation curve will contain two separate steps just as in the binding of protons by succinate dianion. Fig. 1.5.

$$\overline{K} = K_{BX}^2K_t^2\frac{K_{BB}}{K_{AA}} = \frac{K_{BX}^2}{L} \quad \text{...(1.41)}$$

One conformational state dissociated. It may happen that A_2 is a dimer but that B_2 dissociates into monomers because K_{BB} is very small. In such a case binding of X leads to dissociation of the protein. A well-known example is provided by hemoglobin of the lamprey which is a dimer and which dissociates to a monomer upon binding of oxygen. Equation 1.11 simplifies to Eq. 1.42. The reader may wish to consider whether the weakly cooperative binding of oxygen by lamprey hemoglobin is predicted by this equation.

$$2Y = \frac{[BX]}{[A_2]+\frac{1}{2}[BX]} \quad \text{...(1.42)}$$

Look again at the expression for L, the constant determining the relative amounts of a protein in conformations A and B in the absence of ligand. From Eq. 1.36 we see that a large value of L (conformer A favoured) can result either because K_t is very small or because $K_{BB} \ll K_{AA}$. Thus, if $K_t \sim 1$ and L is large, the subunits must associate much more weakly in B_2 than in A_2 and the chances are that binding of X will dissociate the molecule as in the case of lamprey hemoglobin.

On the other hand, if K_t is very small, implying that the molecule is held in conformation A because of some intrinsically more stable folding pattern in that conformation, K_{BB} might exceed K_{AA} very much; if K_{AA} were low enough A_2 could be completely dissociated. Binding of ligand would lead to association and to cooperative binding. This can be verified by writing down the appropriate terms from Eq. 1.35.

Higher Oligomers

Mathematical treatment of binding curves for oligomers containing more than two subunits is complex, but the algebra is straightforward. A computer can be programmed to do necessary calculations. Avoid picking an equation from the literature and assuming that it will be satisfactory. Consider the two tetrameric structures in Fig. 1.22. In the isologous square separate contributions to the free energy of binding can be assigned to the individual pairs of interactions *aj* and *bk*.

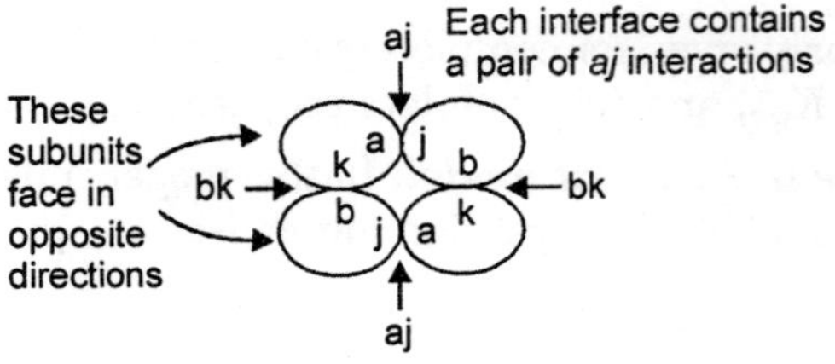

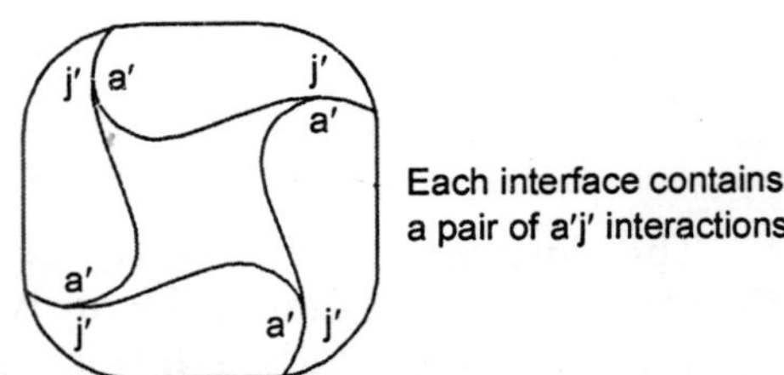

Fig. 1.21. Comparison of the interactions in isologous (dihedral) and heterologous (cyclic) square configurations of subunits.

Thus, following Cornish–Bowden and Koshland for assembly of the tetramer Eq. 1.43:

$$\Delta G_f = 2\,\Delta G_{ajAA} + 2\,\Delta G_{bkAA}$$

$$K_f = K^2_{ajAA} K^2_{bkAA} \qquad ...(1.43)$$

Since Gibbs energies are additive, the formation constant will be the product of formation constants representing the individual interactions; thus, K_{ajAA} represents the formation constant of a dimer in which only the *aj* pair of bonds is formed. In the isologous tetrahedron the third set of paired interactions cl must be taken into account. (However, the third interaction constant will not be an association constant of the type represented by K_{ajAA} and K_{bkAA} but a dimensionless number.)

On the other hand, the heterologous square has only a single interaction constant. the binding of one molecule of X to the isologous tetramer with a conformational change in one subunit. We see that one pair of *aj* interactions and one pair of *bk* interactions have been altered. The equilibrium constant for the binding of X to the tetramer will be in which the 4 is a

$$A_4 + X \rightarrow A_3BX \qquad (1.44)$$

$$K = u\frac{K_{ajAB}K_{bkAB}}{K_{ajAA}K_{bkAA}}K_{BX}K_t \qquad ...(1.45)$$

statistical factor arising from the fact that there are four different ways in which to form A_3BX. When a second molecule of X is added three geometrical arrangements are possible:

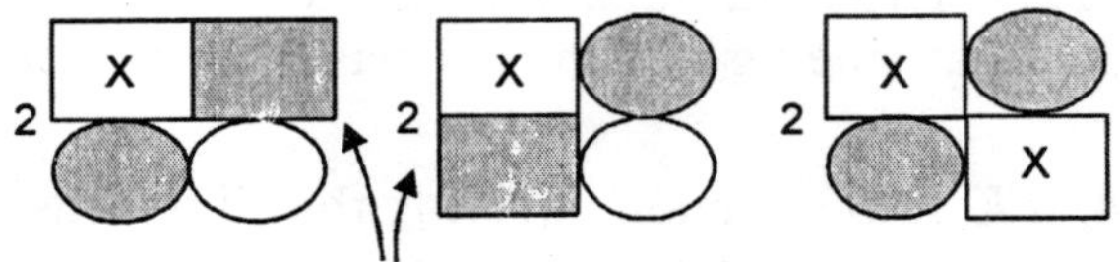

Each one can be formed in two ways. It is a simple matter to write down the microscopic constants for addition of the second molecule of X as the sum of three terms. Because values of the constants for *aj* and *bk* differ, it will be clear that the three ways of adding the second molecule of X are not equally probable.

Thus, the oligomer will show preferred orders of "loading" with ligand X. Two different geometries for the heterologous tetramer are possible in form $A_2B_2X_2$. Again, the different arrangements need not be equally probable and the relative distribution of each will be determined by the specific values of the interaction constants. In the heterologous tetramer A_4 only one type of an interaction is present between subunits.

However, as soon as-a-single molecule ot *X* is bound and one subunit of conformation B is present, two kinds of *aj* interactions exist. (One in which group *a* is present in conformation A and the other in which it is present in conformation B.) Since these interactions always occur in equal numbers they can be lumped together. While the foregoing may seem like an unnecessarily long exercise, it should provide a basic approach which can be applied to specific problems. However, remember that mathematical models require simplification. Real proteins often have more than two stable conformations. The entire outside surface of a protein is made up of potential binding sites for a number of different molecules, both small and large. Filling of almost any of these sites can affect the functioning of a protein.

D. THE OXYGEN-CARRYING PROTEINS

1. Myoglobin and Hemoglobin

The most studied example of a conformational change in a multisubunit protein induced by binding of a small molecule is provided by the cooperative binding of oxygen to hemoglobin. Mammalian hemoglobin is an $\alpha_2\beta_2$ tetramer of ~ 16-kDa subunits, each containing 140 – 150 residues. Within each subunit the peptide chain folds in a characteristic largely α-helical pattern around a single large flat iron-containing ring structure called.

The folding is essentially the same in all hemoglobins, both in the α and β subunits and in the monomeric muscle oxygen storage protein, myoglobiname Amino acid residues are customerily designated by their position in one of the eight helices A–H. The imidazole group of histidine F–8 is coordinated with the iron in the center of the heme on the "proximal" side. The other side of the iron atom (the "distal" side) is the site of binding of a single molecule of O_2. Although the folding of the peptide chain is almost the same in both subunits, and almost identical to that of myoglobin, there are numerous differences in the amino acid sequence. If it were not for these differences, hemoglobin would be a highly symmetric molecule with the bonding pattern indicated in Fig 1.5 with three 2-fold axes of rotation. In fact, hemoglobin has one true axis of rotation and two pseudo-twofold axes. There are two sets of true isologous interactions (those between the two α subunits and between the two β subunits) and two pairs of unsymmetrical interactions (between α and β subunits). The nearly symmetric orientation of different portions of the peptide backbone is clearly seen in the beautiful drawings of Geis. The contact region involved in one pair of interactions in hemoglobin ($\alpha_1\beta_1$) is more extensive than the other. There is close contact between 34 different amino acid side chains and 110 atoms lie within 0.4 nm of each other.

Hydrophobic bonding is the principal force holding the two subunits together, and only a few reciprocal contacts of the type found in a true isologous bond remain. The second contact designated $\alpha_1\beta_2$ involves only 19 residues and a total of 80 atoms. Because this interaction is weaker, hemoglobin dissociates relatively easily into $\alpha\beta$ dimers held together by the $\alpha_1\beta_1$ contacts and motion occurs along the $\alpha_1\beta_2$ contacts during oxygenation. The truly isologous interactions (*i.e.*, αX_2 and ββ) are weak because the identical protomers hardly touch each other.

The binding of oxygen. Curves of percentage oxygenation (Y) vs the partial pressure of O_2 are given in Fig. 1-24 and illustrate the high degree of cooperativity. Depending upon conditions, values of n_{Hill} may be as high as 3. As a result of this cooperativity the hemoglobin, in the capillaries of the lungs at a partial pressure of oxygen of ~ 100 mm of mercury, is nearly saturated with oxygen. However, when the red cells pass through the capillaries of tissues in which oxygen is utilized the partial pressure of oxygen falls to about 5 mm of mercury.

The cooperativity means that the oxygen is more completely "unloaded" than it would be if all four heme groups acted independently. Deoxyhemoglobin has a low affinity for O_2, but the observed cooperativity in binding implies that in the fully oxygenated state the O_2 is held with a high affinity. The monomeric myoglobin also has a high affinity for oxygen, as does the abnormal *hemoglobin H*, which is made up of four β subunits. The latter also completely lacks cooperativity in binding. These results can be interpreted according to the MWC model to indicate that deoxyhemoglobin exists in the T (A) conformation, whereas oxyhemoglobin is in the R (B) conformation.

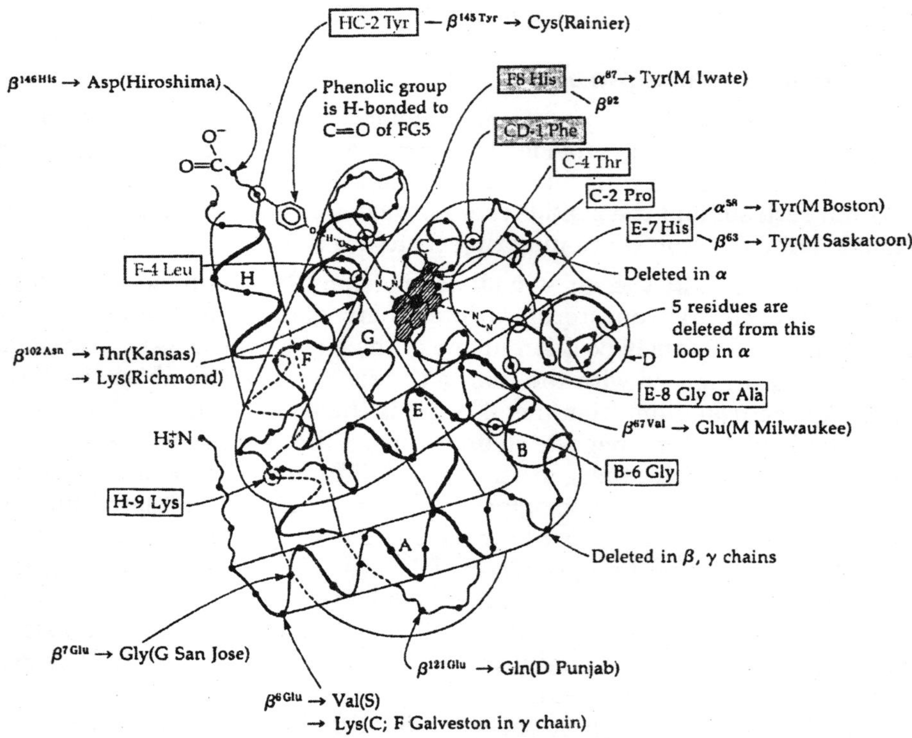

Fig. 1.22. Folding pattern of the hemoglobin monomers. The pattern shown is for the β chain of human hemoglobin. Some of the differences between this and the α chain and myoglobin are indicated. Evolutionarily conserved residues are indicated by boxes, highly conserved, invariant. Other markings show substitutions observed in some abnormal human hemoglobins. Conserved residues are numbered according to their location in one of the helices A-H, while mutant hemoglobins are indicated by the position of the substitution in the entire α and β chain.

Myoglobin stays in the R conformation in *both* states of oxygenation as do the separated α and β chains of hemoglobin. The subunits of hemoglobin H also appear to be frozen in the R conformation, even though the quaternary structure is similar to that of deoxyhemoglobin. Oxygenation curves of hemoglobin are often fitted with the Adair equation. Thus, at pH 7.4 under the conditions given in Table 1.2, Imai found for the successive formation constants $K_1 = 0.004$, $K_2 = 0.009$, $K_3 = 0.002$ and $K_4 = 0.95$ in units of mm Hg^{-1}. From the definition of a formation constant the oxygen pressure Po_2 required for 50 % oxygenation in the first step will be at $Po_2 = 1 / K_f$ or $\log Po_2 = \log (1 /.004) = 2.4$. This is a high oxygen pressure, far to the right side of the oxygenation curve in Fig. 1.24A. However, $\log K_4$ is about 0.02, well to the left on the oxygenation curve.

From these formation constants we can say that after three of the subunits have become oxygenated the affinity of the remaining subunit has increased about 300-fold when the concentration of the effector 2,3-bisphosphoglycerate is present at the normal physiological concentration. However, we must ask what uncertainties are present in the data used to obtain these constants. To extract four successive binding constants from a curve like that in Fig. 1.24A is extremely difficult. This fact has encouraged the widespread use of the simpler MWC model. When the same data were treated by Imai using the MWC model it was found that $L = 2.8 \times 10^6$ and $c = K_{f(T)} / K_{f(R)} = 0.0038$. Changes in enthalpy, entropy, and Gibbs energy are given in Table 1.2. Kintic data as well as O_2-binding measurements with single crystals are partially consistent with the MWC model.

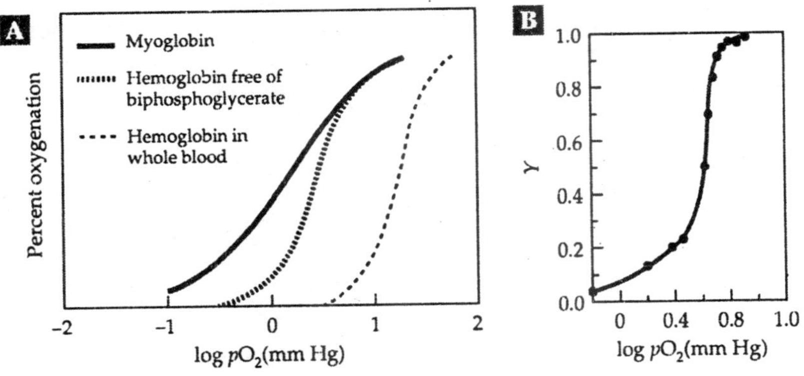

Fig. 1.23. Cooperative binding of oxygen by hemoglobins. (A) Binding curve for myoglobin (noncooperative) and for hemoglobin in the absence and presence (in whole blood) of 2,3-bisphosphoglycerate. Oxygen affinity is decreased by bisphosphoglycerate. (B) Saturation curve for hemoglobin (erythrocruorin) of *Arenicola*, a spiny annelid worm. The molecule contains' 192 subunits and 96 hemes. It shows very strong cooperativity with n_{Hill} ~ 6.

However, the discovery of a third quaternary structure of hemoglobin, similar to the R state but distinct from it, emphasizes the complexity of this allosteric molecule. Hemoglobin tetramers tend to stay tightly associated but some dissociation of oxyhemoglobin into dimers does occur ($K_f = 7 \times 10^5$ M). Deoxyhemoglobin is about 40,000 times more tightly associated. All of the equilibria involved are strongly affected by pH and by the presence of salts such as NaCI. This is in part a reflection of the strong role of ionic interactions in holding together the subunits in the T state as is discussed in the following sections.

Table 1.2. Thermodynamic Functions for Oxygenation of Hemoglobin"

Reaction	*ΔH (kj mol^{-1})*	*ΔS (J°K^{-1} mol^{-1})*	*ΔG (kJ mole^{-1})*	K_f
$T \rightarrow T(O_2)_4$	– 51 ±1	– 154 ±4	–5.0 ±1.7	7.5
$R \rightarrow R(O_2)_4$	– 62 ± 2	– 146 ± 6	– 19 ± 2	2.0×10^3
T→R (unoxygenated)	– 70 ± 7 × 3	–111 ± 25	–37 ±10	3.6×10^{-7}
$T(O_2)_4 \rightarrow R(O_2)_4$ (oxygenated)			– 19	2.1×10^3

Parameters for MWC model

$L = (3.6 \times 10^{-7})^{-1} = 2.8 \times 10^6$

$c = K_{f(T)}/K_{f(R)} = 0.0038$

[a] The measurements were made at pH 7.4 in the presence of 0.1 M chloride ion and 2 mM 2,3-bisphosphoglycerate to mimic physiological conditions. The values of ΔH, ΔS, and ΔG given are per mole of heme, i.e., per monomer unit. They must be multiplied by 4 to correspond to the reactions as shown for the tetramer.

Structural changes accompanying oxygen binding. Perutz and associates, using X-ray crystallography, found *small* but real differences in the conformation of the subunits of deoxy- and oxy-hemoglobin. More striking is the fact that upon oxygenation, both α and β subunits undergo substantial amounts of *rotation,* the net result being that the hemes of the two β subunits move about 0.07 nm closer together in the oxy form than in the deoxy form. Within the $\alpha_1\beta_1$, contacts little change is seen.

On the other hand, contact $\alpha_1\beta_2$, the *"allosteric interface"* is altered drastically. As Perutz expressed it, there is a *"jump in the dovetailing"* of the CD region of the subunit relative to the FG region of the β subunit. The hydrogen-bonding pattern is also changed. A major difference is seen in the hydrogen-bonded salt bridges present at the ends of the molecules of deoxyhemoglobin. The $-NH_3^+$ group of Lys H–10 in each ∝ subunit is hydrogen bonded to the carboxyl group of the C-terminal arginine of the opposite α chain.

The guanidinium group of each C-terminal arginine is hydrogen bonded to the carboxyl group of Asp H-9 in the opposite α chain. It is also hydrogen bonded to an inorganic anion (phosphate or Cl^-), which in turn is hydrogen-bonded to the α amino group of Val 1 of the opposite ∝ chain forming a pair of isologous interactions. At the other end of the molecule, the C-terminal group of His 146 of each β chain binds to the amino group of Lys C–6 of the α chain, while the imidazole side chain binds to Asp FG–1 of the same β chain. These salt bridges appear to provide extra stability to deoxyhemoglobin and account for the high value of the constant L. In deoxyhemoglobin the side chain of the highly conserved Tyr HC–2 lies tucked into a pocket between the H and F helices and is hydrogen bonded to the main chain carbonyl of residue FG-5.

Upon oxygenation this tyrosine in each subunit is released from its pocket; the salt bridges at the ends of the molecules are broken and the subunit shifts into the new bonding pattern characteristic of oxyhemoglobin. Cooperativity in O_2 binding is absent or greatly decreased in mutant hemoglobins with substitutions in the residues involved in these salt bridges or in residues lying in the $\alpha_1\beta_2$. How does the binding of O_2 to the iron of heme trigger the

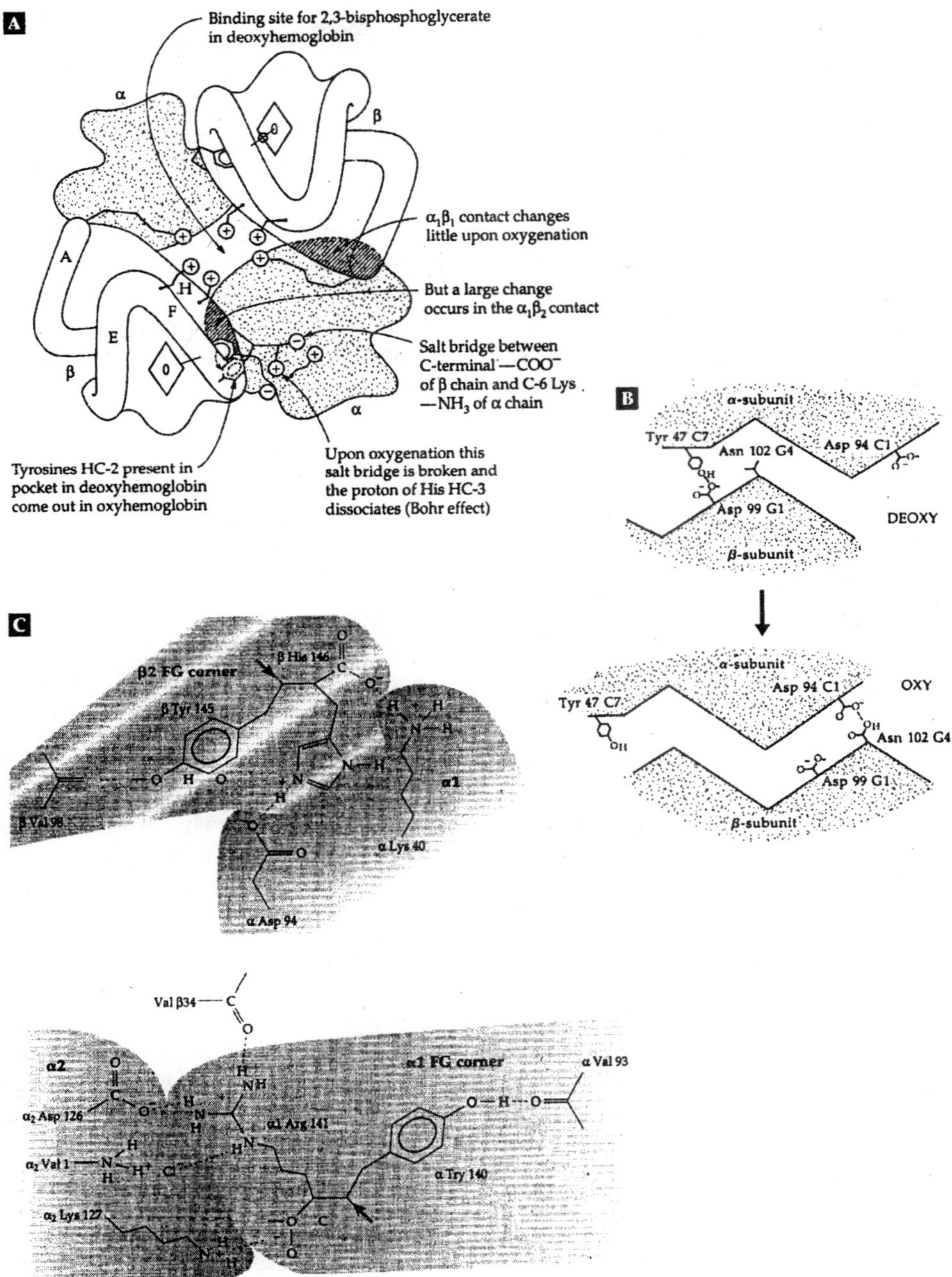

Fig. 1.24. (A) Structural changes occurring upon oxygenation of hemoglobin. (B) "Rotation at the contact $α_1β_2$ causes a jump in the dovetailing of the CD region of α relative to the FG region of β and a switch of hydrogen bonds as shown". (C) Some details of the salt bridges.

conformational change in hemoglobin? An enormous amount of effort by many people has been expended in trying to answer this question.

The iron atom in deoxyhemoglobin lies a little outside the plane of the heme rings. When oxygenation occurs the iron atom moves toward the oxygen and into the plane of the heme. This movement probably amounts to only about 0.05 nm. Nevertheless, this small displacement evidently induces the other structural changes that are observed. The iron pulls the side chain of histidine F-8 with it and moves helix F which is also hydrogen bonded to this imidazole ring.

Because of the tight packing of the various groups this motion cannot occur freely but is accompanied by a movement of the F helix by 0.1 nm across the heme plane. These movements may induce additional structural changes in the irregularly folded FG corners that allow the subunits to shift to the new stable position of the R state. All four subunits appear to change conformation together. This conformational change must also cause the affinity for oxygen of any unoxygenated subunits to rise dramatically, presumably by shifting the iron atoms into the planes of the heme rings.

'This ensures the cooperative loading of the protein by O_2. While there is no doubt that the iron atom moves upon oxygenation, it is not obvious that this will lead to the observed cooperativity. Oxygenated heme has some of the characteristics of an Fe(III)-peroxide anion complex. The iron atom acquires an increased positive charge upon oxygenation by donating an electron for bond formation.

$$Fe^{2+}\text{ (deoxy)} + O_2 \longrightarrow Fe^{3+} - O_2^- \qquad \text{...(1.46)}$$

This may transmit an electronic effect through either His F-8 or the heme ring to the nearby $\alpha_1\beta_2$ interface and also affect the subunit interactions.

The Bohr effect and allosteric regulators. The breaking of the salt bridges at the ends of the hemoglobin molecule upon oxygenation has another important result. The pK_a values of the N-terminal valines of the α subunits and of His HC-3 of the β subunits are abnormally high in the deoxy form because they are tied up in the salt bridges. In the oxy form in which the groups are free, the pK_a values are lower. If hemoglobin is held at a constant pH of 7, these protons dissociate upon oxygenation. This *Bohr effect,* described in 1904 is important because acidification of hemoglobin stabilizes the deoxy form.

In capillaries in which oxygen pressure is low and in which carbon dioxide and lactic acid may have accumulated, the lowering of the pH causes oxyhemoglobin to release oxygen more efficiently. These effects are also strongly dependent on the presence of chloride ions. Just as the conformational equilibria in hemoglobin can be shifted by attachment of oxygen to the heme groups, so the binding of certain other molecules at different sites can also affect the conformation. Such compounds are called *allosteric effectors* or *regulators* because they bind at a site other than the "active site."

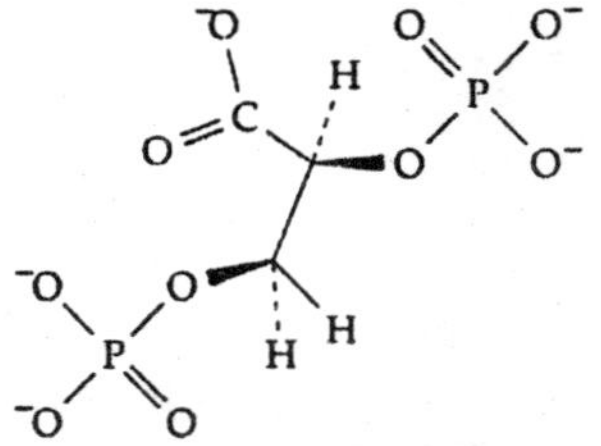

2,3-Bisphosphoglycerate

An important allosteric effector for human hemoglobin is *2,3-bisphosphoglycerate,* a compound found in human red blood cells in a high concentration approximately equimolar with that of hemoglobin. One molecule of bisphosphoglycerate binds to a hemoglobin tetramer in the deoxy form with $K_f = 1.4 \times 10^5$ but has only half this affinity for oxyhemoglobin.159 X-ray crystallography show's that bisphosphoglycerate binds between the two β chains of deoxyhemoglobin directly on the twofold axis. Because of the presence of 2, 3-bisphosphoglycerate in erythrocytes the affinity of oxygen for hemoglobin in whole blood is less than that for isolated hemoglobin. This is important because it allows a larger fraction of the oxygen carried to be unloaded from red corpuscles in body tissues.

The bisphosphoglycerate level of red cells varies with physiological conditions, e.g., people living at higher elevations have a higher concentration. It has been suggested that artificial manipulation of the level of this regulatory substance in erythrocytes may be of clinical usefulness for disorders in oxygen transport.

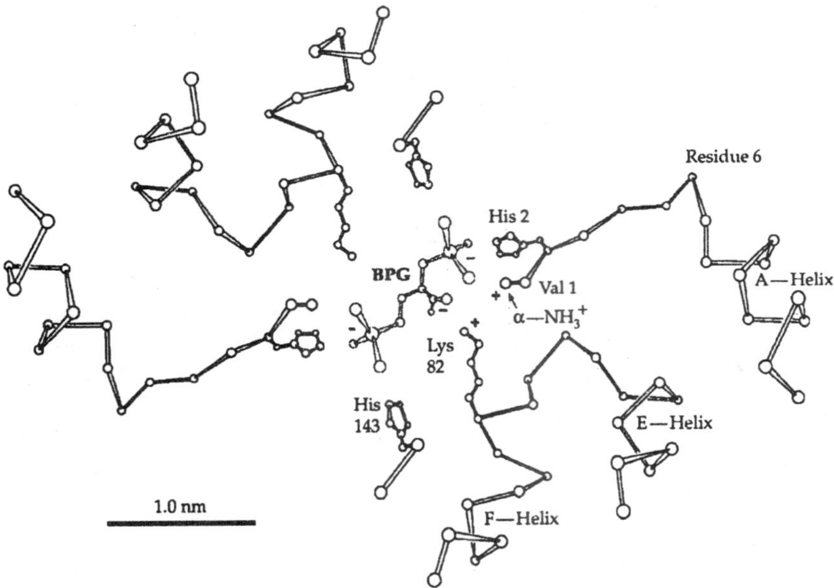

Fig. 1.25. the allosteric effects of 2, 3-bisphosphogly-cerate (BPG) bound to the β chains of human deoxyhemo-globin. The phosphate groups of the BPG form salt bridges with valines 1 and histidines 2 and 143 of both (3 chains and with lysine 82 of one chain. This binding pulls the A helix and residue 6 toward the E helix.

Not all species contain 2,3-bisphosphoglycerate in their erythrocytes. In birds and turtles its function appears to be served by inositol pentaphosphate. In crocodiles the site between the two β chains that binds organic phosphates in other species has been modified so that it binds bicarbonate ion, HCO_3^-, specifically. This ion, which accumulates in tissues as crocodiles lie under water, acts as an allosteric regulator in these animals. It allows the animals to more completely utilize the O_2 from the hemoglobin and to remain under water longer.

'The Bohr effect, which was considered in the preceding section, can be viewed as resulting from the action of *protons* as allosteric effectors that bind to the amino and imidazole groups of the salt linkages. *Carbon dioxide* also acts as a physiological effector in mammalian blood by combining reversibly with NH_2-terminal groups of the α and β subunits to form *carbamino* ($-NH-COO^-$) groups. It is the N-terminal amino groups rather than lysyl side chain groups that undergo this reaction. Because of their relatively low pK_a, values there is a significant fraction of unprotonated $-NH_2$ groups at the pH of blood. The affinity for CO_2 is highest in deoxygenated hemoglobin. Consequently, unloading of O_2 is facilitated in the CO_2-rich respiring tissues. Hemoglobin carries a significant fraction of CO_2 to the lungs, and there the oxygenation of hemoglobin facilitates the dissociation of CO_2 from the carbamino groups. Hemoglobin is also one of the major pH buffers of blood.

$$\text{Protein}-NH_2 + CO_2 \longrightarrow -\underset{\underset{H}{|}}{H}-\overset{\overset{O}{\|}}{C}-O^- \quad + H^+$$

Carbon monoxide, cyanide, and nitric oxide. A danger to hemoglobin and other heme proteins is posed by competing ligands such as CO, CN^-, and NO. All of these are present within organisms and both CO and NO act as hormones. Hemoglobin and myoglobin are partially protected from carbon monoxide by the design of the binding site for O_2.

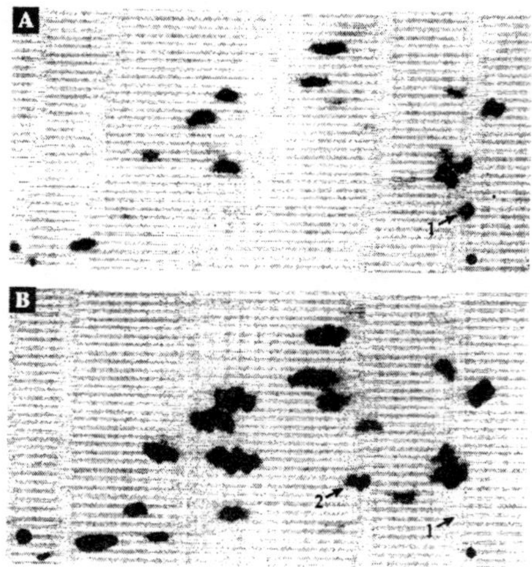

Fig. 1.26. "Fingerprints" of human hemoglobins. The denatured Hemoglobin was digested with trypsin and the 28 resulting peptides were separated on a sheet of paper by electrophoresis in one direction (horizontal in the figures; anode to the left) and by chromatography in the other direction (vertical in the figure). The peptides were visualized by spraying with ninhydrin or with specific reagents for histi-dine or tyrosine residues. Since trypsin cuts only next to lysine, which occurs infrequently, the petide pattern provides a fingerprint, characteristic for any pure protein. (A) The fingerprint of normal adult hemoglobin A. (B) Fingerprint of hemoglobin S (sickle cell hemoglobin). One histidine-containing peptide (1) is missing and a new one (2) is present. This altered peptide contains the first eight residues of the N-terminal chain of the subunit of the protein.

The distal imidazole of histidine E7 hydrogen bonds to O_2 but not to the nonpolar CO. The site also accommodates the geometry of the bound O_2 better than that of CO. Bound CO.can be released from hemes by the action of light. Using X-ray diffraction and X-ray absorption measurements what cryogenic temperatures, it has been possible to observe the motions of both the released CO and the heme in myoglobin, motions which may shed light on the normal oxygen transport cycle. Cooperativity in the binding of CO to hemoglobins has been studied extensively, as has binding to model heme compounds.

Cyanide ions bind most tightly and also cooperatively to the oxidized Fe^{3+} form, which is called *methemoglobin*. Nitric oxide is a reactive, paramagnetic gaseous free radical which is formed in the human body and in other organisms by an enzymatic oxidation of L-arginine. Since about 1980, NO has been recognized as a hormone with a broad range of effects. It binds to the iron of heme groups in either the Fe^{2+} or Fe^{3+} form and also reacts with thiol groups of proteins and small molecules to form S-nitrosothiols (R–S–N=O). It reacts with the heme iron of myoglobin and hemoglobin and, by transfer of one electron, can oxidize the iron of hemoglobin to the Fe^{3+} methemoglobin with formation of the nitroxyl ion NO^-.

This reaction may be a major cause of methemoglobin formation. One of the major effects of NO is to induce the relaxation of smooth muscle of blood vessels, an important factor in the regulation of blood pressure. Hemoglobin can carry NO both on its heme and on the thiol group of cysteine β93. The affinity for NO is high in the T state and low in the R state.

This allows hemoglobin to carry NO from the lungs to tissues, where it can be released and participate in the regulation of blood pressure. A cytoplasmic hemoglobin of the clam *Lucina pectinata* has evolved to carry oxygen to symbiotic chemoautotrophic bacteria located within cells of the host's gills. It is also readily oxidized to the Fe^{3+} methemoglobin form which binds *sulfide ions* extremely tightly and is thought to transport sulfide to the bacteria.

Abnormal Human Hemoglobins

Many alterations in the structure of hemoglobin have arisen by mutations in the human population. It is estimated that one person in 20 carries a mutation that will cause a hemoglobin disorder in a homozygote. There are also many unrecognized and harmless substitutions of one amino acid for another. However, substitutions near the heme group often adversely affect the binding of oxygen and substitutions in the interfaces between subunits may decrease the cooperative interaction between subunits.

One of the most common and serious abnormal hemoglobins is *hemoglobin S.* which is present in individuals suffering from *sickle cell disease.* In Hb S, glutamic acid 6 of the β chain is replaced by valine. Replacement of the same amino acid by lysine leads to *Hb C* and is associated with a mild disease condition . A few of the many other substitutions that have been studied are indicated in Fig. 1.23. The locations of the defects in the hemoglobin structure have been established with the aid of protein *"fingerprinting"*. A group of serious defects are represented by the *hemoglobins M.* Only heterozygotic individuals survive. Their blood is dark because in Hb M the iron in half of the subunits is oxidized irreversibly to the ferric state.

The resulting methemoglobin is present in normal blood to the extent of about 1%. While normal methemoglobin is reduced by a *methemoglobin reductase* system, methemoglobins M cannot be reduced. All of the five hemoglobins M result from substitutions near the heme group. In four of them, one of the heme-linked histidines of either the α or the β subunits is substituted by tyrosine. In the fifth, valine 67 of the β chains is substituted by glutamate. The two hemoglobins M that carry substitutions in the α subunits (M_{Boston} and M_{Iwate}) are frozen in the T (deoxy) conformation and therefore have low oxygen affinities and lack cooperativity.

In hemoglobins Rainier and Nancy the usually invariant C-terminal tyrosine 145 of the β chains is substituted by cysteine and by aspartate, respectively. Oxygen affinity is high and cooperativity is lacking. Hemoglobin Kansas, in which the β asparagine is substituted by threonine, also lacks cooperativity and has a very low oxygen affinity, while hemoglobin Richmond, in which the same amino acid is substituted by lysine, functions normally.

In hemoglobin Creteil the β89 serine is replaced by asparagine with the result that the adjacent C-terminal peptide carrying tyrosine 145 becomes disordered. In hemoglobin Hiroshima the C-terminal histidine in the β chain is replaced by aspartic acid. This histidine is one that donates a Bohr proton and in the mutant hemoglobin the oxygen affinity is increased 3-fold and the Bohr effect is halved. In hemoglobin Suresnes the C-terminal arginines 141 of the ∝ chains are replaced by histidine with loss of one of the anion-binding sites

3. Comparative Biochemistry of Hemoglobin

Even within human beings there are several hemoglobins. In addition to myoglobin, a brain protein neuroglobin, and adult hemoglobin A (*Hb A*, $\alpha_2\beta_2$), there is a minor hemoglobin $A_2(\alpha_2\delta_2)$. Prior to birth the blood contains *fetal hemoglobin,* also called hemoglobin F (*Hb F*, $\alpha_2\beta\gamma_2$)- In the presence of 2, 3-bisphosphoglycerate Hb F has a 6-fold higher oxygen affinity than Hb A as befits its role in obtaining oxygen from the mother's. blood. Hemoglobin F disappears a few months after birth and is replaced by Hb A. Each of the hemoglobins differs from the others in amino acid sequence. In other species the amino acid composition ol hemoglobins varies more, as do the interactions between subunits.

Hemoglobins and myoglobins are found throughout the animal kingdom and even in plants. The *leghemoglobins* are formed in root nodules of legumes and are involved in nitrogen fixation by symbiotic bacteria. Other hemoglobins apparently function in the roots of plants.

Hemoglobins or myoglobins are found in some cyano-bacteria and in many other bacteria. The globin fold of the polypeptide is recognizable in all of these. The quaternary structure of hemoglobin also varies.

Myoglobin is a monomer, as is the leghemoglobin. Hemoglobin of the sea lamprey dissociates to monomers upon oxygenation. The clam *Scapharca inaequivalvis* has a dimeric hemoglobin that binds O_2 cooperatively even though the interactions between subunits are very different from those in mammalian hemoglobins. Hemoglobin of the nematode *Ascaris* is an octamer. It has puzzling properties, including a very high affinity for O_2 and a slow dissociation rate. The distal His E-7 is replaced by glutamine, which has a hydrogen-bonding ability closely similar to that of histidine.

Earthworms, polychaete worms, and leeches have enormous hemoglobin molecules consisting of as many as 144 globin chains arranged into 12 dodecamers and held together by 36-42 linker chains. These hemoglobins are often called *erythrocruorins.* In a few families of polychaetes chloroheme substitutes for heme and the proteins are called *chlorocruorins.* What is common to all of the hemoglobins? The same folding pattern of the peptide chain is always present.

The protein is always wrapped around the heme group in an identical or very similar manner. In spite of this striking conservation of overall structure, when animal hemoglobins are compared, *there are only ten residues that are highly conserved.* They are indicated in Fig. 1.23 by the boxes. The two glycines (or alanine) at B-6 and E-8 are conserved because the close contact between the B and E helices does not permit a larger side chain. Proline C-2 helps the molecule turn a corner.

Four of the other conserved residues are directly associated with the heme group. Histidine E-7 and His F-8 are the "heme-linked" histidines. Tyrosine HC-2, as previously mentioned, plays a role in the cooperativity of oxygen binding. Only Lys H-9 is on the outside of the molecule. The reasons for its conservation are unclear. 'When sequences from a broader range of organisms were determined five residues were found to be highly conserved; *only two are completely conserved.* These are His F-8 and Phe CD-1, which binds the heme noncovalently. Hemoglobins are not the only biological oxygen carriers. The two iron *hemerythrins* are used by *a* few phyla of marine invertebrates, while the copper containing *hemocyanins* are used by many molluscs and arthropods.

E. SELF-ASSEMBLY OF MACROMOLECULAR STRUCTURES

While it is easy to visualize the assembly of oligomeric proteins, it is not as easy to imagine how complex objects such as eukaryotic cilia or the sarcomeres of muscle are formed. However, study of the assembly of bacteriophage particles and other small biological objects has led to the concepts of *selfassembly* and *assembly pathways*, concepts that are now applied to every aspect of the architecture of cells.

1. Bacteriophages

A remarkable example of self-assembly is that of the T-even phage. From genetic analysis at least 22 genes are known to be required for formation of the heads, 21 genes for the tails, and 7 genes for the tail fibers. Many of these genes encode sequences of proteins that are incorporated into the mature virus, but several specify enzymes needed in the assembly process.

Several mutant strains of the viruses are able to promote synthesis of all but one of the structural proteins of the virion. 'Proteins accumulating within these defective bacterial hosts have no tendency to aggregate spontaneously. However, when the missing protein (synthesized by bacteria infected with another strain of virus) is added complete virus particles are formed rapidly. Investigations resulting from this and other observations have led to the conclusion that during assembly *each different protein is added to the growing aggregate in a strictly specified sequence or assembly pathway.*

The addition of each protein creates a binding site for the next protein. In some cases the protein that binds is an enzyme that cuts off a piece from the growing assembly of subunits and thereby creates a site for the next protein to bind. Before considering this complex process further, let's look at the assembly of simpler filamentous bacteriophages and bacterial pili. The filamentous bacteriophages are put together from hydrophobic protein subunits and DNA. After their synthesis the protein subunits are stored within the cytoplasmic membrane of the infected bacteria.

These small, largely α-helical rods can easily fit within the membrane and remain there until a DNA molecule also enters the membrane. Two additional proteins (gene I proteins) also enter the membrane. One has a 348-residue length, while the second is a 107-residue protein formed by translational initiation at a later point in the DNA sequence of the gene. These proteins help to create an assembly site at a place where the inner and outer membrane of the host bacterium are close together. It isn't clear how the process is initiated, but it is likely that each subunit of the viral coat contains a nucleotide binding site that interacts with the DNA.

Adjacent sides of the subunits are hydrophobic and interact with other subunits to spontaneously coat the DNA. As the rod is assembled, the hydrophobic groups become *"buried."* It is postulated that the remaining side chain groups on the outer surface of the virus are hydrophilic and that the formation of this hydrophilic rod provides a driving force for automatic extrusion of the phage from the membrane. Bacterial pili appear to be extruded in a similar manner.

They arise rapidly and may possibly be retracted again into the bacterial membrane. The P pilus in Fig. 1.9A is made up of subunits Pap A, G, F, E, and K which must be assembled in the correct sequence. A chaperonin PapD is also required as is an *"usher protein,"* PapC, and also the disulfide exchange protein DsbA. DsbA helps PapD to form the correct disulfide bridges as it folds and PapD binds and protects the various pilus subunits as they accumulate in the periplasmic space of the host.

The usher protein displaces the chaperonin PapD and *"escorts"* the subunits into the membrane where the extrusion occurs. Because eicosahedra are regular geometric solids and the faces can be made up of hexons and pentons of identical subunits, it might seem that self-assembly of eicosahedral viruses would occur easily. However, the subunits usually must be able to assume three or more different conformations and the shells can easily be assembled incorrectly. Several stategies are employed to avert this problem. Some viruses assemble an empty shell into which the DNA flows, but many others first form an internal *scaffolding* or *assembly core* around which the shell is assembled. An external scaffold may also be needed. The RNA virus MS2 forms its T = 3 capsid by using the RNA molecule as the assembly core. Other viruses may have one or more core proteins which dissociate from the completed shell or are removed by the action of proteases. This is a feature of the small ϕX phage, the tailed phages, and double-stranded RNA viruses including human reoviruses. Bacteriophage PRD1, another virus of *E. coli* and *Salmonella typhimurium,* has a membrane inside the capsid apparently playing a role in assembly. A general concept that seems to hold in all cases is one of "*local rule.*"

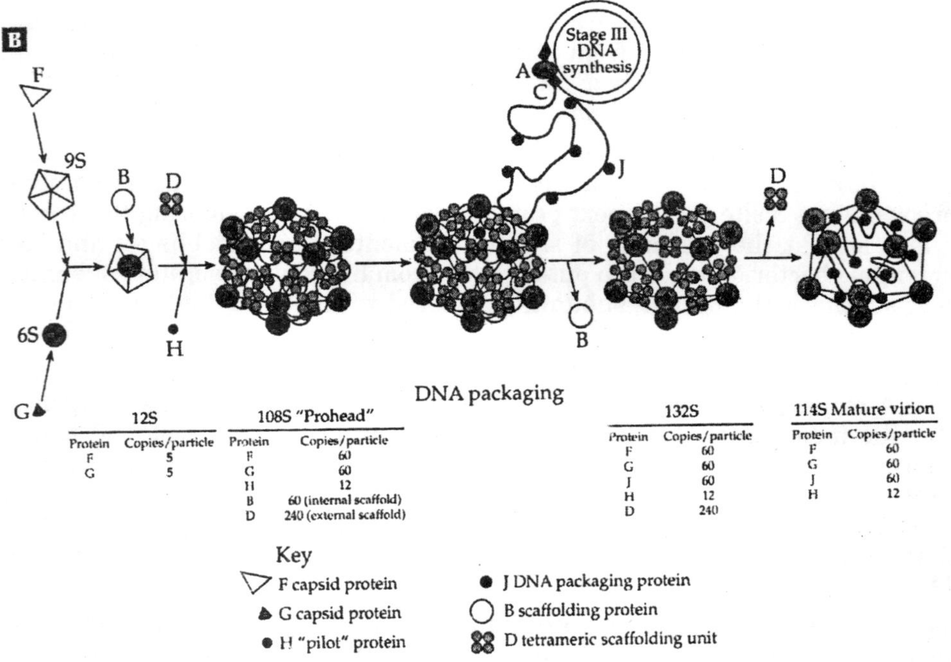

Fig. 1.27. (A) Stereoscopic view of the ϕX174 114 S mature virion viewed down a twofold axis after a cryoelectron microscopy reconstruction. (B) Morphogenesis of ϕ X174. Proteins A and C are required for DNA synthesis.

Several conformers of a virus subunit may equilibrare within a cell. However, *they can associate only through surfaces that are complementary.* A conformer that allows pentons to form cannot assemble into a hexon and only certain combinations of other conformers can give rise to hexons, etc. If there is only one conformation and the shape is right a T = 1 shell will be formed. If there are three conformers a T = 3 shell may arise. Another generalization is that in most cases the *procapsid* **or** *prohead* that is formed initially is fragile. Subunits may still be undergoing conformational changes.

However, a final conformational alteration, which may include chain cleavage by a protease, usually occurs. This often expands the overall dimensions of the capsid and creates new intersubunit interactions which greatly strengthen the mature capsid. Scaffolding proteins are then removed and DNA or RNA enters the capsid, again in a precise sequence. There are many variations and the detail that is known about virus assembly is far too great to describe here. Figure 1.29 illustrates the assembly pathway for the very small ϕX174, a T=l virus. The major capsid protein F is a 426-residue eight-stranded β-barrel. The 175-residue G protein forms pentameric spikes while 60 copies of the internal scaffolding protein and 240 copies of the external scaffolding protein D and 12 copies of the pilot protein are required to form the prohead.

The single-stranded DNA enters along with 60 copies of a DNA packaging protein J. Assembly of the tailed bacteriophages is even more complex. The genome of the viruses is large. The 166-kb circular dsDNA of phage T4 contains ~ 250 genes, many of which encode proteins of the virion or enzymes or chaperonins needed in assembly.

The assembly pathway for the bacteriophage heads requires at least 22 gene products. Seven of these form the assembly core which serves as a scaffolding around which the 840 copies of gp23 and 55 copies of gp24 are added to give the elongated icosahedral prohead I. 'Most of the internal proteins are then dissolved by proteases, one of which is the phage-specified gp21. A protease also cuts a piece from each molecule of gp23 to form the smaller gp23* the major protein of the mature prohead *II*. This cleavage also triggers the conformational change leading to head expansion. The empty proheads are now filled with DNA in a process which is assisted by another series of catalytic proteins. The T4 phage tail is assembled in a separate sequence.

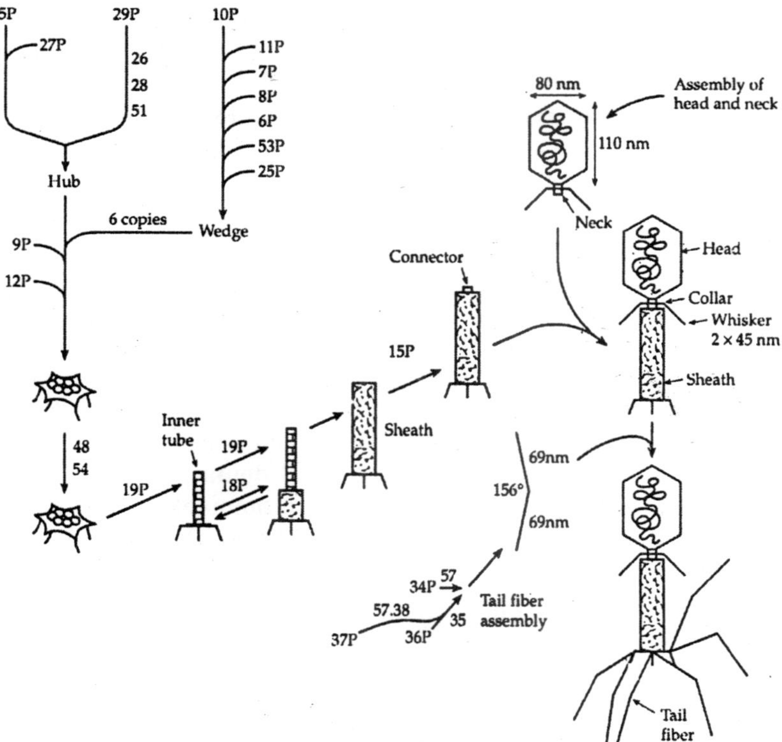

Fig. 1.28. Assembly sequence for bacteriophage T4 with details for the tail. The numbers refer to the genes in the T4 chromosome map. A "P" after the number indicates that the protein gene product is incorporated into the phage tail. Other numbers indicate gene products that are thought to have essential catalytic functions in the assembly process.

Six copies of each of three different proteins form a "*hub*" with hexagonal symmetry. In another assembly sequence, seven different proteins form wedge-shaped pieces, six of which are then joined to the hub to form the hexagonal baseplate. Two more proteins then add to the surface of the base plate and activate it for the growth of the internal tail tube. Only after assembly of the internal tube begins does the sheath begin to grow, and only when both of these tubular structures have reached the correct length, is a cap protein placed on top.

The DNA-filled head is then attached by a special connector and only then do the tail fibers, which have been assembled separately, join at the opposite end. How can each step in this complex assembly process set the stage for the next step? Apparently the structure of each newly synthesized protein monomer is stable only until a specific interaction with another protein takes place. The binding energy of this interaction is sufficient to induce a conformational alteration that affects a distant part of the protein surface and generates complementarity toward a binding site on the next protein that is to be added. Every one of the baseplate proteins must have such self-activating properties!.

Sometimes proteolytic cleavage of a sub-unit is required. If it occurs at an appropriate point in the sequence it provides thermodynamic drive for the assembly process. The induction of a change in one protein by interaction with another protein is a phenomenon that is met also in the construction of microtubules, ribosomes, cilia, and myofibrillar assemblies of muscle. It is basic to the assembly of the many labile but equally real cascade systems of protein-protein interactions such as that involved in the clotting of blood and signaling at membrane surfaces.

2. "Kringles"and Other Recognition Domains and Motifs

The assemble of either transient or long lasting complexes of proteins is often dependent upon the presence of conserved structural domains of 30-100 residues. A similar domain may occur in many different proteins and often two or more times within a single protein. The sequences within such a domain are homologous, allowing it to be recognized from protein or gene sequences alone.

Domains are often named after the protein in which they were first discovered. For example, *EGF-like domains* resemble the 53-residue epidermal growth factor. *SH2* and *SH3* domains are src-homology domains, named after *Src* (c-src), the protein encoded by the *src* protooncogene. The SH2 and SH3 domains are found near the N terminus of this 60-kDa protein. They are also found in many other proteins. An adapter protein called *Grb2,* important in cell signaling, consists of nothing but one SH2 domain and two SH3 domains. The SH2 domains bind to phosphotyrosyl side chains of various proteins, while SH3 domains bind to a polyproline motif. Another phos-photyrosyl binding domain, the plekstrin homology or *PH* domain, is named for the protein in which it was discovered. *Kringle* and *apple* describe the appearances of the folded proteins in those domains.

Strucrural domains often function to hold two proteins together or to help anchor them at a membrane surface by binding to specific protein groupes such as phosphotyrosyl, or calcium ions. Table 1.3 lists a few well-known folding domains and Fig. 1.30 shows three-dimensional structures of two of them. Recognition domains often function transiently. For example, SH2 domains are often found in proteins that interact with phosphotyrosyl groups of "*activated*" cell surface receptors. The receptors become activated by conformational alteration resulting from the binding. The src protein is a tyrosine kinase and, when activated, uses ATP to phosphorylate tyrosyl groups of other proteins, and using its SH2 domains it will bind to such groups forming and passing an intracellular message to them.

Table 1.3. A few Well-Known Structural Domains

Name	Length in amino acid residues	Specific ligands
EGF-like	~ 45	Ca^{2+}
SH2	~ 100	Phosphotyrosine
Structure		
SH3	~ 60	Polyproline, PXXP
Structure		
PTB		Proline-rich sequence
PH (plekstrin homology)		Phosphotyrosine
Structure		
PDZ	80 – 100	C-terminal XS/TXV – COO^-
Immunoglobulin repeat	~ 100	
Kringles, blood clotting proteins	80 – 85	Calcium binding
Apple, Blood clotting Factor X	90	Calcium binding
WW (Trp – Trp)[272]	~ 38	Proline
Serine protease		
P (Trefoil)	~ 50	
TPR (Tetratrico peptide repeat		
ZBD (Zinc-binding domains)		
Zinc finger		

F. THE CYTOSKELETON

The cytoplasm of eukaryotic cells contains a complex network of slender rods and filaments that serve as a kind of internal skeleton. The properties of this *cytoskeleton* affect the shape and mechanical properties of cells. For example, the cytoskeleton is responsible for the biconcave disc shape of erythrocytes and for the ameba's ability to rapidly interconvert gel-like and fluid regions of the cytoplasm.

Three principal components of the cytoskeleton are *microfilaments* of ~ 6 nm diameter, **microtubules** of 23-25 nm diameter, and **intermediate filaments** of –10 nm diameter. A large number of associated proteins provide for interconnections, for assembly, and for disassembly of the cytoskeleton. Other proteins act as *motors that* provide motion. One of these motors is present in **myosin** of muscle. This protein is not only the motor for muscular work but also forms *thick filaments* of 12-16 nm diameter, which are a major structural component of muscle

1. Intermediate Filaments

In most cells the intermediate filaments provide the scaffolding for the cytoskeleton. They may account for only 1% of the protein in a cell but provide up to 85% of the protein in the tough outer layers of skin. Intermediate filament proteins are encoded by over 50 human genes which specify proteins of various sizes, structures, and properties. However, all of them have central 300- to 330-residue α-helical regions through which the molecules associate in parallel pairs to form coiled-coil rods with globular domains at the ends.

Some of these proteins, such as the *keratin of* skin, are insoluble. Others, including the nuclear *laming* and *vimentin* dissociate and reform filaments reversibly. Vimentin is found in most cells and predominates in fibroblasts and other cells of mesenchymal origin. *Desmin* (55-kDa monomer) is found in both smooth and skeletal muscle. In the latter, it apparently ties the contractile myofibrils to the rest of the cytoskeletal network and the individual myofibrils to each other at Z disc. The *glial filaments* from the astroglial cells of the brain are composed mainly of a single type of 55-kDa subunits of the *glial fibrillar acidic protein* but the *neurofilaments* oFmammalian neurons are composed of three distinct subunits of 68–, 150–, and 200–kDa mass. The larger subunits have C-terminal tails that are not required for filament formation but which can be phosphorylated and form bridges to neighboring neurofilaments and other cytoskeletal components and organelles.

'Keratin filaments, which eventually nearly fill the highly differentiated epidermal cells, are also made up of several different subunits. Extensions of the keratin chains are rich in cysteine side chains which form disulfide crosslinkages to adjacent molecules to provide a network that can be dehydrated to form hair and the tough outer layers of skin. Elastin-associated microfibrils are important constituents of elastic tissues of blood vessels, lungs, and skin.

A common architecture of intermediate filaments is a staggered head-to-tail and side-by-side association of pairs of the coiled-coil dimers into 2– to 3–nm proto-filaments and further association of about eight protofilaments to form the 10–nm intermediate filaments.

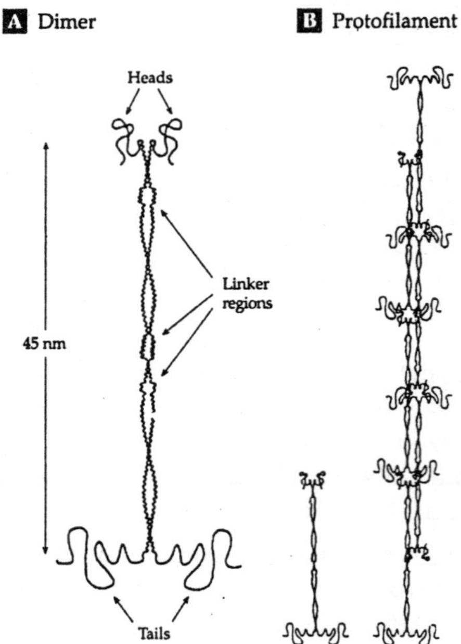

Fig. 1.29. A model for the structure of keratin microfibrils of intermediate filaments. (A) A coiled-coil dimer, 45-nm in length. The helical segments of the rod domains are interrupted by three linker regions. The conformations of the head and tail domains are unknown but are thought to be flexible. (B) Probable organization of a protofilament, involving staggered antiparallel rows of dimers.

2. Microfilaments

The most abundant microfilaments are composed of fibrous actin (F-actin;). *The thin filaments* of F-actin are also one of the two major components of the contractile fibers of skeletal muscle. There is actually a group of closely related actins encoded by a multigene family. At least four vertebrate actins are specific to various types of muscle, while two (β- and γ-actins) are cytosolic. Actins are present in all animal cells and also in fungi and plants as part of the cytoskeleton. The microfilaments can associate to form larger arrays and actin often exists as thicker "cables," some of which form the *stress fibers* seen in cultured cells adhering to a glass surface. In the red blood cells the *spectrin-actin* meshwork, which lies directly beneath the plasma membrane, together with the proteins that anchor it to the membrane, form the cytoskeleton. Its mechanical properties appear to be responsible for the biconcave disc shape of the cell. "The *acrosomal process* of some invertebrate sperm cells is an actin cable that sometimes forms almost instantaneously by polymerization of the actin monomers and shoots out to penetrate the outer layers of the egg during fertilization.

The stereocilia, the *"hairs"* of the hair cells in the inner ear, contain bundles of actin filaments. Motion of the stereocilia caused by sound produces changes in the membrane potential of the cells initiating a nerve impulse. In certain lizards each hair cell contains about 75 stereocilia of lengths up to 30 μm and diameter 0.8 μm and containg more than 3000 actin filaments in a semicrystalline array

Microvilli contain longitudinal arrays of actin filaments. In every instance, groups of microfilaments are held together by other proteins. Stress fibers of higher eukaryotes contain the *"muscle proteins" tropomyosin, α-actinin,* and myosin, although the latter is usually not in fibrillar form. *Filamin* (250-kDa) and a 235-kDa protein are associated with actin in platelets

The high-molecular-weight *synemin* crosslinks vimentin and desmin filaments, while the smaller, highly polar *filaggrin* provides a matrix around the keratin filaments in the external layers of the skin. Postsynthetic modifications of cytoskeletal micro-filaments can also occur. For example, epidermal keratin has been found to contain lanthionine, (*Y*-glutamyllysine) and lysinoalanine, both presumably arising from crosslinkages.

3. Microtubules

A prominent component of cytoplasm consists of microtubules which appear under the electron microscope to have a diameter of 24 ±2 nm and a 13 - to 15-nm hollow core. However, the true diameter of a hydrated microtubule is about 30 nm and the microtubule may be further surrounded by a 5-20 nm low density layer of associated proteins. Microtubules are present in the most striking form in the flagella and cilia of eukaryotic cells. The *stable microtubules* of cilia are integral components of the machinery causing their motion. *Labile microtubules,* which form and then disappear, are often found in cytoplasm in which motion is taking place, for example, in the pseudopodia of the ameba.

The mitotic spindle consists largely of microtubules which function in the movement of chromosomes in a dividing cell. Microtubules in the long axons of nerve cells function as *"rails"* for the *"fast transport"* of proteins and other materials from the cell body down the axons. In fact, microtubules appear to be present throughout the cytoplasm of virtually all eukaryotic cells and also in spirochetes.

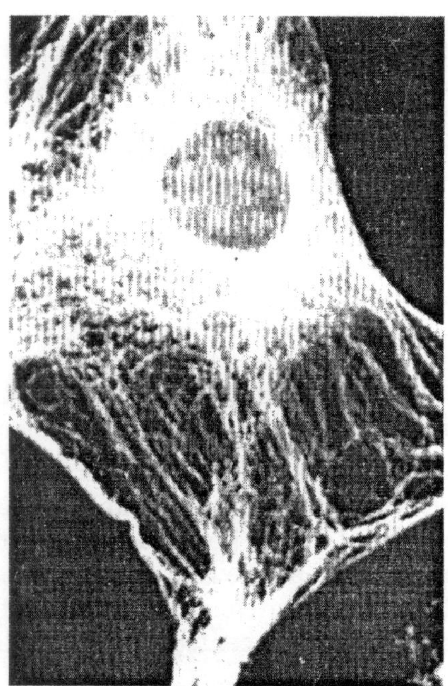

Fig. 1.30. Micrograph of a mouse embryo fibroblast was obtained using indirect immunofluorescence techniques. The cells were fixed with formaldehyde, dehydrated, and treated with antibodies (formed in a rabbit) to microtubule protein. The cells were then treated with fluorescent goat antibpdies to.rabbit γ-globulins and the photograph was taken by fluorescent light emission.

Motion in microtubular systems depends upon motor proteins such as *kinesin,* which moves bound materials toward what is known as the *"negative"* end of the microtubule, *dyneins* which move toward the positive end. These motor proteins are driven by the Gibbs energy of hydrolysis of ATP or GTP and in this respect, as well as in some structural details, resemble the muscle protein myosin. Dynein is present in the arms of the microtubules of cilia whose motion results from the sliding of the microtubules driven by the action of this protein. Microtubules are assembled from ~ 55-kDa *tubulins,* which are mixed dimers of *a* subunits (450 residues) and β subunits (445 residues) with 40% sequence identity.

The αβ dimers, whose structure is shown in Fig. 1.33 are thought to be packed into an imperfect helix. The structure can also be regarded as an array of longitudinal protofilaments. Naturally formed microtubules usually have precisely 13 protofilaments and a discontinuity in the helical stacking of subunits as shown in Fig 1.34. When grown in a laboratory the microtubules usually have 14 protofilaments and rarely 10 or 16 protofilaments with regular helical packing.

Microtubules of some "months and also of male germ cells of *Drosophila* have 16-protofilament microtubules without a discontinuity, an architecture that is specified by the geometry of a specific p-tubulin isoform. Each tubulin dimer binds one molecule of GTP strongly in the ∝ subunit and a second molecule of GTP or GDP more loosely in the β subunit. In this respect, tubulin resembles actin, whose subunits are about the same size. However, there is little sequence similarity. Labile microtubules of cytoplasm can be formed or disassembled very rapidly. GTP is essential for the fast growth of these microtubules and is hydrolyzed to GDP in the process.

However, nonhydrolyzable analogs of GTP, such as the one containing the linkage $P–CH_2–P$ between the terminal and central phosphorus atoms of the GTP, also support polymerization. Since microtubules have a distinct *polarity* the two ends have different tubulin surfaces exposed, and polymerization and depolymerization can occur at different rates at the two ends. As a consequence, microtubules often grow at one end and disassemble at the other. Such *"treadmilling"* may be important in movement of chromosomes in neuronal migration and in fast axonal transport of macromolecules.

During mitosis the minus ends of the microtubules are believed to be tightly anchored at the centrosome while subunit exchanges can occur at the plus ends where the β subunits are exposed. Using a phage display system it could be shown that the N termini of the ∝ subunits are exposed at the minus ends. Kinesin can bind to the β subunits all along the microtubule. Microtubules are formed by growth from microtubule nucleation sites in *microtubule organizing centers* found in centrosomes, spindle poles, and other locations.

Several proteins, including *-γ-tubulin,* are required. proposed assembly pathway is illustrated in Fig 1.34 isolated microtubules always contain small amounts of larger ~300-kDa *microtubule-associated proteins-*(MAPS). These elongated molecules may in part lie in the grooves between the tubulin subunits and in part be extended outward to form a low-density layer around the tubule. Nerve cells that contain stable microtubules have associated stabilizing proteins. A family of proteins formed by differential splicing of mRNA are known as *tau.* The tau proteins are prominent components of the cytoskeleton of neurons. They not only interact with microtubules but also undergo reversible phosphorylation.

Hyperphosphorylated tau is the primary component of the paired helical filaments found in the brains of persons with Alzheimer disease.

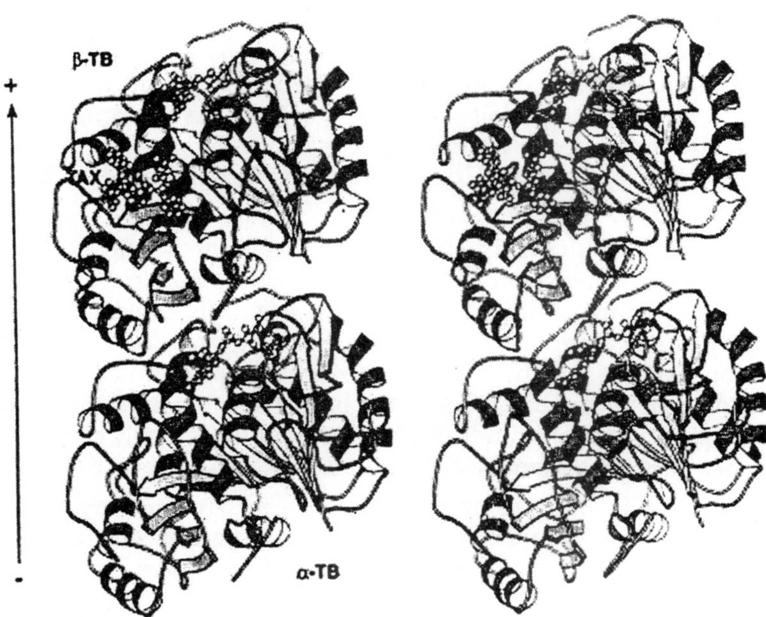

Fig. 1.31. **Stereoscopic ribbon diagram of the tubulin dimer with α-tubulin with bound GTP at the top and β-tubulin with bound GDP at the bottom. The β-tubulin subunit also contains a bound molecule of taxotere which is labeled TAX. This model is based upon electron crystallography of zinc-induced tubulin sheets at 0.37-nm resolution and is thought to approximate closely the packing of the tubulin monomers in microtubules. The arrow at the left points toward the plus end of the microtubule.**

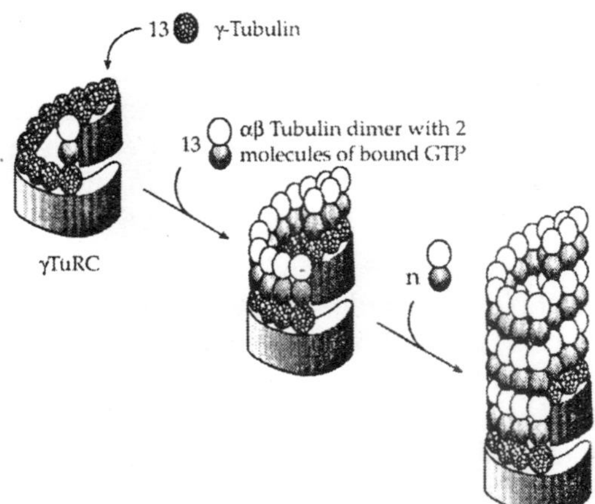

Fig. 1.32. **Growth of a microtubule from a γ-tubulin ring complex (γTURC). The helical γ-tubulin rings are formed in the microtubule organizing centers which, in animal cells, are the centrosomes. Thirteen γ-tubulin subunits are shown in a hypothetical array formed together with a base of other molecules of unknown structure. The microtubule grows by addition of successive layers of α/β-tubulin dimmers, each a split rign of 13 dimers with the β-tubulin subunits toward the base, the negative end, and the α-tubulin subunits toward the growing positive end.**

LIFE AND DEATH FOR PROTEINS: CHAPERONINS AND PROTESOMES

In 1968, a tiny cylindrical particle, which appeared to be a stack of 11–nm rings, was observed by electron microscopy of an extract of erythrocytes. Later, a similar particle was found in both the nucleus and the cytoplasm of other cells of many organisms. The particles were soon recognized as *a* new type of protein–hydrolyzing enzyme, a large 700–kDa particle consisting of 20–30 subunits of several different types which came to be known as the *multicatalytic protease* or *20S proteasome*.

Electron microscopy and X–ray diffraction showed that the parficlels formed from four stacked rings, each of which con-sists of seven subunits whose molecular masses range from 21 to 31 kDa. Proteasomes are strikingly similar in architecture, though not in peptide sequences, to another particle found in both bacteria and eukaryotes: a molecular "chaperone" or *chaperonin*. The chaperonins, of which there are several types, protect proteins while they fold or undergo translocation within cells. One of the best studied members is the *E. coli* protein *GroEL*, which is also composed of double rings of 14 subunits with seven–fold rotational symmetry and with two of these assemblies associated back–to–back with dihedral symmetry.

The dimensions ot GroEL and 20S proteasomes are nearly the same. However, GroEL has only two rings of ~60–kDa subunits, more than twice the size of proteosomal subunits. The accompanying sketch illustrates this fact and also the basic structural similarity of 20S proteosomes with GroEL. The αβ pairs of the proteosome, correspond to single subunits of the chap-eronin, but these subunits have three distinct domains–apical, intermediate, and equatorial (labeled A, I, and E , respectively, in the drawing). After a protein, whether correctly, incorrectly, or only partially folded, enters a cavity in GroEL, a second protein *GroES* of smaller size (~10 kDa) but with seven–fold symmetry binds to one end of the chaperonin. Seven molecules of ATP also bind to sites on the GroEL ring to which GroES binds (the *cis* ring).

The binding of the ATP and GroES evidently induces a major conformational change in the GroEL subunits which caxises the binding cavity to expand to over twice the original volume. This change (see drawing) also causes hydrophobic surfaces of the cavity to become buried and hydro-philic side chains to be exposed. The cavity surface was initially largely hydrophobic and able to bind many proteins nonspecifically, but upon expansion it becomes hydrophilic and less likely to bind. This releases the encased protein to complete its folding or to partially unfold and refold without interference from other proteins. While a proteirils adjusting its folding in the *cis* compartment another protein molecule may become trapped in the *trans* compartment.

After some time the bound ATP molecules are hydrolyzed. As in the contraction of muscle, which is discussed in Chapter 19, the loss of inorganic phosphate (P_i) and ADP from the active site can be accompanied by movement. In the chapenonin this involves a conformational switch so that the ES heptamer is released and the conformation of the *trans* ring of EL is switched to that of the initial *cis* ring and vice versa. The new *cis* ring is ready to receive an ES cap and the new *trans* ring can release the folded protein. A variety of experimental approaches are being used in an effort to further understand the action of GroEL. The chaperonin may function repeatedly before a protein becomes properly folded.[t] While chaperonins assist proteins to fold correctly proteasomes destroy unfolded chains by partial hydrolysis, cutting the chains into a random assortment of pieces from 3 to 30 residues in length with an average length of ~8 residues.

Proteasomes destroy not only unfolded and Improperly folded proteins but also proteins marked for destruction by the ubiquitin system described in Box 10–C. It has been hard to locate true proteosomes in most bacteria. However, they do contain pro tease particles with similar characteristics and archaeons, such as *Thermoplasma acidophilum,* have proteasomes similar to those of eukaryotes. The *Thermoplasma* proteasome contains only two kinds of subunits, α and β, which have similar amino acid sequences. These form α_7 and β_7 rings which associate in $\alpha_7\beta_7$ pairs with two of these double rings stacked back–to–back with dihedral D7 symmetry: $\alpha_7\beta_7\beta_7\alpha_7$. The crystal structure has been determined for this 20S proteasome from T. *acidophilum*[g,dd] and for the corresponding proteasome from yeast *(Sacchammyces cerevisiae.* The accompanying drawings

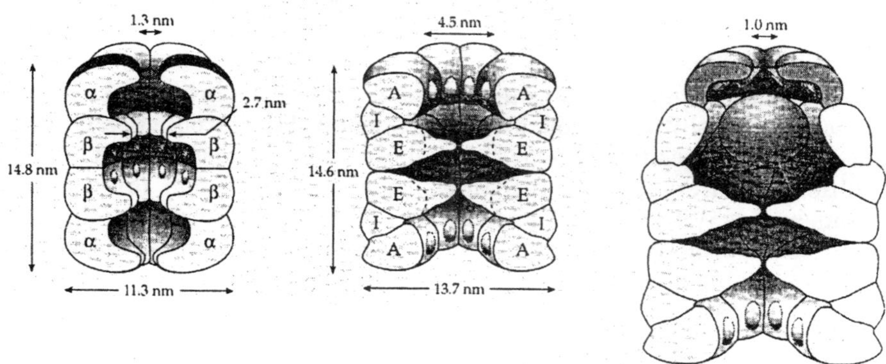

The accompanying drawings illustrate top and side views of the *T. acidophilum* proteasome. The particle contains three internal cavities. The outer two are formed between the α_7 and β_7 rings and the inner is formed between the two β_7 rings. A channel only. 1.3 nm in diameter permits the entrance of peptide chains into the compartmetns. The active sites of the enzymes are located in the β subunits in the central cavity.[dd] While the yeast and human proteasomes are similar to those of *Thermoplasma,* the β subunits consist of seven different protein–hydrolyzing enzymes.

There are also seven different a subunits, all of whose sequences are known. To make the story more complex, additional subunits, some of which catalyze ATP hydrolysis, form a 600- to 700-kDa cap which adds to one or both ends of a 20S proteosome to give a larger *26S proteasome.* These larger proteasomes carry out an ATP–dependent cleavage of proteins selected for degradation by the ubiquitin system. Some of the short peptide segments formed by proteasomes may leave cells and participate in intercellular communication.

For example, pieces of antigenic peptides are used by cells of the immune system for "antigen presentation" an important process by which the immune system recognizes which cells are "*self*" and which are foreign or malignant and must be killed. The structure of caps on the 26S proteasome ends is complex. At least 20 different regulatory subunits have been identified.

SICKLE CELL DISEASE, MALARIA, AND BLOOD SUBSTITUTES

Many person, especially if they are of west. African descent, suffer from the crippling and often lethal sickle cell disease. In 1949 , Pauling, Itano, and associates discoveredx that hemoglobin from such individuals migrated unusually rapidly upon electrophoresis. Later, Ingram divised themethod of protein fingerprinting illustrated in Fig. 1.27 and applied it to hemoglobin. He split the hemoglobin molecule into 15 tryptic pepties which he separated by electrophoresis and chromatograpy. From these experiments the abnormality in sickle cell hemoglobin (hemoglobin S; Hb S) was located at position 6 in the β chain. The glutamic acid present in this position in hemoglobin A was replaced by valine in Hb S. This was the first instance in which a genetic disease was traced directly to the presence of a single amino acid substitution in a specific protein. The DNA of the normal gene for the β globin chain has since been sequenced and found to have the glutamic acid codon GAG at position 6. A single base change to GTG causes the sickle cell mutation. Persons homozygous for this altered gene have sickle cell disease, while the much more numerous heterozygotes have, at most, minor problems. When HbS is deoxygenated it tends to "crystallize" in red blood cells, which contain 33% by weight hemoglobin.

The crystallization (actually gel forma-tion) distorts the cells into a sickle shape and these distorted corpuscles are easily destroyed, leading to anemia. The introduction of the hydrophobic valine residue in Hb S at position 6 near the end of the molecule helps form a new bonding domain by which the hemoglobin tetramers associate to form long semicrystalline microfilamentous arrays. Why is there such a high incidence of the sickle cell gene, estimated to be present in three million Americans? The occurrence and spread of the gene in Africa was apparently the result of a balance . between its harmful effects and a beneficial effect under circumstances existing there.

The malaria parasite, the greatest killer of all time, lives in red blood cells during part of its life cyle Red cells that contain Hb S as well as Hb A are apparently less suitable than cells containing only Hb A for growth of the malaria organism. Thus, het-erozygotic carriers of the sickle cell gene survived epidemics of malaria but at the price of seeing one-fourth of their offspring die of sickle cell disease. What is the outlook for the many (50,000 in the United States alone) sufferers of sickle cell disease today? Careful medi-cal care, including blood transfusion, can prolong life greatly and intense efforts are under way to find drugs that will prevent Hb S from crystallizing.

The problem arises from a hydrophobic interaction of valine B6 with phenylalanine B85 and leucine B88 of another molecule in the filaments of Hb S. The latter two residues are on the outside surface of helix F. It is difficult to modify one of these residues chemically but various alterations at the nearby N-termini of the β chains do inhibit sickling. Cyanate does

β Chain
Cyanate
β Chain

so by specifically carbamoylating these amino groups. However, although it was tested in humans, cyanate is too toxic for use.' Another approach employs an aldehyde that will form

Schiff bases with the same amino groups. A third approach is to use an acylating reagent. For example, methylacetyl phosphate acetylates the same β Lys 82 amino groups that react with bisphosphoglycerate and with cyanate.

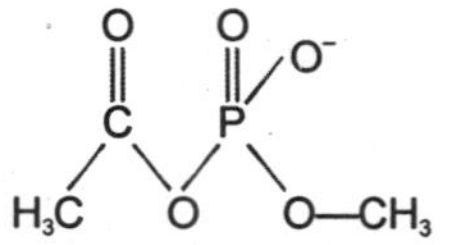

$$H_3C-C(=O)-O-P(=O)(O^-)-O-CH_3$$

Aspirin (2-acetoxybenzoic acid) is also a mild acetylating reagent and "*two-headed*" aspirins such as the following react specifically to crosslink hemoglobin β chains.

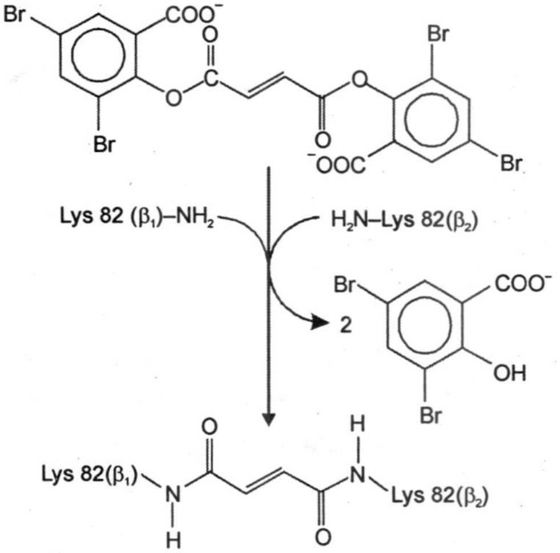

Lys 82 (β_1)–NH_2 + H_2N–Lys 82(β_2) → 2 (3,5-dibromosalicylate, COO^-, OH, Br, Br)

Lys 82(β_1)–NH–CO–CH=CH–CO–NH–Lys 82(β_2)

These compounds bind into the bisphospho-glycerate binding site and prevent the chains from spreading apart as far as they normally do in the deoxy (T) state. Since it is only the latter that crystallizes, the compounds have a powerful antisickling action.

Various other crosslinking reagents have been developed and more than one could be used together. New drugs that serve as allosteric modifiers in the same fashion as bisphosphoglycerate mayjdso be useful. A touth approach to treatment of sickle cell disease is gene therapy. This might allow patients to produce, in addition to Hb S, an engineered hemoglobin with compensating mutations that would mix with the Hb S and prevent gelling. This is impractical at present but there is another approach. Persons with sickle cell disease some-times also have the disorder of hereditary persis-tence of fetal hemoglobin. They continue to make Hb F into adulthood.

Great amelioration of sickle cell disease is observed in patients with 20-25% Hb F.' Hydroxyurea stimulates a greater production of Hb F and in patients with hereditary persistence of Hb F hydroxyurea may raise its level in erythocytes to ~50% of the total hemoglobin. Crosslinking ot the *alpha* chaTnsof normal deoxyhemoglobin through lysines 99 yields a hemoglobin with normal oxygen-binding behavior and an increased stability. It makes a practical emer-gency blood substitute, whereas unmodified hemo-globin is unsatisfactory. Unless encapsulated in an erythrocyte the hemoglobin tetramers tend to disso-ciate to dimers, losing cooperativity and escaping through kidneys. Sviitable crosslinking helps to solve this problem. Both a and p chains can be produced from cloned genes and reassembled to form hemoglobin. This will probably allow genetic engineering to form more stable but suitably cooperative hemoglobins that can be used to avoid hazards of transmission of viruses by transfusion.

THE T-EVEN BACTERIOPHAGES

Among the most remarkable objects made visible by the electron microscope are the T-even bacteriophage (T2, T4, and T6) which attack E. *coli* While it is not often evident how a virus gains access to a cell, these *"molecular syringes"* literally inject their DNA through a hole dissolved in the cell wall of the host bacterium. The viruses, of length ~200. nm and mass -225 × 10^6 Da each, contain 130 × 10^6 Da of DNA in a 100 × 70 nm head of elongated icosahedral shape. The head surface appears to be formed from ~840 copies of a 45-kDa protein known as gp23 (gene product 23; it is encoded by gene 23; it its encoded by gene 23.

These protein molecules are arranged as 140 hexamers (hexons) and together with ~ 55 copies of protein gp24, arranged as 11 pentamers (pentons), make up the bulk of the shell. The head also contains at least nine other proteins, including three internal, basic proteins that enter the bacterium along with the DNA. Addi-tional proteins form the *neck, collar*, and *whiskers*. The phage *tail,* which fastens to the collar via a *connector* protein,[g] contains an *internal tube* with a 2.5-nm hole, barely wide enough to accommodate the flow of the DNA molecule into the bacterium. The tube is made up of 144 subunits of gp19. The 8 × 10^6-Da *sheath* that surrounds the tail tube is made up of 144 subunits of gp!8, each of mass 55 kDa, arranged in the form of 24 rings of six subunits sheath

The sheath has contractile properties. After the virus has become properly attached to the host it shortens from ~80 to ~30 nm, forcing the inner tube through a hole etched in the wall of the bacterium. At the end of the tail is a *baseplate,* a hexagonal structure bearing six short *pins*, each a trimer of a 55 kDa zinc metalloprotein.

One of the ten proteins known to be present in the baseplafels the enzyme T4 lysozyme. The baseplate also contains six molecules of the coenzyme 7,8-dihydropteroyl- hexaglutamate. Six elongated, *"jointed"* " *tail fibers* are attached to the baseplate. The proximal segment of each fiber is a trimer of the 1140-kDa protein gp34. A globular domain attaches it to the baseplate. The distal segment is composed of three subunits of the 109-kDa gp37, three subunits of the 23-kDa gp36, and a single copy of the 30-kDa gp35.[k] The C-terminal ~140 residues of the 1026-residue gp37 are the specific *adhesjnjhat* binds to a lipopolysac-charide of E. *coli* type B cells or to the outer mem-brane protein OmpC? Among the smaller molecules present in the virus are the polyamines *putrescine* and *spermidine* which neutralize about 30% of the basic groups of

How is infection by a T-even virus initiated? f Binding of the tail fibers to-specific receptor sites on the bacterial surface triggers a sequence of confor-mational changes in the fibers, baseplate, and sheath. The lysozyme is released from the baseplate and etches a hole in the bacterial cell wall. Contraction of the sheath is initiated at the baseplate and continues to the upper end of the sheath. The tail tube is forced into the bacterium and the DNA rapidly flows through the narrow hole into the host cell. During contraction the subunits of the sheath undergo a remarkable rearrangement into a structure containing 12 larger rings of 12 subunits each.h

Thus, a kind of mutual *"intercalation"* of subunits occurs. In its unidirectional and irreversible nature the shortening of the phage tail differs from the contraction of muscle. The protein subunits of the sheath seem to be in an unstable high energy state when the tail sheath of the phage is assembled. The stored energy remains available for later contraction.

Mitosis, Tetraploid Plants and Anticancer Drugs

Microtubules in cells undergoing mitosis are the target of several important drugs. One of these is the alkaloid *colchicine* which is produced by various members of the lily family and has been used since ancient Egyptian times for the alleviation of the symptoms of gout.

Colchicine

Taxol (pacliaxel)

This compound, with its tropolone ring system, binds specifically and tightly and prevents assembly of microtubules, including those of the mitotic spindle. Colchicine forms a complex with soluble tubulin, perhaps a dimeric αβ complex of the two subunits. Dividing cells treated with colchicine appear to be blocked at metaphase and daughter cells with a high degree of polyploidy are formed. This has led to the widespread use of colchicine in inducing formaiton of tetraploid varieties of flowering plants. Similar effects upon microtubules are produced by the antitumor agents *vincristine* and *vinblastine,* alkaloids formed by the common plant *Vinca* (periwinkle), and also by a variety of other drugs.

The more recently discovered Taxol (paclitaxel) was extracted from the bark of the Pacific yew. It *stabilizes* microtubules, inhibiting their disassembly. Taxol also blocks mitosis and causes the cells which fail to complete mitosis to die. Taxol has been synthesized and is a promising drugh that is being used in treatment of breast, ovarian, and other cancers. Binding sites for the compound have been located in β tubulin subunits. Attempts are being made to develop "taxoids" and other drugs more effective than taxol against cancer cells.

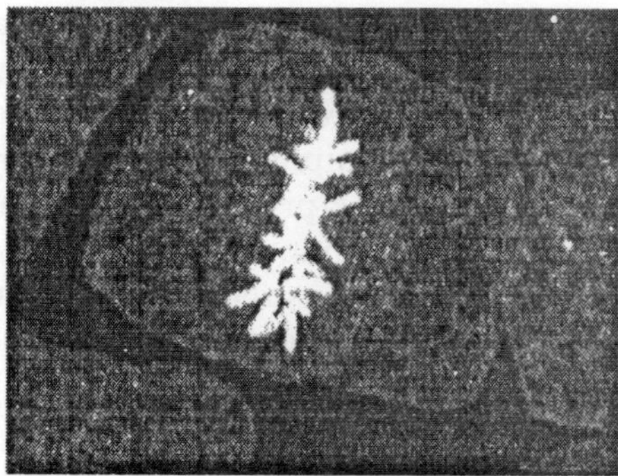

Another group of drugs that bind to microtubules are *benzimidazole* and related compounds. These have been used widely to treat infection by parasitic nematodes in both humans and animals. Unfortunately resistance has developed rapidly. In a nematode that infects sheep a single tyrosine to phenylalanine mutation at position 200 in the β-tubulin subunit confers resistance.

2 INTRAGASTRIC INTUBATION

Fetal alcohol spectrum disorder (FASD) is the leading known cause of preventable mental retardation in the Western world and animal models have been instrumental in isolating alcohol as a teratogen and describing behavioural and neural deficits induced by alcohol. Animal models of FASD are currently being used to conduct translational research directed towards delineating possible behavioural (*e.g.*) and/or pharmacological treatments (e.g.) of FASD and mechanisms of alcohol-induced damage (*e.g.*).

Alcohol administration during development in rodents (predominantly mice and rats but also including guinea pigs and ferrets) has been accomplished using a variety of different methods, and each of these methods has different strengths and weaknesses. The most common models use rats or mice. The gestational period in rats and mice is only equivalent to the first two trimesters in the human with respect to brain growth and so, to target a period equivalent to all three trimesters, alcohol administration in both the prenatal and postnatal period must occur. An ideal method of alcohol administration in rodents would allow complete control over the actual alcohol exposure, entail no stress or handling of the animal, mimic the pharmacological time course of blood alcohol concentrations in the human fetus, be technically easy to do, and control for nutritional effects resulting from intoxication from alcohol. Such a method does not exist, but it is useful to go through the different methods and evaluate them along the different criteria.

The most commonly used methods include vapour inhalation, liquid diets, artificial rearing, and intragastric intubation. All of these methods can be considered stressful to the animal although the nature of the stressor varies across methods. Vapour inhalation and artificial rearing require the purchase of expensive equipment, whereas the other methods do not. In contrast to the other methods, the liquid diet procedure does not allow close experimental control of the dose given to the animal. This chapter will focus on a three trimester model of FASD that uses intragastric intubation during the prenatal and postnatal period. This procedure is relatively easy to do, gives excellent control over the dose, and allows administration across both the prenatal and postnatal period.

The alcohol is administered orally so that the time course of blood alcohol levels is appropriate and there are no detectable nutritional effects using the doses described here. However, the administration procedure of intubation is stressful to the animals. As a result, it is important to include a control for the stress and allow a comparison with a nontreated control group. It is also very important to make efforts to minimize the stress; these efforts are emphasized in the Methods section. This alcohol administration method does not require expensive equipment and is not technically demanding. In addition, a method to evaluate blood alcohol concentrations is also included since this is typically done in many alcohol studies. The procedure described here uses doses in dams and pups that, when given in a single bolus, results in equivalent peak blood alcohol concentrations between 300 and 400 mg/dL in both dams and pups.

MATERIALS

Alcohol Administration During the Prenatal Period

1. Feeding needles, curved, 18 gage (7.62cm, 2.25-mm ball; VWR).
2. 10-mL syringes, slip tip connection.
3. Maltose-dextrin (Bio-Serv; 3.89kcal/g) is dissolved in water to give a solution that is isocaloric with the ethanol solution. This solution must be heated and stirred to get the maltose-dextrin into solution.
4. Ethanol (Acros) (100%, 0.7893 g/mL or 95%, 0.7498 g/mL, 7 kcal/g) is dissolved in distilled water to give a solution of 0.225 g of ethanol per millilter. When injected in a volume of 20mL/kg, this will give a dose of 4.5 g/kg.

Alcohol Administration During the Postnatal Period

1. *Intramedic* tubing, PE 10 Clay Adams and PE 50 Clay Adam (VWR).
2. 1-mL syringes and 23-gage needles.
3. A milk solution that is similar in composition to rat milk is made *(25)*. This solution is made under sterile conditions. A mineral mix is first made combining 1.2g of $ZnSO_4$, 1.2g of $CuSO_4$, 1.2g of $FeSO_4$, 20g of $MgCl_2$, and 20 g of KC_1 in 500mL of sterile H_2O. It is a slurry mix that must be stirred before adding to the milk solution. A specially formulated vitamin mix is ordered from Bioserv; it is listed as the custom mix for the University of Iowa. The milk solution is made by homogenizing 1500mL of evaporated milk (purchased at a grocery store), 450 mL of sterile H_2O, 70 g of Supro 710 protein power, 130 mL of corn oil, 2 g of methionine, 1 g of tryptophan, 10 g of vitamin mix, 11 g of calcium phosphate dibasic, 0.2 g of deoxycholic acid, and 50 mL of the mineral mix.
4. After homogenizing, the solution is put in bottles of a measured amount (usually 50 mL), and the bottle is sealed a rubber topper. The milk is then pasteurized by putting the bottles into an oven at 60 to 65 °C for 30 min. The milk is then cooled rapidly and stored at -80°C, where it is stable indefinitely. When milk solutions are needed, a bottle of milk is thawed and used alone for the second injection for the pups. A separate bottle is used as the base of the ethanol intubation and is made such that 3.0g/kg of ethanol is given when 0.0278 mL/g is given to the pup. The milk solutions that are being used are stored in the refrigerator (4°C) and will be stable for approximately 2 weeks.

Measurement of Blood Alcohol Concentrations (BACs)

1. Heparinized capillary tubes (10 μL, Drummond Scientific special ordered through VWR, vendor part 1-000-0100-H).
2. 0.53 *N* perchloric acid (Sigma).

Table 2.1. Evaluation of Alcohol Administration Methods for Rats During Development

Method	*Period of exposure*	*Experimenter control of dose*	*Nutritional control*	*Stress*	*Pharmacological time course*	*Technical difficulty and equipmment*
Vapour inhalation	Prenatal	Yes	No, but no observed body weight differences	Yes, restiricted to breathing ethanol fumes	Very swift increase phase compared with oral ingestion	Easey, requires inhalation chambers
Vapour inhalation	Postnatal	Yes	No, but no observed body weight differences	Yes, restricted to breathing ethanol fumes	Very swift increase phase compared with oral ingestion	Easy, requires inhalation chambers
Liquid diet	Prenatal	No	Yes, observed effects	Yes, diet restriction	Results from oral ingestion	Easy, requires graduated liquid containers
Artificial rearing	Postnatal	Yes	Yes, observed effects	Yes, handling and maternal and sib-ling deprivation	Results from oral ingestion by the pup	Difficult, requires infusion pumps Easy, requires feeding tubes
Intragastric intubation	Postnatal	Yes	Yes, no observed body weight differences in dams	Yes, intragastric intubation and handling	Results from oral ingestion by dam	Moderate, requires PE tubing
Intragastric intubation	Postnatal	Yes	No, but no observed body weight differences in pups	Yes, intragastric intubation and handling	Results from oral ingestion by the pup	Moderate, requires PE tubing

3. 0.30M potassium carbonate (Sigma).
4. Alcohol dehydrogenase (Sigma) is dissolved in distilled water to get a solution of 89.25 units/mL. The ADH solution is frozen at -4°C in 5-mL aliquots and thawed before the assay.
5. β-Nicotinamide adenine dinuculeotide (NAD) (Grade III, Sigma) is dissolved in 0.5 M TRIS buffer (Sigma 7-9) to give a solution of 1.875 mM NAD. This solution is kept refrigerated (4°C) until time of assay.
6. Ethanol standards are made such that there are solutions of 0, 50, 100, 200, 300, 400, 500, and 50 mg/dL of ethanol dissolved in water. These are kept refrigerated (4°C) until time of assay.

METHODS

Subjects in our experiments are Long-Evans rats purchased from Harlan and housed in an animal colony with a 12-h light -dark cycle (7:00 am lights on), temperature at 22°C, and humidity at 20%. Female animals are bred before alcohol administration in this model. The female rats are at least 90d of age and are immediately put on breeder blocks (Purina, a diet specifically designed for breeding rats) and allowed a week of recovery after shipment from Harlan. They are housed in groups of two or three during this period.

For breeding, they are put in groups of four or five overnight (from 5 pm on) with a proven male breeder. The next morning at 08:00 AM vaginal smears are taken and examined for the presence of sperm. If sperm is detected, that day is designated gestational day (GD) 1 and the female rat is assigned to one of three experimental groups, which are ET (Ethanol), 1C (intubated control), or NC (nontreated control; The experimental dams are singly housed in polypropylene cages on GD 1.

Alcohol Administration During the Prenatal Period

1. Each ET dam is allowed free access to rat chow (breeder blocks) and water; the amount of food intake is measured daily in order to provide data for pair-feeding of the 1C dams.
2. From GD 1 through GD 22, the ET dam is weighed and then intubated with 4.5 g/kg of ethanol in a volume of 20 mL/kg via intragastric intubation. The ethanol solution is drawn up into a 10-mL syringe and then attached to an feeding needle. Intragastric intubations are given by first dipping the stainless steel feeding tube in corn oil to provide lubrication; the tube is then inserted down the esophagus of the rat. Handling time should be minimized. Intubation of ethanol is done in the late afternoon in order to minimize effects on circadian rhythms.
3. For 1C dams, the treatment is similar except that food is restricted and the intubation is of isocaloric maltose-dextrin solution. On GD 1, the 1C dam is matched by body weight to an ET dam of similar weight that has successfully given birth to a litter. The 1C dam is then only given the amount of rat chow consumed by the matched ET dam on that particular gestational day. The 1C dam is weighed daily and immediately intubated with a maltose-dextrin solution that is isocaloric with the ethanol solution in a volume of 20mL/kg. The intubation process is the same as for the ET dams.

4. The NC dams are weighed on GD 1 and GD 22 to minimize handling during gestation. In earlier studies, NC dams were weighed on a daily basis in order to conclusively show that there were no dam weight differences, but it is felt that there are enough data showing this finding to warrant fully controlling handling to allow maximal detection of handling and stress effects.

Alcohol Administration During the Postnatal Period

1. The day of birth (typically GD 23) is designated postnatal day 1 (PD 1) and the dams and pups do not receive any intubations on this day Litters are culled to 10 pups with as close to an even number of males and females as possible. On PD 2 through PD 10, pups from the ET and IC groups are revmoved from their litter, one at a time and weighed. They are then given their first intragastric intubation and then 2h later, they are given their second intragastric intubation. This is done on PD 2 through PD 10.
2. All intubations given to the pups are administered using PE 10 *Intramedic* tubing attached to a 1-mL syringe via a short piece of PE 50 *Intramedic* tubing. A small amount of waterproof glue from a hot glue gun makes the connection between the PE 10 and PE 50 tubing tight. The PE 10 tubing is dipped in corn oil prior to the intubation in order to facilitate the procedure. ET pups receive a 3.0 g/kg dose of ethanol in a volume of 0.0278 mL/g milk solution (PD 2–10). Two hours after the first intubation. ET pups are intubated a second time with the milk solution only (0.0278 mL/g). This procedure does not result in deficits in body weight when this dose of alcohol of alcohol is used. The IC pups receive the same procedure (two intubations) as the ET pups except that solutions are not given. NC pups are weighed on PD 2 and PD 10 but not treated in any other way.
3. On PD 2, pups can be permanently paw-marked with India ink for identification purposes. India ink is injected subcutaneously using a 1-mL syringe and 26-gage needle. The coding system resulting from this injection is such that the two forepaws represent the numbers 1 and 2 and the two hindpaws represent 4 and 8. With this numbering system, addition of the numbers represented by the different paws can number pups from 1 through 15. Alternatively, a rat tattooing system can be used to tattoo the paws of the rats.

BACs

1. On GD 20, 10μL of blood is taken up using a heparinized capillary tube from a nick to the tail from the ET and 1C dams 3 h after the intubation procedure. On PD 10, 10μL of blood is taken using a heparinized capillary tube from a nick to the tail from the ET and IC pups 2 h after the alcohol intubation and just before the second intubation of milk only. The blood from the pups can be encourage to flow by holding the pup so that its tail is hanging, and then slowly moving your fingers along the tail to encourage the blood drops to come out. It is important not to put too much pressure on the tail.

   The blood from the ET dams and pups is used to measure BACs via a colorimetric enzymatic assay as described below. The alcohol doses are chosen because they produce similar peak BACs prenatally and postnatally and the time points are chosen to assay peak BACs which has been shown to be a critical determinant of the teratogenic effects of ethanol.

2. The blood alcohol concentrations are measured as described in Dudek and Abbott. The 10 μL of blood of the experimental animal and 10 μL of distilled water are added to 190 μL of 0.53 *N* perchloric acid. Then, *200 μL* of 0.30M potassium carbonate is added. The solution is vortexed and then centrifuged in a refrigerated (4°C) centrifuge for 15 min at 12,000rpm (14.4 g). On the same day, standards are made.

   Ethanol standards of 0, 50, 100, 200, 300, 400, 500, and 600 mg/dL (ethanol in distilled water) are made and stored at 4°C. Ten microliters of the standard and 10μL of blood from a nontreated, non-experimental animal are added to 190μL of 0.53 *N* perchloric acid. The standards are then treated as the blood samples from the experimental animals. At this stage, the samples and standards can be frozen together at –8.0°C until the time of assay or the assay can be conducted immediately.

3. If the samples and standards are frozen, they should be thoroughly thawed and centrifuged for 15 min in a refrigerated (4°C) centrifuge at 12,000rpm. The ADH solution should also be thoroughly thawed. Into a glass culture tube, 400 μL of the NAD-Tris solution and 50μL of ADH should be combined. Then, 50 μL of the samples and standards should be added to each tube.

   The tubes should be vortexed and then incubated for one hour at room temperature. Each solution should then be read for absorbance of a light of 340 nm wavelength in a spectro photometer. The standards should be used to construct a linear standard curve (with a correlation of greater than 0.95) and then the unknown samples are calculated from that curve.

NOTES

1. This procedure uses 4.5g/kg and 3.0g/kg of ethanol in dams and pups respectively, delivered in one bolus. Others have used higher doses and found that loss of body weight in the pups occurs. If higher doses or two ethanol intubations of a higher dose are used, careful pilot work and consideration of more than one supplemental milk injection should be considered. In addition, it is likely that the peak blood alcohol concentration will occur at a different point with two intubations.
2. The Supro 710 protein powder can be purchased from Purina only in bulk (40kg). It is possible to get free samples from some of the sales representatives.
3. A common problem with this assay will appear as a flat standard curve. This occurs because the aliquots of ADH have been thawed and frozen too many times, resulting in a degradation of the enzyme and a failure of the reaction.
4. In general, the earlier the smears are taken, the better the detection of sperm.
5. Animals are assigned in cohorts such that all three groups are represented. Because survival of the litter tends to be less from ET dams because of neglect of the pups or spontaneous resorptions of the litters, more ET dams are designated than IC or NC dams.
6. Learning to do intragastric intubations in dams can be tricky, and it is highly recommended that extensive training take place prior to being allowed to intubate experimental animals. Training can be facilitated by first training with lightly anesthetized rats. Protective gloves should be thin enough to enable a good touch; we use gardening gloves. A good intubation should take less than 1 min.

7. Very rarely a litter is born either early or late. If this occurs, postnatal day is determined from the point of conception; postnatal day 2 then is always gestational day 24.
8. Intubating rat pups is the most demanding part of this technique and requires practice. Marking the PE 10 intubation tube to indicate the length that would reach the stomach helps gauge how deep the tube should be inserted. The tube should slide down the esophagus with little resistance although it is sometimes a tight fit in rats at PD 2. If resistance occurs, the tube should be removed and the procedure tried again. For training, starting with rat pups at PD 4 or 5 tends to result in more success.
9. Initially with this model, it was thought that the IC control animals should receive intubations of milk solution in attempt to match the ET groups. However, pups do not regulate their food intake well, and the intubation of milk resulted in IC control animals weighing considerably more that NC control animals; this has been shown in pilot studies in our laboratory and in a published study.

GENE EXPRESSION IN A COMPLEX TISSUE

The changing pattern of estradiol (E) and progesterone (P) secretion during the primate menstrual cycle governs the hormonal regulation of endometrial growth, differentiation, shedding, and reconstruction that is an essential component of continued reproductive competence. The focus of our laboratory has been on P-dependent regulation of endometrial response in the rhesus monkey, an appropriate model for human endometrial function. P is expected to regulate a wide variety of endometrial genes that include growth factors and their receptors, extracellular matrix proteins, and enzymes involved in cellular metabolism.

These features of P action most likely involve a cascade of signal transduction pathways that control the global maturation of the endometrium. One important mechanism that regulates the factors in such pathways is the activation or repression of their respective gene products at the transcriptional/translational level; a fate ultimately determined by specific promoter-binding transcription factors and associated coactivators/repressors including interference RNA (iRNA).

PRIMATE ENDOMETRIUM

The rhesus endometrium has been characterized by Bartelmez, using histologic criteria, as composed of four horizontal zones : the transient functionalis is composed of zone I, the luminal epithelia and densely packed stroma, and zone II, the upper third segment of the glands; the germinal basalis is composed of zone III, the middle third of the glands, and zone IV, the deepest portion of the glands adjacent to the myometrium.

A similar zonation of the endometrium is also apparent in the human. One of the most striking characteristics of the primate endometrium is its remarkable regenerative capacity. Hartman showed that in the rhesus monkey, the endometrium could regenerate completely after an endometrectomy that would leave only a few endometrial cells on the surface of the myometrium. This regenerative capacity of the endometrium is perhaps not surprising because of the central role it plays in menstruating primate reproduction. This feature of the primate endometrium has allowed us to use these very valuable animals several times depending on their health and an appropriate postsurgical recovery period.

Spatiotemporal Regulation

In addition to the morphologic zonation described above, the endometrium's complexity is further defined by the number of different cell types that it harbors within these zones. These cell types include luminal and glandular epithelia, stromal fibroblasts, vascular smooth muscle cells, endothelial cells, and cells of the lymphocytic system. It has become increasingly clear that different cells or cell types within a target tissue can respond dissimilarly to the same hormonal milieu.

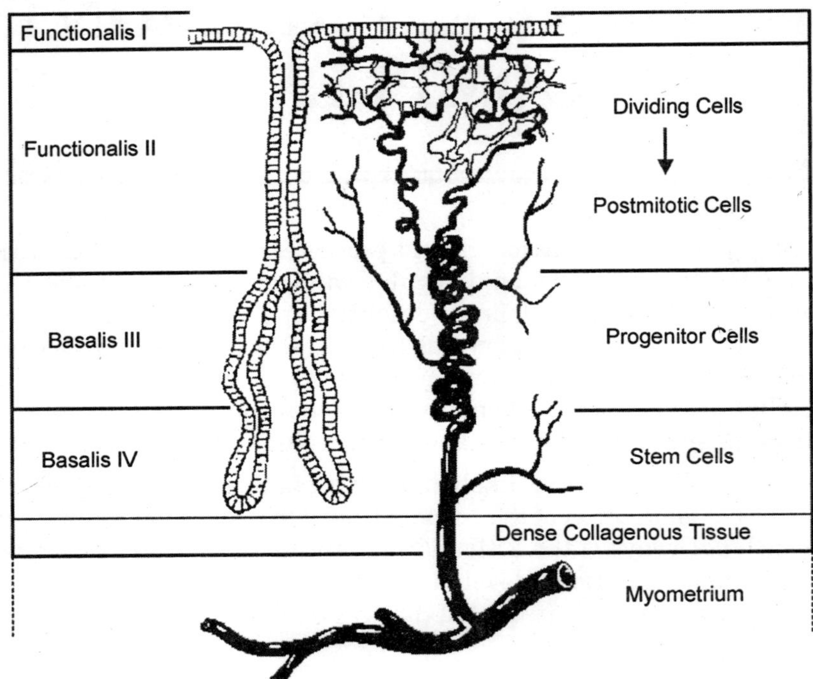

Fig. 3.1. Zonation of the primate endometrium

In our laboratory and others, it has been shown by immunohistochemical techniques that P inhibition of nuclear estrogen receptor (ER) is most pronounced in zones I, II, and III while strong positive staining of glandular epithelia in zone IV is retained. Stromal cells in zones I, II, and III are also more rapidly affected by P demonstrating a temporal regulation of this inhibition. A recent study in the human has also shown that stromal cells are more sensitive to P downregulation of ER.

It has also been shown that there are zonal dependent differences in proliferation during both E- or P-dominance during both natural menstrual cycles and artificial menstrual cycles in the rhesus monkey. These studies and others strongly support the concept that the primate endometrium contains distinctive microenvironments that can respond differentially to the same hormonal stimulation, and different cell types within these regions of the endometrium are also differentially responsive. This important concept is essential to the understanding of responses stimuli in complex tissues such as the endometrium.

ANIMAL MODEL

The development and use of artificial menstrual cycles in the rhesus monkey was first described by Hodgen. These studies showed that simulation of the menstrual cycle by the timed insertion and removal of silastic implants of estradiol (E) or progesterone (P) was sufficient to allow the endometrium to support implantation and eventual delivery (IVF and surrogate transfer). Luteal-phase defects in women are purported to be the most common endocrinopathy in infertility and recurrent abortion wherein low secretory P levels are not sufficiently elevated to achieve appropriate endometrial maturation. Both short and inadequate luteal phases similar to those found in women have been described in the rhesus monkey.

These latter studies also provide support for the usefulness of the rhesus monkey as a model for luteal phase defects in women wherein low secretory P levels lead to retarded endometrial maturation. We have created our inadequate secretory phase P levels in accord with the levels determined to be inadequate based on these previous studies. Our previously published studies describe in detail the protocols for creation of adequate and inadequate cycles. These studies showed that the hormone levels produced by these protocols are coincident with for those observed in the natural menstrual cycle as well as inadequate secretory phases. The profiles of serum E and P observed using the above protocols are shown in Fig 3.2 adequate (A) and inadequate (B) secretory phases.

Laser Capture Microdissection

Traditionally, cell-type-specific and region-specific regulation of genes and gene products has been analyzed using *in situ* hybridization and immunohistochemistry, respectively. Although these techniques remain important tools in our experimental arsenal, a new technology, laser capture microdissection (LCM), initiated by the National Institutes of Health (NIH), has substantially expanded our ability to examine cell-type-specific and region-specific responses. Microdissection using this technique allows the retrieval of specific cells from specific morphologic units within the tissue of interest.

Material harvested using this approach can subsequently be used for a variety of genetic or proteomic analyses. Since the introduction of LCM technology, laboratories in a wide variety of scientific areas have capitalized on its power to harvest specific cell types within their tissue of interest. The heterogeneity of most tissues (e.g., cell types and regional areas) makes this technology particularly amenable to many studies. Perhaps not surprisingly, some of the first applications of LCM were in the area of molecular diagnostics particularly in cancers because the cancer cells could constitute less than 5% if the total tissue and render analyses of whole tissue confusing or uninformative. Some of the initial studies focused on cancerous lesions such as the prostate, breast, colon, lung, and follicular lymphoma as examples. Other applications have included studies on spermatogenesis and the development of the fetal testis as well as in neuroscience research.

Although studies on mammalian systems have dominated the literature, there have also been investigations on selective retrieval of plant cell types/tissues using this technology. Our laboratory has used LCM to study the primate endometrium, a complex heterogeneous structure (see above) whose components are difficult at best or impossible to study in isolation. For example, separation of endometrial stroma from epithelia has been shown to dramatically alter epithelial response to hormonal signals.

Microdissection can overcome limitations of traditional means of analysis and allow the application of powerful molecular methods of analysis on specific cell types within specific morphologic units in the tissue of interest. Our studies described herein will be used as an example of the power and application of this technology. We have used this approach in combination with PCR analysis, differential display and microarray analysis to study differential gene expression in different cell types within different regions of the primate endometrium

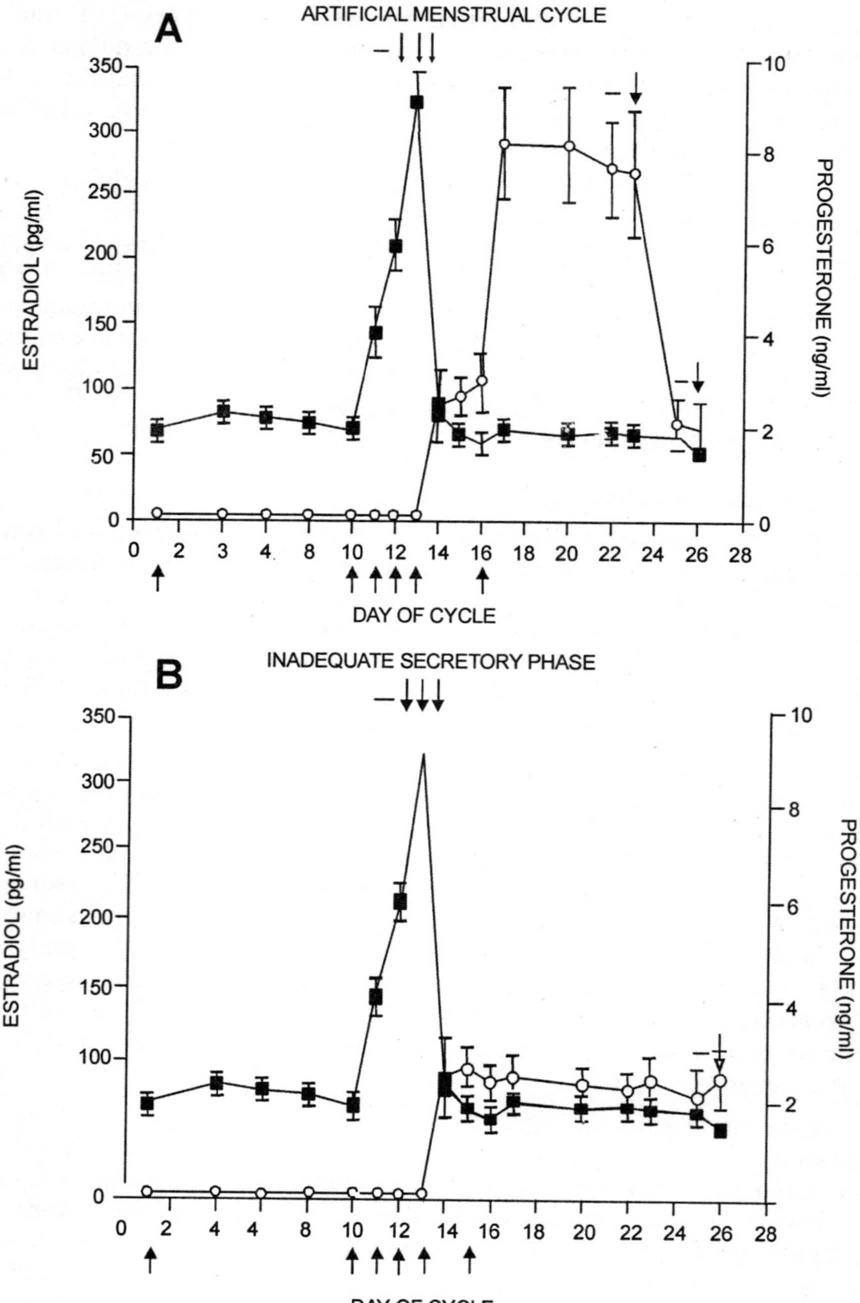

Fig. 3.2. Serum E and P levels during simulated (A) adequate or (B) inadequate menstrual cycles in the rhesus monkey. Closed squares and closed arrowheads represent the mean ± SEM (n = 6 to 8) of serum E levels and implant insertion and removal, respectively. Open circles and open arrowheads represent the mean ± SEM (n = 6 to 8) of serum P levels and implant insertion and removal, respectively.

LCM Analysis of Endometrial Tissue

Details of our LCM procedures can be found in our previous publications (see above). We provide below the steps we have used to obtain appropriate cellular targets from primate endometrial tissue. Additional information on procedures and Arcturus LCM equipment can be found at www.moleculardevices.com.

Procedure

1. Endometrial tissue is obtained at laparotomy by endometrectomy from rhesus monkeys.
2. The tissue is oriented in a an ice-cooled, small aluminum foil cup and frozen immediately in Tissue Tek OCT embedding compound and stored desiccated at –80°C prior to further processing.

   (**Note :** It is important to orient the tissue according to the plane of interest subsequent to cryostat sectioning. For example, we orient the tissue so the myometrium will be part of our sections in order to clearly establish different zones/regions within the endometrium Dessication of the samples can limit ice formation upon frequent opening and closing of the freezer and the resulting condensation that can affect tissue integrity.)
3. After mounting the tissue block, we obtain cryostat (–25°C) sections (6 μm) to determine proper tissue orientation microscopically using Toluidine Blue O (0.5% aqueous) staining (5 to 10 s, water wash). Once proper orientation or reorientation of the tissue has been established, tissue sections for LCM [cryostat (–25°C), 6 μm] are placed on untreated plain glass slides and immediately fixed in 70% ethanol (10 min).

   (**Note :** Slides are precooled to cryostat temperature prior to sectioning and very briefly mounted (finger warmth) within the cryostat. This procedure minimizes the condensation that can occur when sections are mounted outside of the cryostat. In addition, the fixation in 70% ethanol is also done within the cryostat to also limit water condensation.)
4. Slides are air-dried for 20 min to promote adhesion of the section.
5. Slides are stained with hematoxylin and eosin using the following sequential solutions (10 to 30 s each): Mayers hematoxylin, distilled water, bluing reagent, 70% ethanol, 95% ethanol, and Eosin Y.
6. Slides are then dehydrated with two 10-s washes using 95% ethanol, and two 10-s washes using 100% ethanol. Finally, the slides are placed in xylene twice for at least 5 min and then dried in a vacuum desiccator for 15 to 20 min prior to LCM.

   (**Note :** A fresh bottle of 100% ethanol and fresh xylene should be used routinely.)
7. Luminal or glandular epithelia or stromal cells from appropriate regions of the endometrium are identified morphologically and harvested using a Pix Cell II LCM System.

   (**Note :** The lOx objective is used for tissue orientation and the 20× and 40× for specific cell-type harvesting depending on the application. Photographic documentation prior to and after microdissection of the cells and region of interest should be done routinely.)

8. The 15- or 30-μm-diameter beams are used depending on the application. Amplitude and pulse duration ranged from 35 to 50 mw and 3 to 5 ms, respectively.
9. Cells are collected on TF-100 caps (Arcturus Engineering) containing the transfer film from a minimum of two to three sections for each sample. Tissue samples from at least three different animals are analyzed individually or pooled for subsequent analyses where appropriate depending on the application.
10. Excess cellular/tissue debris, not part of the laser etched area, is removed by a brief treatment with a Capture Sure Pad.

    (**Note :** The removal of this material can be Visualized microscopically and documented photographically if required.)
11. The TF-100 caps are placed on ice-cooled Microfuge tubes of the appropriate size that contained the RNAqueous lysis/binding buffer (100 μL) (RNAqueous-4PCR Kit).
12. The tubes are inverted several times to remove the captured tissue form the transfer cap, vortexed, briefly centrifuged, and processed for RNA using the RNAqueous protocol (see below).

Four different cDNA populations were prepared from endometrial tissue harvested by LCM from adequate secretory cycles (days 21 to 23). These populations were glandular epithelia and stroma from the functionalis (FG and FS, respectively) and glandular epithelia or stroma from the basalis (BG and BS, respectively). One of the first objectives of our studies was to assess the quality and potential usefulness of endometrial tissue harvested in this manner for subsequent gene expression studies.

There are numerous steps in the preparation of suitable genetic material from laser microdissected tissue any one of which could compromise the quality of a sample. Because of the time and effort that is required for an analysis of gene expression using this approach, it is useful to have some guide to the relative quality of a sample.

Synthesis and Amplification of cDNA Populations

Tissue limitations for our laboratory and others may not allow traditional means of analysis of RNA integrity (e.g., relative size of the mRNA population by agarose gel electrophoresis). In an effort to overcome this drawback we have used an adapter-specific primer amplification approach to allow visualization of a cDNA smear. This approach coupled with the detection of an appropriate housekeeping gene(s) can serve as a useful guide to estimate sample quality for those investigators faced with limited tissue/cells. The above approach also provides considerable material (cDNA) from a single round of amplification (approximately 75-fold) that will subsequently allow a number of comparative studies on gene expression to be performed. RNA extraction was performed using the RNAqueous-4PCR Kit (Ambion).

The LCM transfer caps were placed in Brinkmann microcentrifuge tubes with RNAqueous lysis/binding solution, which contained guanidinium thiocyanate. After vortexing and centrifuging the tubes, the caps were removed and RNA was isolated using the RNAqueous protocol. The DNAse I treatment that is part of this protocol has been shown previously to effectively remove genomic DNA. The Superscript Choice System (Life Technologies, Rockville, MD), was used for first strand cDNA synthesis using a mixture of both oligo(dT) and random hexamer primers according to the manufacturer's protocol. Second strand cDNA synthesis and adaptor ligation with EcoRI *(Not* I) adapters were performed using the same kit. cDNA populations were purified in Qiaquick spin columns and amplified by PCR in 100 µL containing 0.5-µm LINK-CUA primer, 0.25 mM dNTPs, 1.5 mM $MgCl_2$, Ix buffer, and 2 units Taq polymerase in a thermal cycler (94°C, 1 min; 50°C, 1 min; 72°C, 2 min) for 30 cycles. The LINK-CUA primer (5′-CUACUACUACUAAATTCGCGGCCGCGTCGAC-3′) is complementary to the EcoRI adaptor.

After amplification of the cDNA populations, one-fiftieth (2 µL or 40 ng) was used as template in PCR reactions using primers specific for human 18S ribosomal RNA (Ambion) and G3PDH. The results obtained with 18S ribosomal RNA were used to normalize the cDNA populations. Details of the above approach can be found in our previous publications. In addition to cDNA smears, the relative size range of our cDNA populations was also estimated using. the above primers for both 18S ribosomal RNA (product, 324 bp) and G3PDH (product, 983 bp). Our data suggested that the relative size range of a cDNA smear can influence the subsequent detection of these housekeeping genes (see below).

cDNA Amplification Strategy

With the use of primers for a housekeeping gene, it would be expected that its presence at the correct fragment size would provide evidence for the suitability of the genetic material

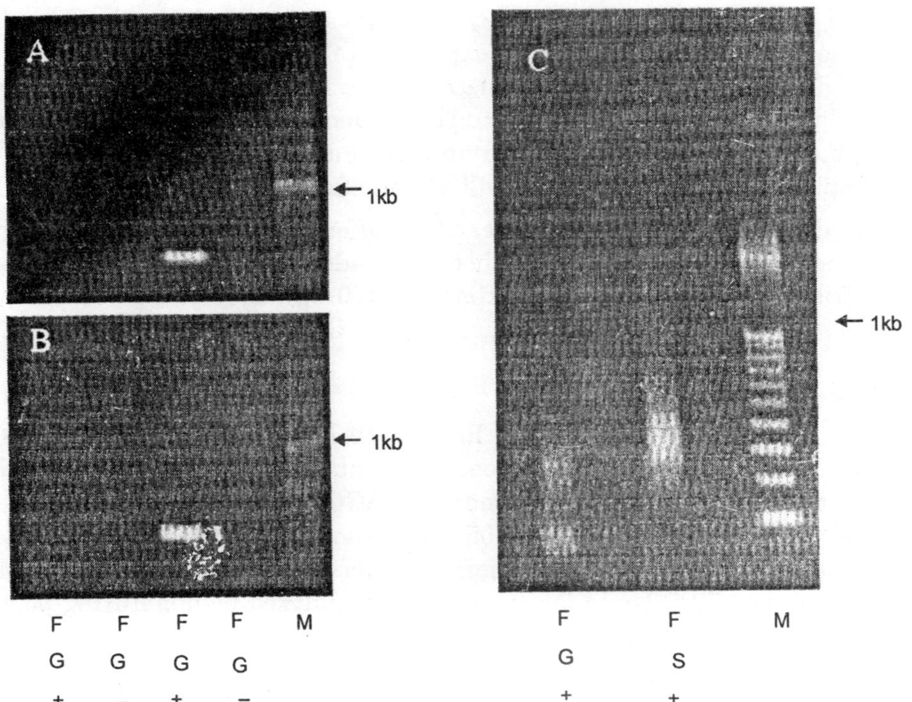

Fig. 3.3. Detection of 18S ribosomal RNA expression before and after amplification of cDNA. (A) Results after 30 cycles of PCR, (B) results after 40 cycles of PCR, (C) corresponding cDNA smears after amplification of adaptor-ligated cDNA. "+" signs indicate cDNA that has been amplified after adaptor ligation, and "–" signs indicate double-stranded cDNA before amplification. FG, cDNA from functionalis glandular epithelia; FS, cDNA from functionalis stromal cells. A 100–bp ladder is shown with 1 kb band indicated by an arrow.

for further analysis. Whereas this may in part be true, the absence of an appropriate PCR product could mean either that the preparation is poor (e.g., degraded) or that there is insufficient material for detection (false negative). Our data for 18S ribosomal RNA expression prior to and after amplification show that a false negative can be detected).

For example, a detectable band of the correct size is only apparent following amplification for either 30 or 40 cycles (FS+ vs FS). The use of this amplification strategy cannot only allow detection of a false-negative result but also can provide some additional information with regard to the quality of the genetic material through cDNA smears. For example, the cDNA populations prior to or after amplification for glands from the functionalis did not show an 18S ribosomal RNA band (324 bp) despite amplification (FG+,).

These data would suggest that the quality of this material rather than the quantity is most likely at fault. The very low molecular weight smear of this cDNA population correlates with this result. cDNA smears after amplification of these laser microdissected samples can also be useful in the design of an appropriate expected fragment size for a given housekeeping gene or other gene of interest.

DIFFERENTIAL DISPLAY

Differential display is an approach for the identification of mRNAs that show differences in expression level between two or more experimental groups or that are unique to a cell type, tissue or developmental stage. The method allows random cDNAs to be amplified by a pair of short arbitrary PCR primers (10-mer). The sequence of the primers dictates which panel of cDNA fragments of the total will be amplified: only those gene fragments containing sequences complementary to the primers will be amplified.

The use of different primer sets will result in different patterns of gene fragments that are amplified, and we have used this method to broaden the scope of genes to be analyzed. Because the fragments are small (approximately 400 bp), they can be cloned and quickly sequenced and subsequently compared by homology to GenBank database entries.

Cloning and Sequencing

DDRT-PCR was performed using the RNAimage kit (GenHunter, Nashville, TN). Two nanograms of cDNA is amplified in 20 μL reactions containing 1X buffer, 2.0 μM dNTPs, 20 Ci/mmol alpha-[^{33}P]ATP, 0.2 μM HT_{11} A primer (5'-AAGCTTTTTTTTTTA–3′), 0.2 μM H-AP1 primer (5'-AAGCTTGATTGCC–"), and 0.05 units Taq polymerase (Qiagen, Valencia, CA). Reactions were carried out in a PTC-200 thermal cycler (MJ Research, Waltham, MA) at 94°C/1 min, 40°C/2 min, and 72°C/1 min for 40 cycles and analyzed by denaturing polyacrylamide gel electrophoresis, omitting the fixing stage.

The autoradiogram and gel were aligned by needle punctures, and individual bands were carefully excised from the gel with a razor blade. Gel slices attached to filter paper were eluted by boiling in 100 μL water for 10 min, spun to remove debris, and the supernatant precipitated with glycogen. DNA fragments thus isolated were reamplified as described above except with 250 μM dNTPs in the absence of radiolabel. Products were directly cloned into the plasmid vector pCR 2.1-TOPO^{R} (Invitrogen, Carlsbad, CA) and sequenced (UMass Medical School Nucleic Acid Facility). Homology searches were performed against GenBank entries using BLAST programs (NCBI). Alternatively, the following arbitrary primers were used to provide additional patterns of gene expression:

Arbitrary Primers
ARB-1: CTGATCCATG
ARB-2: CTTGATTGCC
ARB-4: GTTGCGATCC
ARB-5: GACCGCTTGT
ARB-7: CTTTGGTCAG
ARB-8: CAAGCGAGGT
ARB-9: AACGCGCAAC

Window of Receptivity

The different patterns (bands) of expression that are observed after electrophoresis and autoradiography of the radioactive fragments must be subsequently confirmed by PCR analysis (false positives are common) and, if appropriate, sequenced. We had previously hypothesized different patterns of gene expression during the secretory phase in the rhesus endometrium based on our work and other studies in the literature. The results of our studies (using whole endometria) are shown in Fig. 3.4, and the hypothetical patterns of expression are described in the legend (see also the composite panel E).

Of particular interest are the fragments that displayed a restricted expression during the expected window of receptivity (panel D). Two fragments exhibited high homology to previously characterized human genes, syncytin (the envelope gene) and secretory leukocyte protease inhibitor (SLPI). A third fragment, similar to BAT2 (KIAA1096), also displayed a restricted pattern of expression during the window of receptivity. Syncytin is a highly fusogenic membrane glycoprotein that appears to be expressed specifically in the placenta.

The protein induces the formation of giant syncytia and can mediate fusion of cytotrophoblasts into the syncytiotrophoblast layer, which is essential for pregnancy maintenance. Endogenous viral syncytin appears to have been sequestered to serve an important physiologic role during pregnancy and placental morphogenesis Our results are the first to describe the expression of this gene in the primate endometrium. Its potential role in the non-pregnant uterus, however, remains to be elucidated. We speculate that syncytin could be a P-induced endometrial decidualization factor (it induces syncytia, or multinucleated cells). Interestingly, the expression of HERV-K (closely related to HERV-W) is stimulated in cultured human tumor cells by sequential E and P treatment, most likely mediated by the presence of a P receptor binding site situated in its long terminal repeat (LTR).

Because differentiation of P–induced endometrial stromal cells into decidual cells is essential for embryo implantation and placentation, expression of syncytin during the window of receptivity of the primate menstrual cycle is likely to play an important role in preparation of the endometrium for successful implantation. SLPI is a neutrophil elastase inhibitor that also has antibacterial and antinflammatory properties. In addition to its antiprotease activity, secretory leukocyte protease inhibitor (SLPI) has been shown to regulate intracellular enzyme synthesis, epithelial cell growth by activation and repression of distinct growth-regulatory genes, and mediate normal wound healing.

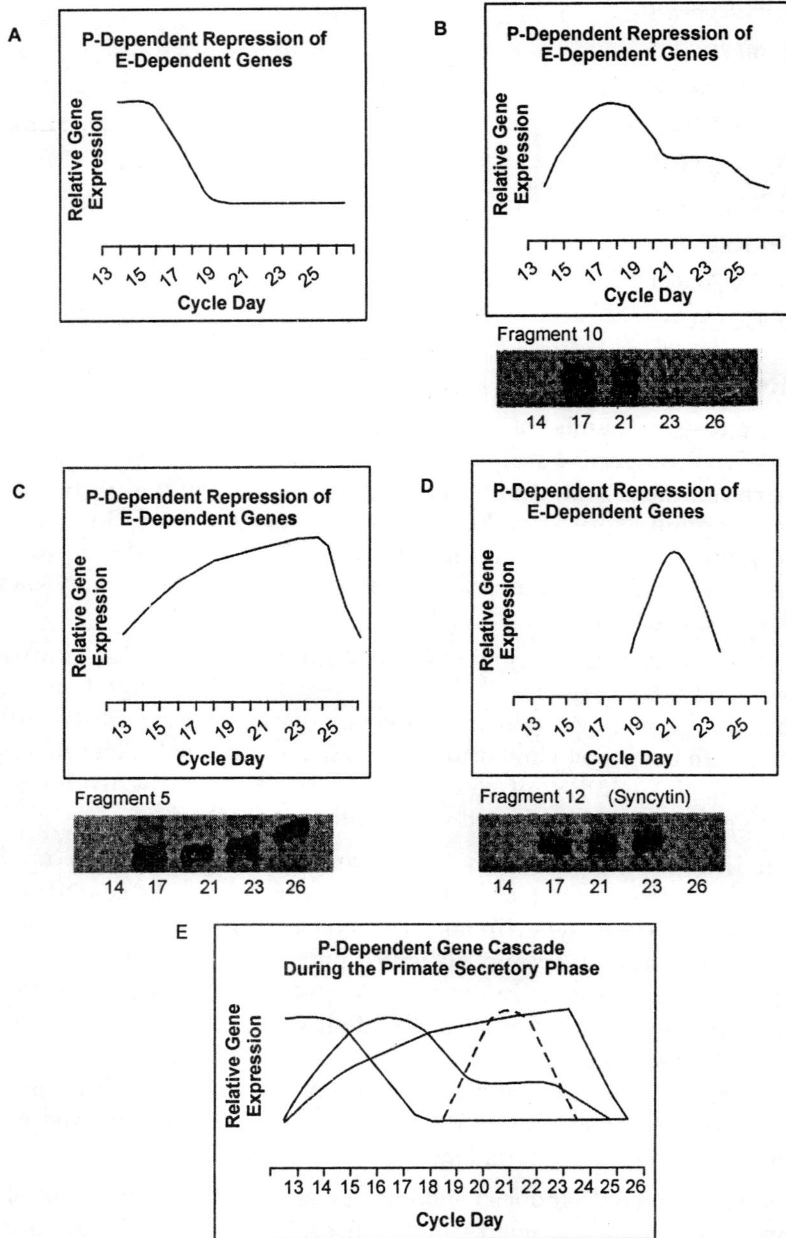

Fig. 3.4. A schematic representation together with examples of relative gene expression patterns in the rhesus endometrium during the secretory phase. (A) Progesterone (P) repression of estradiol (E)-dependent genes. (B) Autologous downregulation of P-dependent genes (e.g., fragment 10). (C) P-induction of genes during the secretory phase (e.g., fragment 5). (D) P-induction of gene expression during the window of receptivity (e.g., fragment 12, syncytin). (E) A composite cascade of gene expression patterns (A) to (D). The associated panel (B-D) for each fragment shows the relevant temporal portion of the gel after PCR analyses.

King et al. have shown by immunohistochemistry that SLPI is expressed in glandular epithelium of human endometrium from the mid to late secretory phase suggesting a direct or indirect regulation by progesterone.' These studies are in agreement with our results that show increased expression of this gene from days 17 to 23 of adequate secretory phases in the rhesus monkey. Leukocytes infiltrate the endometrium prior to menstruation and may in part be responsible for the rise in secretory SLPI during the secretory phase. SLPI has been found to have antibacterial effects, and one region of the molecule has 37% homology with defensins, a family of antibacterial proteins. SLPI also inhibits the NF kappa-B signal transduction pathway involved in inflammatory response.

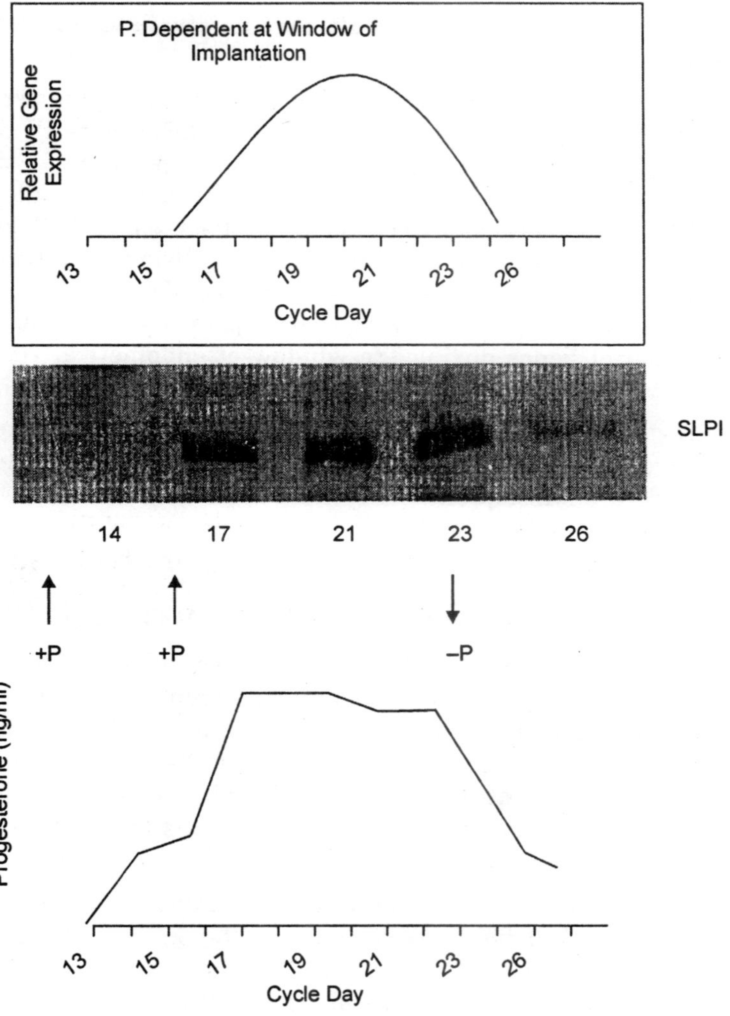

Fig. 3.5. SLPI expression and hormonal levels of progesterone (P) during the expected window of receptivity in the rhesus monkey endometrium. Upper panel shows the hypothetical expression pattern and the associated temporal pattern of SLPI expression as determined by PCR analyses. The lower panel shows the associated serum progesterone levels during this timeframe.

The most likely role of SLPI is as a natural antibiotic and antinflammatory molecule. Infection ascending through the cervix could pose a threat to the implanting and developing conceptus. The third fragment initially represented a predicted human protein, KIAA1096, based on the open reading frame, which was subsequently shown to be BAT2. BAT2 shares similarities with some transcriptional regulatory proteins containing zinc finger motifs and proline- or glutamine rich regions. As noted above, BAT2 also contains motifs typical of the integrin receptor family. Integrins (receptors) have been implicated in successful implantation in the human. BAT2 has also been mapped within the class III region of the major histocompatibility complex (MHC), which encodes genes involved in immune function or MHC-associated disease susceptibility.

Serum P Level

An example of the relationship between the temporal change in serum P level (adequate cycle) and restricted gene expression is shown in Fig 3.5 for SLPI during the window of receptivity in an adequate secretory phase. Note that the pattern of rising and falling serum P is paralleled by the rise and fall of SLPI expression. Similar profiles for the genes (gene fragments) described above also displayed such a pattern (e.g., BAT2, syncytin, WFDC2). We have also studied the expression of these genes during inadequate secretory phases (cannot support implantation, see above) and have shown that SLPI, BAT2, and WFCD2 genes are strikingly underrepresented compared with their expression in adequate secretory phases. A restricted expression of genes during the window of endometrial receptivity/implantation provides support for the potential importance of these genes in maturation of the endometrium.

All levels of regulation described in our working model (see above) are, however, considered important because of their potential to be linked. That is, preceding gene expression patterns may control and direct subsequent gene expression that will allow proper endometrial maturation necessary for embryo implantation. It is anticipated that refinements or additions to this working model will result from future studies in our laboratory and others.

GENE MICROARRAY ANALYSIS

We, as well as other investigators, have used the powerful but often perplexing technique of gene expression (microarray) profiling to identify differentially expressed genes in the human and non-human primate endometrium. A complete discussion of these data is beyond the scope of this chapter, and our approach and methodology can be found in our previous publication. Our studies focused on the identification of differentially expressed endometrial genes during a normal secretory phase (progesterone-dominant) versus the proliferative phase (estrogen-dominant) in the rhesus monkey.

Specifically, our data confirmed the elevated expression of SLPI (see above) and identified another member of the SLPI family of secretory proteins, whey acidic protein four-disulfide core domain 2 (WFDC2). Both WFDC2 and SLPI belong to a family of 14 WAP proteins (whey acidic proteins) that have duplicated on chromosome 20 over evalution. Interestingly, WFDC2 was highly upregulated (27.5-fold) in our microarray studies. The expression profile of WFDC2 was confirmed by semiquantitative PCR in parallel with SLPI. In addition, WFDC2 showed a restricted expression similar to that of SLPI during the window of receptivity.

Although there is limited literature on WFDC2, it is interesting to speculate that one or more of these uncharacterized family members could provide functional redundancy regarding endometrial development in the female reproductive tract. Our data on SLPI and WFDC2 confirm the microarray analysis and provide an important avenue for future studies on the potential importance of this gene family in endometrial function.

INADEQUATE SECRETORY PHASES

As shown above, the regulation of syncytin, SLPI, BAT2, and WFDC2 displayed a restricted expression during the expected window of receptivity during adequate secretory phases. We extended our studies to compare both adequate and inadequate secretory phases. Our results clearly showed that the expression of both SLPI and WFDC2 are not temporally elevated during an inadequate secretory phase (2.8- and 4.8-fold decrease, respectively). In addition, we have also shown that both syncytin and BAT2 are also temporally underrepresented during an inadequate secretory phase (2.3-and 3.0-fold decrease, respectively). Together, these data suggest that these four genes may play important roles in endometrial maturation and function during the receptive period of an adequate secretory phase

LCM AND DIFFERENTIAL DISPLAY

In order to couple LCM with differential display, we prepared cell type-specific cDNA populations (glandular epithelia or stroma) from both the functionalis and basalis of adequate midsecretory endometria as described above. After differential display analysis, we selected six fragments that showed a putative cell-type-specific and/or a region-specific expression. The differential display gels and patterns of expression of these fragments are shown in

Although differential display reverse transcriptase-polymerase chain reaction (DDRT-PCR) is a potentially powerful and important approach, a drawback as noted above can be the appearance of false positives. After cloning and sequencing, specific primers for each of these fragments were designed and used to verify their expression patterns. Although three of these fragments were shown to be false positives with regard to their regional or cell-type specificity, they remain potentially important because of their elevated expression during an adequate secretory phase. The three fragments identified in laser microdissected samples showing the expected regulatory pattern were F1 (highly expressed in the glands and stroma of the functionalis), BG-1 (highly expressed in the glands of the basalis), and FS-1 (highly expressed in the stroma of the.functionalis).

Although BG–1 and FS–1 are currently uncharacterized gene fragments, F1 showed a 94% homology to a known gene, namely, the human leukotrienre B4 receptot. Leukotriene B4 (LTB_4) is one of the most potent chemoattractant mediators, acting mainly on neutrophils but also on related granulocytes, macrophages, and endothelial cells. LTB_4 activates inflammatory cells by binding to its cell surface receptors BLTR1 and BLTR2 and has been implicated in a number of inflammatory diseases. Interestingly, levels of leukotrienes were elevated in the endometrium of women with primary dysmenorrhea and endometriosis. Levels of LTB_4 have also been shown to increase in the rat uterus during the peri-implantation phase implicating a role for this cytokine in uterine receptivity and implantation.

Importantly, the expression of this gene was localized to the endometrial functionalis, the target for blastocyst invasion/implantation. To our knowledge, this is the first time expression of a BLTR2 receptor ortholog (F1) has been shown in the endometrium. Further studies will be required to identify the role of this receptor in proper maturation of the primate endometrium.

LCM and SLPI, WFDC2, BAT2, and Syncytin

We have also used LCM to study cell-type and regional differences in expression of SLPI, WFDC2, BAT2, and syncytin (see above) in the rhesus endometrium. SLPI and BAT2 showed an increased expression primarily in the stromal compartment of the functionalis, whereas WFDC2 displayed elevated expression in both stroma and epithelia of the functionalis compared with the basalis.

No detectable differences for syncytin were observed with these LCM samples. These studies demonstrate a differential expression of several known genes during the window of receptivity that appears to be primarily localized in the functionalis of the primate endometrium, the primary target for blastocyst invasion. Preliminary studies (data not shown) showed that SLPI and WFDC2 are both upregulated in LCM harvested luminal epithelia during the receptive window of an adequate secretory phase. Further studies are necessary to understand the functions and roles of these genes in primate endometrial maturation and receptivity.

MICROARRAY ANALYSIS OF ENDOMETRIAL CELLS

cDNA populations representing functionalis glandular epithelium (FG) and stroma (FS) isolated by LCM were constructed as previously described . The two populations were used to prepare biotin-labeled probes by random primer extension and separately hybridized to Affymetrix (Santa Clara, CA) HGU95Av2 oligonucleotide microarrays. When compared, the results from our initial experiment showed that >50 genes were upregulated >5-fold in FG, while >100 genes were upregulated in FS. Likewise, these upregulated genes were

Table 3.1. Genes Upregulated in the Functionalis Glands (FG), Secretory Phase.

Accession no.	*Fold-up*	*Gene*	*Biological function (NCBI)*	*Regulation in endometrium*
L07648	10.3	Max interacting factor (MXI1)	DNA binding, transcription corepressor activity	
M38449	5.5	Transforming growth factor beta 1 (TGFB1)	Contrpl of Proliferation and differentiation	Upregulated in secretory phase glands
AC005175	5.1	Thromboxane A2 receptor (TBXA2R)	Muscle contraction	Expressed in gland
D14838	4.5	Fibroblast growth factor 9 (FGF9)	Embryonic development, cell growth, morphogenesis, tissue repair	
M32334	4.4	Intercellular adhesion molecule 2 (ICAM2)	Cell-cell adhesion, integrin binding	Expressed in vascular endothelium

downregulated in the comparative cDNA population. Examples of some of these data are presented in Table 3.1 and Table 3.2. Genes upregulated in FG included transforming growth factor beta 1 (TGFBl), MAX interacting protein 1 (MXI1), fibroblast growth factor 9 (FGF9), and intercellular adhesion molecule 2 (ICAM2). TGFB1 has previously been shown by others expressed at elevated levels in secretory phase glandular epithelium of the human endometrium.

Table 3.3. Genes Upregulated in Functionalis Stroma (FS), Secretory Phase.

Accession no.	*Fold-up*	*Gene*	*Biological function (NCBI)*	*Regulation in endometrium*
AF084367	46	Inversin (INV)	Embryonic development, left-right axis determination	
D14043	39	Sialomucin (CD 164 antigen)	Regulation of Cell adhesion	
AA401397	28	Kallikrein 13 (KLK13)	Proteolysis and Peptidolysis	
D83402	15	Prostacyclin (PTGIS)	Prostaglandin biosynthesis	Expressed in functionalis stroma
AB015051	12	Death-associated protein 6 (DAP6)	Regulation of transcription, DNA–dependent apoptosis	

Genes upregulated in FS included inversin (INV), sialomucin (CD 164), kallikrein 13 (KLK13), and death-associated protein 6 (DAP6). Expression of these genes in the endometrium has not previously been documented. These data provide support for the use of combined LCM and microarray analyses.

CONCLUSION

The use of LCM affords an important and powerful tool to characterize and analyze gene or protein expression patterns in tissues composed of multiple cell types within different regions. LCM coupled with techniques such as PCR analysis, differential display, and microarray analysis further augments the usefulness of this approach. Indeed, our data described above support the applicability of these approaches in combination with LCM. Our results demonstrate that regional and cell-type differences in gene expression are a property of a complex tissue such as the primate endometrium and that these approaches can be applicable to the analyses of other complex tissues. The application of the above tools to answer complex biological questions and other techniques yet to be developed will help lead us down new and exciting pathways of discovery and knowledge.

4 ANALYZING GENE EXPRESSION

Alcoholism is generated from long term exposure and the development of this disease may be due to corresponding changes in molecular signaling events. Critical molecular events underlying this neuronal plasticity are thought to include changes in gene expression. High-density DNA microarrays allow for simultaneous, nonbiased measurement of the expression of thousands of genes and the identification of patterns of genes with similar function or regulation. Instead of quantifying single genes, microarray analysis can focus on entire gene networks and their associated biological pathways and functions.

Such an approach allow a nonbiased analysis of complex biological events involved in brain responses to ethanol or other drugs of abuse and can potentially generate novel hypotheses about mechanisms and treatment of these diseases. Expression profiling with DNA microarrays has been used to identify molecular network responses to ethanol in cell culture, animal models, and humans. Most of this work has been in neural cells or brain tissue. This enormous amount of complex data has grown to the point of allowing sophisticated meta-analyses of expression patterns across multiple laboratories and the compilation of microarray databases across large genetic models of inbred mouse strains.

Despite this progress, the precision and sensitivity of microarray data analysis are critical determinants on whether downstream analysis and interpretation will yield significant results. Because of the inherent complexity of the central nervous system, expression profiling in neurobiology is even more challenging. Studies of whole brain neglect the elegant construction of distinct brain regions, yet microdissection can introduce errors from casual cross-contamination and dissection variability as well as the inclusion of multiple cell types in any dissected tissue.

Studies of the central nervous system are further complicated by the need for accurate detection of low-abundance genes that are postulated to have significant impact on neuronal phenotype. Regardless of the aforementioned complications that are involved in performing microarray studies on brain tissue, more than 1000 citations are currently listed in a PubMed search for "*brain*" and "*microarray*". A growing number of these involve studies on ethanol and other drugs of abuse.

Although the inherent complexity of brain regional and cellular makeup cannot be avoided, careful attention to experimental design and technical issues can greatly improve the quality and impact of microarray studies on brain tissue. This methods review will focus largely on techniques for oligonucleotide microarray analysis of brain responses to ethanol in the hope of encouraging future work in this challenging research area.

MATERIALS

DNA Microarrays

The two principle types of deoxyribonucleic acid (DNA) microarrays are spotted complementary (c) DNA and oligonucleotide arrays. Oligonucleotide arrays can be spotted by robotic liquid handlers, as with spotted cDNA arrays, synthesized *in situ* or immobilized on specialized bead supports. The spotted cDNA or oligonucleotide arrays can be synthesized in-house or acquired commercially, and many research universities now have core facilities that offer these services. Although in-house spotted cDNA arrays are relatively inexpensive and allow the researcher to dictate the sequences to be probed on the array, commercial oligonucleotide arrays offer higher standards of quality assurance and consistency, and are rapidly declining in price.

Major commercial sources of oligonucleotide arrays are Affymetrix, Agilent, and Illumina, and each of these has its own unique analysis features. Additionally, 15 National Institute of Health divisions have provided microarray core facilities that allow grantees to have analyses done at greatly reduced costs with several different microarray platforms including Affymetrix and Illumina. A description of feature and use of multiple array types is beyond the scope of this chapter. Therefore, we will focus on the use of Affymetrix GeneChip™ oligonucleotide arrays in studies on ethanol.

This type of commercial microarray is the most commonly used and offers considerable resources in terms of microarray availability, annotation of sequences on the arrays, standard reagents and reaction protocols, and availability of multiple statistical tools for array analysis. In particular, because the reagents and protocols for preparation of probes and hybridization of Affymetrix arrays are highly standardized, we will largely limit discussion to experimental design features and analysis of Affymetrix arrays. It should be stressed that alternative approaches to Affymetrix microarrays can produce comparable data, often at a considerably reduced price. Some caveats about Affymetrix microarray should be considered.

All Affymetrix produced microarrays are manufactured through a novel photo-lithography technique for simultaneous synthesis of very high density (currently up to >600,000 probes/array) short oligonucleotide (approx 25 nucleotides in length) probes. In general, there are now two major types of "expression arrays" from Affymetrix, 3′-end biased arrays and exon-arrays. The former type uses multiple oligonucleotides derived mainly from the 3′-untranslated end of a target messenger ribonucleic acid (mRNA). This produces a composite picture of the abundance for most transcripts from a given gene.

However, because many genes produce multiple mRNA species either through alternative splicing or use of gene expression can be obtained through measurement of transcript abundance for individual exons. The extremely high density of the photolithography arrays allows production of chips sampling most exons for the entire expressed genome. Another crucial feature of using Affymetrix microarrays for gene expression studies concerns the need to produce a complementary (c)RNA probe for hybridization to the arrays.

This process is a three-step protocol where total RNA is first converted to a double-stranded cDNA containing a T7-polymerase recognition site at the 3′-end of the molecule. This allows production of large amounts of biotinlabeled cRNA from the cDNA (general yields are 5–10X the initial amount of input total RNA). This amplification process is linear and produces a highly faithful representation of the original mRNA abundance. Moreover, the product cRNA can actually be re-converted to double-stranded cDNA and the process repeated to generate adequate amounts of cRNA for microarray hybridization even with sub-microgram amounts of starting total RNA. This two-step amplification process, however, biases the product cRNA toward the 3′–end of the mRNA molecules due to incomplete extension during the cycles of cDNA and cRNA synthesis. Thus, such two-cycle approaches do not routinely allow use of exon chips since more distal exons would be severely under-represented in the hybridization mixture.

Reagents

Because of the complex hybridization kinetics and multiple steps entailed in preparation of probes and array hybridization, reagents recommended by the microarray manufacturer should be used. These reagents can either be assembled by buying individual reagents from various suppliers or by using various kits developed and quality control checked by Affymetrix. Because of the sensitivity of the cRNA synthesis and labeling steps, reagents lots should not be mixed during an experiment. Doing so will produce large *"batch effects"* such that the cRNA reagent kit defines more variance of the array results than resulting from the desired experimental variables themselves.

Thus, arrays from different samples but run with one batch of cRNA reagents could appear more similar to one another than arrays run on identical biological samples but with differing reagent lots. Statistical approaches for minimizing such batch effects can be used, but it is preferable to avoid as many such confounds as possible during the experimental design stage. If more cRNA samples are to be prepared than can be made from a single lot of reagents, multiple reagent kits should be combined at the beginning of the experiment so that enough reagents are available as a single *"batch."*

METHODS

Experimental Design

The experimental design step of a microarray experiment is crucial to the generation and analysis of data and overall success of the experiment. Even if an array experiment produces high quality data, improper design can make the experiment extremely difficult to analyze and interpret, so much so that the results may be inconclusive and the hypothesis untested. Because microarray experiments are so expensive in terms of labour hours, reagents, biological materials, and the chips themselves, care should be exercised from the start to ensure success. Extensive discussions of statistical considerations of microarray experimental design have been published. The following factors should be considered during the design of microarray experiments:

1. Number of replicates

As with any experiment, an number of independent biological replicates are required to ensure statistical power for microarray data analysis. Traditional power analysis of microarray experiments suggests that upwards of ten replicates is required for each group, in part because of statistical issues caused by multiple testing when studying the entire genome simultaneously. Although the more replicates the better, in practice far fewer replicates are necessary to achieve statistical significance and quality data.

In addition, the number of replicates required is partly dictated by the type of experiment because the needs of a time course or dose response experiment might differ from an experiment on a panel of recombinant inbred strains. Anything less than three replicates makes downstream statistical filtering methods nearly useless, except for particular experimental designs such as QTL mapping or the application of advanced approaches such as the Bioconductor "Timecourse" package (http://www.bioconductor.org). In our experience four biological replicates produces much higher quality data for statistical filtering and clustering than three replicates, and also allows for the occasional loss of a sample caused by error or chance. It is also important to distinguish biological samples derived from separate animals or cultures, from technical replicates, which are repeated measurements (microarrays) of the same sample (mRNA prep).

Due to the very high reproductibility of Affymetrix microarrays, technical replicates run on the same biological sample are usually avoided in favour of running single arrays on more biological replicates. Extensive statistical discussions on the use of technical versus biological replicates in microarray experiments have been published *(24,26)*.

2. Amount of Biological Material

Experimental design should account for the amount of material necessary for arrays and subsequent validation experiments. Affymetrix array protocols call for a minimum of 10 μg of total RNA for standard methods of producing cRNA for chip hybridization. However, as mentioned previously, 2-cycle amplification protocols exist and are useful for instances where much lower amounts of starting RNA are available, such as when using laser capture microdissection.

Although commercial RNA isolation kits and reagents specify a minimum amount for starting material, high-quality RNA is more easily obtainable when processing larger amounts of biological sample. Thus, pooling of animals or cultures may be necessary to obtain sufficient material for RNA isolation and subsequent microarray analysis. This is .particularly important for isolated brain region RNA samples from small experimental animals such as mice, where dissections of nucleus accumbens or prefrontal cortex which may be 30–50 mg in mass from a single animal. Pooling of animals samples also has the advantage of reducing the influence of random environmental influences that can complicate microarray studies on ethanol with whole animals or cell cultures.

Obviously, multiple biological replicates of "pooled" samples are necessary for statistical analysis as mentioned above. This pooling approach can decrease the statistical power of some experimental designs where it is desired to correlate a phenotypic variable (e.g., behaviour) with gene expression. In such cases, analysis of individual animals and a relatively larger number of microarrays are needed. In general, 1 μg of total RNA is obtained from 1 mg (wet weight) of brain tissue. Different tissues or cell cultures will produce different yields.

Human autopsy brain tissue will generally produce lower yields. For single-cycle cRNA synthesis protocols, 5-10 μg of starting total RNA is needed. Thus, as a general rule for best results with brain tissue, this requires starting with >50mg of tissue (wet weight). As mentioned previously, two-cycle cRNA amplification protocols can use much less starting material. In addition, our laboratory has obtained high quality microarray data starting with as little as 1 μg of total RNA and using a single-cycle cRNA protocol but such designs are not used routinely.

If a given sample contains a very 'low amount of total RNA (but has passed RNA integrity checks) we routinely use that amount of input RNA for all cRNA reactions. Keeping the amount of input total RNA constant across the cRNA reactions is an important factor in avoiding variability due to cRNA synthesis. Similarly, if less than the recommended amount of cRNA is available for array hybridization, then it is critical that all arrays are at least hybridized with the same amount of cRNA, In extreme cases, samples with greatly reduced RNA or cRNA yields have to be discarded since they likely have issues with RNA integrity.

3. Experimental Technique

Improper handling of samples can result in changes in gene expression because of factors unrelated to the experiment, including animal handling, tissue dissection, and sample preparation. For example, a change in an animal's environment or handling can increase biological variance or noise and thus greatly complicate experiment analysis. The impact of laboratory environment of ethanol behavioural studies is well documented.

Because whole-genome microarrays are extremely sensitive to environmental factors, extraneous variables must be reduced so that observed changes in gene expression reflect experimental perturbations and not unrelated environmental factors. Thus, factors such as lab chow, light-dark cycle, and animal handler should remain constant during the course of an experiment as much as possible. For cell culture experiments, vendor and lot of reagents should be maintained constant and all cultures must be handled identically with very rapid approaches used for going from cell culture dish to frozen cell pellet or RNA extraction.

Seemingly trivial factors such as position within the incubator or having a mildly prolonged incubation on ice before freezing or homogenization can produce systematic patterns of gene expression changes when monitoring the expression of more than 40,000 transcripts.

4. RNA Quality Control

High-quality total RNA is critical for successful labeled-cRNA synthesis and array hybridizations. As with any RNA work, a RNAse-free environment must be maintained at each step from tissue extraction through labeled-cRNA synthesis. This can be assisted by liberal application of anti-RNAse reagents such as RNAseZap (Ambion, Austin, TX) on bench surfaces, gloves, pipettes and sample tubes and the use of rapid protocols for going from intact tissue/ cells to homogenization in denaturing RNA isolation buffers. Traditionally, RNA quality has been accessed by 260/280 UV absorbance ratios. and RNAse-free agarose gel electrophoresis.

More advanced techniques are now available to characterize RNA quality. The Agilent Bio Analyzer and Bio Rad Experion microfluidics systems are automated and require much less material (approx 0.5μg) than gel electrophoresis (approx 5μg). We routinely perform a size distribution analysis on both total RNA and the product cRNA. For total RNA, all samples should show near 1:1 molar ratios for 28S and 18S ribosomal RNA. For cRNA synthesis, size distribution patterns should be very similar across all samples and the highest molecular weight products should extend above 3000 nucleotides unless autopsy material or two-cycle amplification protocols are being used.

5. Randomization

As referred to previously, expression profiling with DNA micro arrays is highly sensitive for detecting variation in gene expression due to seemingly trivial environmental or experimental design features. Although the source of these effects is sometimes unclear, it is known that array background and gene expression can correlate with reagent lot, day/time of RNA isolation, order of labeled-cRNA synthesis and hybridization, sample age, and technician, among other factors).

In addition to controlling experimental conditions as much as possible, the use of randomization schemes at multiple levels throughout the experimental design can be very useful method for minimizing systematic bias in microarray experiments. Thus, it is critical to use a supervised randomization of control and treatment samples at RNA isolation, cRNA synthesis and array hybridization steps to eliminate systematic variance due to uncontrolled environmental factors.

Protocol

Tissue isolation or cell harvesting are highly individualized protocols dependent on the particular experimental system at study. As a general rule, however, very rapid harvesting of tissue/cells must be used to reduce ischemic, traumatic or cold stress induced changes in gene expression. For example, the mouse brain dissection protocol utilized in our laboratory requires the combined efforts of three to four laboratory members and takes tissue from seven to eight brain region microdissections to being frozen in liquid nitrogen within 5–7 min per mouse. RNA isolation, labeled-cRNA synthesis, and array hybridization protocols are extensive and therefore not within the scope of this review.

There are several total RNA isolation kits available with similar guanidine denaturation/ phenol-chloroform extraction protocols; we have used the Stat 60 reagent with very reproducible results. More recent protocols for fatty tissue such as brain have been developed, for example the BioRad Aurum™ Total RNA Fatty and Fibrous Tissue Kit.

These newer kits offer advantages either in a reduced number of steps or column steps for nucleic acid recovery, as opposed to ethanol precipitation in traditional methods. The best approach is to practice each step until high quality RNA can be obtained in the laboratory.

In general, except for points mentioned above regarding cRNA synthesis, Affymetrix protocols are generally strictly followed for cDNA and cRNA synthesis and microarray hybridization.

Microarray Quality Control

After hybridization and scanning, arrays must be rigorously evaluated for quality and consistency across arrays. Arrays that fail these standards will confound downstream analysis efforts and that sample should either be re-hybridized or discarded from the analysis. Microarray quality control includes metrics specific to the Affymetrix GeneChip platform and approaches that are applicable to all array types.

Affymetrix-Specific Quality Control Metrics

These parameters are shown on the metrics tab of the Affymetrix software and can be exported to a text file. These should be uniform across arrays and within threshold guidelines described below.

1. Scaling Factor (SF)

This is a measure of background and depends on the target average intensity (TGT) used during scanning. For TGT 190, used for older arrays, SF should be ≤3. For newer arrays TGT 500 is recommended; SF should be ≤10. Consistency within 3. SD units of the mean SF across all arrays of a batch is used as a cutoff for outliers.

2. Statistical Difference Threshold (SDT)

This is a measure of noise and is calculated from SF and RawQ. Standards for SDT are also dependent on TGT used for scanning. For TGT = 190 this value should be in the range 12–30; for TGT = 500 this value may be considerably greater. The critical feature of SDT is that it be consistent over all arrays and outliers are excluded as mentioned above.

3. Per Cent-present

The Affymetrix MAS and GCOS software will calculate the percentage of probesets (genes) present in the sample based on probe intensities corrected for background. Arrays with 40–60% present are generally accepted. Arrays outside this range are generally due to errors in cRNA synthesis or array hybridization. Again, array-to-array consistency is the most crucial factor and outliers are excluded as for the SF and SDT measures.

4. *3′-5′* GAPDH Control Expression Ratios

Arrays should only be accepted that have signals for the 3′–and 5′ end of the gene for glyceraldehyde phosphate dehydrogenase (GAPDH) within an accepted range. The Affymetrix chips have probesets with oligonucleotides addressing the 5′ end or 3′ end of GAPDH. Since the cRNA probes are synthesized starting at the 3′ end, it is normal to have lower signal produced against 5′ end probes but an excessive 375′ ratio is indicative of RNA or cRNA degradation or synthesis failure. GAPDH 375' ratios near or below 3.0 are acceptable but these are generally below 2.0 for good quality RNA/cRNA.

Generalized Quality Control Measures

These descriptive statistical studies are standard approaches for assessing the quality of data from any array platform.

1. Scatter Plot

Scatter plots of the log transformation of probe or probeset intensities between two arrays should be visually linear. Plots of arrays that curve at high or low intensity values, or other abnormal non-linear shapes, reflect abnormal saturation or background in one or more arrays. The shape of the scatter plot is a better indicator of array quality than Pearson correlations, which should be high ($R > 0.96$). Excessive scatter between closely related samples is also indicative of problems with one of the arrays and is accompanied by decreases in the Pearson correlation coefficients.

Pairwise scattergrams between all arrays within an experiment should appear very similar. An alternative to the straight scatter plot is the so-called "M vs. A" plot. This transformation is visually somewhat easier to interpret than the normal scatter plot. In this case the x-axis is the average log intensity for each gene/probe across the two arrays being compared. The y-axis displays the log of the ratio of intensities between the two chips. Thus, if a gene shows no change, the ratio should be 1 with a log value of 0. The M vs. A plot will display a cigar-shaped scatter of points centered around 0 on the *y*-axis. Deviations from linearity at the high or low abundance range can be easily detected.

2. Box-and-whisker Plot

The box plot function of the Bioconductor Affy Package will plot the median, 25th and 75th percentiles, and extremes of probe intensities. This function is useful to compare the relative background and scale of multiple arrays in an experiment, and may indicate which arrays should be removed prior to normalization and summarization.

Primary Analysis

Primary analysis refers to the production of intensity values for genes from the raw fluorescence measurements derived from scanning microarrays. For Affymetrix oligonucleotide arrays, this expression intensity determination is not as straightforward as one might think. The short oligonucleotides on these arrays are prone to nonspecific hybridization or lack of signal because of failed hybridization kinetics. Fortunately, the very high density of these arrays allowed the designers to incorporate two features aimed at circumventing the difficulties of using short oligonucleotides. First, each specific oligonucleotide on the array (termed a "perfect match" or PM probe) is paired with a "mismatch" or MM probe that has a single base substitution in the center of the 25 nucleotide oligonucleotide, aimed at disrupting specific hybridization.

These MM probes are meant to detect non-specific hybridization. Secondly each gene on the Affymetrix arrays is represented by a "probeset," which encompasses 8-15 different PM probes and their paired MM probes. The exact number of PM/MM probe pairs depends on the particular array design. Ideally, each different PM for a given gene should have the same hybridization intensity, but in actuality this rarely occurs due to variable oligonucleotide synthesis kinetics, hybridization kinetics, alternative splicing and a number of other factors. Thus, the first step in low-level analysis of Affymetrix arrays is. to derive a single intensity value from all the PM/MM probes making up a given probeset.

A very large number of low-level analysis methods have evolved that produce robust estimates of gene expression from the individual probe data. The manufacturer's most recent algorithm, entitled MASS or GCOS, uses a Tukey bi-weight estimate of modeled oligonucleotide signals with greatly improved performance compared to the initial algorithms, and is particularly useful for low abundance genes. However, more advanced and robust gene expression summarization methods, for instance RMA, MBEI, and PDNN, have been developed, and the use of at least one of these methods is often preferred.

Some reports have suggested the use of multiple algorithms with the contention that different algorithms work better for different genes or that the use of multiple algorithms for statistical filtering produces a more robust set of selected genes. An additional approach developed by our laboratory involves a ratio method, the S-score algorithm, that compares expression of two arrays and produces a summary statistic reflecting the probability of their being a difference in expression for a given gene (probeset) between the two arrays compared.

This method is unique in that it compares individual probe intensities between two arrays rather than comparing probeset summary statistics. By performing S-score analyses across biological replicates and using statistical approaches to find reproducible changes in expression (see below), we have found the S-score algorithm to be very specific and sensitive for detecting small changes in expression. An additional advantage of the S-score is the use of this approach for experiments having small numbers of replicates.

Statistical Filtering

As mentioned in the discussion about replicates, statistical analysis of microarray data is complicated by multiple-testing errors affecting calculation of statistical significance. As a first step before the actual statistical analysis of any array platform a filtering is performed to remove genes that are "*absent*" or expressed at levels below the linear range of detection for the microarray. Several algorithms exist to make decisions about which genes are absent, but a common approach is to simply define a cutoff as 3 standard deviation units above the background signal (defined either in an area with no probes or with cross-species probes known not to cross-hybridize (e.g., bacterial genes).

Affymetrix arrays have a decision metric built into the low-level analysis software that defines a gene as "*absent*," "*present*," or "*maybe*". Some investigators choose to require that all genes are called "*present*" on *all* arrays for inclusion in statistical analysis. Our laboratory takes a more inclusive approach, to avoid generating excessive false negatives (see discussion below), by just eliminating genes judged to be "*absent*" in *all* samples. Detailed discussions of statistical approaches for analysis of microarray data have been previously published. Although there are vocal arguments about the best statistical approach to take with microarray data, some investigators feel that "*statistical filtering*" is not meant as a "*final analysis*" but rather, only a method for reducing false positives.

Because multivariate analysis is frequently performed after statistical filtering, this secondary "*network level*" analysis serves to further reduce false positives. Such an approach, with a somewhat relaxed statistical filtering followed by multivariate analysis, serves to minimize the false negative rate and thus aids potential network analysis. It is difficult to study gene networks if you eliminate all but a handful of genes through a rigorous statistical filtering. The most widely used primary approaches for statistical filtering with microarray data currently involve some type of permutation or re-sampling analysis to produce a false-discovery rate calculation to identify genes whose expression change is likely to not have occurred by chance alone.

Many types of software are available for such statistical filtering. Perhaps the most popular application is the Significance Analysis of Microarrays (SAM) tool which can be used as a plug-in application for Excel spreadsheets on Windows based PCs or within the TIGR Multi Experiment Viewer (TMEV) multi-functional analysis platform. Other popular statistical approaches include ANOVA models (also present in the TMEV software) and the Timecourse package within the Bio-Conductor suite of array analysis programs. This latter algorithm and the SAM Excel Plug-in software are useful particularly for analysis of time-course data.

Multivariant Analysis

Multivariant analysis groups genes with statistically similar expression patterns across all samples in an experiment. Such data reduction algorithms have proven to be exceedingly useful in analysis of microarray data for two major reasons. First, grouping genes into "*clusters*" serves to reduce the complexity of the analysis and make overall interpretation easier. Second, use of techniques such as hierarchical and k-means clustering of microarray data has been shown to group the genes together in functionally meaningful ways. A cluster can thus provide hints as to the biological function of the gene regulatory events themselves. For example, upregulation of a group of genes involved in glycolysis might implicate a demand for energy by whatever biological process is being studied.

Additionally, clustering of an unknown or unannotated gene tightly with a group of known genes all having a coherent biological function can provide clues as to the function of the unknown genes . The TMEV software again provides a comprehensive selection of multivariate analysis tools. These include additional clustering methods such as self-organizing maps, and post-hoc association methods such as Pearson-correlation template matching. Each clustering algorithm has unique features and a number of adjustable parameters resulting in a very large number of options for this approach. Generally, several methods are used and the results evaluated for biological relevance. Thus, for a standard microarray analysis, after running quality control studies on the microarray data and eliminating genes consistently called "*absent*" we will import the RMA or S-score data from an experiment, into TMEV and perform statistical filtering with SAM or ANOVA using permutation analysis.

The statistical cutoff depends upon the source and quality of the data. Genes passing~statistial filtering will be studied by multivariate analysis using the TMEV hierarchical or k-means clustering. The latter algorithm requires "guessing" the number of clusters and we frequently use principal components analysis to first estimate the number of clusters. The individual clusters are then subjected to bioinformatics analysis.

Bioinformatics and Network Analysis

Bioinformatics methods are used to identify key regulated biological networks from lists of regulated genes. The biological significance or mechanism of regulation of selected genes can be explored by three general approaches: (1) Over-representation analysis: the identification of a common gene ontology, functional group assignment, or biological pathway for an abnormally large number of genes within a gene list or cluster; 2 literature association analysis: the identification of pairings of genes within the biodmedical literature either directly or indirectly in a network of literature associations; and (3) biological network mining such as finding common transcription factor binding sites within a large number of genes from a microarray gene list or cluster. All of these approaches take advantage of specialized software, forms of which are found as open source freeware.

In particular, we generally perform over-representation analysis of gene ontology or other functional groupings with web-based tools such as DAVID (http://david.abcc.ncifcrf.gov/) or WebGestalt (http://bioinfo. vanderbilt.edu/webgestalt/). For literature association analyses we generally use free-ware such as Chilibot (http://www.chilibot.net) or the Agilent Literature plug-in for Cytoscape. Networks constructed from functional and regulatory relationships between genes can be studied using commercial programs such as BiblioSphere or the Ingenuity Pathway Analysis platform. Finally, a number of academic databases can provide extremely important resources for microarray analysis. The WebQTL tool in GeneNetwork is such a resource that is very frequently used by our laboratory in array analysis.

GeneNetwork provides a searchable compilation of behavioural and gene expression QTL data from a variety of rodent models and other species. This allows the identification of loci controlling the expression of target genes either in cis or trans, the correlation of gene expression across rich recombinant inbred strain batteries, and the correlation of gene expression and behaviour across these same rodent lines (generally mouse). A host of other analysis tools and links makes WebQTL a powerful tool for identify gene networks and gene-behaviour interactions from microarray data.

Validation of Candidate Genes

When microarray analysis was first introduced, it was generally deemed necessary to confirm array results in-bulk by identical experiments using more accepted forms, or arguably more reliable or more sensitive methods of mRNA quantification, that is, "gold standard" methods such as Northern blotting and quantitative real-time rtPCR (QPCR). It was not uncommon for a publication of microarray data to be accompanied by QPCR data for dozens of genes, irrespective of the significance of such genes within the results and conclusions of the report.

With the acceptance of microarray hybridization as a reliable technology, expression validation solely for the sake of reproducing data and increasing confidence in array results no longer seems a necessity (except for an occasional obstinate reviewer!). However, in practice we tend to validate genes that are representative of gene networks networks or gene groupings selected by the bioinformatics analysis above. If sufficient materials exist, we prefer Northern blotting but QPCR confirmation can obviously also be used.

NOTES

1.Detailed descriptions and instructions regarding spotted cDNA arrays are available at the website of the Brown Lab at Stanford. Popular commercial oligonucleotide array vendors include Affymetrix, Agilent, and Illumina and more detailed information can be obtained at their corporate websites.

2. We usually dilute 1 μL of RNA sample in 99 μL of TE buffer for UV measurements. Ratios 260, 280 ≥ 2.0 .and ≤ 2.2 are best. Alternatively, if using ddH_2O as diluent, ratios should be between 1.8 and 2.0.

3. A very large number of primary analysis algorithms for the Affymetrix platform have been developed. Many of these are available through the Bioconductor packages for the R programming environment (www.bioconductor.org).

4. Software is readily available for normalization, primary analysis, and both generating and viewing cluster diagrams. These programs include TIGR Multi-Experiment Viewer the Bioconductor package for the R statistical computing environment, and cluster analysis and visualization tools Cluster and Tree view.

5 SENSORY SYSTEMS

Our senses provide us with means for detecting a diverse set of external signals, often with incredible sensitivity and specificity. For example, when fully adapted to a darkened room, our eyes allow us to sense very low levels of light, *down to a limit of less than ten photons.* With more light, we are able to distinguish millions of colors. Through our senses of smell and taste, we are able to detect thousands of chemicals in our environment and sort them into categories: pleasant or unpleasant? Healthful or toxic? Finally, we can perceive mechanical stimuli in the air and around us through our senses of hearing and touch. How do our sensory systems work? How are the initial stimuli detected? How are these initial biochemical events transformed into perceptions and experiences?

We have previously encountered systems that sense and respond to chemical signals—namely, receptors that bind to growth factors and hormones. Our knowledge of these receptors and their associated signal-transduction pathways provides us with concepts and tools for unraveling some of the workings of sensory systems. For example, 7TM receptors (seven-transmembrane receptors, play key roles in olfaction, taste, and vision. Ion channels that are sensitive to mechanical stress are essential for hearing and touch.

In this chapter, we shall focus on the five major sensory systems found in human beings and other mammals: olfaction (the sense of smell; *i.e.*, the detection of small molecules in the air), taste or gestation (the detection of selected organic compounds and ions by the tongue), vision (the detection of light), hearing (the detection of sound, or pressure waves in the air), and touch (the detection of changes in pressure, temperature, and other factors by the skin).

Each of these primary sensory systems contains specialized sensory neurons that transmit nerve impulses to the central nervous system. In the central nervous system, these signals are processed and combined with other information to yield a perception that may trigger a change in behaviour. By these means, our senses allow us to detect changes in our environments and to adjust our behaviour appropriately.

Color Perception. The photoreceptor rhodopsin (bottom), which absorbs light in the process of vision, consists of the protein opsin and a bound vitamin A derivative, retinal. The amino acids (shown in red) that surround the retinal determine the color of light that is most efficiently absorbed. Individuals lacking a light-absorbing photoreceptor for the color green will see a colorful fruit stand (top) as mostly yellows (middle).

ORGANIC COMPOUNDS DETECTED BY OLFACTION

Human beings can detect and distinguish thousands of different compounds by smell, often with considerable sensitivity and specificity. Most odorants are relatively small organic compounds with sufficient volatility that they can be carried as vapors into the nose. For example, a major component responsible for the smell of almonds is the simple aromatic compound benzaldehyde, whereas the sulfhydryl compound 3-methylbutane-l-thiol is a major component of the smell of skunks. What properties of these molecules are responsible for their smells? First, *the shape of the molecule rather than its other physical properties is crucial.* We can most clearly see the importance of shape by comparing molecules such as those responsible for the smells of spearmint and caraway. These compounds are identical in essentially all physical properties such as hydrophobicity because they are exact mirror images of one another.

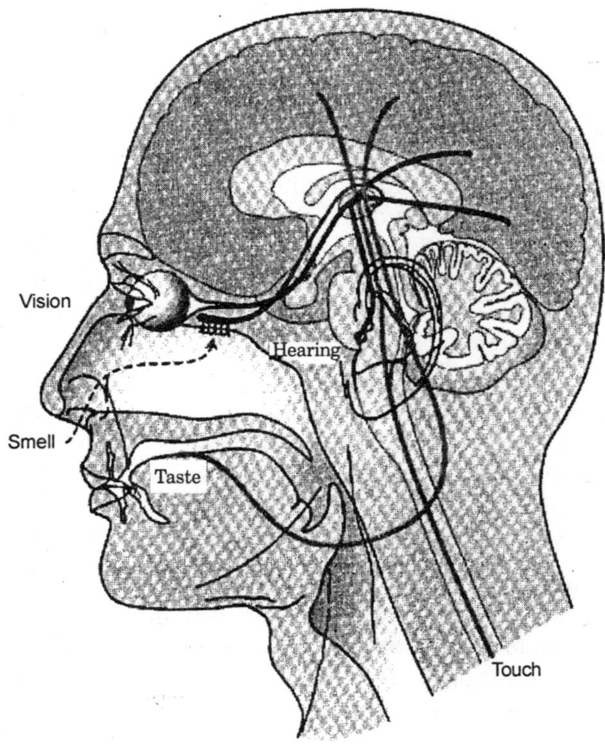

Fig. 5.1. Sensory Connections to the Brain. Sensory nerves connect sensory organs to the brain and spinal cord.

Benzaldehyde (Almond)

3-Methylbutane-1-thiol (Skunk)

Geraniol (Rose)

Zingiberene (Ginger)

Thus, the smell produced by an odorant depends not on a physical property but on the compound's interaction with a specific binding surface, most likely a protein receptor. Second, some human beings (and other animals) suffer from *specific anosmias;* that is, they are incapable of smelling specific compounds even though their olfactory systems are otherwise normal. Such anosmias are often inherited. These observations suggest that mutations in individual receptor genes lead to the loss of the ability to detect a small subset of compounds.

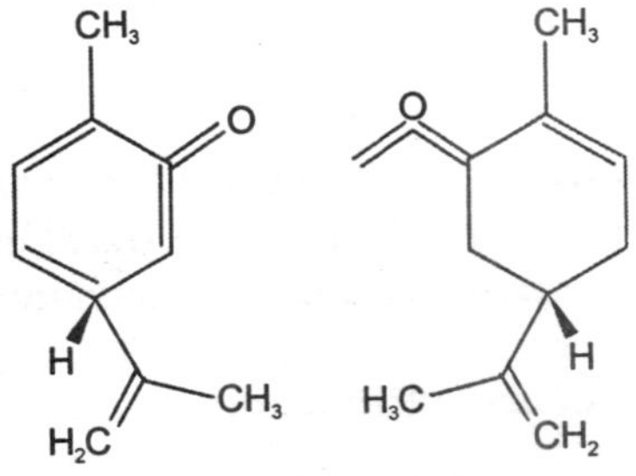

R-Carvone (Spearmint) S-Carvone (Caraway)

Seven-Transmembrane-Helix Receptors

Odorants are detected in a specific region of the nose, called the *main olfactory epithelium,* that lies at the top of the nasal cavity. Approximately 1 million sensory neurons line the surface of this region. Cilia containing the odorant-binding protein receptors project from these neurons into the mucous lining of the nasal cavity. Biochemical studies in the late 1980s examined isolated cilia from rat olfactory epithelium that had been treated with odorants. Exposure to the odorants increased the cellular level of cAMP,'and this increase was observed only in the presence of GTP. On the basis of what was known about signal-transduction systems, *the participation of cAMP and GTP strongly suggested the involvement of a G protein and, hence, 7TM receptors.*

Indeed, Randall Reed purified and cloned a G protein χ subunit, termed $G_{(olf)}$, which is uniquely expressed in olfactory cilia. The involvement of 7TM receptors suggested a strategy for identifying the olfactory receptors themselves. cDNAs were sought that (1) were expressed primarily in the sensory neurons lining the nasal epithelium, (2) encoded members of the 7TM receptor family, and (3) were present as a large and diverse family to account for the range of odorants. Through the use of these criteria, cDNAs for odorant receptors from rats were identified in 1991 by Richard Axel and Linda Buck. The *odorant receptor* (hereafter, OR) family is even larger than expected: *more than 1000 OR genes are present in the mouse and the rat, whereas the human genome encodes between an estimated 500 and 750 ORs.* The OR family is thus one of the largest gene families in human beings.

However, more than half the human odorant receptor genes appear to be pseudogenes that is, they contain mutations that prevent the generation of a full-length, proper odorant receptor. In contrast, essentially all rodent OR genes are fully functional. Further analysis of primate OR genes reveals that the fraction of pseudogenes is greater in species more closely related to human beings. Thus, we may have a glimpse at the evolutionary loss of acuity in the sense of smell as higher mammals presumably became less dependent on this sense for survival. The OR proteins are typically 20% identical in sequence to the 5-adrenergic receptor and from 30 to 60% identical with each other.

Several specific sequence features are present in most or all OR family members. The central region, particularly transmembrane helices 4 and 5, is highly variable, suggesting that this region is the site of odorant binding. That site must be different in odorant receptors that bind distinct odorant molecules. What is the relation between OR gene expression and the individual neuron? Interestingly, *each olfactory neuron expresses only a single OR gene,* among hundreds available. Apparently, the precise OR gene expressed is determined largely at random. The mechanism by which all other OR genes are excluded from expression remains to be elucidated. The binding of an odorant to an OR on the neuronal surface initiates a signal-transduction cascade that results in an action potential.

The ligand-bound OR activates $G_{(olf)}$, the specific G protein mentioned earlier. $G_{(olf)}$ is initially in its GDP-bound form. When activated, it releases GDP, binds GTP, and releases its associated δ_i subunits. The χ subunit then activates a specific adenylate cyclase, increasing the intracellular concentration of cAMP. The rise in the intracellular concentration of cAMP activates a nonspecific cation channel that allows calcium and other cations into the cell. The flow of cations through the channel depolarizes the neuronal membrane and initiates an action potential. This action potential, combined with those from other olfactory neurons, leads to the perception of a specific odor.

Combinatorial Mechanism

An obvious challenge presented to the investigator by the large size of the OR family is to match up each OR with the one or more odorant molecules to which it binds. Exciting progress has been made in this regard. Initially, an OR was matched with odorants by overexpressing a single, specific OR gene in rats. This OR responded to straight-chain aldehydes, most favourably to *n*-octanal and less strongly to *n*-heptanal and *n*-hexanal. More dramatic progress was made by taking advantage of our knowledge of the OR signal-transduction pathway and the power of PCR

A section of nasal epithelium from a mouse was loaded with the calcium–sensitive dye Fura–2. The tissue was then treated with different odorants, one at a time, at a specific concentration. If the odorant bound to and activated an OR, that neuron could be detected under a microscope by the change in fluorescence caused by the influx of calcium that occurs as part of the signal-transduction process. To determine which OR was responsible for the response, cDNA was generated from mRNA that had been isolated from single identified neurons. The cDNA was then subjected to PCR with the use of primers that are effective in amplifying most or all OR genes. The sequence of the PCR product from each neuron was then determined and analyzed. Using this approach, investigators analyzed the responses of neurons to a series of compounds having varying chain lengths and terminal functional groups.

The results of these experiments appear surprising at first glance. Importantly, there is not a simple 1:1 correspondence between odorants and receptors. *Almost every odorant activates a number of receptors* (usually to different extents) and *almost every receptor is activated by more than one odorant.* Note, however, that each odorant activates a unique combination of receptors. In principle, this combinatorial mechanism allows even a relative by small array of receptors to distinguish a vast number of odorants. How is the information about which receptors have been activated transmitted to the brain? Recall that each neuron expresses only one OR and that the pattern of expression appears to be largely random.

A substantial clue to the connections between receptors and the brain has been provided by the creation of mice that express a gene for an easily detectable colored marker in conjunction with a specific OR gene. Olfactory neurons/that express the OR-marker protein combination were traced to their destination in the brain, a structure called the *olfactory bulb.* The processes from neurons that express the same OR gene were found to connect to the same location in the olfactory bulb.

Moreover, this pattern of neuronal connection was found to be identical in all mice examined. Thus, *neurons that express specific ORs are linked with specific sites in the brain.* This property creates a spatial map of odorant-responsive neuronal activity within the of factory bulb. Can such a combinatorial mechanism truly distinguish many different odorants? An "*electronic nose*" that functions by the same principles provides compelling evidence that it can. The receptors for the electronic nose are polymers that bind a range of small molecules. Each polymer binds every odorant, but to varying degrees. Importantly, the electrical properties of these polymers change on odorant binding. A set of 32 of these polymer sensors, wired together so that the pattern of responses can be evaluated, is capable of distinguishing individual compounds such as *n*-pentane and *n*-hexane as well as complex mixtures such as the odors of fresh and spoiled fruit.

Magnetic Resonance Imaging

Can we extend our understanding of how odorants are perceived to events in the brain? Biochemistry has provided the basis for powerful methods for examining responses within the brain. One method, *called functional magnetic resonance imaging (fMRI),* takes advantage of two key observations. The first is that, when a specific part of the brain is active, blood vessels relax to allow more blood flow to the active region. Thus, a more active region of the brain will be richer in oxyhemoglobin. The second observation is that the iron center in hemoglobin undergoes substantial structural changes on binding oxygen.

These changes are associated with a rearrangement of electrons such that the iron in deoxyhemoglobin acts as a strong magnet, whereas the iron in oxyhemoglobin does not. The difference between the magnetic properties of these two forms of hemoglobin can be used to image brain activity. Nuclear magnetic resonance techniques detect signals that originate primarily from the protons in water molecules but are altered by the magnetic properties of hemoglobin. With the use of appropriate techniques, images can be generated that reveal differences in the relative amounts of deoxy- and oxyhemoglobin and thus the relative activity of various parts of the brain. These noninvasive methods reveal areas of the brain that process sensory information. For example, subjects have been imaged while breathing air that either does or does not contain odorants.

When odorants are present, the fMRI technique detects an increase in the level of hemoglobin oxygenation (and, hence, brain activity) in several regions of the brain. Such regions include those in the primary olfactory cortex as well as other regions in which secondary processing of olfactory signals presumably takes place. Further analysis reveals the time course of activation of particular regions and other features. Functional MRI shows tremendous potential for mapping regions and pathways engaged in processing sensory information obtained from all the senses; Thus, *a seemingly incidental aspect of the biochemistry of hemoglobin has yielded the basis for observing the brain in action.*

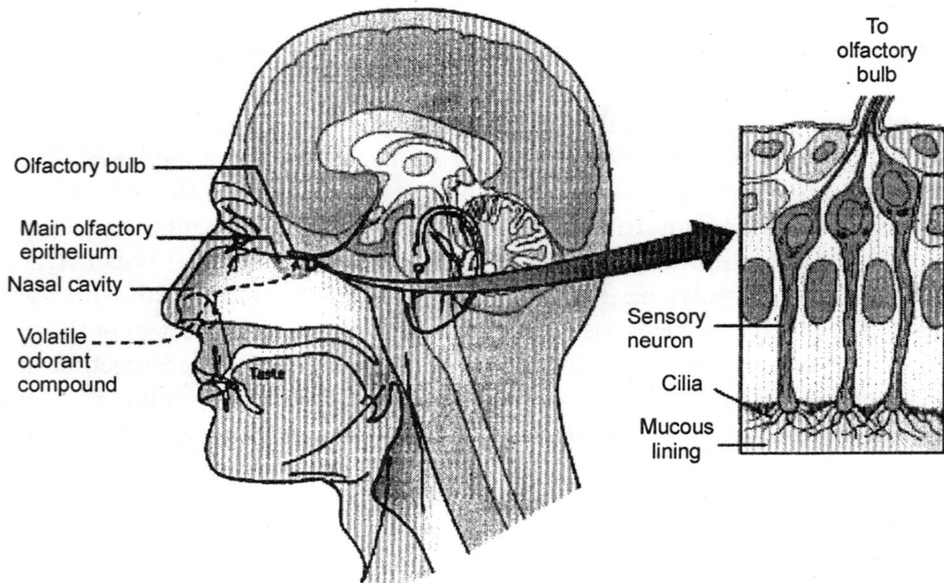

Fig. 5.2. The Main Nasal Epithelium. This region of the nose, which lies at the top of the nasal cavity, contains approximately 1 million sensory neurons. Nerve impulses generated by odorant molecules binding to receptors on the cilia travel from the sensory neurons to the olfactory bulb.

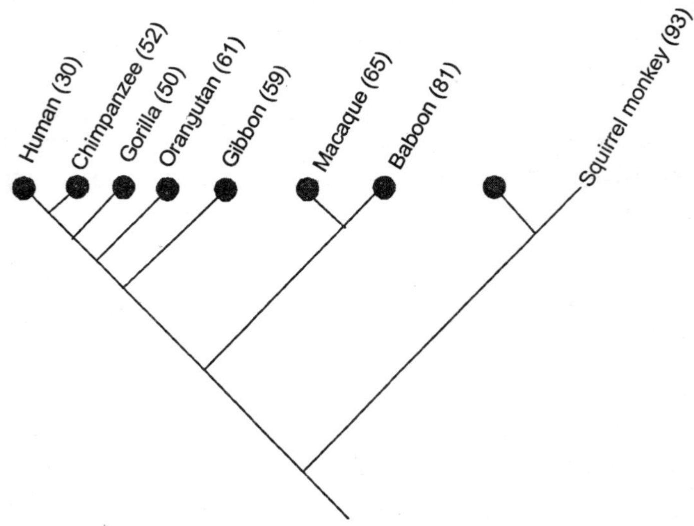

Fig. 5.3. Evolution of Odorant Receptors. Odorant receptors appear to have lost function through conversion into pseudogenes in the course of primate evolution. The percentage of OR genes that appear to be functional for each species is shown in parentheses.

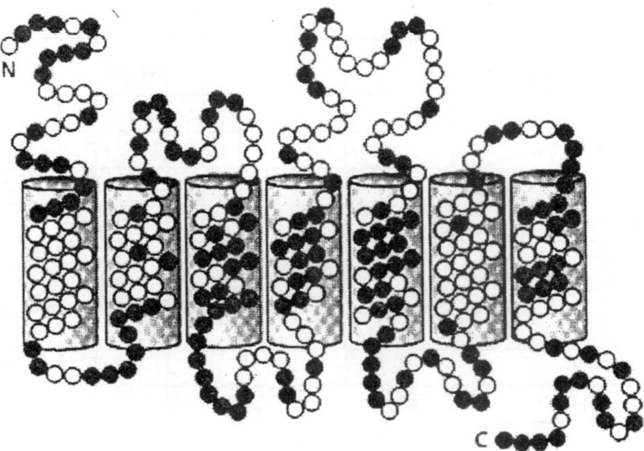

Fig. 5.4 Conserved and Variant Regions in Odorant Receptros. Odorant receptors are members of the 7TM receptor family. The green cylinders represent the seven presumed transmembrane helices. Strongly conserved residues characteristic of this protein family are shown in bule, whereas highly variable residues are shown in red.

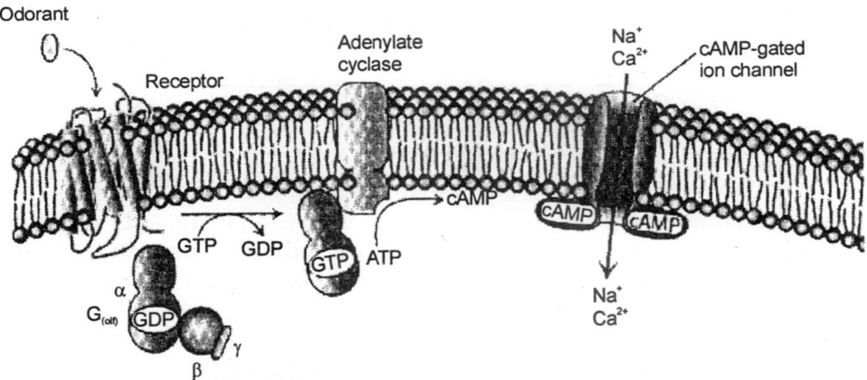

Fig. 5.5 The Olfactory Signal-Transduction Cascade. The binding of odorant to the olfactory receptor activates a signaling pathway similar to those initiated in response to the binding of some hormones to their receptors. The final result is the opening of cAMP-gated ion channels and the initiation of an action potential.

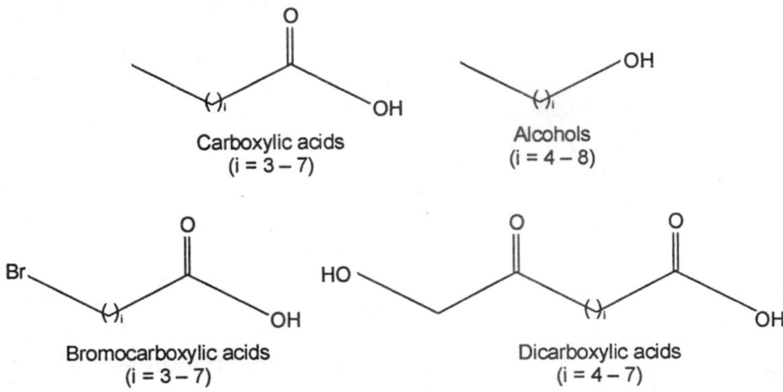

Fig. 5.6. Four Series of Odorants Tested for Olfactory Receptor Activation.

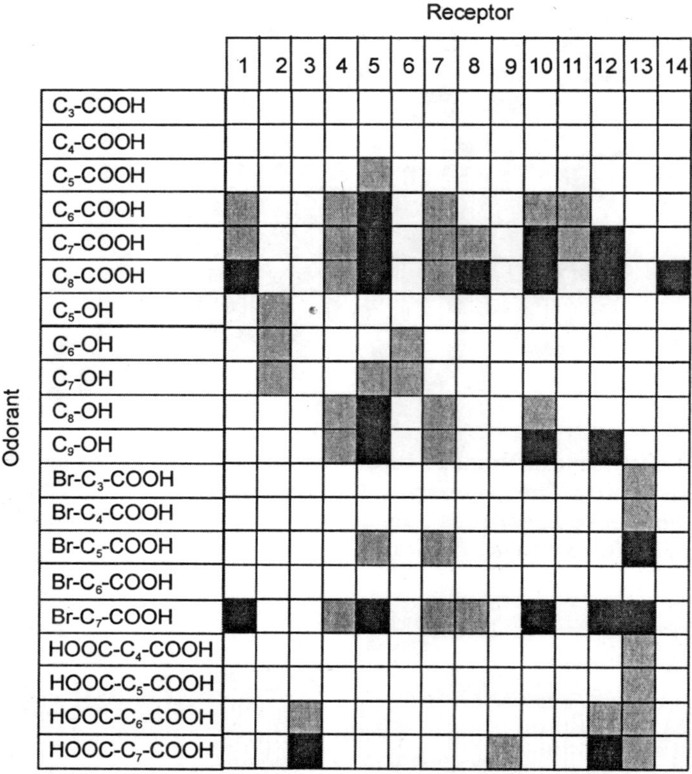

Fig. 5.7. Patterns of Olfactory Receptor Activation, Fourteen different receptors were tested for responsiveness to the compounds. A colored box indicates that the receptor at the top responded to the compound at the left. Darker colors indicate that the receptor was activated at a lower concentration of odorant.

COMBINATION OF SENSES

The inability to taste food is a common complaint when nasal congestion reduces the sense of smell. Thus, smell greatly augments our sense of taste (also known as *gustation),* and taste is, in many ways, the sister sense to olfaction. Nevertheless, the two senses differ from each other in several important ways. First, we are able to sense several classes of compounds by taste that we are unable to detect by smell; salt and sugar have very little odor, yet they are primary stimuli of the gustatory system. Second, whereas we are able to discriminate thousands of odorants, discrimination by taste is much more modest.

Five primary tastes are perceived: *bitter, sweet, sour, salty,* and *umami* (the taste of glutamate from the Japanese word for *"deliciousness"*). These five tastes serve to classify compounds into potentially nutritive and beneficial (sweet, salty, umami) or potentially harmful or toxic (bitter, sour). *Tastants* (the molecules sensed by taste) are quite distinct for the different groups. The simplest tastant, the hydrogen ion, is perceived as sour. Other simple ions, particularly sodium ion, are perceived as salty. The taste called umami is evoked by the amino acid glutamate, often encountered as the flavor enhancer monosodium glutamate (MSG). In contrast, *tastants perceived as bitter or sweet are 'extremely diverse.* Many bitter compounds are alkaloids or other plant products of which many are toxic. However, they do not have any common structural elements or other common properties.

Carbohydrates such as glucose and sucrose are perceived as sweet, as are other compounds including some simple peptide derivatives, such as aspartame, and even some proteins.

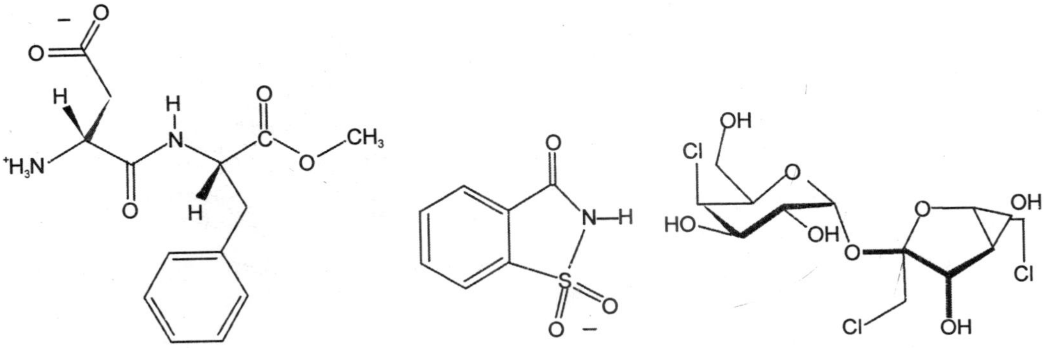

Aspartame Saccharin Sucrolose

These differences in specificity among the five tastes 'are due to differences in their underlying biochemical mechanisms The sense of taste is, in fact, a number of independent senses all utilizing.the same organ, the tongue, for their expression. Tastants are detected by specialized structures called *taste buds,* which contain approximately 150 cells, including sensory neurons. Fingerlike projections called *microvilli,* which are rich in taste receptors, project from one end of each sensory neuron to the surface of the tongue. Nerve fibers at the opposite end of each neuron carry electrical impulses to the brain in response to stimultation by tastants. Structures called *taste papillae* contain numerous taste buds.

Bitter Receptors

Just as in olfaction, a number of clues pointed to the involvement of G proteins and, hence, 7TM receptors in the detection of bitter and sweet tastes. The evidence included the isolation of a specific G protein χ subunit termed *gustducin,* which is expressed primarily in taste buds. How could the 7TM receptors be identified? The ability to detect some compounds depends on specific genetic loci in both human beings and mice. For instance, the ability to taste the bitter compound 6–*n*–propyl–2–thiouracil (PROP) was mapped to a region on human chromosome 5 by comparing DNA markers of persons who vary in sensitivity to this compound.

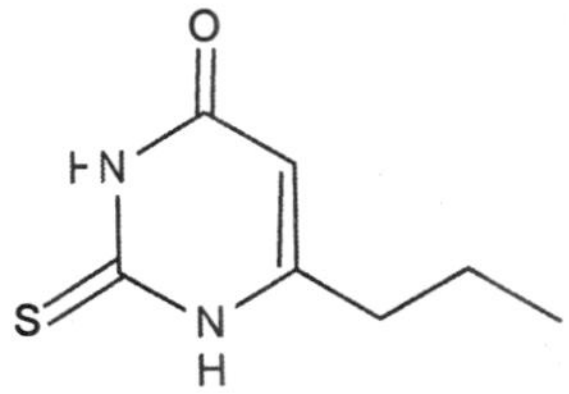

6-*n*-Propyl-2-thiouracil
(PROP)

This observation suggested that this region might encode a 7TM receptor that responded to PROP. Approximately 450 kilobases in this region had been sequenced early in the human genome project. This sequence was searched by computer for potential 7TM receptor genes, and, indeed, one was detected and named *T2R-1.* Additional database searches for sequences similar to this one detected 12 genes encoding full-length receptors as well as 7 pseudogenes within the sequence of the human genome known at the time. The encoded proteins were between 30 and 70% identical with T2R–1. *Further analysis suggests that there are from 50 to 100 members of this family of 7TM receptors in the entire human genome.* Similar sequences have been detected in the mouse and rat genomes. Are these proteins, in fact, bitter receptors?

Several lines of evidence suggest that they are. First, their genes are expressed in taste-sensitive cells—in fact, in many of the same cells that express gustducin. Second, cells that express individual members of this family respond to specific bitter compounds. For example, cells that express a specific mouse receptor (mT2R-5) responded when exposed specifically to cycloheximide. Third, mice that had been found unresponsive to cycloheximide were found to have point mutations in the gene encoding mT2R-5. Finally, cycloheximide specifically stimulates the binding of GTP analogs to gustducin in the presence of the mT2R-5 protein. Importantly, each taste receptor cell expresses many different members of the T2R family.

This pattern of expression stands in sharp contrast to the pattern of one receptor type per cell that characterizes the olfactory system. The difference in expression patterns accounts for the much greater specificity of our perceptions of smells compared with tastes. *We are able to distinguish among subtly different odors because each odorant stimulates a unique pattern of neurons. In contrast, many tastants stimulate the same neurons.* Thus, we perceive only "bitter" without the ability to discriminate cycloheximide from quinine.

Sweet Compounds

Most sweet compounds are carbohydrates, energy rich and easily digestible. Some noncarbohydrate compounds such as saccharin and aspartame also taste sweet. The structural diversity among sweet-tasting compounds, though less than that among bitter compounds, strongly suggested that a family of receptors detects these compounds. The observation that mice in which the gene for gustducin was disrupted lost much of their ability to sense sweet, as well as bitter, compounds strongly suggested that the sweet receptors would belong to the 7TM receptor superfamily.

Recently, a small group of 7TM receptors that respond to sweet compounds has been identified. Interestingly, simultaneous expression of two members of the family in the same cell is required for the cells to respond to sweet compounds. The biochemical explanation for this observation remains to be elucidated.

Salty Tastes

Salty tastants are not detected by 7TM receptors. Rather, they are detected directly by their passage through ion channels expressed on the surface of cells in the tongue. Evidence for the role of these ion channels comes from examining known properties of sodium channels characterized in other biological contexts. One class of channels, characterized first for their role in salt reabsorption, are thought to be important in salt taste detection because they are sensitive to the compound *amiloride,* which mutes the taste of salt and significantly lowers sensory neuron activation in response to sodium.

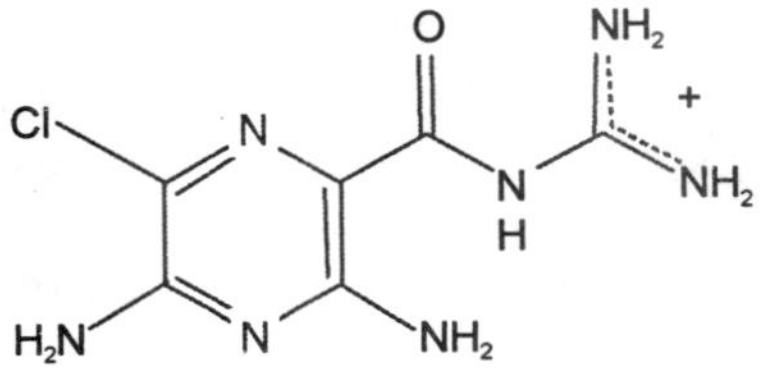

Amiloride

An *amiloride-sensitive sodium channel* comprises four subunits that may be either identical or distinct but in any case are homologous. An individual subunit ranges in length from 500 to 1000'amino acids and includes two presumed membrane-spanning helices as well as a large extracellular domain in between them. The extracellular region includes two (or, sometimes, three) distinct regions rich in cysteine residues (and, presumably, disulfide bonds).

A region just ahead of the second membrane-spanning helix appears to form part of the pore in a manner analogous to the structurally characterized potassium channel. The members of the amiloride-sensitive sodium-channel family are numerous and diverse in their biological roles. We shall encounter them again in the context of the sense of touch. Sodium ions passing through these channels produce a significant transmembrane current. Amiloride blocks this current, accounting for its effect on taste. However, about 20% of the response to sodium remains even in the presence of amiloride, suggesting that other ion channels also contribute to salt detection.

Effects of Hydrogen Ions

Like salty tastes, *sour tastes are also detected by direct interactions with ion channels,* but the incoming ions are hydrogen ions (in high concentrations) rather than sodium ions. For example, in the absence of high concentrations of sodium, hydrogen ion flow can induce substantial transmembrane currents through amiloride-sensitive sodium channels. However, hydrogen ions are also sensed by mechanisms other than their direct passage through membranes. Binding by hydrogen ions blocks some potassium channels and activates other types of channels. Together, these mechanisms lead to changes in membrane polarization in sensory neurons that produce the sensation of sour taste.

Glutamate Receptor

Glutamate is an abundant amino acid that is present in protein-rich foods as well as in the widely used flavour enhancer monosodium glutamate. This amino acid has a taste, termed *umami,* that is distinct from the other four basic tastes. Adults can detect glutamate at a concentration of approximately 1 mM. Glutamate is also a widely used neurotransmitter, and thus, not surprisingly, several classes of receptors for glutamate have been identified in the nervous system. One class, called *metabotrophic glutamate receptors,* are 7TM receptors with large amino-terminal domains of approximately 600 amino acidis.

Sequence analysis reveals that the first half of the aminoterminal region is most likely a ligand-binding domain, because it is homologous to such domains found in the Lac represser and other bacterial ligand-binding proteins. One glutamate receptor gene, encoding a protein called the metabotrophic glutamate receptor 4 (mGluR4), has been found to be expressed in taste buds. Further analysis of the mRNA that is expressed in taste buds reveals that this mRNA lacks the region encoding the first 309 amino acids in brain mGluR4, which includes most of the high-affinity glutamate-binding domain. The glutamate receptor found in taste buds shows a lowered affinity for glutamate that is appropriate to glutamate levels in the diet. Thus, *the receptor responsible for the perception of glutamate taste appears to have evolved simply by changes in the expression of an existing glutamate-receptor gene.* We shall consider an additional receptor related to taste, that responsible for the "hot" taste of spicy food, when we deal with mechanisms of touch perception.

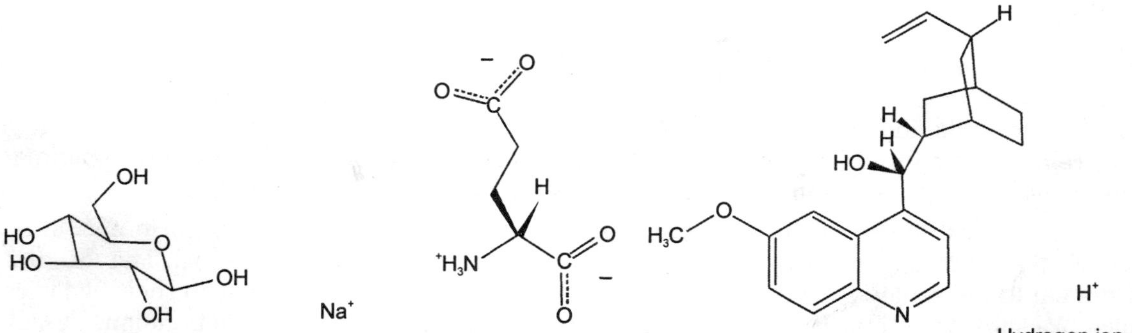

Glucose (sweet) — Na^+ Sodium ion (salty) — Glutamate (umami) — Quinine (bitter) — H^+ Hydrogen ion (sour)

Fig. 5.8. Examples of Tastant Molecules. Tastants fall into five groups: sweet, salty, umami, bitter, and sour.

PHOTORECEPTOR MOLECULES

Vision is based on the absorption of light by photoreceptor cells in the eye. These cells are sensitive to light in a relatively narrow region of the electromagnetic spectrum, the region with wavelengths between 300 and 850 nm. Vertebrates have two kinds of photoreceptor cells, called *rods* and *cones* because of their distinctive shapes. Cones function in bright light and are responsible for color vision, whereas rods function in dim light but do not perceive color. A human retina contains about 3 million cones and 100 million rods. Remarkably, a rod cell can respond to a single photon, and the brain requires fewer than 10 such responses to register the sensation of-a flash of light.

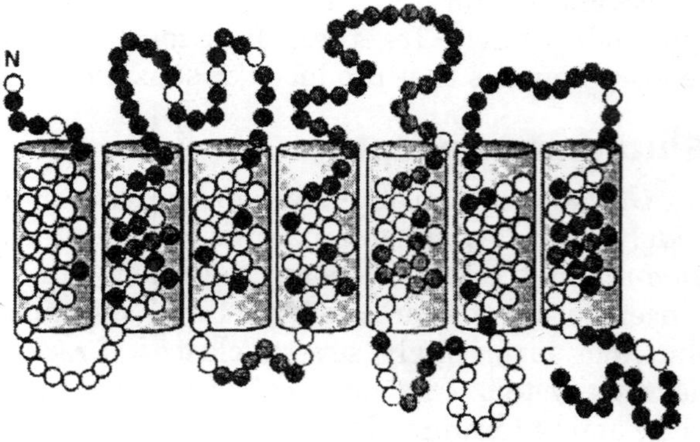

Fig. 5.9. Conserved and Variant Regions in Bitter Receptors. The bitter receptors are members of the 7TM receptor family. Strongly conserved residues characteristic of this protein family are shown in blue, and highly variable residues are shown in red.

Rhodopsin

Rods are slender elongated structures; the outer segment is specialized for photoreception. It contains a stack of about 1000 discs, which are membrane-enclosed sacs densely packed with photoreceptor molecules. The photosensitive molecule is often called a *visual pigment* because it is highly colored owing to its ability to absorb light. The photoreceptor molecule in rods is *rhodopsin,* which consists of the protein *opsin* linked to 11-cis-*retinal,* a prosthetic group.

11-*cis-Retinal*

Rhodopsin absorbs light very efficiently in the middle of the visible spectrum, its absorption being centered on 500 nm, which nicely matches the solar output. A rhodopsin molecule will absorb a high percentage of the photons of the correct wavelength that strike it, as indicated by the extinction coefficient of 40,000 M^{1} cnr^{1} at 500 nm. The extinction coefficient for rhodopsin is more than an order of magnitude greater than that for tryptophan, the most efficient absorber in proteins that lack prosthetic groups.

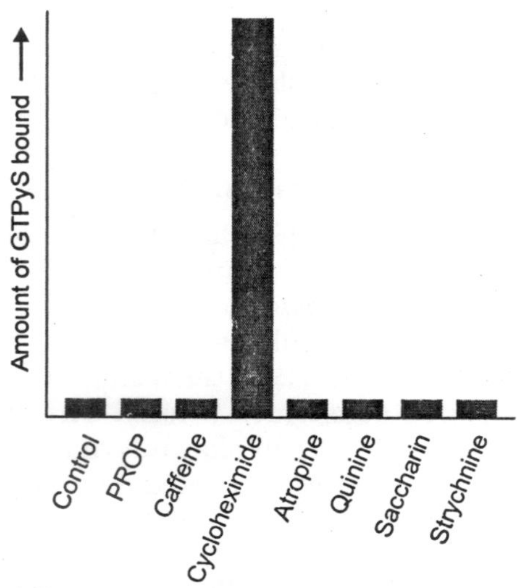

Fig. 5.10. Evidence that T2R Proteins Are Bitter Taste Receptors. Cycloheximide uniquely stimulates the binding of the GTP analog GTP i S to gustducin in the presence of rhe mT2R protein.

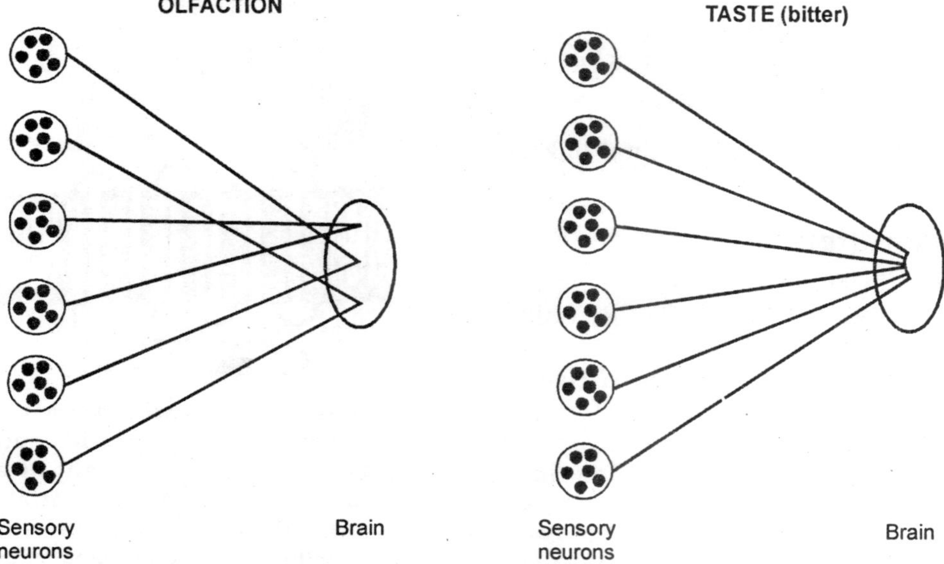

Fig. 5.11. Differing Gene Expression and Connection Patterns in Olfactory and Bitter Taste Receptors. In olfaction, each neuron expresses a single OR gene, and the neurons expressing the same OR converge to specific sites in the brain, enabling specific perception of different odorants. In gustation, each neuron expresses many bitter receptor genes, so the identity of the tastant is lost in transmission.

Opsin, the protein component of rhodopsin, isa member of the 7TM receptor family. Indeed, rhodopsin was the first member of this family to be purified, its gene was the first to be cloned and sequenced, and its three-dimensional structure was the first to be determined.

The color of rhodopsin and its responsiveness to light depend on the presence of the light-absorbing group *(chromophore)* 11-*cis*-retinal. This compound is a powerful absorber of light because it is a polyene; its six alternating single and double bonds constitute a long, unsaturated electron network. Recall that alternating single and double bonds account for the chromophoric properties of chlorophyll. The aldehyde group of 11-cis-retinal forms a Schiff base with the γ-amino group of lysine residue 296, which lies in the center of the seventh transmembrane helix. Free retinal absorbs maximally at 370 nm, and its unprotonated Schiff-base adduct absorbs at 380 nm, whereas the protonated Schiff base absorbs at 440 nm or longer wavelengths. Thus, *the 500-nm absorption maximum for rhodopsin strongly suggests that the Schiff base is protonated;* additional interactions with opsin shift the absorption maximum farther toward the red. The positive charge of the protonated Schiff base is compensated by the negative charge of glutamate 113 located in helix 2; the glutamate residue closely approaches the lysine-retinal linkage in the three-dimensional structure of rhodopsin.

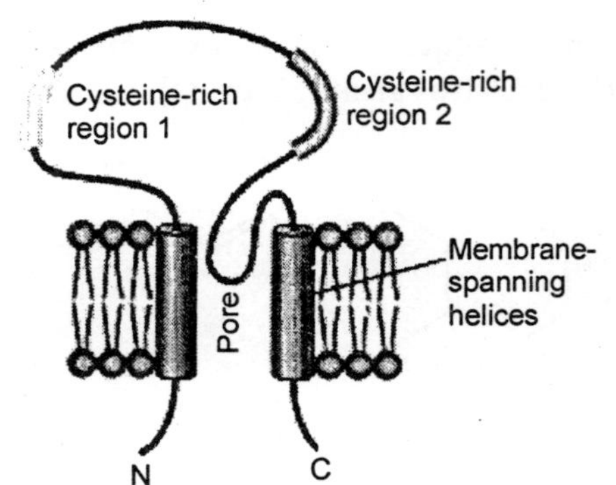

Fig. 5.12. Schematic Structure of the Amiloride-Sensitive Sodium Channel. Only one of the four subunits that constitute the functional channel is illustrated. The amiloride-sensitive sodium channel belongs to a superfamily having common structural features, including two hydrophobic membrane-spanning regions, intracellular amino and carboxyl termini; and a large, extracellular region with conserved cysteine-rich domains.

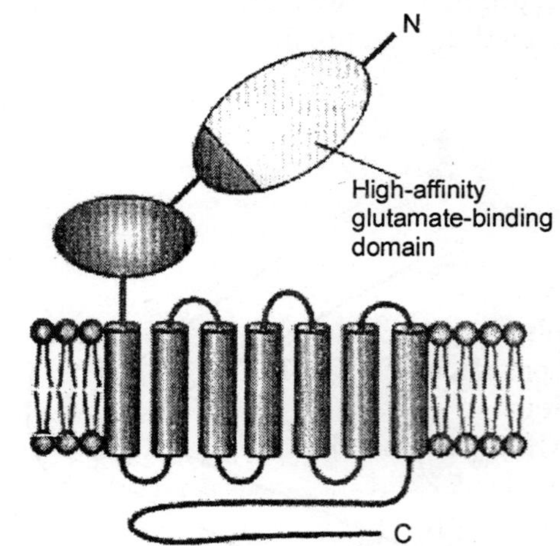

Fig. 5.13. Schematic Structure of a Metabotrophic Glutamate Receptor. The umami receptor is a variant of a brain glutamate receptor. A substantial part of the high-affinity glutamate-binding domain (shown in yellow) is missing in the form expressed in the tongue.

Light Absorption

How does the absorption of light by the retinal Schiff base generate a signal? George Wald and his coworkers discovered that *light absorption results in the isomerization of the 11-cis-retinal group of rhodopsin to its all-*transform. This isomerization causes the Schiff-base nitrogen atom to move approximately 5 Å, assuming that the cyclohexane ring of the retinal group remains fixed. In essence, *the light energy of a photon is converted into atomic motion.* The change in atomic positions, like the binding of a ligand to other 7TM receptors, sets in train a series of events that lead to the closing of ion channels and the generation of a nerve impulse. The isomerization of the retinal Schiff base takes place within a few picoseconds of a photon being absorbed.

The initial product, termed *bathorhodopsin,* contains a strained all-*trans*-retinal group. Within approximately 1 millisecond, this intermediate is converted through several additional intermediates into *metarhodopsin II.* In metarhodopsin II, the Schiff base is deprotonated and the opsin protein has undergone significant reorganization. Metarhodopsin II is analogous to the ligand-bound state of 7TM receptors such as the δ_2-adrenergic receptor and the odorant and tastant receptors heretofore discussed. Like these receptors, this form of rhodopsin activates a heterotrimeric G protein that propagates the signal. The G protein associated with rhodopsin is called *transducin.* Metarhodopsin II triggers the exchange of GDP for GTP by the χ subunit of transducin.

On the binding of GTP, the δ_i subunits of transducin are released and the χ subunit switches on a *cGMP phosphodiesterase* by binding to and removing an inhibitory subunit. The activated phosphodiesterase is a potent enzyme that rapidly hydrolyzes cGMP to GMP. The reduction in cGMP concentration causes cGMP-gated ion channels to close, leading to hyperpolarization of the membrane and neuronal signaling. *At each step in this process, the initial signal—the absorption of a single photon—is amplified so that it leads to sufficient membrane hyperpolarization to result in signaling.*

Light-induced the Calcium Level

As we have seen, the visual system responds to changes in light and color within a few milliseconds, quickly enough that we are able to perceive continuous motion at nearly 1000 frames per second. To achieve a rapid response, the signal must also be terminated rapidly and the system must be returned to its initial state. First, activated rhodopsin must be blocked from continuing to activate transducin. *Rhodopsin kinase* catalyzes the phosphorylation of the carboxyl terminus of R* at multiple serine and threonine residues. *Arrestin,* an inhibitory protein then binds phosphorylated R* and prevents additional interaction with transducin. Second, the χ subunit of transducin must be returned to its inactive state to prevent further signaling.

Like other G proteins, the χ subunit possesses built-in GTPase activity that hydrolyzes bound GTP to GDP. Hydrolysis takes place in less than a second when transducin is bound to the phosphodiesterase. The GDP form of transducin then leaves the phosphodiesterase and reassociates with the δ_i subunits, and the phosphodiesterase returns to its inactive state. Third, the level of cGMP must be raised to reopen the cGMP-gated ion channels. *The action of guanylate cyclase accomplishes this third step by synthesizing cGMP from GTP.* Calcium ion plays an essential role in controlling guanylate cyclase because it markedly inhibits the activity of the enzyme.

In the dark, Ca^{2+} as well as Na^+ enter the rod outer segment through the cGMP-gated channels. Calcium ion influx balanced by its efflux through an exchanger, a transport system that uses the thermodynamically favourable flow of four Na+ ions into the cell and one K^+ ion out of the cell to extrude one Ca^{2+} ion. After illumination, the entry of Ca^{2+} through the cGMP-gated channels stops, but its export through the exchanger continues. Thus, the cytosolic Ca^{2+} level drops from 500 nM to 50 nM after illumination. This drop markedly stimulates guanylate cyclase, rapidly restoring the concentration of cGMP to reopen the cGMP-gated channels.

<u>Activation</u> | <u>Recovery</u>

[cGMP] ↓ ⟶ Ion channels closed ⟶ [Ca^{2+}] ↓ ⟶ Guanylate cyclase activity increased ⟶ [cGMP] ↑

By controlling the rate of cGMP synthesis, *Ca^{2+} levels govern the speed with which the system is restored to its initial state.*

Colour Vision

Cone cells, like rod cells, contain visual pigments. Like rhodopsin, these photoreceptor proteins are members of the 7TM receptor family and utilize 11-*cis*-retinal as their chromophore. In human cone cells, there are three distinct photoreceptor proteins with absorption maxima at 426, 530, and ~ 560 nm. *These absorbances correspond to (in fact, define) the blue, green, and red regions of the spectrum.* Recall that the absorption maximum for rhodopsin is 500 nm. The amino acid sequences of the cone photoreceptors have been compared with each other and with rhodopsin.

The result is striking. Each of the cone photoreceptors is approximately 40% identical in sequence with rhodopsin. Similarly, the blue photoreceptor is 40% identical with each of the green and red photoreceptors. The green and red photoreceptors, however, are > 95% identical with each other, differing in only 1 5 of 364 positions. These observations are sources of insight into photoreceptor evolution. First, the green and red photoreceptors are clearly products of a recent evolutionary event.

The green and red pigments appear to have diverged in the primate lineage approximately 35 million years ago. Mammals, such as dogs and mice, that diverged from primates earlier have only two cone photoreceptors, blue and green. They are not sensitive to light as far toward the infrared region as we are, and they do not discriminate colours as well. In contrast, birds such as chickens have a total of six pigments: rhodopsin, four cone pigments, and a pineal visual pigment called *pinopsin.* Birds have highly acute colour perception. Second, the high level of similarity between the green and red pigments has made it possible to identify the specific amino acid residues that are responsible for spectral tuning.

Three residues (at positions 180, 277, and 285) are responsible for most of the difference between the green and red pigments. In the green pigment, these residues are alanine, phenylalanine, and alanine, respectively; in the red pigment, they are serine, tyrosine, and threonine. A hydroxyl group has been added to each amino acid in the red pigment. The hydroxyl groups can interact with the photoexcited state of retinal and lower its energy, leading to a shift toward the lower-energy (red) region of the spectrum.

Green and Red Pigments

The genes for the green and red pigments lie adjacent to each other on the human X chromosome. These genes are more than 98% identical in nucleotide sequence, including introns and untranslated regions as well as the protein-coding region. Regions with such high similarity are very susceptible to unequal homologous recombination. Recombination can take place either between or within transcribed regions of the gene.

If recombination takes place between transcribed regions, the product chromosomes will differ in the number of pigment genes that they carry. One chromosome will lose a gene and thus may lack the gene for, say, the green pigment; the other chromosome will gain a gene. Consistent with this scenario, approximately 2% of human X chromosomes carry only a single colour pigment gene, approximately 20% carry two, 50% carry three, 20% carry four, and 5% carry five or more.

A person lacking the gene for the green pigment will have trouble distinguishing red and green colour, characteristic of the most common form of colour blindness. Approximately 5% of males have this form of colour blindness. Recombination can also take place within the transcription units, resulting in genes that encode hybrids of the green and red photoreceptors. The absorption maximum of such a hybrid lies between that of the red and green pigments. A person with such hybrid genes who also lacks either a functional red or a functional green pigment gene does not discriminate colour well.

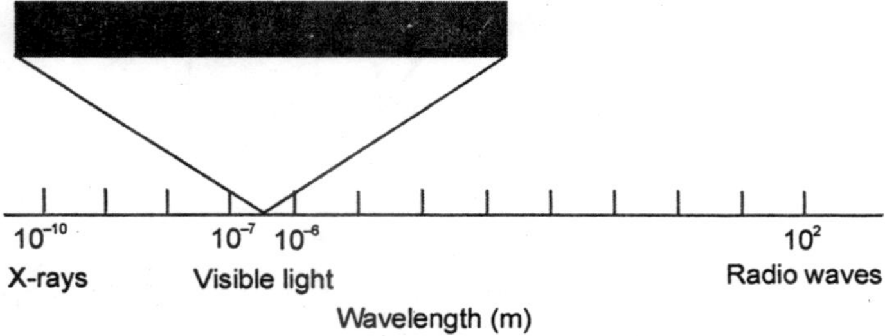

Fig. 5.14. The Electromagnetic Spectrum. Visible light has wavelengths between 300 and 850 nanometers.

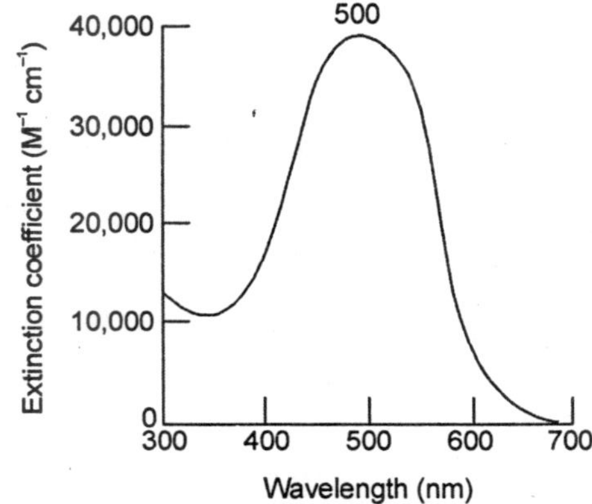

Fig. 5.15. Rhodopsin Absorption Spectrum.

Detection of Mechanical Stimuli

Hearing and touch are based on the detection of mechanical stimuli. Although the proteins of these senses have not been as well characterized as those of the senses already discussed, anatomical, physiological, and biophysical studies have elucidated the fundamental processes. *A major clue to the mechanism of hearing is its speed.* We hear frequencies ranging from 200 to 20,000 Hz (cycles per second), corresponding to times of 5. to 0.05 ms. Furthermore, our ability to locate sound sources, one of the most important functions of hearing, depends on the ability to detect the time delay between the arrival of a sound at one ear and its arrival at the other.

Given the separation of our ears and the speed of sound, we must be able to accurately sense time differences of 0.7 ms. In fact, human beings can locate sound sources associated with temporal delays as short as 0.02 ms. This high time resolution implies that hearing must employ direct transduction mechanisms that do not depend on second messengers. Recall that, in vision, for which speed also is important, the signal-transduction processes take place in milliseconds.

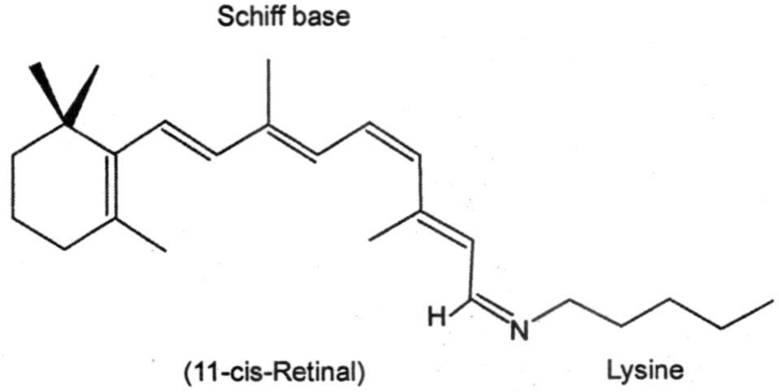

Schiff base

H N

(11-cis-Retinal) Lysine

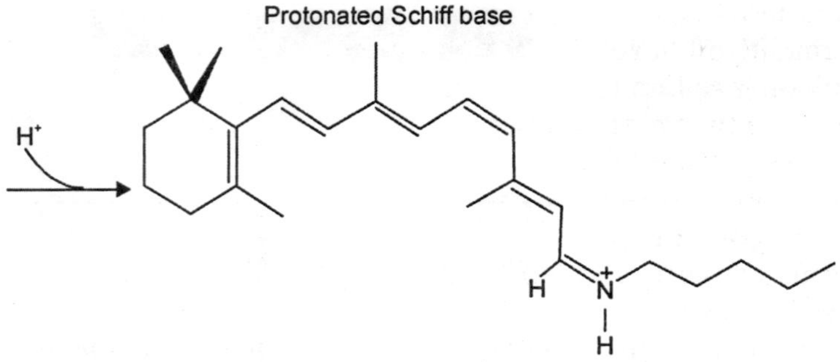

Protonated Schiff base

Fig. 5.16. Retinal-Lysine Linkage. Retinal is linked to lysine 296 in opsin by a Schiff-base linkage. In the resting state of rhodopsin, this Schiff base is protonated.

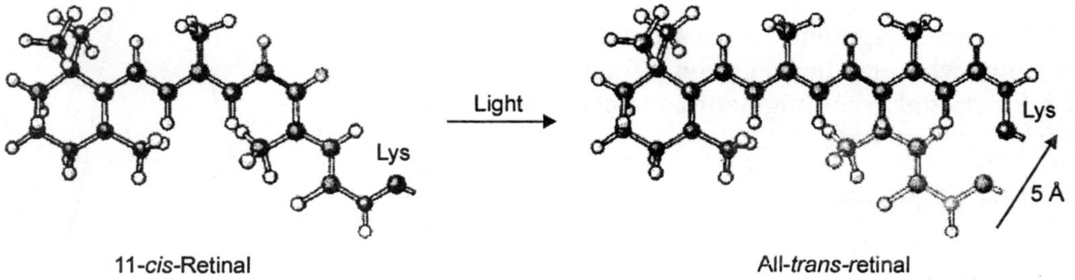

Light

Lys

5 Å

11-*cis*-Retinal

All-*trans*-retinal

Fig. 5.17 Atomic Motion in Retinal. The Schiff-base nitrogen atom moves 5 Å as a consequence of the light-induced isomerization of 11-cis-retinal to all-trans-retinal by rotation about the bond shown in red.

Stereocilia

Sound waves are detected inside the cochlea of the inner ear. The *cochlea* is a fluid-filled, membranous sac that is coiled like a snail shell. The primary detection is accomplished by specialized neurons inside the cochlea called *hair cells*. Each cochlea contains approximately 16,000 hair cells, and each hair cell contains a hexagonally shaped bundle of 20 to 300 hairlike projections called *stereocilia*.

These Stereocilia are graded in length across the bundle. Mechanical deflection of the hair bundle, as occurs when a sound wave arrives at the ear, creates a change in the membrane potential of the hair cell. Micromanipulation experiments have directly probed the connection between mechanical stimulation and membrane potential. Displacement toward the direction of the tallest part of the hair bundle results in depolarization of the hair cell, whereas displacement in the opposite direction results in hyperpolarization.

Motion perpendicular to the hair-length gradient does not produce any change in resting potential. Remarkably, *displacement of the hair bundle by as little as 3 Å (0.3 nm results in a measurable (and functionally important) change in membrane potential.* This motion of 0.003 degree corresponds to a 1-inch movement of the top of the Empire State Building. How does the motion of the hair bundle create a change in membrane potential? The rapid response, within microseconds, suggests that the movement of the hair bundle acts on ion channels directly.

An important observation is that adjacent stereocilia are linked by individual filaments called *tip links*. The presence of these tip links suggests a simple mechanical model for transduction by hair cells. The tip links are coupled to ion channels in the membranes of the stereocilia that are gated by mechanical stress. In the absence of a stimulus, approximately 15% of these channels are open.

When the hair bundle is displaced toward its tallest part, the stereocilia slide across one another and the tension on the tip links increases, Causing additional channels to open. The flow of ions through the newly opened channels depolarizes the membrane. Conversely, if the displacement is in the opposite direction, the tension on'the tip links decreases, the open channels close, and the membrane hyperpolarizes. *Thus, the mechanical motion of the hair bundle is directly converted into current flow across the hair-cell membrane.*

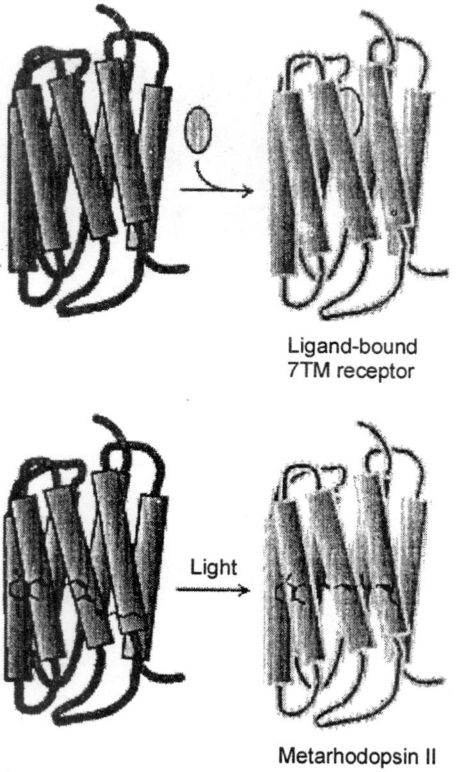

Ligand-bound 7TM receptor

Light

Metarhodopsin II

Fig. 5.18. Analogous 7TM Receptors. The conversion of rhodopsin into metarhodopsin II activates a signal-transduction pathway analogously to the activation induced by the binding of other 7TM receptors to appropriate ligands.

Mechanosensory Channels

Although the ion channel that functions in human hearing has not been identified, other mechanosensory channels in other organisms have been. *Drosophila* have sensory bristles used for detecting small air currents. These bristles respond to mechanical displacement in ways similar to those of hair cells; displacement of a bristle in one direction leads to substantial transmembrane current. Strains of mutant fruit flies that show uncoordinated motion and clumsiness have been examined for their electrophysiological responses to displacement of the sensory bristles. In one set of strains, transmembrane currents were dramatically reduced.

The mutated gene in these strains was found to encode a protein of 1619 amino acids, called NompC for *no* mechanoreceptor potential. The carboxyl-terminal 469 amino acids of NompC resemble a class of ion channel proteins called TRP (transient receptor potential) channels. This region includes six putative transmembrane helices with a porelike region between the fifth and sixth helices. The amino-terminal 1150 amino acids consist almost exclusively of 29 *ankyrin repeats*. Ankyrin repeats are structural motifs formed by 33 amino acids folded into a hairpin loop followed by a helix-turn-helix.

Importantly, in other proteins, regions with tandem arrays of these motifs mediate protein-protein interactions, suggesting that these arrays couple the motions of other proteins to the activity of the Nomp channel. Prokaryotes such as *E. coli* have ion channels in their membranes that open in response to mechanical changes. These channels play a role in regulating the osmotic pressure within the bacteria.

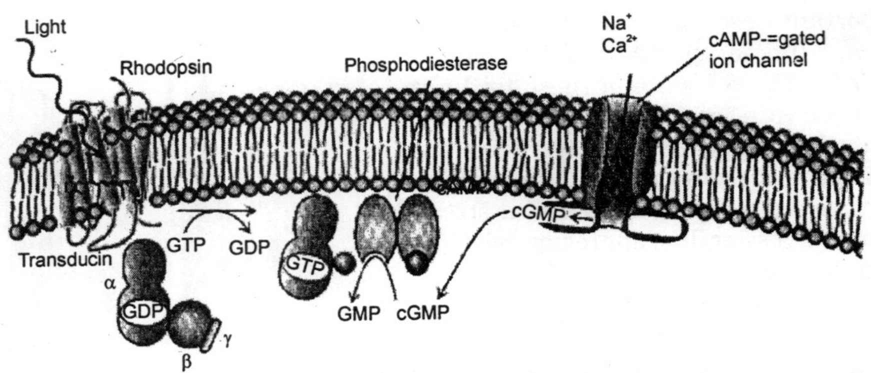

Fig. 5.19. Visual Signal Transduction. The light-induced activation of rhodopsin leads to the hydrolysis of cGMP, which in turn leads to ion channel closing and the initiation of an action potential.

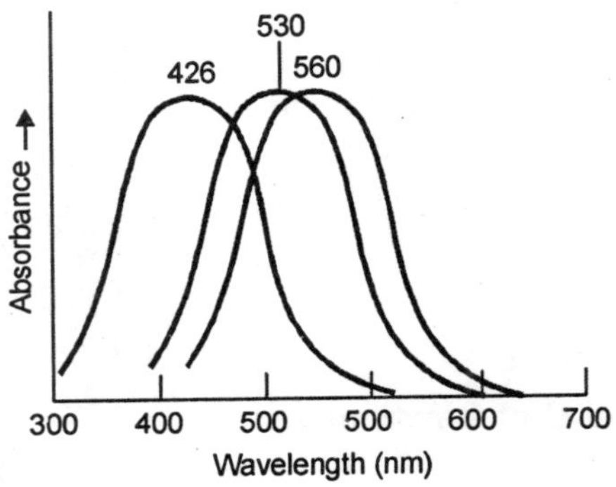

Fig. 5.20 Cone-Pigment Absorption Spectra. The absorption spectra of the cone visual pigment responsible for color vision.

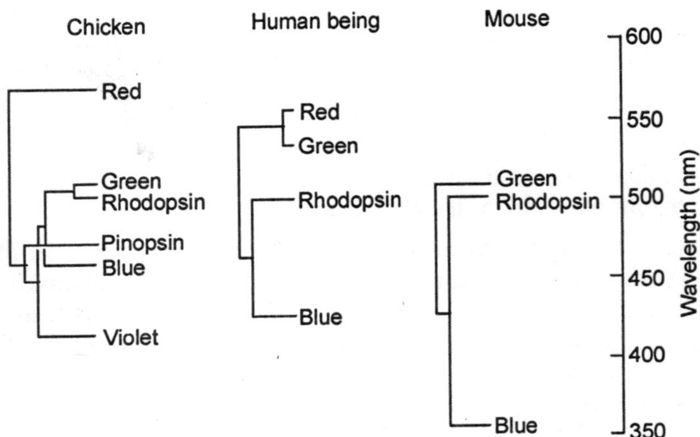

Fig. 5.21 Evolutionary Relationships among Visual Pigments. Visual pigments have evolved by gene duplication along different branches of the animal evolutionary tree. The branch lengths of the "trees" correspond to the percentage of amino acid divergence.

(A) Recombination between genes (B) Recombination with in genes

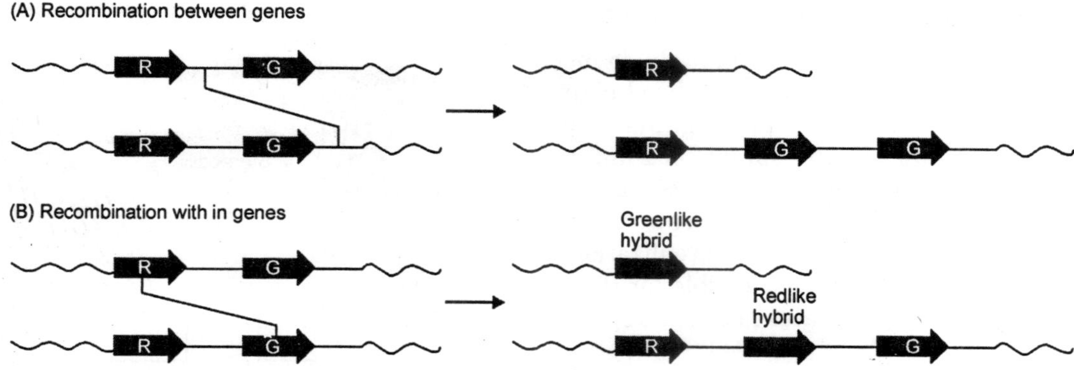

Fig. 5.22. Recombination Pathways Leading to Colour Blindness. Rearrangements in the course of DNA replication may lead to (A) the loss of visual pigment genes or (B) the formation of hybrid pigment genes that encode photoreceptors with anomolous absorption spectra. Because the amino acids most important for determining absorption spectra are in the carboxyl-terminal half of each photoreceptor protein, the part of the gene that encodes this region most strongly affects the absorption characteristics of hybrid receptors.

The three-dimensional structure of one such channel, that from *Mycobaterium tuberculosis,* has been determined. The channel is constructed of five identical subunits arranged such that an alpha helix from each subunit lines the inner surface of the pore. Further studies should reveal whether the transduction channel in hearing is homologous to either of these clases of mechanosensory channels or represents a novel class.

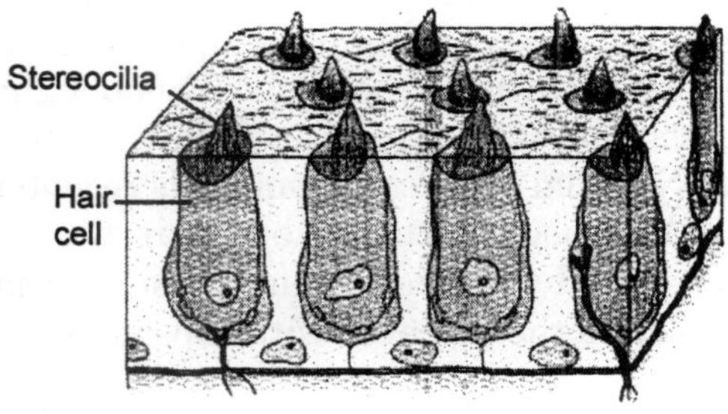

Fig. 5.23. Hair Cells, the Sensory Neurons Crucial for Hearing.

TOUCH SENSING

Like taste, touch is a combination of sensory systems that are expressed in a common organ — in this case, the skin. The detection of pressure and the detection of temperature are two key components. Amiloride-sensitive sodium channels, homologous to those of taste, appear to play a role. Other systems are responsible for detecting painful stimuli such as high temperature, acid, or certain specific chemicals. Although our understanding of this sensory system is not as advanced as that of the other sensory systems, recent work has revealed a fascinating relation between pain and taste sensation, a relation well known to anyone who has eaten "spicy" food.

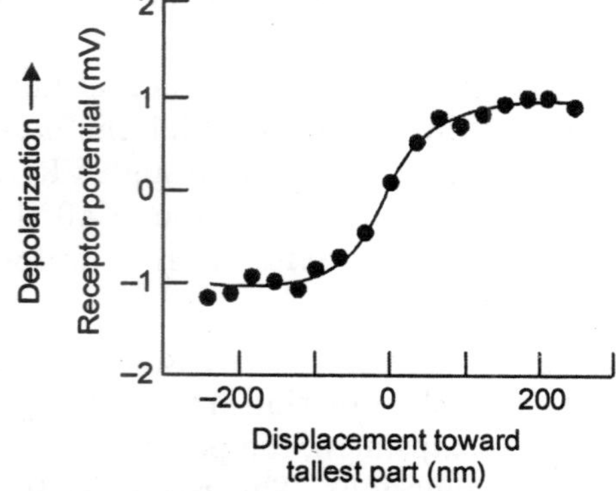

Fig. 5.24. Micromanipulation of a Hair Cell. Movement toward the tallest part of the bundle depolarizes the cell as measured by the microelectrode. Movement toward the shortest part hyperpolarizes the cell. Lateral movement has no effect.

Calo Receptors and Other Painful Stimuli

Our sense of touch is intimately connected with the sensation of pain. Specialized neurons, termed *nociceptors,* transmit signals to pain-processing centers in the spinal cord and brain in response to the onset of tissue damage. What is the molecular basis for the sensation of pain? An intriguing clue came from the realization that *capsaicin,* the chemical responsible for the "hot" taste of spicy food, activates nociceptors.

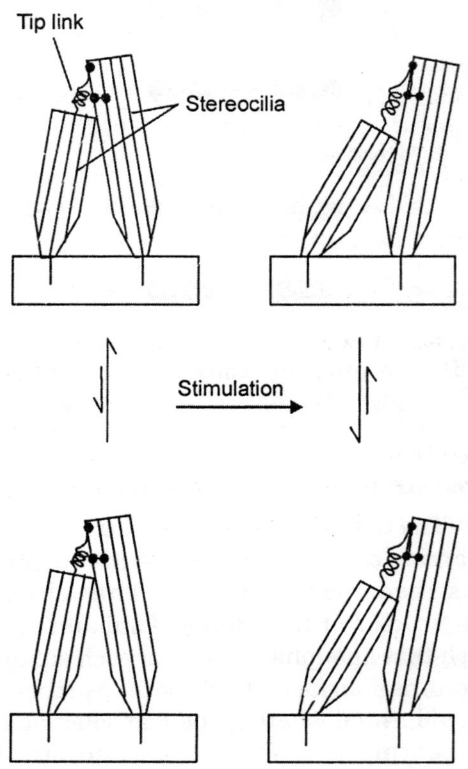

Fig. 5.25. Model for Hair-Cell Transduction. When the hair bundle is tipped toward the tallest part, the tip link pulls on and opens an ior channel. Movement in the opposite direction relaxes the tension in the tip link, increasing the probability that any open channels will close.

Early research suggested that capsaicin would act by opening ion channels that are expressed in nociceptors. Thus, a cell that expresses the *capsaicin receptor* should take up calcium on treatment with the molecule. This insight led to the isolation of the capsaicin receptor with the use of cDNA from cells expressing this receptor. Such cells had been detected by their fluorescence when loaded with the calcium-sensitive compound Fura-2 and then treated with capsaicin or related molecules.

Cells expressing the capsaicin receptor, which is called VR1 (for vanilloid receptor 1), respond to capsaicin below a concentration of 1 oM. The deduced 838-residue sequence of VR1 revealed it to be a member of the TRP channel family. The ammo-terminal region of VR1 includes three ankyrin repeats. Currents through VR1 are also induced by temperatures above 40°C and by exposure to dilute acid, with a midpoint for activation at pH 5.4.

Temperatures and acidity in these ranges are associated with infection and cell injury. The responses to capsaicin, temperature, and acidity are not independent. The response to heat is greater at lower pH, for example. Thus, *VR1 acts to integrate several noxious stimuli.* We feel these responses as pain and act to avoid the potentially destructive conditions that caused the unpleasant sensation.

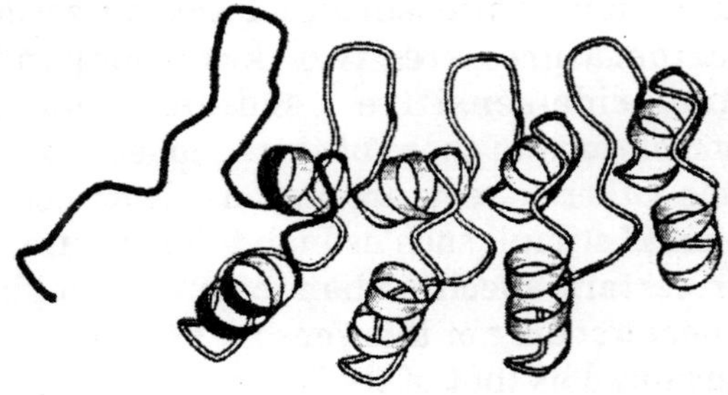

Fig. 5.26. Ankyrin Repeat Structure. Four ankyrin repeats are shown with one shown it red. These domains interact with other proteins, primarily through their loops.

Mice that do not express VR1 suggest that this is the case; such mice do not mind food containing high concentrations of capsaicin and are, indeed, less responsive than control mice to normally noxious heat. Plants such as chili peppers presumably gained the ability to synthesize capsaicin and other "hot" compounds to protect themselves from being consumed by mammals. Birds, which play the beneficial role of spreading pepper seeds into new territory, do not appear to respond to capsaicin. Because of its ability to simulate VR1, capsaicin is used in pain management for arthritis, neuralgia, and other neuropathies. How can a compound that induces pain assist in its alleviation? Chronic exposure to capsaicin overstimulates pain-transmitting neurons, leading to their desensitization.

Subtle Sensory Systems

In addition to the five primary senses, human beings may have counterparts to less-familiar sensory systems characterized in other organisms. These sensory systems respond to environmental factors other than light, molecular shape, or air motion. For example, some species of bacteria are magnetotactic; that is, they move in directions dictated by Earth's magnetic field. In the Northern Hemisphere, Earth's magnetic field points northward but also has a component directed downward, toward Earth's center.

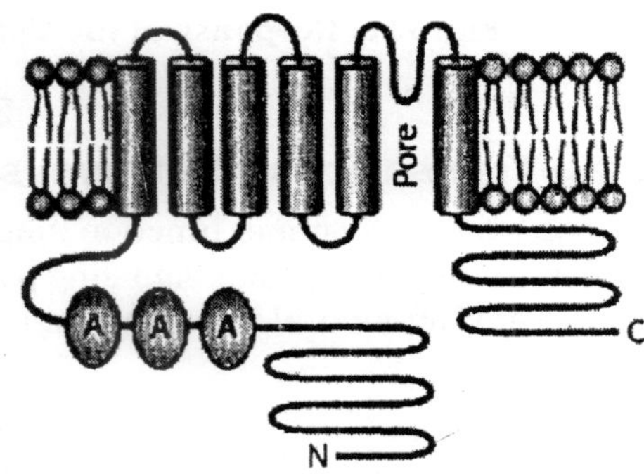

Fig. 5.27. The Membrane Topology Deduced for VR1, the Capsaicin Receptor. The proposed site of the membrane pore is indicated in red, and the three ankyrin (A) repeats are shown in orange. The active receptor comprises four of these subunits.

Magnetotactic bacteria not only swim northward but also swim downward, away from the surface and the presence of high levels of oxygen, toxic to these bacteria. Remarkably, these bacteria synthesize intracellular chains of small particles containing a magnetic ore called magnetite Fe_3O_4) that run through the center of each bacterium. Such chains are called *magnetosomes*. The magnetic force exerted by these particles is sufficiently strong in relation to the size of the bacterium that it causes the bacterium to become passively aligned with Earth's magnetic field. Intriguingly, similar magnetite particles have been detected in the brains of birds, fish, and even human beings, although their role in sensing magnetic fields has not yet been established. There may exist other subtle senses that are able to detect environmental signals that then influence our behaviour.

The biochemical basis of these senses is now under investigation. One such sense is our ability to respond, often without our awareness, to chemical signals called pheromones, released by other persons. Another is our sense of time, manifested in our daily (circadian) rhythms of activity and restfulness. Daily changes in light exposure strongly influence these rhythms. The foundations for these senses have been uncovered in other organisms; future studies should reveal to what extent these mechanisms apply to human beings as well.

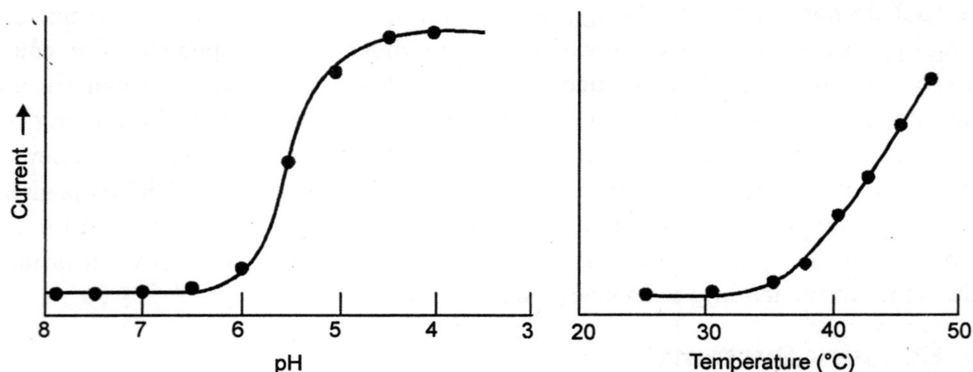

Fig. 5.28. Response of the Capsaicin Receptor to pH and Temperature.

SUMMARY

Signal–Transduction Pathways

These sensory systems function similarly to the signal-transduction pathways for many hormones. These intercellular signaling pathways appear to have been appropriated and modified to process environmental information.

Olfaction

The sense of smell, or olfaction, is remarkable in its specificity—it can, for example, discern stereoisomers of small organic compounds as distinct aromas. The 7TM receptors that detect these odorants operate in conjunction with $G_{(olf)}$, a G protein that activates a cAMP cascade resulting in the opening of an ion channel and the generation of a nerve impulse. An outstanding feature of the olfactory system is its ability to detect a vast array of odorants. Each olfactory neuron expresses only one type of receptor and connects to a particular region of the olfactory bulb. Odors are decoded by a combinatorial mechanism—each odorant activates a number of receptors, each to a different extent, and most receptors are activated by more than one odorant.

Taste

We can detect only five tastes: bitter, sweet, salt, sour, and umami. The transduction pathways that detect taste are, however, diverse. Bitter and sweet tastants are experienced through 7TM receptors acting through a special G protein called gustducin. Salty and sour tastants act directly through membrane channels. Salt tastants are detected by passage though sodium channels, whereas sour taste results from the effects of hydrogen ions on a number of types of channels. The end point is the same in all cases — membrane polarization that results in the transmission of a nerve impulse. Umami, the taste of glutamate, is detected by a receptor that is a modified form of a brain receptor that responds to glutamate as a neurotransmitter rather than as a tastant.

Photoreceptor

Vision is perhaps the best understood of the senses. Two classes of photoreceptor cells exist: cones, which respond to bright lights and colours, and rods, which respond only to dim light.

The photoreceptor in rods is rhodopsin, a 7TM receptor that is a complex of the protein opsin and the chromophore 1 1-cis-retinal. Absorption of light by 1 1 -*cis*-retinal changes its structure into that of all-*trans*-retinal, setting in motion a signal-transduction pathway that leads to the breakdown of cGMP, to membrane hyperpolarization, and to a subsequent nerve impulse. Colour vision is mediated by three distinct 7TM photoreceptors that employ 11-*cis*-retinal as a chromophore and absorb light in the blue, green, and red parts of the spectrum.

Mechanical Stimuli

The immediate receptors for hearing are found in the hair cells of the cochleae, which contain bundles of stereocilia. When the stereocilia move in response to sound waves, cation channels will open or close, depending on the direction of movement. The mechanical motion of the cilia is converted into current flow and then into a nerve impulse.

Touch

Touch, detected by the skin, senses pressure, temperature, and pain. Specialized nerve cells called nociceptors transmit signals that are interpreted in the brain as pain. A receptor responsible for the perception of pain has been isolated on the basis of its ability to bind capsaicin, the molecule responsible for the hot taste of spicy food. The capsaicin receptor, also called VRI, functions as a cation channel that initiates a nerve impulse.

6 EXCRETORY SYSTEM

Because it is found in so many compounds and can exist in several oxidation states, nitrogen has a complex metabolism. The inorganic forms of nitrogen found in our surroundings range from the highly oxidized nitrate ion, in which N has an oxidation state of +5, to ammonia, in which the oxidation state is –3. Living cells both reduce and oxidize these inorganic forms. The organic forms of nitrogen are most often derived by incorporation of *ammonium ions* into amino groups or amide groups. Once it has been incorporated into an organic compound, nitrogen can be transferred into many other carbon compounds.

Certain compounds including glutamic acid, aspartic acid, glutamine, asparagine, and carbamoyl phosphate are especially active in these transfer reactions. They constitute a *nitrogen pool* from which nitrogen can be withdrawn and to which it can be returned. In addition to the pathways for synthesis and degradation of nitrogenous substances, many organisms have specialized metabolism for incorporation of excess nitrogen into relatively nontoxic excretion products. 'All of these aspects of nitrogen metabolism will be dealt with in this and the following chapter. We will look first at the reactions by which organic nitrogen compounds are formed from inorganic compounds, then at the reactions of the nitrogen pool. After that we will examine the specific reactions of synthesis and catabolism of individual nitrogenous compounds.

FIXATION OF N_2 AND THE NITROGEN CYCLE

Most of the nitrogen of the biosphere exists as the unreactive N_2, which makes up 80% of the molecules of air. The "fixation" of N_2 occurs principally by the action of a group of bacteria known as diazotrophs and to a lesser extent by lightning, which forms oxides of nitrogen and eventually nitrate and nitrite.

Human beings also contribute a smaller but significant share through production of chemical fertilizer by the Haber process. These reactions are an important part of the *nitrogen cycle.* Quantitatively even more important are the biochemical processes of *nitrification,* by which ammonium ions from decaying organic materials are oxidized to NO_2^- and NO_3^- by soil bacteria (Fig. 6.1), and reactions of reduction and *assimilation* of nitrate and nitrite by bacteria, fungi, and green plants. Another reductive process catalyzed by *denitrifying bacteria* returns N_2 to the atmosphere (Fig. 6.1).

Reduction of Elemental Nitrogen

One of the most remarkable reactions of nitrogen metabolism is the conversion of dinitrogen (N_2) to ammonia. It was estimated that in 1974 this biological nitrogen fixation added 17×10^{10} kg of nitrogen to the earth (compared with 4×10^{10} kg fixed by chemical reactions).

The quantitative significance can be more easily appreciated by the realization that one square meter of land planted to nodulated legumes such as soybeans can fix 10–30 g of nitrogen per year. Fixation of N_2 by *Clostridium pasteurianum* and a few other species was recognized by Winogradsky in 1893. Subsequent nutritional studies indicated that both iron and molybdenum were required for the process.

Inhibition by CO and N_2O was observed. While ammonia was the suggested product, the possibility remained that more oxidized compounds such as hydroxylamine were the ones first incorporated into organic substances. When cell-free preparations capable of fixing nitrogen were obtained in 1960 rapid progress became possible.[5] It was discovered that nitrogen-fixing bactria are invariably able to reduce acetylene to ethylen, a catalytic ability that goes hand in hand with the ability to reduce N_2.

A simple, sensitive *acetylene reduction test* permits easy measurement of the nitrogen-fixing potential of cells. Application of this test revealed that nitrogen fixation is not restricted to a few species, but is a widespread ability of many prokaryotes. Most studied are *Azotobacter vinelandii,* Winogradsky's C. *pasteurianum, Klebsiella pneumoniae* (a close relative of E. *coli),* and several species of *Rhizobium,* the symbiotic bacterium of root nodules of legumes. The latter deserves special attention?

Although some free-living rhizobia reduce N_2; the reaction usually takes place only in nodules developed by infected roots. Within these nodules the bacteria degenerate into *bacteroids* and the special hemoglobin *leghemoglobin,*whose sequence is specified by a plant gene, synthesized. Legumes are not the only plants with nitrogen-fixing symbionts. Some other angiosperms are hosts to nitrogen-fixing actinomycetes and some gymno-sperms contain nitrogen-fixing blue-green algae.

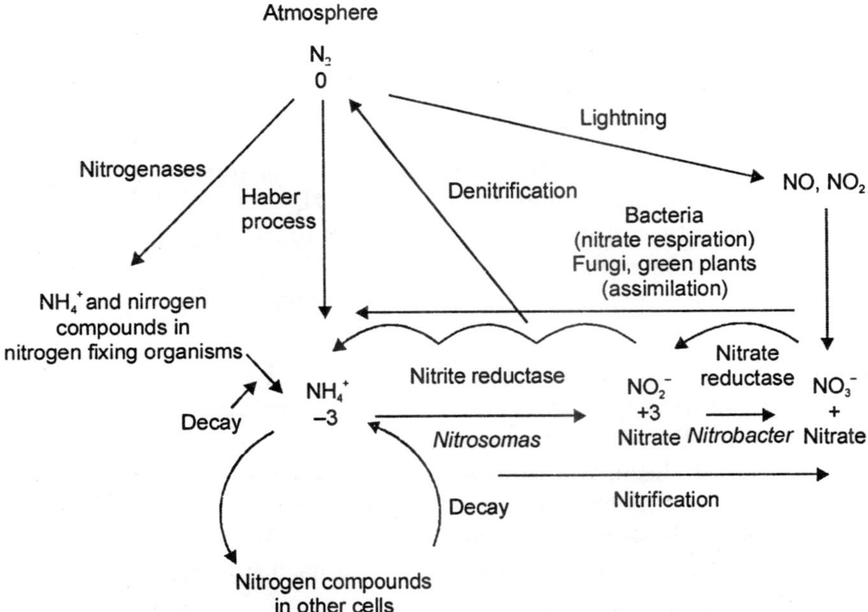

Fig. 6.1. The nitrogen cycle. Conversion of N_2 (oxidation state 0) to NH_4^+ by nitrogen-fixing bacteria, assimilation of NH_4^+ by other organisms, decay of organic matter, oxidation of NH_4^+ by the nitrifying bacteria *Nitrosomas* and *Nitrobacter*, reduction of NO_3^- and NO_2^- back to NH_4^+, and release of nitrogen as N_2 by denitrifying bacteria are all part of this complex cycle.

Leaf nodules of certain plants infected with *Klebsiella* fix nitrogen. While the nutritional significance is uncertain, nitrogen-fixing strains of *Klebsiella* have also been found in the intestinal tracts of humans in New Guinea. Of the free-living nitrogen-fixing organisms, cyanobacteria appear to be of most importance quantitatively. For example, in rice paddy fields cyanobacteria may fix from 2.4 to 10 g of nitrogen per square meter per year. Cyanobacteria in the oceans fix enormous amounts of nitrogen.

Nitrogenases

Cell-free nitrogenases have been isolated from a number of organisms. These enzymes all share the property of being inactivated by oxygen, a fact that impeded early work. Apparently nitrogen fixation occurs in anaerobic regions of cells. Leghemoglobin may protect the nitrogen-fixing enzymes in root nodules from oxygen. It probably also functions to deliver O_2 by facilitated diffusion to the aerobic mitochondria of the bacteroids at a stable, low partial pressure.

Some bacteria utilize protective proteins to shield the nitrogenase molecules when the O_2 pressure is too high. Nitrogenases catalyze the six-electron reduction of N_2 to ammonia and are also able to reduce many other compounds. For example, the reduction of acetylene to

$$N_2 + 6H^+ + 6e^- \rightarrow 2NH_3 \quad ...(6.1)$$

ethylene is a two-electron process. Azide is reduced to N_2 and NH_4^+ in another two-electron reduction. Cyanide ions yield methane and ammonia. Alkyl nitriles as well as N_2O and carbonyl sulfide (COS) are also reduced. Carbon dioxide is reduced slowly to CO, and nitrogenases invariably catalyze reduction of protons to H_2

$$HC \equiv CH + 2\,H^+ + 2\,e^- \rightarrow H_2C = CH_2 \quad ...(6.2)$$

$$N{=}N{=}N^- + 4\,H^+ + 2\,e^- \rightarrow N_2 + NH_4^+ \quad ...(6.3)$$

$$^-C \equiv N + 8\,H^+ + 6\,e^- \rightarrow CH_4 + NH_4^+ \quad ...(6.4)$$

$$2H^+ + 2\,e^- \rightarrow H_2 \quad ...(6.5)$$

In early experiments it was found that sodium pyruvate was required for fixation of N_2 in cell-free extracts, and that large amounts of CO_2 and H_2 accumulated. Investigation showed that cleavage of pyruvate supplies cells with two important products: ATP and reduced ferredoxin. Pyruvate can be replaced by a mixture of ATP plus Mg^{2+} and reduced ferredoxin (Fd_{red}). Furthermore, the nonbiological reductant dithionite ($S_2O_4^{2-}$) can replace the reduced ferredoxin.

Since ADP is inhibitory to the nitrogenase system, it is best in laboratory studies to supply ATP from an ATP-generating system such as a mixture of creatine phosphate, creatine kinase, and a small amount of ADP. The commonest type of nitrogenase can be separated easily into two components. One of these, the *iron protein* (dinitrogenase reductase, azoferredoxin, or component II), is an extremely oxygen-sensitive iron-sulfur protein. It consists of two identical ~ 32-kDa peptide chains; those of *A. vinlandii* each contain 189 amino acid residues. The three-dimensional structure of the dimeric protein shows that each subunit forms a nucleotide-binding domain with an ATP-binding site.

About 2 nm away from this site is *a single Fe_4S_4 cluster which is shared symmetrically by the two subunits of the protein.* Each subunit contributes two thiolate groups from Cys 97 and Cys 132 as well as three N–H--S hydrogen bonds from NH groups at helix ends. The other component, the *molybdenum-iron protein* (dinitrogenase, molybdoferredoxin, or component I), contains both iron and molybdenum as well as labile sulfide. It is a mixed ($\alpha_2\beta_2$) tetramer of ~ 240-kDa mass and an analytical metal ion composition ~$Mo_2Fe_{30}S_{26}$. However, the X-ray structure suggests the composition $Mo_2Fe_{34}S_{36}$. The MoFe protein is a symmetric molecule in which each αβ subunit contains two types of complex metal clusters.

The active sites for N_2 reduction, which are embedded in the α subunits, contain the *FeMo-coenzyme* molecules, each with the metal composition $MoFe_7S_9$ and also containing a molecule of *homocitrate.* The other clusters, known as *P-clusters,* are shared between the α and β subunits, which for *A. vinlandii* contain 491 and 522 amino acid residues, respectively. Each P-cluster is actually a *joined pair* of cubane-type clusters, one Fe_4S_4 and one Fe_4S_3 with two bridging cysteine –SH groups and one iron atom bonded to three sulfide sulfur atoms.

The FeMo-coenzyme can be released from the MoFe-protein by acid denaturation followed by extraction with dimethylformamide. While homocitrate was identified as a component of the isolated coenzyme, the three-dimensional structure of FeMo-co was deduced from X-ray crystallography of the intact molybdenum-iron protein. When the Fe-protein is reduced an EPR signal at $g = 1.94$, typical of iron-sulfur proteins, is observed. This signal is altered by interaction with Mg and ATP, whereas ATP has no effect on the complex EPR signals produced upon reduction of MoFd. These are among the observations that led to the concept that the Fe-protein is an electron carrier responsible for reduction of the molybdenum in the MoFe-protein.

The Mo(IV) or two atoms of Mo(III) formed in this way could then reduce N_2 in three two-electron steps with formation of Mo(VI). Three successive two-electron steps are required to completely reduce N_2 to two molecules of ammonia. An unexpected feature of nitrogenase action is that there is inevitably what was once regarded as a side reaction, the reduction of protons to H_2. The amount of H_2 formed is variable and may be much greater than that of N_2 reduced. However, at high pressures of N_2 the ratio of H_2 formed to N_2 reduced is 1 : 1. This led to the suggestion that H_2 formation is not a side reaction but an essential step in preparing the active site for the binding of N_2.

Two protons that are bound somewhere on "the reduced MoFe-protein could be reduced to H_2 in an obligatory step that would, for example, cause a conformational change required for binding of N_2. If no reducible substrate (N_2, C_2H_2, etc.) is present, H_2 would still be formed slowly. Reducible substrates inhibit H_2 formation. However, addition of N_2 or any other reducible substrate causes an initial "burst" of H_2 to be released. This can be measured readily when a slow substrate such as CN^- is used as the inhibitor. The amount of H_2 released in the burst is stoichiometric with one H_2 per Mo being formed. The overall stoichiometry for reduction of one N_2 becomes:

$$N_2 + 8e^- + 8H^+ + 16\ MgATP \longrightarrow 2NH_3 + H_2 + 16MgADP + 16\ P_i \quad ...(6.7)$$

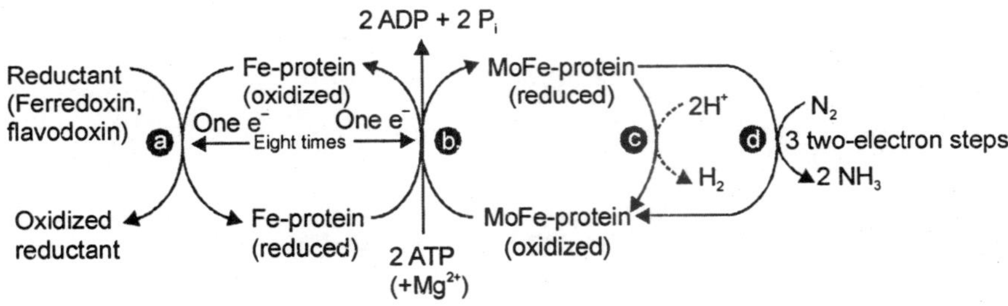

A second remarkable feature of nitrogenase is a requirement for hydrolysis of MgATP that is coupled to reduction of the MoFe-protein. Two molecules of ATP are hydrolyzed to ADP and inorganic phosphate for each electron transferred. This large ATP requirement seems surprising in view of the fact that reduction of N2 by reduced ferredoxin 6.8 is thermodynamically (eq. 6.8) spontaneous:

$$N_2 + 6\ Fd_{red} + 8\ H^+ \rightarrow 2\ NH_4^+ + 6\ Fd_{ox} + \Delta G'\ (pH\ 7) = -89.3\ kj\ mol^{-1}$$

However, N_2 is exceedingly unreactive. In the conmercial Haber process high pressure and temperature are needed to cause H_2 and N_2 to combine.

Evidently cleavage of 16 molecules of ATP must be coupled to the nitrogenase reduction system to overcome the very high activation energy. Not only are two molecules of ATP hydrolyzed to pump each electron, but the Fe-protein must receive electrons from a powerful (low E°) reductant such as reduced ferredoxin, reduced flavodoxin, or dithionite. *Klebsiella pneumoniae* contains a *pyruvate:flavodoxin oxidoreductase* that reduces either flavodoxin or ferredoxin to provide the low potential electron donor: In some bacteria, e.g., the strictly aerobic *Azotobacter,* NADPH is the electron donor for reduction of N_2.

The Fe-protein is thought to accept electrons from a chain that includes at least the ordinary bacterial ferredoxin (Fd) and a special one-electron-accepting *azotoflavin,* a flavoprotein that is somewhat larger than the flavodoxins and appears to play a specific role in N_2 fixation. In *Clostridium* and *Rhizobium* reduced ferredoxins generated by cleavage of pyruvate reduce nitrogenase directly.

The mechanism of nitrogenase action. The one-electron reduction of the Fe_4S_4 cluster of the Fe-protein initiates the action. This reaction occurs before the Fe-protein forms a complex with the MoFe-protein. Following this initial reduction step the two molecules of ATP required for step *b* of Eq 6.6 bind to the Fe-protein. One is bound to each subunit of this protein but neither is immediately adjacent to the shared Fe_4S_4 cluster. The binding to MgATP appears to induce a conformational change that permits the "docking" of the Fe-protein with the MoFe-protein to form the complex in which the electron transfer of step *b* occurs. Abundant evidence indicates that electron transfer does not occur without the binding of MgATP.

The electron transfer is coupled to the hydrolysis of the ATP, but the two reactions appear to be consecutive events. In a deletion mutant of the Fe-protein (lacking Leu 127) the hydrolysis of ATP does not occur, but the complex between Fe-protein and MoFe-protein is formed and electron transfer to the MoFe-protein takes place. The binding of the MgATP causes the midpoint redox potential to drop from –0.42 V to –0.62 V, assisting the transfer. X–ray crystallographic studies reveal a distinct conformational change similar to those observed with G-proteins and involving movement of the Fe_4S_4 center into a better position for electron transfer. After electron transfer the complex of the two proteins is thought to be tightly bonded when unhydrolyzed ATP is present.

This has allowed the direct observation and imaging of the complex at low resolution (~1.5 nm) using rapid synchrotron X-ray scattering measurements. The ATP is hydrolyzed, and the Fe-proton is released from the complex. Only one electron is transferred to the MoFe-protein in each catalytic cycle of the Fe-protein. Thus, the cycle must be repeated eight times to accomplish the reduction of $N_2 + 2H^+$. Where in the MoFe-protein does a transferred electron go? EPR spectroscopic and other experiments with incomplete and catalytically inactive molybdenum coenzyme have provided a clear answer.

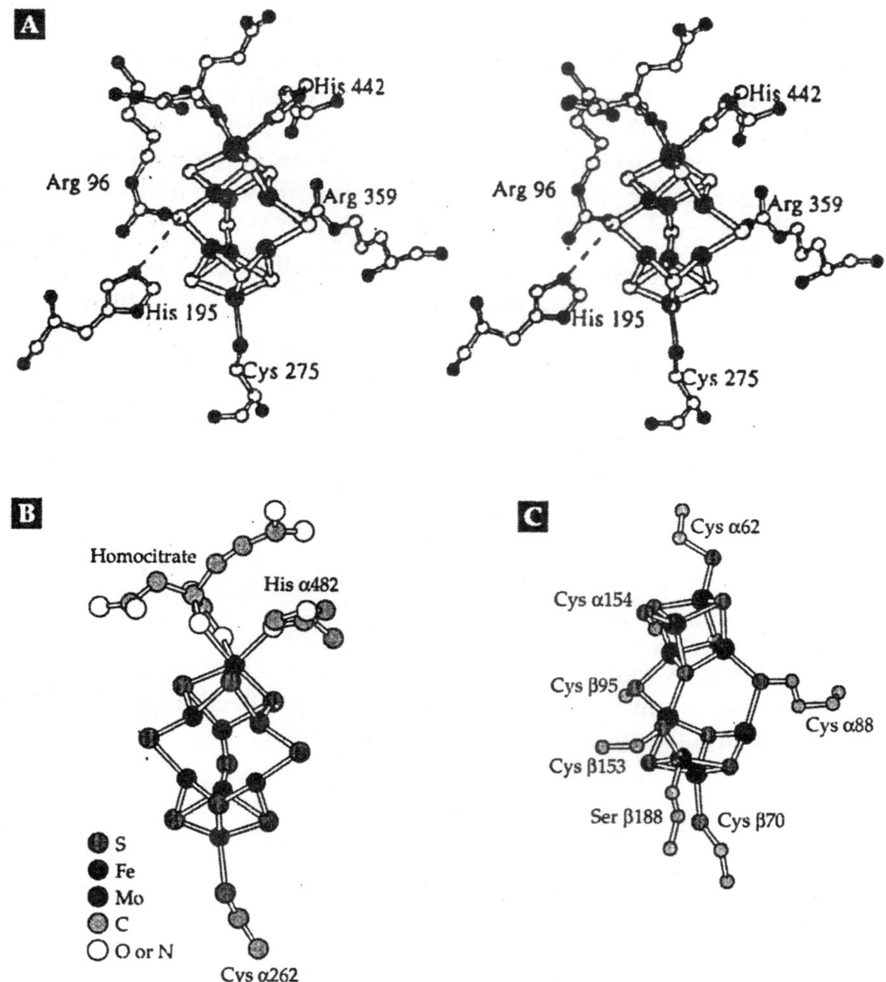

Fig. 6.2. Structures of the metal-sulfide clusters of the MoFe-protein. (A) Stereoscopic view of the FeMo-co coenzyme with interacting side chains from the MoFe-protein of *A. vinlandii*. (B) FeMo-co with atom labels, (C) The structure of the oxidized form of the P-cluster.

The electron is transferred first to one of the two P-clusters, both of which are close to the Fe_4S_4 cluster of the Fe-protein. The transfer causes an observable change both in the spectroscopic properties and in the three-dimensional structure of the P-cluster. Since protons are needed at the active site for the reduction reactions (the FeMo-coenzyme), it is probable that hydrolysis of ATP in the Fe-protein is accompanied by transport of protons across the interface with the MoFe-protein. The electron transfer from the P-cluster on to the FeMo-co center would be assisted by a protic force resulting from ATP cleavage.

With defective FeMo-co (apparently lacking ho-mocitrate) no reduction of N_2, acetylene, or protons is observed. If intact FeMo-co is present, reduction of the cofactor can be observed. An s = 3/2 EPR signal arising from the Mo is seen and EXAFS measurements reveal decreased Mo–

Fe distances as the coenzyme is reduced. The molybdenum is probably present as Mo(VI) in the oxidized state of nitrogenase, but after reduction it isn't clear whether it is Mo(lll) or Mo(IV).

Isolated FeMo-co exists in three identified oxidation states related by $E^{\circ\prime}$ values of –0.17 and –0.465 V. Only the middle state is EPR–active, but it is the most reduced state that is involved in N_2 reduction. With its P-cluster and FeMo-co center each αβ unit of the MoFe-protein could store several electrons. Two or more might be stored in a P-cluster, and Mo(VI) could, in principle, accept three electrons to form Mo(III). However, it is a little hard to imagine storage of the eight electrons needed to reduce both N_2 and H_2.

The reduction of N_2 may begin before all eight electrons have been transferred into the Mofe-protein. Another uncertainty lies in the mode of binding of N_2 and other substrates. Does N_2 bind end-on to Mo, does it slide between Fe atoms within the coenzyme, or does it bind in some other way? While N_2 is unreactive, it forms nitrides with metals and complexes with some metal chelates. These complexes are generally of an end-on nature, *e.g.*, N≡N–Fe. Stiefel suggested that N_2 first forms a complex of this type with an iron atom of the MoFe-protein. Then an atom of Mo(IV) could donate two electrons to the N_2 to form a complex of N_2 and Mo(VI).

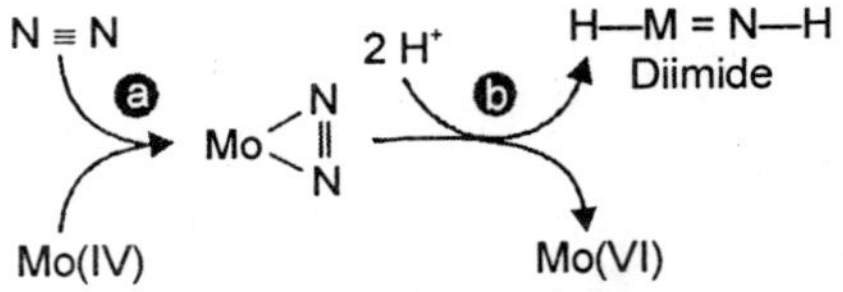

N ≡ N + Mo(IV) —(a)→ Mo(N≡N) —(b) 2 H⁺→ H—M = N—H (Diimide) + Mo(VI) ...(6.9)

Addition of two protons would yield a molecule of *diimide*, which would stay bound at the iron site while the molybdenum underwent another round of reduction. The diimide could be reduced to hydrazine and finally to ammonia.

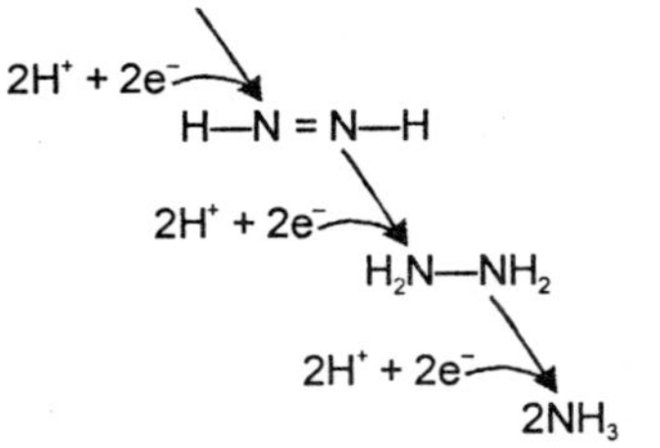

$$\xrightarrow{2H^+ + 2e^-} H\text{—}N = N\text{—}H \xrightarrow{2H^+ + 2e^-} H_2N\text{—}NH_2 \xrightarrow{2H^+ + 2e^-} 2NH_3 \qquad \text{...(6.10)}$$

Mo(VI) attracts electrons sufficiently strongly that protons bound to surrounding ligands, such as H_2O, tend to dissociate completely. Thus, the molybdate ion MoO_4^{2-} is not protonated. The same would be true of nitrogenous ligands of a protein that might be coordinated with the bound molybdenum.

On the other hand, reduction to Mo(IV) would tend to favour protonation of-ligands such as the His 442 imidazole seen in Fig. 6.3A. Concurrently with the electron transfer from molybdenum to N_2 these protons could be transferred to the N_2 molecule. The fact that strictly *cis*-dideuteroethylene is formed from acetylene in the presence of $2H_2O$ is in accord with this idea. However, looking at the FeMo-co molecule and the crowded surroundings of the Mo atom it may be more likely that reduction of N_2 occurs while it is bound to iron.

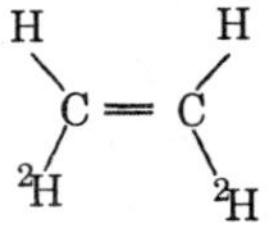

H H
C = C
²H ²H

Theoretical calculations as well as experimental data support this possibility. Recent crystallographic studies at a resolution of 0.12 nm revealed the presence of an atom, probably N, coordinated to six Fe atoms of FeMo-Co. This suggets, as previously proposed by Thorneley and Lowe, that a nitride ion (N_3^-) may be an intermediate in the formation of N_2.

Many mutant forms of nitrogenase have been investigated. Substitutions of His 195, Lys 191, and Gly 69 of the α chain affect reactions with various substrates. For example, the mutant obtained by substitution of His 195, whose imidazole forms an N–H--S hydrogen bond to a central bridging sulfide atom of FeMo-co, with glutamine (H195Q mutant) reduces N_2 only very slowly.

However, it still reduces both acetylene and protons. Thus, it may be that different modes of subsstrate binding are needed for the individual steps of Eq. 6.10. Because of the practical significance to agriculture there is interest in devising better nonenzymatic processes for fixing nitrogen using nitrogenase models that mimic the natural biological reaction. One interesting catalyst is the following molybdenum complex Mo(III)$(NRAr)_3$ where R = $C(C_2H_3)_2CH_3$ and Ar = 3,5-dimethylphenyl. Many other synthetic complexes have been studied including cubic $MoFe_3S_4$ clusters. However, no exact chemical model for the FeMo-co coenzyme has been developed, and the rates of reaction for all of the model reactions are much slower than those of nitrogenases.

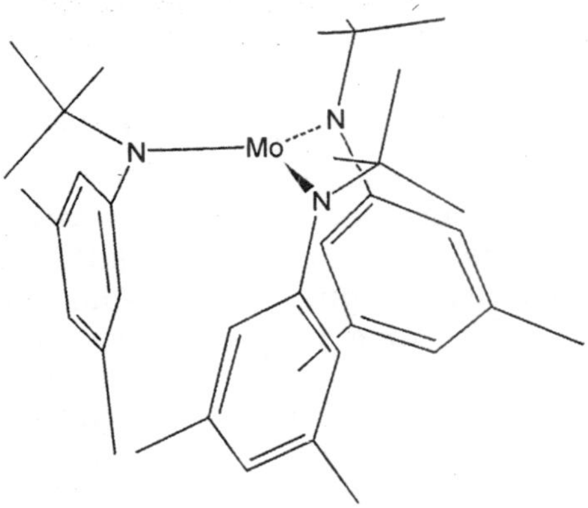

Nitrogen fixation genes. At least 17 genes needed for nitrogen fixation are present in the 23-kb *nif* region of the *Klebsiella* chromosome. A similar gene cluster in *A. vinlandii* contains five polygenic transcriptional units and one monogenic unit. The nitrogenase structural genes are *nifK, D,* and H as is indicated in Fig. 6.4. The *nifF* and *J* genes encode associated electron-transport proteins. *NifM* is needed to activate the Fe-protein in an unknown fashion. *NifS* encodes a cysteine desulfhydrase needed for assembly of Fe-S clusters in the nitrogenase and the nifU and nifY proteins assist the assembly.

The chaperone GroEL is also required. *Nif Q, B, V,* X, *N, ei,* and *H* are needed for synthesis of FeMo-co and for its incorporation into the MoFe-protein. *Nif A* is an activator gene for the whole cluster including the *nifL* gene product, which is altered by the presence of O_2 or of glutamine. Accumulation of the latter in cells strongly represses transcription of the nitrogenase genes.

Legume nodules and cyanobacterial heterocysts. Nitrogen fixation requires an anaerobic environment. Free-living bacteria fix nitrogen only when anaerobic. However, Rhizobia produce their own anaerobic environment by symbiotic association with the roots of legumes. Formation of root nodules is a genetically determined process, several nodulation *(nod)* genes of the bacterium being required along with an unknown number of plant genes.

Initiation of nodulation results from a two-way molecular conversation between root hairs of the plant and bacterial cells. The roots secrete *flavonoid compounds* which are recognized by bacterial sensors and induce transcription of the *nod* genes. Several of these genes encode enzymes required for synthesis of *Nod factors,* small β-linked N-acetyl-D-glucosamine oligosaccharides containing 3-5 sugar residues and an N-linked long-chain fatty-acyl substituent at the nonreducing terminus *(lipo-chitooligosaccharides).* Genes *nodA, B, C* specify enzymes needed for synthesis of the oligosaccharide core present in all Nod factors. Other Nod genes provide for modifications that restrict infection to specific species of legumes. For example, *nodS* encodes a methyltransferase and *nodU a* carbomoyltransferase. NodH is a sulfotransferase. NodD is a transcriptional activator that binds to DNA and induces the synthesis of the other Nod factors needed to initiate nodulation.

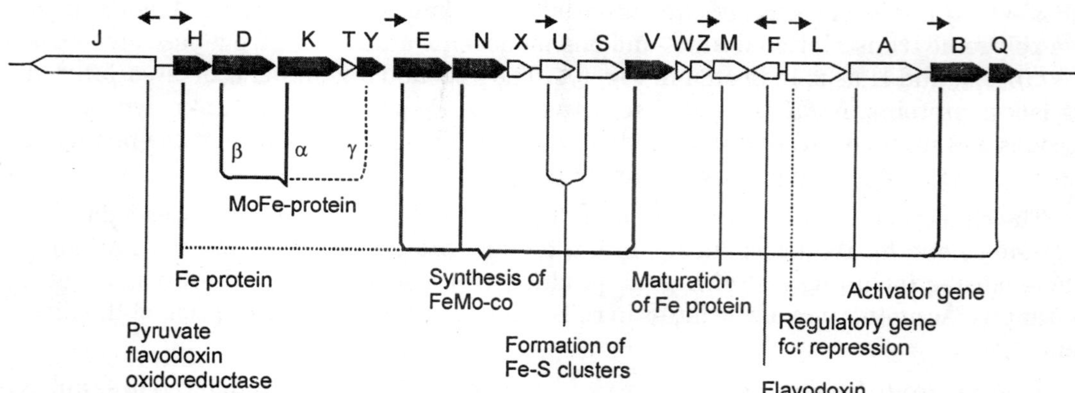

Fig. 6.4. Sequence of nif genes of *Klebsiella pneumoniae.* These precede, the *his* operon directly at the right side. The nitrogenase structural genes are marked with green.

When an appropriate Nod factor is recognized, the root hairs on the legume curl around the bacteria to initiate nodulation: However, there are other factors. Infecting bacteria must reach a region of low oxygen in the plant. A hemoprotein *FixL* is a sensor kinase that regulates phosphorylation of transcription factor *FixJ*. This two-component system induces transcription of *nif A* and others.

Nitrogen-fixing nodules, which are filled with the bacteroids derived from the infecting bacteria, synthesize leghemoglobin. The polypeptide chain of this protein is encoded by the plant, but its heme may be synthesized by bacteroid enzymes. In at least one strain of *Rhizobium* the *nod* genes as well as *the fix* and *nif genes* are all carried on a 536-kb plasmid, which is almost as large as the whole 580-kb genome of *Mycoplasma genitalium.*

This arrangement seems to have allowed these bacteria to form an unusually large number of Nod factors and to colonize a wider variety of hosts including a non-leguminous tree. The H_2 that is produced in Eq 6.6 (step *c)* may be used by bacteroids or by the plant cells. Some nodules evolve H_2, but in others it is utilized by hydrogenases as a source of energy. From Eq 3.6 it can be seen that up to 1/4 of the ATP utilized can, ideally, be recovered by use of the H_2 in this manner. In cyanobacteria nitrogen fixation occurs in the *heterocysts,* specialized cells with thickened cell envelopes. They supply NH_4^+ to other cells in the filament of which they are a part.

The cell envelopes prevent rapid diffusion of O_2 into the cells but do permit rapid enough entry of N_2 to maintain the observed rate of fixation of N_2. In actinomycetes of the genus *Frankia,* which forms root nodules with woody plants, nitrogen fixation occurs in vesicles that are sheathed by multiple layers of *hopanoid lipids.*

Genetic engineering. Because of the high cost of nitrogen fertilizers there is intense interest in improving biological nitrogen fixation. Ideas range from increasing the efficiency of nitrogenase by using fewer molecules of ATP, by limiting excessive evolution of H_2, or transferring the whole *nif* region of a bacterial genome into nonleguminous plants. The last proposal has generated much publicity, but it will probably be difficult because of the need to create an anaerobic environment suitable for nitrpgen fixation.

A crop plant engineered in this way might not resemble the hoped-for product. It would have an enormous energy requirement for nitrogen fixation, which would have to be met by photosynthesis. At present genetic engineering on *nif* genes to increase efficiency seems most likely to succeed.

Other nitrogenases. Although the well-characterized Mo-containing nitrogenase is responsible for most of their nitrogen fixation, bacteria often have alternative nitrogen fixation systems. *Azotobacter vinlandii* produces three different nitrogenases in response to varying metal compositions in its surroundings. When the molybdenum level is adequate nitrogenase 1 is formed with its FeMo-coenzyme.

In a low-molybdenum environment containing vanadium nitrogenase 2 is formed with an FeV-coenzyme. If both molybdenum and vanadium are lacking, the bacteria form nitrogenase-3, which has an iron-only FeFe-coenzyme. An unusual nitrogenase is formed by the chemolithotrophic *Streptomyces thermoautotrophicus,* which obtains energy from reduction of CO_2 or CO by H_2. These organisms form a MoFe nitrogenase that utilizes a manganese-containing superoxide dismutase to generate superoxide anion radicals'. The latter transfer electrons to the MoFe protein in an ATP-dependent process. Electrons for generation of superoxide are formed using another molybdenum enzyme, a CO dehydrogenase containing molybdpterin cytosine dinucleotide and Fe-S centers. The two systems function together as indicated by Eq 6.11.

H_2O | $4\,ATP + 4\,P_i$
CO → CO_2 (Four times; Mo-pterin Fe–S) → 2 e^-, 2H^+
O_2^- ⇌ O_2 (Eight times; SOD)
Mn(III) ⇌ Mn(II)
MoFe nitrogenase: 4 MgATP; N_2, 2H^+ → H_2, 2 NH_3

Interconversion of Nitrate, Nitrite, and Ammonium Ions

As is indicated in Fig. 6.1 the interconversions of nitrate and nitrite with ammonia and with organic nitrogen compounds are active biological processes. Two genera of nitrifying soil bacteria, oxidize ammonium ions to nitrate. *Nitrosomas* carries out the six-electron oxidation to nitrite and *Nitrobacter* the two-electron oxidation of nitrite to nitrate. The opposite sequence, reduction of nitrate and nitrite ions, provides a major route of acquisition of ammonia for incorporation into cells by bacteria, fungi, and green plants.

Assimilatory (biosynthetic) *nitrate reductases* catalyze the two-electron reduction of nitrate to nitrite. This is thought to occur at the molybdenum atom of the large ~900-residue highly regulated molybdopterin-dependent enzyme. In green plants the reductant is usually NADH while in fungi it is more often NAD-PH. In all cases the cofactors FAD, heme, and molybdopterin are bound to a single polypeptide chain with the molybdopterin domain near the N terminus, and the heme in the middle. The electron-accepting FAD domain is near the C terminus and is thought to transfer the two electrons through the following chain.

$$\text{NAD(P)H} \rightarrow \text{FAD} \rightarrow \text{Cyt } b \rightarrow \text{Mo-pterin} \rightarrow NO_3^- \quad ...(6.12)$$

Bacterial assimilatory nitrate reductases have similar properties. In addition, many bacteria, including *E. coli,* are able to use nitrate ions as an oxidant for *nitrate respiration* under anaerobic conditions. The *dissimilatory nitrate reductases* involved also contain molybdenum as well as Fe-S centers. The *E. coli* enzyme receives electrons from reduced quinones in the plasma membrane, passing them through cytochrome *b,* Fe-S centers, and molybdopterin to nitrate. The three-subunit αβγ enzyme contains cytochrome *b* in one subunit, an Fe_3S_4 center as well as three Fe_4S_4 clusters in another, and the molybdenum cofactor in the third. Nitrate reduction to nitrite is also on the pathway of denitrification, which can lead to release of nitrogen as NO, N_2O, and N_2 by the action of *dissimilatory nitrite reductases.*

Assimilatory nitrite reductases of plants, fungi, and bacteria carry out me six-electron reduction of nitrite to ammonium ions using electron donors such as reduced ferredoxins or NADPH.

$$NO_2^- + 6\,e^- + 8\,H^+ \rightarrow NH_4^+ + 2\,H_2O \quad ...(6.13)$$

The enzymes from green plants and fungi are large multifunctional proteins, which may resemble assimilatory sulfite reductases. These contain *siroheme*, which accepts electrons from "either reduced ferredoxin (in photosynthetic organisms) or from NADH or NADPH. FAD acts as an intermediate carrier. It seems likely that the nitrite N binds to Fe of the siroheme and remains there during the entire six-electron reduction to NH_3. Nitroxyl (NOH) and hydroxylamine (NH_2OH) may be bound intermediates as is suggested in steps *a-c* of Eq. 6.14.

$$NO_2^- \xrightarrow[2H^+ \quad HO^-]{2e^- \ (a)} [HNO] \text{ (Nitroxyl)} \xrightarrow[2H^+]{2e^- \ (b)} [NH_2OH] \xrightarrow[2H^+ \quad H_2O]{2e^- \ (c)} NH_3 \qquad ...(6.14)$$

INCORPORATION OF NH_3 INTO AMINO ACIDS AND PROTEINS

Prior to 1935, amino acids were generally regarded as relatively stable nutrient building blocks. That concept was abandoned as a result of studies of the metabolism of $^{15}NH_3$ and of ^{15}N-containing amino acids by Schoenheimer and Rittenberg and more recent studies using ^{13}N by Cooper *et al.* These investigations showed that nitrogen could often be shifted rapidly between one carbon skeleton and another. This confirmed proposals put forth earlier by Braunstein, Meister, and others who had pointed out that the C_4 and C_5 amino acids, aspartate and glutamate, which are closely related to the tricarboxylic acid cycle, are able to exchange their amino groups rapidly with those of other amino acids via transamination, step *d*).

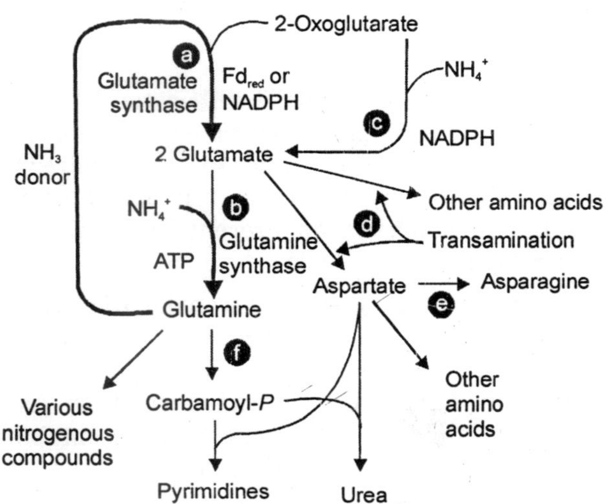

Fig. 6.5 Major pathways of incorporation of nitrogen from ammonium ions into organic compounds, traced by green arrows.

Since ammonia can be incorporated readily into glutamate, (step *a;* see next section), a general means is available for the biosynthesis of amino acids. The citric acid cycle is able to provide any needed amount of 2-oxoglutarate for the synthesis of both glutamate and glutamine. Glutamine, and to a lesser extent asparagine, act as soluble, n.ontoxic carriers of additional ammonia in the form of their amide groups. An active synthase converts glutamate and ammonia to glutamine, (step *d*) and another enzyme transfers the amide nitrogen into aspartate, in an ATP-dependent reaction, to form asparagine (step *e*).

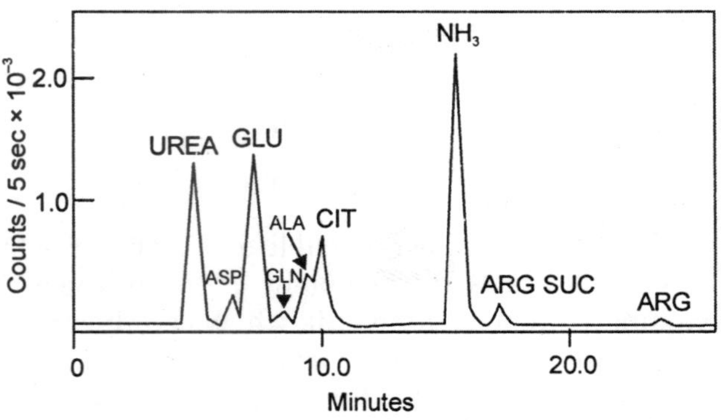

Fig. 6.6 Elution profile of N-containing metabolites extracted from liver 15 s after injection of $^{13}NH_3$ into the portal vein of an anesthetized adult male rat. CIT, citrulline; ARG SUC, argininosuccinate.

The amide nitrogen of glutamine is incorporated in a similar way into a great variety of other biochemical compounds, including carbamoyl phosphate, glucosamine NAD^+, *p*-aminobenzoate, histidine, CTP, and purines. These reactions are catalyzed by a family of *amidotransferases*, which hydrolyze the glutamine to glutamate and NH_3. The last is entrapped until it reacts with the second substrate. *Asparagine synthetase* apparently first binds ATP and aspartate, which probably react to form β-aspartyl adenylate (β-aspartyl-AMP). Glutamine then

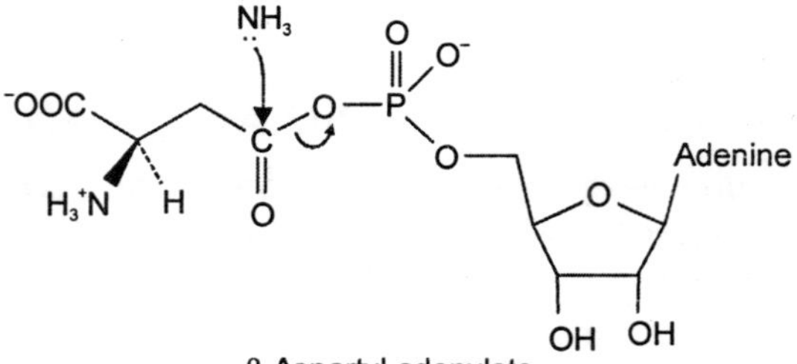

β-Aspartyl adenylate

binds and is hydrolyzed. The liberated NH_3 can attack the β-aspartyl-AMP as indicated in the accompanying diagram to form asparagine. However, it is also possible that NH_3 is transferred via covalently bonded complexes and is never free NH_3. An asparagine synthetase that utilizes free NH_3 as a nitrogen donor is also present in many organisms. A. third mechanism of synthesis, which was only recently recognized, appears to provide the sole source of asparagine for many bacteria.

The asperagine-specific transfer RNA tRNA^{Asn} is "mischarged" with aspartic acid to form Asp-tRNA^{Asn}. This compound is then converted to the properly am tRNA^{Asn} by a gl͜tamine-dependent amidotransferase. The entire ATP-dependent sequence is shown in Eq. The activated asparaginyl grpup is then transferred from Asn-tRNA^{Asn} into proteins as they are synthesized. Most green plants transport nitrogen from roots to growing shoots as asparagine. However, in peanuts *β-methyleneaspartate* is the major nitrogen carrier, and in some legumes, including soybeans, *allantoin* and *allantoate* play this role. Allantoin arises from hydrolysis of purines, which are synthesized in root nodules of nitrogen-fixing plants. Glutamate, glutamine, and aspartate also play central roles in *removal* of nitrogen from organic compounds.

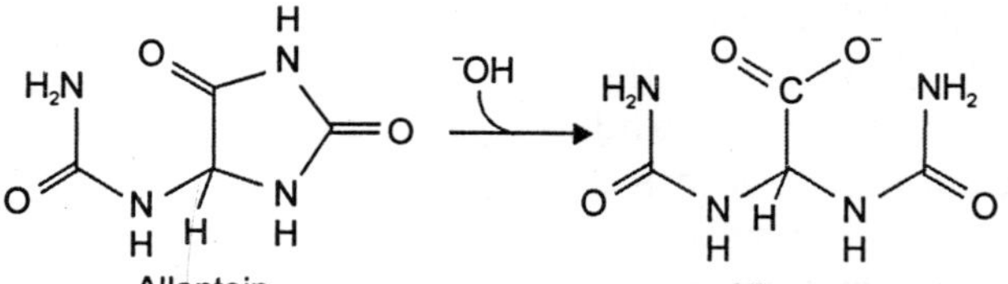

Allantoin Allantoate

Transamination is reversible and is often the first step in catabolism of excess amino acids. 2-Oxoglutarate is the recipient of the nitrogen, and the glutamate that is formed can be deaminated to form ammonia which can then be incorporated into glutamine. Glutamate can also donate its nitrogen to form aspartate. In the brain glutamate is a major neurotransmitter but is toxic in excess. The astrocyte glial cells take up glutamate from the synaptic clefts between neurons, converting it to glutamine, which is then released into the extracellular space for reuptake by neurons.

In the animal body both aspartate and glutamine (via carbamoyl phosphate) are precursors of *urea*, the principal nitrogenous excretion product. These relationships are also summarized in Fig. 6.5 and details are provided in later sections. While reductive amination of glutamate via glutamate synthase appears to be the major pathway for incorporation of nitrogen into amino groups, some direct amination of pyruvate and other 2-oxoacids in reactions analogous to that of glutamate dehydrogen-ase occurs in bacteria. Another bacterial enzyme catalyzes reversible addition of ammonia to fumarate to form asparate. An initially surprising conclusion drawn from the studies of Schoenheimer and Rittenberg was that proteins within cells are in a continuous steady state of synthesis and degradation. The initial biosynthesis, the processing, oxidative and hydrolytic degradative reactions of peptides, and further catabolism of amino acids all combine to form a series of metabolic loops. Within cells some proteins are degraded much more rapidly than others, an important aspect of metabolic control. This is accomplished with the aid of the ubiquitin system and proteasomes. Proteins secreted into extracellular fluids often undergo more rapid turnover than do those that remain within cells.

Uptake of Amino Acids by Cells

While cells of autotrophic organisms can make all of their own amino acids, other organisms utilize many preformed amino acids. Human beings and other higher animals require several *essential amino acids* in their diets. Additional amounts of "nonessential" amino acids are also needed. It is true that amino groups can be transferred from one carbon skeleton to another among most of the amino acids. However, the body must take in enough amino groups to supply its need for all of the 20 amino acid components of proteins. Because of an unfavourable equilibrium constant, and the normally low concentration of NH_4^+, glutamate dehydrogenase does not normally synthesize glutamate in the animal body. Its function is to deaminate excess glutamate.

Furthermore, cells of some tissues take up amino acids that are made in other tissues. In mammals the absorption of amino acids takes place through epithelial cells of the intestinal tract, kidney tubules, and the brain (blood-brain barrier). Both Na^+ -dependent transport and Na^+ -independent processes occur. Among the latter is the proposed *γ-glutamyl cycle.*

The cycle makes use of the γ-carboxyl group of glutamate, the same carboxyl that carries ammonia in the form of glutamine. Glutathione supplies the activated γ-glutamyl group. The amino acid to be transported reacts on the membrane surface by *transpeptidation* to form a γ-*glutamylamino acid* which enters the cytoplasm. It releases the free amino acid through an internal displacement by the free amino group of the glutamyl group. The natural tendency of the 5-carbon glutamate to undergo cyclization is used to provide the driving force for release of the bound amino acid.

The cyclic product 5-oxoproline is then opened hydrolytically in an ATP-requiring reaction. Cysteinyl-glycine formed in the initial transpeptidation is hydrolyzed by a peptidase, and glutathione is regenerated in two ATP-dependent steps. The significance of the γ-glutamyl cycle is not fully understood. However, the finding of a mentally retarded individual who excretes 25-50 g / day of 5-oxoproline in the urine (possibly because of a defective 5-oxoprolinase) suggests that the pathway is a very active one.

A few persons deficient in γ-glutamyl transpeptidase have been found. They excrete glutathione and have a variety of medical problems.

Glutamate Dehydrogenase and Glutamate Synthetase

In animal tissues and in some bacteria the *glutamate dehydrogenase* reaction provides a means of incorporating ammonia reversibly into glutamic acid. In eukaryotic cells the allosteric enzyme is found largely in the mitochondria. Glutamate dehydrogenase is also found in chloroplasts where it may function in glutamate synthesis when ammonia is present in excess. The action of aminotransferases, both within and without mitochondria, distributes nitrogen from glutamate into most of the other amino acids.

Especially active is aspartate aminotransferase which equilibrates asparatate and oxaloacetate with the 2-oxoglutarate–glutamate couple. However, the body obtains glutamate, as well as other amino acids, from foods, the initial source being largely green plants. In plants as well as in E. *coli* and many other bacteria most glutamate is formed *by glutamate synthase,* which carries out reductive amination of 2-oxoglutarate. Glutamate synthase (also called *GOGAT*) utilizes the amide side chain of glutamine as the nitrogen donor. It is one of the previously mentioned amidotransferases in which glutamine is hydrolyzed to glutamate and NH_3 within the active site of the enzyme.

Formation of a Schiff base and reduction probably occurs. However, one of the two glutamate molecules formed in reaction *a* of Fig. 6.5 must be reconverted by glutamine synthase to glutamine with the utilization of a molecule of ATP. Because of this coupling of ATP cleavage to the reaction the equilibrium in reaction *a* lies far toward the synthesis of glutamate. The low value of K_m for NH_4^+ that is characteristic of glutamine synthase favours glutamate synthesis even when little nitrogen is available. Bacterial glutamate synthases are large oligomeric proteins containing flavin and Fe-S centers. That of *Azospirillum brasilense* consists of αβ units in which the 53-kDa β chains contain FAD and an NADPH binding site. The NADPH evidently transfers electrons to the FAD, which transfers them to an Fe_3S_4 center in the large 162-kDA α subunit.

A molecule of bound FMN receives the electrons and reduces the iminoglutarate to glutamate. The site of binding and hydrolysis of glutamine is also present in subunit α. Chloroplasts of higher, plants contain two glutamate synthases. One resembles the bacterial enzyme and utilizes NADPH as the reductant. The other requires reduced ferredoxin. Bacteria utilize both D-alanine and D-glutamate in the synthesis of their peptidoglycan layers. Both D-amino acids are formed by racemases. That of alanine uses PLP but *glutamate race mase* does not. It may be able to remove the α-H of glutamate by utilizing the –COOH of the substrate, rather than the PLP ring, as an electron sink. Small amounts of D-amino acids occur also in animals. Animal livers and kidneys contain *D-amino acid oxidase* and *D-aspartate oxidase,* which apparently function to metabolize D-amino acids from foods or those formed by brain activity or by aging.

Glutamine Synthetase

The formation of glutamine from glutamate also depends upon a coupled cleavage of ATP: Glutamine synthase, as isolated from E. *coli,* contains 12 identical 51.6-kDa subunits arranged in the form of two rings of six subunits each with a center-to-center spacing of 4.5 nm. The units in one layer lie almost directly above those in the next, the center-to-center spacing between the two layers is also 4.5 nm, and the array has 622 dihedral symmetry. The enzyme displays complex regulatory properties. The enzyme exists in two forms *Active glutamine synthetase* requires Mg^{2+} in addition to the three substrates glutamate, NH_4^+, and ATP. If the glutamate precursor, 2-oxoglutarate, is present in excess, the enzyme tends to remain in the active form because conversion to a modified form is inhibited; when the oxoglutarate concentration falls to a low value and glutamine accumulates, alteration is favoured.

The modifying enzyme *adenylyltransferase* (AT) in its active form AT_A transfers an adenylyl group from ATP to a tyrosine hydroxyl on glutamine synthase to give an adenylyl enzyme (GS-AMP). This *modified enzyme* requires Mn^{2+} instead of Mg^{2+} and is far more sensitive than the original enzyme to feedback inhibition by a series of end products of glutamine metabolism. All nine of the feedback inhibitors (serine, alanine, glycine, histidine, tryptophan, CTP, AMP, carbamoyl-P, and glucosamine 6-*P*) seem to bind to specific sites on the enzyme surface and to exert a cumulative inhibition. (Serine, alanine, and glycine appear to be competitive inhibitors at the glutamate binding site. Relaxation of adenylylated glutamine synthase to the unmodified form is not catalyzed by a separate hydrolase but is promoted by a modified form of the adenylyltransferase AT_D. The active enzyme AT_A is actually a complex AT PII containing the regulatory protein PII. Subunit PII can be uridylylated on a tyrosine side chain by action of a 95-kDa *uridylyltrans-ferase* (UT) to form the modified glutamine synthgtase AT• PII-UMP or AT_D.

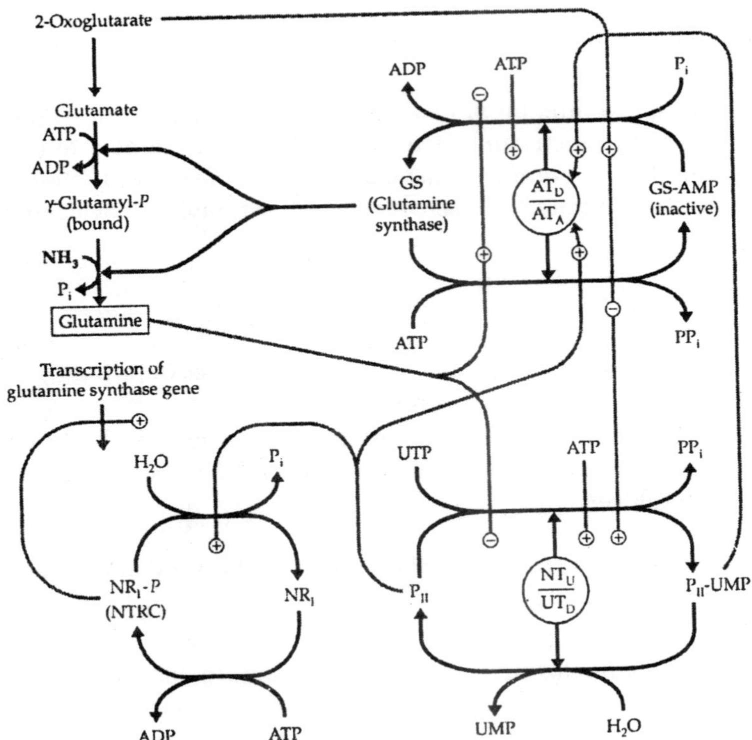

Fig. 6.7. Regulation of glutamine synthase of *E. coli* using activation (+) and inhibition (–). Glutamine synthase (GS upper center) is converted by adenylylation of Tyr 397 into an inactive form GS-AMP by the action of an adenylyltransferase AT_A in complex with regulatory protein PII. PII is uridylylated at up to four sites by action of uridylyltransferase UT_U, which resides in the same polypeptide chain as a uridylyl removing enzyme UT_D (or UR). When PII carries a uridylyl group (PII-UMP), AT_A is transformed to AT_D, which reconverts the inactive GS-AMP to active GS by phosphorolytic removal of the adenylyl group. The ratios of AT_A/AT_D and UT_U / UT_D are controlled by the concentrations of the metabolites 2-oxoglutarate, a precursor, and glutamine, the immediate product. The amount of GS formed is controlled at the transcriptional level by an enhancer-binding transcription factor called NRI or NtrC (lower left). It is active when phosphorylated. Dephosphorylation of NRI-P is catalyzed by yet another protein and is stimulated by PH. Thus, PII both decreases synthesis of GS and promotes conversion of GS to its inactive form.

This form catalyzes phosphorolytic deadenylylation of glutamine synthetase, P_i displacing the adenylyl group to form ADP. Removal of the uridylyl group from PII-UMP is catalyzed by a fourth enzyme, UT_D (or UR), which is part of the same polypeptide chain as UT_U. The cycle of interconversions of PII catalyzed by the UT_U and UT_P activities is shown at the lower right side of Fig 6.7. From the allosteric modification reactions indicated by the gray lines, it is seen that glutamine not only promotes the adenylylation of glutamine synthetase but also inhibits the uridylylation of PII, thereby preventing AT_D from removing the adenylyl group from the synthetase.

Furthermore, it allosterically inhibits the deadenylylation reaction itself. On the other hand, 2-oxoglutarate acts in the opposite way. The glutamine synthase regulatory system has another important function. Protein PII stimulates the dephosphorylation of the enhancer-binding transcriptional regulator NRI-*P* (NtrC-*P*). This slows transcription of the glutamine synthase gene as well as a variety of other genes including those for the nitrogenase proteins in organisms that have them. As a consequence, a deficiency of glutamine turns on a number of genes involved in nitrogen metabolism.

Accumulation of glutamine promotes PII accumulation, modification of the synthase and loss of gene activation. Nitrogen can be transferred from glutamine into many other substrates. Several antibiotic analogs of glutamine have been useful in studying these processes. Examples are the streptomyces antibiotics *L-azaserine* and 6-diazo-5-oxo-L-norleucine (DON).

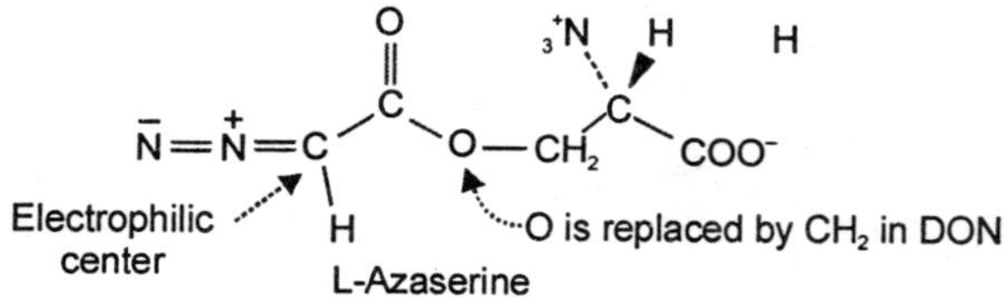

$$\bar{N}=\overset{+}{N}=CH-C(=O)-O-CH_2-C(H)(\overset{+}{N}H_3)-COO^-$$

Electrophilic center

O is replaced by CH_2 in DON

L-Azaserine

These compounds act as alkylating agents; N_2 is released and a nucleophilic group from the enzyme becomes attached at the carbon atom indicated. Other inhibitors bind noncovalently to form dead-end complexes.

Catabolism of Glutamine, Glutamate, and Other Amino Acids

Glutamine is hydrolyzed back to glutamate by glutaminases that are found both in eukaryotic tissues and in bacteria. Liver, contains an isozyme whose function appears to be to release NH_3 from glutamine for urea synthesis. Glutamate dehydrogenase deaminates excess glutamate back to 2-oxoglutarate, which is degraded to succinyl-CoA and via β oxidation to malate, pyruvate, and acetyl-CoA. The last can reenter the citric acid cycle and be oxidized to CO_2. In fact, in mammalian tissues glutamate is essentially in equilibrium with 2-oxoglutarate and other citric acid cycle intermediates

$$\text{Glu} \rightarrow \text{2-oxoglutarate} \rightarrow \text{succinyl-CoA} \rightarrow \rightarrow$$
$$\rightarrow \text{malate} \rightarrow \text{pyruvate} \rightarrow \text{acetyl-CoA} \rightarrow CO_2 \quad \text{...(6.17)}$$

Many other amino acids are degraded in similar ways. In most cases the sequence is initiated by transamination to the corresponding 2-oxoacid. Beta oxidation and breakdown to such compounds as pyruvate and acetyl-CoA follows.

Catabolism initiated by decarboxylation. An alternative pathway for glutamate degradation is through the γ-aminobutyrate shunt. This pathway is initiated by a PLP-dependent decarboxylation rather than by a deamination or transamination. Since decarboxylases are known for most amino acids, there are usually alternative breakdown pathways, initiated by decarboxylation. In many cases these pathways lead to important products.

For example, γ-aminobutyrate functions in the brain as an important neurotransmitter. Dihydroxyphenylalanine is converted to noradrenaline and adrenaline, tryptophan to serotonin, and histidine to histamine. All of these are neurotransmitters and/or have other hormonal functions. A calmodulin-dependent glutamate decarboxylase occurs in higher plants, which accumulate γ-aminobutyrate in response to a variety of stresses. However, the significance of this accumulation is unclear.

Fermentation of glutamate. Special problems face anaerobic bacteria subsisting on amino acids. Their energy needs must be met by balanced fermentations. For example, glutamate may be converted to CO_2, ammonia, acetate", and butyrate" according to the reactions of Fig. 6.8. The end result is described by Eq. 6.18.

$$2\ \text{Glutamate}^- + 2\ H_2O + H^+ \rightarrow 2\ CO_2 + 2\ NH^{4+} + 2\ \text{acetate}^- + \text{butyrate}^-\ \Delta G'\ (\text{pH } 7) = -131\ \text{kJ}$$

The sequence begins with the γ-aminobutyrate shunt reactions but succinic semialdehyde is reduced to γ-hydroxybutyric acid using the NADH generated in the trans-deamination process of step *c*. With the aid of a CoA-transferase (step *d)* two molecules of the CoA ester of this hydroxy acid are formed at the expense of two molecules of acetyl-CoA. Use is then made of a β,γ elimination of water (step *e),* analogous to that involved in the formation of vaccenic acid

Isomerization (perhaps by the same enzyme that catalyzes elimination) forms crotonyl-CoA (step *f*). The latter undergoes dismutation, one-half being reduced to butyryl-CoA and one-half being hydrated and oxidized to acetoacetyl-CoA in the standard β-oxidation sequence. Acetoacetyl-CoA is cleaved to regenerate-the two molecules of acetyl-CoA. The organism can gain one molecule of ATP through cleayage of the butyryl-CoA. Perhaps a second can be gained by oxidative phosphorylation between the NADH produced in the formation of acetoacetyl-CoA and the reduction of crotonyl-CoA to butyryl-CoA.

The two processes take place at sufficiently different redox potentials to permit this kind of coupling. Another fermentation of glutamate is initiated by the vitamin B_{12}-dependent isomerization of glutamate to β-methylaspartate. This rearrangement of structure permits α,β elimination of ammonia (step *b*), a process not possible in the original glutamate. Hydration to *citramalate* (step *c)* and aldol cleavage yields acetate and pyruvate.

Acetate is one of the usual end products of the fermentation. The pyruvate can be cleaved to H_2, CO_2, and acetyl-CoA by the pyruvate-formate-lyase system and cleavage of the acetyl-CoA can provide ATP. Alternatively, two molecules of acetyl-CoA can be coupled and reduced to butyryl-CoA. The reducing power generated in the cleavage of pyruvate is used to reduce crotonyl-CoA rather than being released as H_2. Still other fermentation mechanisms are used by some clostridia to degrade glutamate.

SYNTHESIS AND CATABOLISM OF PROLINE, ORNITHINE, ARGININE, AND POLYAMINES

The 5-carbon skeleton of glutamic acid gives rise directly to those of proline, ornithing and arginine. The reactions are outlined in Fig. 6.9 Arginine , in turn, is involved in the urea cycle, which is shown in detail in Fig. 6.10. Arginine is also a biosynthetic precursor of the polyamines. Another important biosynthetic product of glutamate metabolism is δ-aminotevulinate, a precursor to porphyrins in some organisms.

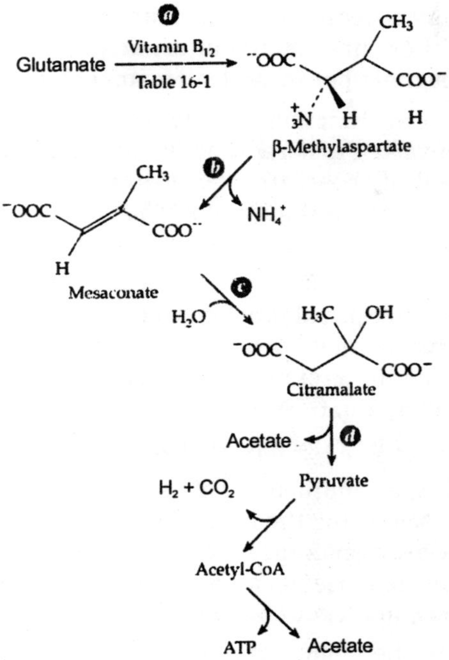

Glutamate → (a) Vitamin B_{12}, Table 16-1 → β-Methylaspartate → (b) NH_4^+ → Mesaconate → (c) H_2O → Citramalate → (d) Acetate + Pyruvate → $H_2 + CO_2$ → Acetyl-CoA → ATP, Acetate

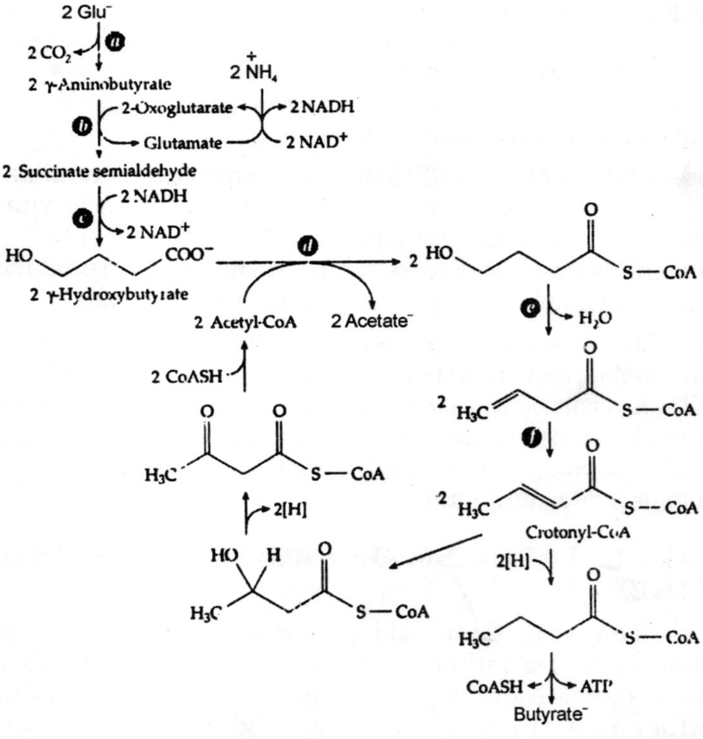

2 Glu⁻ → (a) 2 CO_2 → 2 γ-Aminobutyrate → (b) 2-Oxoglutarate → Glutamate; 2 $\overset{+}{N}H_4$, 2 NADH, 2 NAD^+ → 2 Succinate semialdehyde → (c) 2 NADH → 2 NAD^+ → 2 γ-Hydroxybutyrate → (d) 2 Acetyl-CoA → 2 Acetate⁻ → 2 HO–CH₂CH₂CH₂–CO–S—CoA → (e) H_2O → 2 H_3C=…–CO–S—CoA → (f) 2 H_3C–CH=CH–CO–S—CoA (Crotonyl-CoA) → 2[H] → H_3C–CH₂CH₂–CO–S—CoA → CoASH, ATP → Butyrate⁻

Crotonyl-CoA → H_3C–CH(OH)–CH₂–CO–S—CoA → 2[H] → H_3C–CO–CH₂–CO–S—CoA → 2 CoASH → 2 Acetyl-CoA

Fig. 6.8. Fermentation of glutamate by Clostridium ammobutylicum.

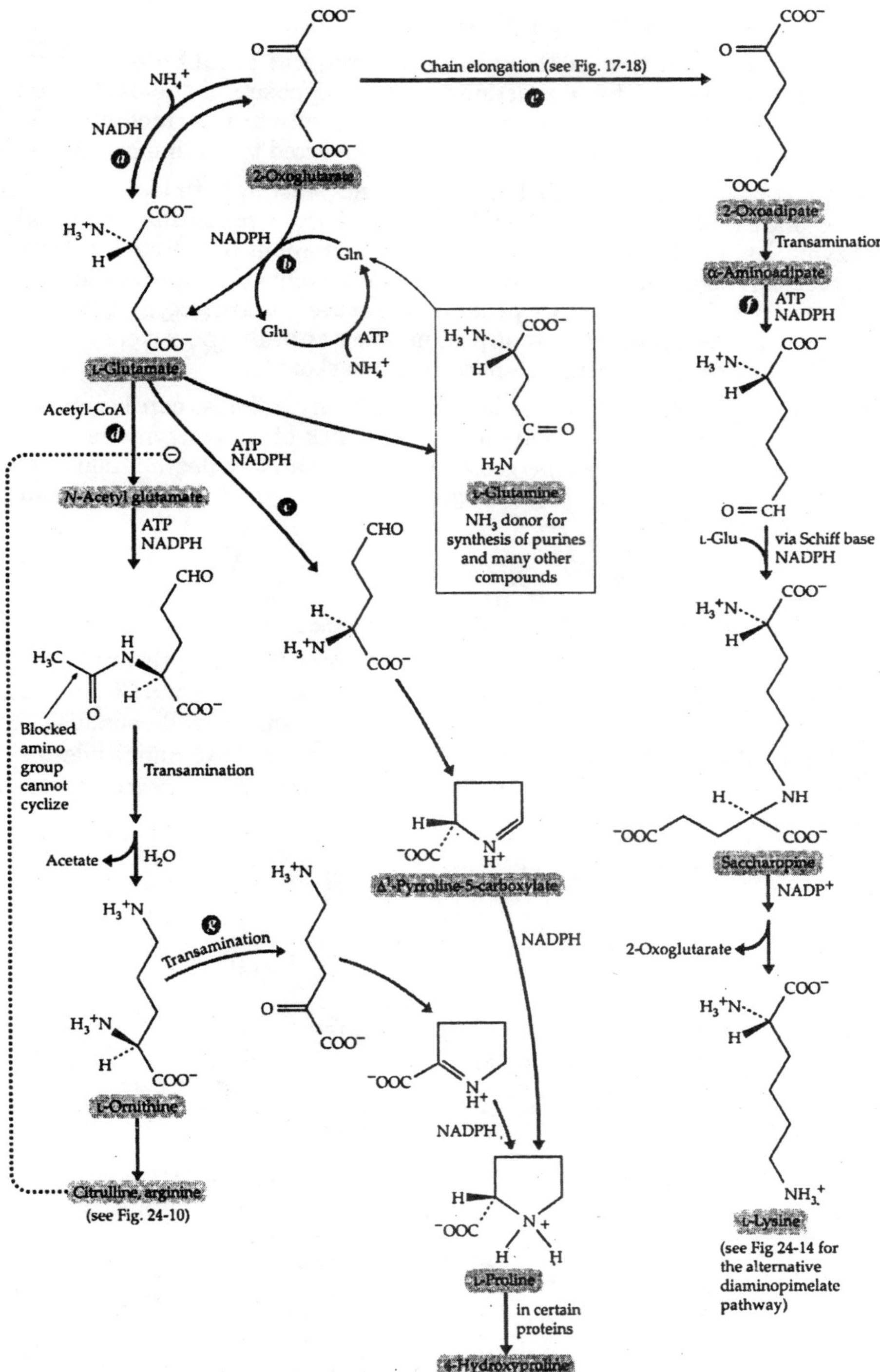

Fig. 6.9. Biosynthesis of glutamate, glutamine, praline, and lysine from 2-oxoglutarate.

Synthesis and Catabolism of Proline

The ATP-dependent reduction of the γ-carboxyl group of glutamate to an aldehyde by NADPH is of a standard biosynthetic reaction type, the opposite of the oxidation reaction of Fig. Like the latter it is thought to occur via an acyl phosphate intermediate. The oxidation product, *glutamate semialdehyde*, cyclizes and can be converted to proline by further reduction.

The pathway has been well established in bacteria and yeast by both biochemical and genetic experiments. In plants both the initial reduction and the cyclization are catalyzed by a bifunctional enzyme. An alternative pathway important in animals is initiated by transamination of ornithine to the corresponding 2-oxoacid, spontaneous cyclization, and reduction to proline. Selected prolines in collagen and in plant glycoproteins are oxygenated to form 4-hydroxyproline. One route of catabolism of proline is essentially the reverse of its formation from glutamate. *Proline oxidase,* yields Δ^1–pyrroline 5-carboxylate.

The corresponding open-chain aldehyde, formed by hydrolysis, can be oxidized back to glutamate by pyrroline 5-carboxylate dehydrogenase. Lack of this enzyme is associated with the human genetic deficiency causing *hyperprolinemia.* Alternatively, degradation can be initiated by oxidation on the other side of the ring nitrogen to form Δ^1pyrroline 2-carboxylate.

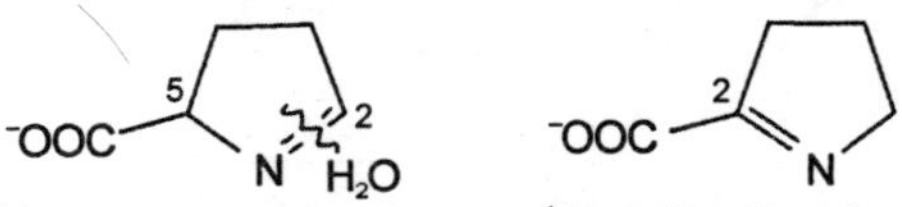

$^-$OOC 5 2 N H_2O $^-$OOC 2 N

Δ^1-Pyrroline 5-carboxylate Δ^1-Pyrroline 2-carboxylate

The metabolic fate of this compound is uncertain. A Corresponding pathway for break down of 4-hydroxy-L-proline of collagen yields glyoxylate and pyruvate or malate and CO_2. Oxidation on the other side of the ring nitrogen of hydroxyproline is utilized by some pseudomonads to convert the amino acid into 2-oxoglutarate. Anaerobic bacteria may reduce proline to 5-aminovalerate and couple this reaction to the oxidative degradation of another amino acid (Stickland reaction).

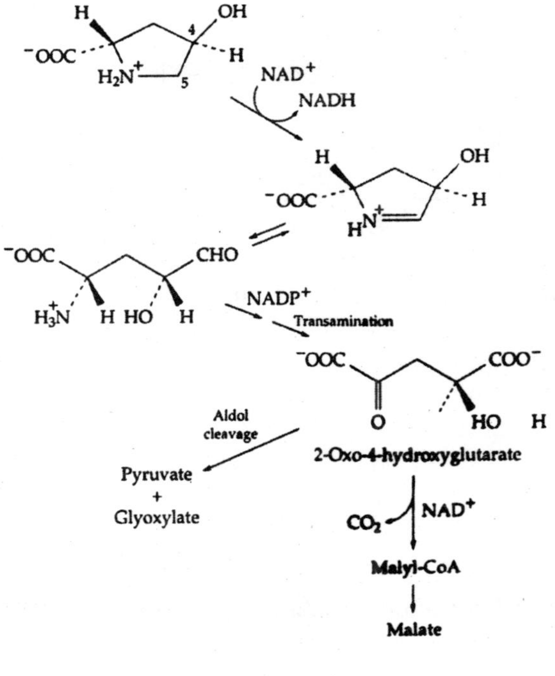

H OH $^-$OOC 4 H H_2N^+ 5

NAD^+ NADH

H OH $^-$OOC H HN^+

$^-$OOC CHO H_3N^+ H HO H

$NADP^+$ Transamination

$^-$OOC COO^- O HO H

2-Oxo-4-hydroxyglutarate

Aldol cleavage

Pyruvate + Glyoxylate

CO_2 NAD^+

Malyl-CoA

Malate

Synthesis of Arginine and Ornithine and the Urea Cycle

If the amino group of glutamate is blocked by acetylation prior to the reduction to the semialdehyde cyclization is prevented. The γ-aldehyde group can be transaminated to an amino group and the acetyl blocking group removed to form *ornithine.* Ornithine is not usually a constituent of proteins, but it is sometimes formed by hydrolytic modification of arginine at specific sites in a protein. A 67-kDa urate-binding glycoprotein of plasma is reported to contain 43 residues of ornithine. It is postulated that a special arginase is needed to form these residues, and that it may be lacking in some cases of gout in which the urate-binding capacity of blood is impaired.

Ornithine appears to be present in specific sites in a few other proteins as well. Neurospora grown in a minimal medium accumulates large amounts of both ornithine and arginine, over 98% of which is sequestered in vesicles within the cytoplasm. This appears to.be a way of accumulating a store of arginine that is protected from the active catabolism of that amino acid by the fungus. However, accumulation of ornithine in the human body, as a result of lack of *ornithine aminotransferase* causes gyrate atrophy of the choroid and retina, a disease that results in tunnel vision and blindness. A major interest in arginine metabolism asises from its role in formation of urea in the human body. Study of arginine biosynthesis in bacteria has also been important in developing our understanding of regulation of gene expression.

The urea cycle. In 1932, Krebs and Henseleit proposed that urea is formed in the liver by a cyclic process in which ornithine is converted first to *citrulline,* and then to arginine. The hydrolytic cleavage of arginine produces the urea and regenerates ornithine. Subsequent experiments fully confirmed this proposal. Urea is the principal nitrogenous end product of metabolism in mammals and many other organisms, but the urea cycle reactions have other functions. As with the citric acid cycle, products other than urea can be withdrawn in any needed quantity.

Most notably, the reactions of Fig. 6.10 provides for the biosynthesis of arginine in all organisms. Also of physiological importance is the fact that the urea cycle involves both mitochondria and cytosolic enzymes. Let us trace the entire route of nitrogen removed by the liver from excess amino acids. Transaminases transfer nitrogen to 2-oxoglutarate to form glutamate. Since urea contains two nitrogen atoms, two molecules of glutamate must donate their amino groups. One molecule is deaminated directly by glutamate dehydrogenase to form ammonia. This ammonia is combined with bicarbonate to form carbamoyl phosphate, which transfers its carbamoyl group onto ornithine to form citrulline. The second molecule of glutamate transfers its nitrogen by transamination to oxaloacetate (reaction *e)* to form aspartate.

The aspartate molecule is incorporated intact into *argininosuccinate* by reaction with citrulline. Undergoing a simple elimination reaction, the 4-carbon chain of argininosuccinate is converted to fumarate (step g) with arginine appearing as the elimination product. Finally, the hydrolysis of arginine (step *h)* yields urea and regenerates ornithine.

Carbamoyl phosphate synthetases. The first of the individual steps in the urea cycle is the formation of carbamoyl phosphate. Carbon dioxide and ammonia equilibrate spontaneously

$$CO_2 + NH_4^+ \rightleftharpoons H_2N\text{—}C(=O)\text{—}OH + H^+ \quad ...(6.21)$$

Carbamic acid

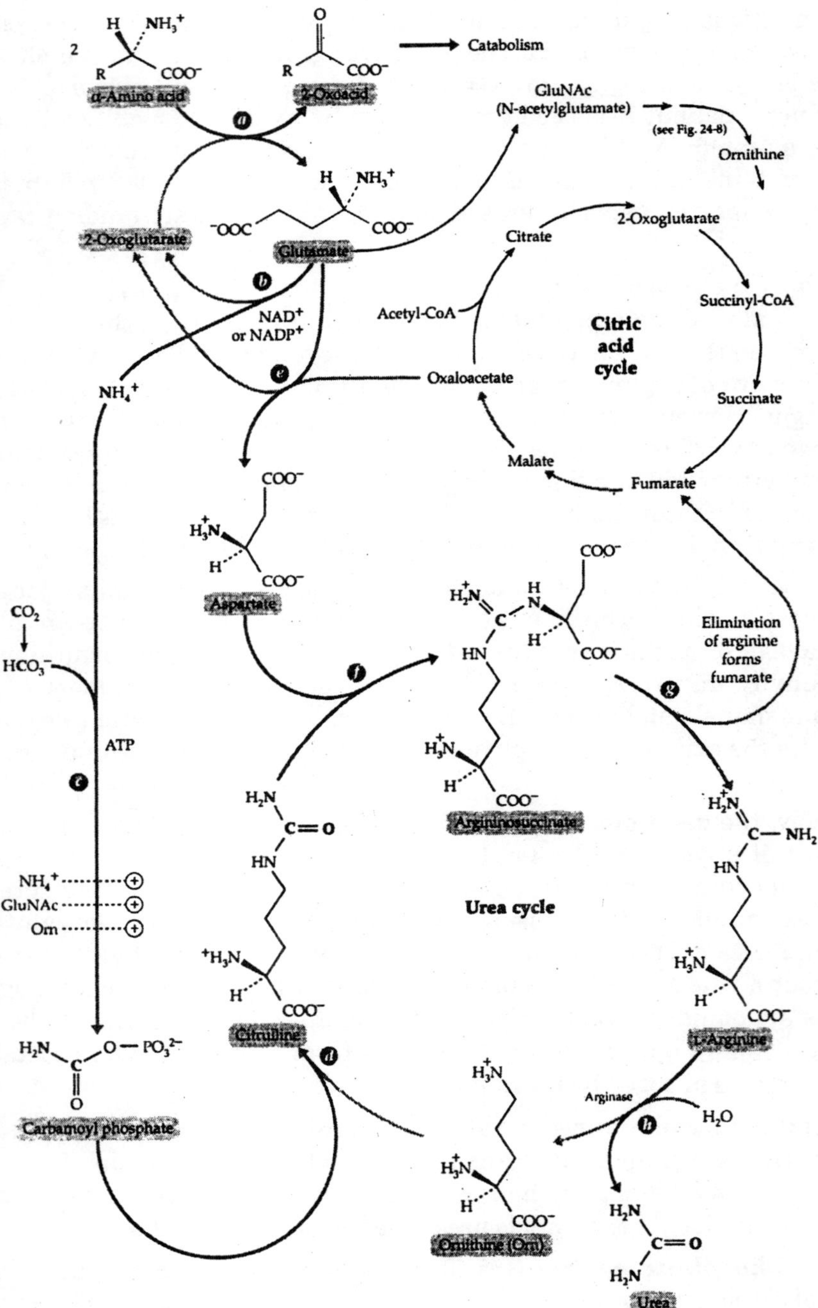

Fig. 6.10. Biosynthesis of citrulline, arginine, and urea. The green arrows indicate reactions directly involved in deamination of amino acids and the synthesis of urea. N from amino acids and C from CO_2 are traced in green.

with carbamic acid: Some bacteria have a kinase able to convert carbans ate into carbamoyl phosphate starting with step *a* of Eq. 6.22 However, the equilibrium constant is low (0.04 at pH 9,10°C), and it is now believed that carbamate kinase functions in the opposite direction, providing a means of synthesis of ATP for bacteria degrading arginine.

The biosynthetic carbamoyl phosphate synthases harness the cleavage of *two* molecules of ATP to formation of one molecule of carbamoyl phosphate. In bacteria such as *E. coli,* a single synthase provides carbamoyl phosphate for biosynthesis of both arginine and pyrimidines. However, fungi and higher animals have at least two carbamoyl-P synthases. Synthase I provides substrate for formation of citrulline from ornithine, while carbamoyl-P synthase II, which is part of a larger multifunctional protein, functions in pyrimidine synthesis. Synthase I is found in mitochondria and synthase II in the cytoplasm. Mammalian carbamoyl phosphate sythase I consists of a single 160-kDa peptide.

A powerful allosteric effector for the liver synthase is *N-acetyl-glutamate*, a precursor of ornithine. The enzyme from certain marine elasmobranchs, such as the spiny dogfish *Squalus acanthias,* have carbamoyl-P synthase III, an enzyme with somewhat different molecular properties. It probably functions in synthesis of urea, which is used by these animals to regulate osmotic pressure. Synthase I utilizes only free NH_3. The others are amidotransferases and prefer glutamine as the ammonia donor. Carbamoyl-P synthase from *E. coli* consists of two subunits (~42 and 118 kDa, respectively) and can utilize *either* free ammonia or glutamine. The light subunit has *glutaminase* activity; i.e., it is able to hydrolyze glutamine to ammonia. All of these synthatases presumably act by first phosphorylating bicarbonate to an enzyme-bound carboxyl phosphate, which can then undergo a displacement of phosphate by NH_3 to give enzyme bound carbamate. Phosphorylation of the latter by ATP completes the reaction. In the single-chain enzymes the amidotransferase domain is at the N terminus. Crystallographic study of a mutant form of the *E. coli* enzyme unable to act rapidly on glutamine showed that the latter released its ammonia to form a thioester with cysteine 269, suggesting a mechanism resembling that of serine proteases or papain for the glutaminase action.

$$^{-}O{-}C({=}O)OH \xrightarrow[a]{ATP \rightarrow ADP} {}^{2-}O_3P{-}O{-}C({=}O)O^{-} \text{ (Carboxyl phosphate)}$$

$$Gln + H_2O \rightarrow Glu + NH_3$$

$$\xrightarrow[b]{NH_3,\ -P_i} H_2N{-}C({=}O)O^{-} \xrightarrow[c]{ATP \rightarrow ADP} H_2N{-}C({=}O){-}O{-}PO_3^{2-} \text{ (Carbamoyl phosphate)} \qquad ...(6.22)$$

The X-ray crystallography also showed that the released NH_3 must travel 4.5 nm through the interior of the protein to the site of carbamate formation. The carbamate must travel Fig.

another ~ 4.5 nm to the site from which carbamoyl phosphate is released. The C-terminal regions of the synthases undergo allosteric modification by a number of effectors.163-173c Both ornithine and IMP are activators for the E. *coli* enzyme, whereas UMP, a pyrimidine end product, exerts feedback inhibition. Phosphoribosyl pyrophosphate activates synthase II, and N-acetylglutamate activates the mammalian liver synthase I by binding near the C terminus.

Citrulline and argininosuccinate. One NH_3 and one HCO_3^- for urea formation are provided by the carbamoyl group, which is transferred from carbamoyl-P to ornithine to form citrulline. The second nitrogen atom is transferred from glutamate to asparatate into argininosuccinate. The equilibrium constant for ornithine transcarbamoylase (reaction *d)* is very high so that ornithine is completely converted to citrulline. The trimeric human enzyme is a trimer of 36-kDa subunits whose structural gene is on the X chromosome.

Like many other mitochondrial matrix enzymes it is synthesized as a larger (40 kDa) precursor, which enters the mitochondria in an energy-dependent process. A genetic defect in this sex-linked gene is often lethal to boys, and even girls, heterozygous for the defect, sometimes have serious problems with accumulation of ammonia in the brain. The conversion of citrulline to argininosuccinate and the subsequent breakdown to fumarate and arginine take place in the cytosol. The ureido group of citrulline is activated by ATP for the argininosuccinate synthase reaction. Thus, ^{18}O present in this group is transferred into AMP. A citrulline adenylate intermediate (center) is likely. *Argininosuccinate lyase* catalyzes the elimination of arginine with formation of fumarate.

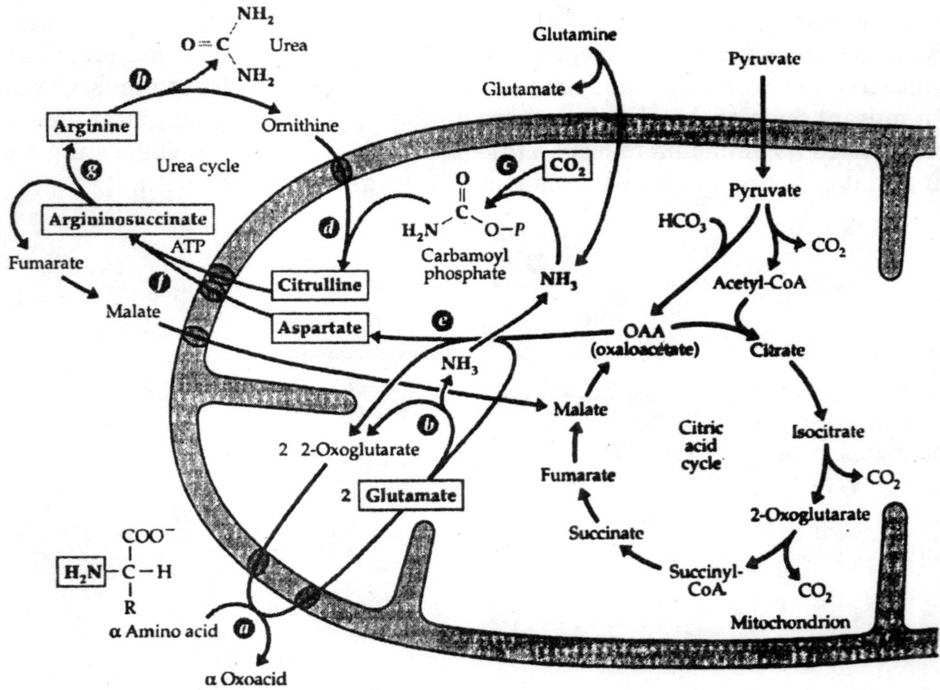

Fig. 6.11. Integration of the urea cycle with mitochondrial metabolism. Green lines trace the flow of nitrogen into urea upon deamination of amino acids or upon removal of nitrogen from the side chain of glutamine.

It is entirely analogous to the bacterial aspartase that eliminates ammonia from aspartate to form fumarate. Like the latter enzyme and fumarate hydratase argininosuccinase promotes a trans elimination. The fumarate produced can be reconverted through reactions of the citric acid cycle to oxaloacetate, which can be reaminated to aspartate. Aspartate is used to introduce amino groups in an entirely similar way in other metabolic sequences such as in the formation of adenylic acid cleavage of arginine to ornithine and urea by the Mn^{2+}-containing *arginase* converts the biosynthetic route to arginine into a cycle for the synthesis of urea. This cyclic pathway is unique to organisms that excrete nitrogenous wastes as urea, but the biosynthetic path is nearly ubiquitous.

Human adults excrete approximately 20 g of urea nitrogen per day. If this rate decreases, ammonia accumulates in the blood to toxic levels. Normally, plasma contains 0.03 mM ammonia, and only 2–3 times this level is required to produce toxic symptoms. Therefore, it is not surprising that five different well-documented hereditary enzyme deficiencies affect the urea cycle. One of the most common, *argininosuccinic aciduria,* is a deficiency of the breakdown of argininosuccinic acid. Both lethal and nonlethal variants of this disease are known.

Human argininosuccinate lyase consists of two subunits. Defects may occur in either subunit but considerable genetic heterogeneity exists and intragenic complementation between the two subunits accounts for many of the nonlethal forms of the disorder: A common feature of all of the hereditary defects of the urea cycle is an intolerance to high protein intake and mental symptoms. Toxic accumulation of ammonia in blood is often seen also in *alcoholic liver cirrhosis* as a result of a decreased capacity of the liver for synthesis of urea. For some urea cycle defects a combination of a low-protein diet together with an arginine supplement prevents the ammonia intoxication while allowing normal growth. In other cases it is necessary to replace the natural dietary protein with a mixture of essential amino acids or with the corresponding 2-oxoacids, which can be converted to amino acids in the body with utilization of endogenous ammonia. A specific treatment for lack of *N*-acetylglutamate synthetase, which forms the carbamoyl phosphate synthase activator *N*-acetylglutamate, is administration of the analog *N*-carbamoylglutamate. This also activates carbamoyl phosphate synthase and is not cleaved by acylases that would prevent the natural activator from being supplied artifically via the blood. Although the primary function of the urea cycle is usually regarded as the removal of NH_4^+ from the body, it also removes HCO_3^- in equal amounts. This is essential for maintenance of neutral pH,

$$2\,HCO_3^- + 2\,NH_4^+ \longrightarrow H_2NCONH_2 + CO_2 + 3H_2O \qquad ...(6.24)$$

and Atkinson and Bourke suggested that removal of HCO_3~ is as important a function of the cycle as removal of NH_4^+. However, there are strong arguments against this concept.

Excretion of ammonia. While mammals excrete urea many invertebrate organisms that live in water as well as some fishes simply excrete NH_3. Other organisms hydrolyze urea to NH_3. Even green plants recycle nitrogen via urea and the Ni^{2+}-dependent urease. Two compounds that can be hydrolyzed by cells to urea and glyoxylate are allantoin and allantoic acid. If cells of *Saccharomyces cerevisiae* are grown on either of these compounds as a sole source of nitrogen, they make a biotin-dependent *urea carboxylase*. This enzyme facilitates the hydrolysis of urea by conversion to the more easily degraded allophanate.

Catabolism of arginine. Arginine can also be converted back to glutamate and 2-oxoglutarate. The initial step is removal of the guanidino group to form ornithine. This occurs in the urea cycle and also in many bacteria by the action of arginase. A parallel pathway involving conversion of arginie to N[1]-succinylarginine, then on to succinyl-glutamate, and to free glutamate and succinate is used by some pseudomonads. The alternative *arginine dihydrolase* pathway, used by some bacteria and a few protozoa such as *Giardia,* is initiated by a different hydrolase that cleaves arginine to citrulline and ammonia.

Phosphorolysis of citrulline yields carbamoyl phosphate whose breakdown to CO_2 and ammonia (catalyzed by carbamate kinase can be utilized for generation of ATP by microorganisms that subsist on arginine. Degradation of L-arginine by *Streptomyces griseus* is initiated by a hydroxylase that causes decarboxylation and conversion of the amino acid into an amide, a reaction analogous to that catalyzed by the flavin-dependent lysine oxygenase. The product formed from arginine is γ-guanidinobutyramide, which is further degraded by the hydrolysis of the amide gioup and cleavage of the guanidino group to form urea and γ-aminobutyrate. *Pseudomonas putida* initiates degradation of arginine by decarboxylation to the corresponding 2-oxoacid and oxidative decarboxylation with a thiamin diphosphate-requiring enzyme to γ-guanidiriobutyraldehyde.

Dehydrogenation and hydrolysis lead, again, to γ-aminobutyrate. Specific arginine residues in proteins are methylated on their guanidino groups to give monomethylated and both sym metrically and asymmetrically dimethylated derivatives. These methylated arginines also occur free in various mammalian tissues, where they may serve as endogenous regulators of nitric oxide synthases. A Zn^{2+}-containing dimethylarginase hydrolyzes the monomethyl and dimethyl arginines to cirullene and monomethyl or dimethyl amines.

Insecticidai analogs of arginine. The toxic amino acid *L-canavanine* is synthesized by more than 1500 species-of legumes including alfalfa and cloyer. It is structurally similar to arginine, the 5-CH_2 group being replaced by O. However, the guanidino group is much less basic than in arginine. Canavanine is a natural insecticide, which in some plants accumulates to a level of 13% of the total dry matter. Plants that store canavanine hydrolyze it to *canaline* and urea, which they use as a nitrogen source. Canaline is a toxic derivative of hydroxylamine and forms oximes with 2-oxoglutarate, other oxoacids, and PLP-cohtaimng enzymes.

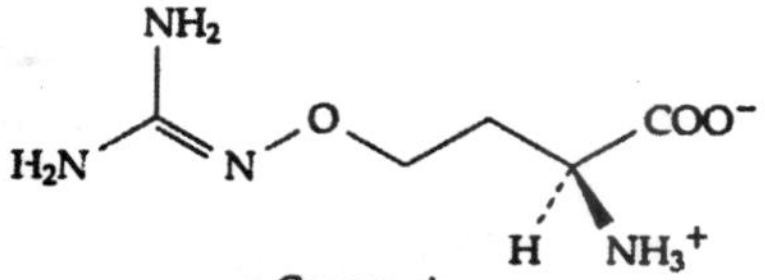

L-Canavanine

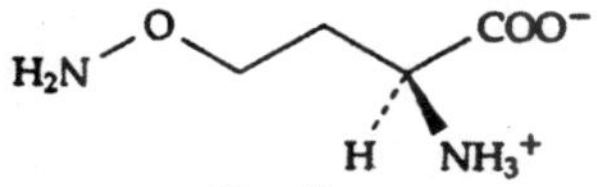

L-Canaline

Although canavanine and canaline are effective insecticides, some beetles are adapted to these compounds to the extent that they feed exclusively on canavanine-containing seeds. The tobacco budworm is likewise resistant to these toxins and produces a *canavanine hydrolase* that converts canavanine to *i*-homoserine, a normal intermediate in ths threonine, isoleucine, methionine biosynthetic pathway, and *hydroxyguanidine.* The latter undergoes NADH-dependent reduction to guanidine which can be catabolized.

Amidino Transfer and Creatine Synthesis

The terminal amidino group of arginine is transferred intact to a number of other substances in simple displacement reactions. An example is the formation of *guanidinoacetic acid.* The amidino group appears to be transferred first to the SH group of cysteine 407 then to glycine in a double displacement mechanism. Transmethylation from S-adenosylmethionine converts guanidinoacetic acid to *creatine,* a compound of special importance in muscle. Creatine kinase reversibly transfers the phospho group of ATP to creatine to form the N-phosphate. *Creatine phosphate.* and in some invertebrates phosphoarginine, serves as an important "*energy buffer*" for muscular contraction.

Through the reversible action of creatine kinase it is able rapidly to transfer its phospho group back onto ADP as fast as the latter is formed during the hydrolysis of ATP in the contraction process. An end product of creatine phosphate metabolism is the anhydride creatinine formed from creatine phosphate as is indicated in step *e* as well as directly from creatine. The urinary creatinine excretion for a given individual is extremely constant from day to day, the amount excreted apparently being directly related to the muscle mass of the person. Another example of the transfer of amidino groups from arginine is found in the synthesis of streptomycin.

A cyclic analog of creatine, *cyclocreatine,* when fed to animals, accumulates in large amounts in muscle, heart, and brain and is a long-acting phosphagen.

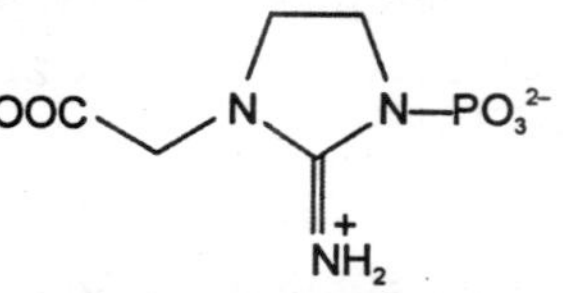

Cyclocreatine phosphate

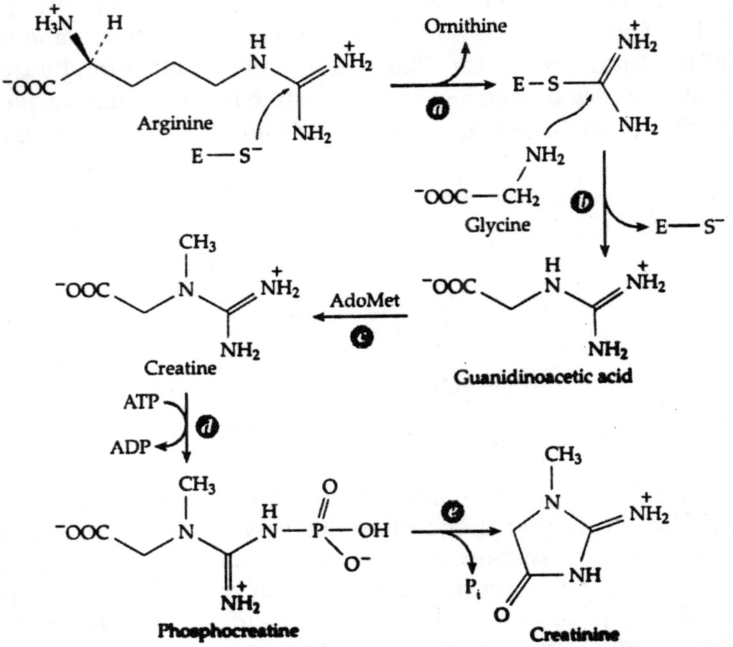

The Polyamines

A series of related polyamino compounds, which are derived in part from arginine, are present in all cells in relatively high, often millimolar, concentrations. The content of polyamines in cells tends to be stoichiometric with that of RNA, and the polyamines are concentrated in the ribosomes and also in the nucleus. Two moles of polyamine are usually present per mole of any isolated tRNA. The first satisfactory crystals of tRNA for X-ray structure determination were obtained in the presence of spermine. Spermidine is associated with RNA in the turnip yellow mosaic virus.

The T-even bacteriophage and most bacteria contain polyamines in association with DNA. Polyamines are able to interact with double helical nucleic acids by bridging between strands, the positively charged amino groups interacting with the phosphates of the nucleic acid backbones. Tsuboi suggested that the tetramethylene portion of the polyamine lies in the minor groove bridging three base pairs, and the trimethylene portions (one in spermidine, two in spermine) bridge adjacent phosphate groups in one strand.

Polyamines may also stabilize supercoiled or folded DNA. The structures of polyamines are shown here as di-and tri-cations, but it should be realized that there are multiple positions for protonation and therefore various tautomers. Also, polyamines show extreme anticooperativity in proton binding, i.e., successive pKa values range from very low to very high for the last proton to leave. Polyamines are thought to have several functions. They can substitute to some extent for cellular K+ and Mg2+, and they may play essential controlling roles in nucleic acid and protein synthesis.

A specific role of spermidine in cell division seems likely. An absolute requirement for polyamines has been demonstrated for some bacteria such as *Hemophilus parainfluenzae* and for mutants of *Asper-gillus* and *Neurospora*. Polyamines are also essential for mammalian cells. Polyamines activate some enzymes including the serine/threonine protein kinase CK2. Mutants of *E. coli* have been constructed in which enzymes of all known bio-synthetic pathways for polyamines are blocked by deletion of the genes for arginine decarboxylase *(SpeA)*, agmatine ure ahydrolase *(SpeB)*, ornithine decarboxylase *(SpeC)*, and adenosylmethionine decarboxylase (*SpeD*).

Even though polyamines cannot be detected in these cells they grow at one third the normal rate. However, yeast cells require both putrescine and spermidine or spermine for growth. Another effect is seen in strains of yeast carrying the *"killer plasmid"* a 1500-kDa double-stranded RNA plasmid that encodes a toxic protein, which is secreted and kills other susceptible strains of yeast. Yeast cells carrying the killer plasmid lose it when made deficient in polyamines. The bacterial outer membrane porins OmF and OmC bind polyamines, especially spermine, and inhibit passage of ions. Polyamines may also modulate ion channels of heart, muscle, and neurons. Both prokaryotic and eukaryotic cells have transporters that allow uptake of polyamines from their surroundings.

Putrescine (1,4-diaminobutane)

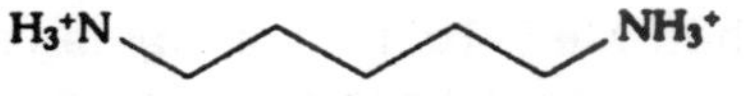

Cadaverine

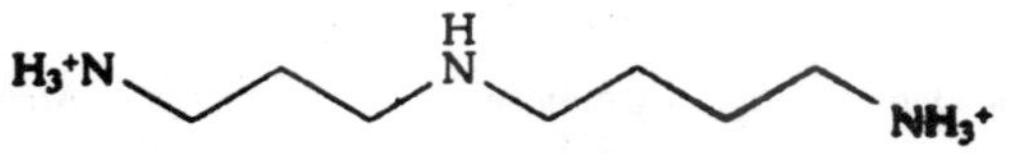

Spermidine

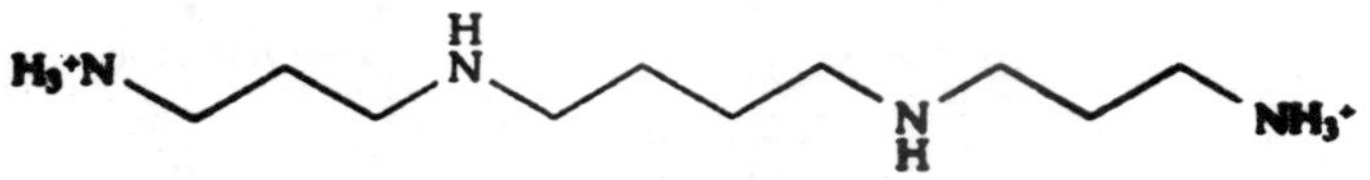

Spermine

Major naturally occurring polyamines (as dications)

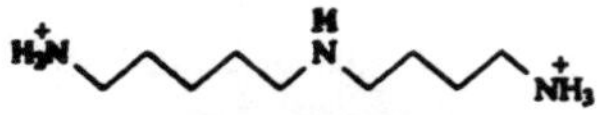

4-Aminobutylcadaverine

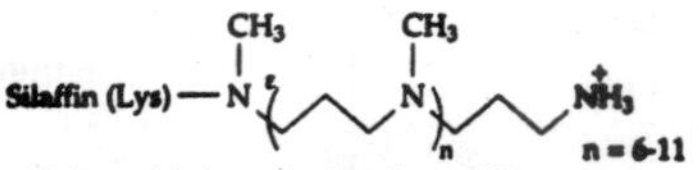

Polypeptide-bound polyamines of diatoms

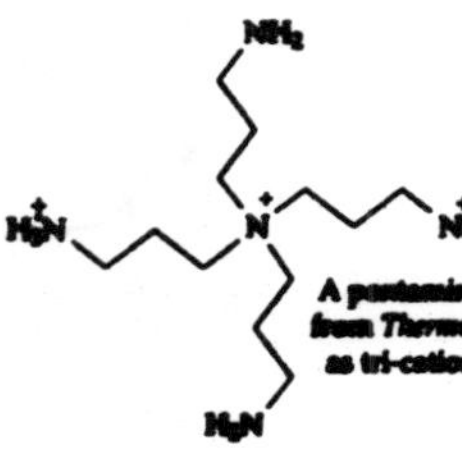

A pentamine from *Thermus*, as tri-cation

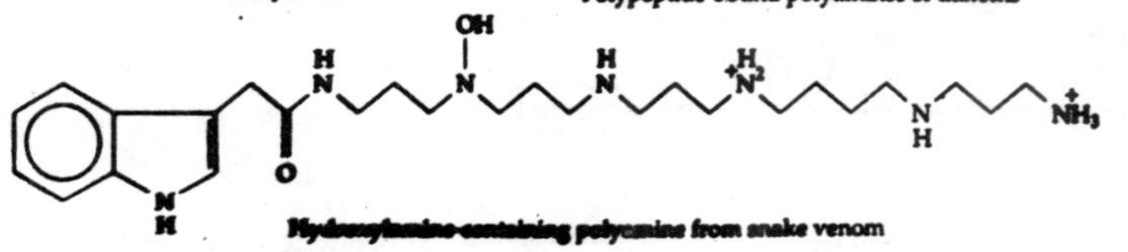

Hydroxylamine-containing polyamine from snake venom

Biosynthesis. The 4-carbon putrescine arises most directly by decarboxylation of ornithine but it can also be formed by decarboxylation of arginine to agmatine followed by hydrolysis of the latter. An alternative pathway utilizes an *"agmatine cycle"* in which agmatine is first hydrolyzed to ammonium ions and *N*-carbamoylputrescine. The latter transfers its ureido group to ornithine to form citrulline and releases free putrescine. The citrulline is reconverted to arginine. This pathway appears to be important in plants.

Putrescine is normally present in all cells, and all cells are able to convert it on to spermidine. This is accomplished by decarboxylation of S-adenosylmethionine and transfer of the propylamine group from the resulting decarboxylation product onto an amino group of putrescine. The more complex spermine is found only in eukaryotes. It is formed by transfer of a second propylamine group onto spermidine. A historical note is that Anthony von Leeuwenhoek with one of his first microscopes observed crystals of the phosphate salt of spermine in human semen in 1678. The 5-carbon diamine cadaverine arises from decarboxylation of lysine. The extremely thermophilic bacterium *Thermus thermophilis* produces several additional polyamines including a pentamine a quaternary nitrogen compound. Many other polyamines are known.

Among these are a 4-aminobutylcadaverine isolated from root nodules of the adzuki bean and very long partially aromatic hydroxylamine derivatives from venom of common funnel-web spiders (structures at top of page). Cationic polypeptides called *silaffins,* with masses of ~3 kDa, apparently initiate the growth of the silica cell walls of diatoms. These peptides contain polyamines consisting of 6 to 11 repeated N-methylpropylamine units covalently attached to lysine residues and also many phosphoserines. The synthesis of polyamines is tightly regulated. The PLP-dependent ornithine decarboxylase is present in very low concentrations and apparently has the shortest half-life (~10 min) of any mammalian enzyme. Its concentration increases rapidly in most species with the onset of rapid growth, transformation to a neoplastic state, or initiation of cell differentiation.

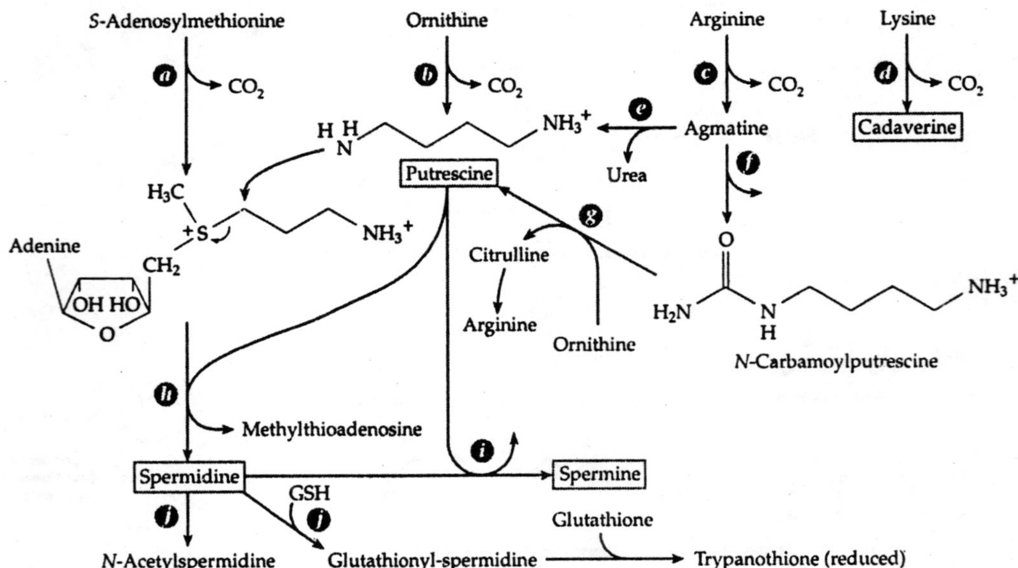

Fig. 6.12. Pathways of biosynthesis of polyamines. Also shown is the formation by trypanosomes of trypantohione.

The rate of synthesis of the enzyme appears to be regulated by feedback repression by spermidine and by inactivation in response to a buildup of putrescine. One mechanism of inactivation is the synthesis of a 26-kDa specific inhibitor called an *antizyme* in response to the presence of putrescine, spermidine, or spermine. The antizyme is ubiquitous in both prokaryotes and eukaryotes and keeps most of the ornithine decarboxylase bound and inactive and also promotes its degradation by 26S proteosomes. A polyamine-dependent protein kinase in *Physarum* phosphorylates the decarboxylase thereby inhibiting its activity.

Breakdown. The catabolism of polyamines is less well understood than is their biosynthesis. Oxidative cleavages of spermine to spermidine and of the latter to 1,4-diaminobutane appear to occur in the animal body, and a substantial amount of this diamine is excreted in the urine. Spermidine is acetylated on N^1 by acetyl-CoA and a spermidine N-acetyltransferase. The resulting N^1-acetylspermidine is more readily cleaved by hepatic polyamine oxidase than is free spermine; again 1,4-diaminobutane is reformed together with an N-acetylaminopropionaldehyde. This and other aldehydes formed from polyamines are very toxic but they may play essential roles in regulation of metabolism.

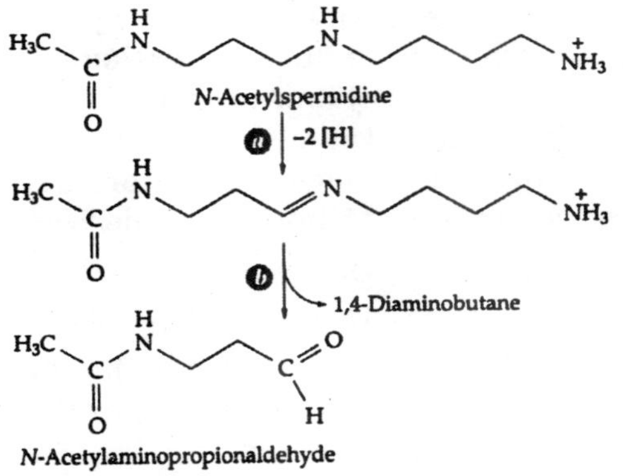

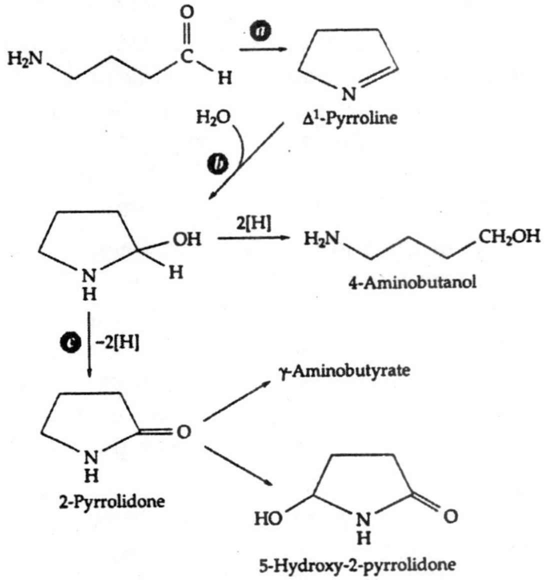

N-Acetylspermidine

(a) −2 [H]

(b) → 1,4-Diaminobutane

N-Acetylaminopropionaldehyde

(a)

Δ^1-Pyrroline

H_2O (b)

2[H]

4-Aminobutanol

(c) −2[H]

γ-Aminobutyrate

2-Pyrrolidone

5-Hydroxy-2-pyrrolidone

Transamination of 1,4-diaminobutane yields γ-aminobutyraldehyde which cyclizes. Diamine oxidases of animal tissues oxidize 1,4-diaminobutane with formation of the same products. Further metabolism of Δ^1-pyrroline yields γ-aminobutyrate, which can undergo transamination and oxidative metabolism. Other products are also indicated (in Eq. 6.29 Metabolism of other polyamines also begins by oxidation at the primary amino termini. Formation of β-alanine, needed for synthesis of pantothenic acid, can also occur by oxidation of spermine. When *E. coli* cells enter the stationary phase of the growth curve, most of the spermidine is converted to *glutathionylspermidine* (γ-glutamyl-cysteinylglycylspermidine) in which glutathione and spermidine are joined by an amide linkage.

Trypanosomes join a second glutathione at the other end of the spermidine to form reduced *tryptathione* N^1-γ-Glutamylspermidine and related compounds have been found in proteolytic digests of certain proteins, suggesting that polyamines may be physiological substrates for transglutaminases. Portions of polyamines are incorporated into a variety of products including *nicotine* and the unusual amino acid *hypusine*.

Ornithine decarboxylase is speifically inhibited by the enzyme-activated inhibitor ∞-difluoromethyl-ornithine, which can cure human infection with *Trypanosoma brucei* (African sleeping sickness) by interfering with polyamine synthesis. In combination with inhibitors of spermidine synthase or S-adenosylmethionine decarboxylase, it can reduce polyamine levels and growth rates of cells. Another powerful inhibitor that acts on both ornithine and adenosylmethionine decarboxylases is the hydroxylamine derivative l-aminooxy-3-aminopropane.

$$H_2N—O—CH_2CH_2CH_2—NH_2$$

l-Aminooxy-3-aminopropane

Like difluoromethylornithine the compound at low concentrations is not toxic to cells but inhibits growth. It is hoped that adequate inhibition of growth of normal cells may allow more aggressive chemotherapeutic treatment of cancer.

COMPOUNDS DERIVED FROM ASPARTATE

The 4-carbon aspartate molecule is the starting point for synthesis of *pyrimidines* and of the amino acids *lysine*, *methionine*, *threonine isoleucine* and *asparagine*. There are several branch points, and aspartate can be converted directly to asparagine, to carbamoylaspartate (the precursor of pyrimidines), or to β-aspartyl phosphate and aspartate semialdehyde. The latter can be converted in one pathway to lysine and in another to homoserine. Homoserine can yield either homocysteine and methionine or threonine.

Although threonine is one of the end products and a constituent of proteins, it can also be converted further to 2-oxobutyrate, a precursor of isoleucine: Most of the chemistry has been considered already. The reduction of aspartate via β-aspartyl phosphate and aspartate β-semialdehyde is a standard one. Conversion to methionine can occur in two ways. In *E. coli* homoserine is succinylated with succinyl-CoA. The γ-succinyl group is then replaced by the cysteine molecule in a PLP-dependent γ-replacement reaction. The product *cystathionine* undergoes elimination to form homocysteine. A similar pathway via O-phospho-homoserine occurs in chloroplasts of green plants.

A more direct γ replacement of the hydroxyl of homocysteine or O-phosphohomoserine by a sulfide ion has also been reported for both *Neurospora* and green plants. Methylation of homocysteine to methionine has been considered previously, as has the conversion of homoserine to threonine by homoserine kinase and the PLP-dependent *threonine synthase.* A standard PLP-requiring β elimination converts threonine to *2-oxobutyrate,* a precursor to isoleucine. Formation *asparagine* has been discussed in section B.

Asparagine synthase of *E. coll* cleaves ATP to AMP and PP_i rather than to ADP via an aspartyladenylate intermediate. In higher animals glutamine serves as the ammonia donor for synthesis of asparagine, but NH_4^+ can also function. *L-Asparaginase,* a bacterial hydrolase, is an experimental antileukemic drug. It acts to deprive fastgrowing tumor cells of the exogenous asparagine needed for rapid growth. Tissues with a low asparagine synthase activity are also damaged, limiting the clinical usefulness. Aspartate can be decarboxylated either to α-alanine by a PLP-dependent enzyme or to β-alanine by a pyruvoyl group-containing enzyme

Beta-alanine is not only a component of the vitamin pantothenic acid but is found in the dipeptides carno-sine (β-alanylhistidine) and anserine (β-alanyl-$N^δ$-methylhistidine) present in vertebrate muscles. It is a crosslinking agent in insect cuticle. Aspartate can be deaminated to fumarate by bacterial *L-aspartate oxidase."* This flavoprotein is structurally and mechanistically related to succinate dehydrogenase and can function as a soluble fumarate reductase. However, its main function appears to be to permit the intermediate iminoaspartate to react with dihydroxyacetone-*P* to form quinolinate, which can be converted to NAD.

Control of Biosynthetic Reactions of Aspartate

In *E. coli* there are three *aspartokinases* that catalyze the conversion of aspartate to β-aspartyl phosphate. All three catalyze the same reaction, but they have very different regulatory properties. Each enzyme is responsive to a different set of end products. The same is true for the two *aspartate semialdehyde reductases*, which catalyze the third step. Both repression of transcription and feedback inhibition of the enzymes are involved.

Two of the aspartokinases of *E. coli* are parts of bifunctional enzymes, which also contain the homoserine dehydrogenases that are needed to reduce aspartate semialdehyde in the third step. These aspartokinase-homoserine dehydrogenases I and II are encoded by *E. coli* genes *thrA* and *metL,* respectively, and have homologous sequences. The N-terminal portions are also homologous to the lysine-sensitive aspartokinase III which is encoded by the *lys*C gene. In *Bacillus subtilis* the lysine-sensitive enzyme is known as aspartokinase II. It has an $\alpha_2\beta_2$ oligomeric structure and both α and β chains are encoded within a single gene. There is no associated homoserine dehydrogenase. Both genetic organization and processing of the synthesized protein are thus different in these two bacteria.

Lysine Diaminopimelate, Dipicolinic Acid, and Carnitine

Lysine cannot be made at all by animals but is nutritionally essential. There are two distinct pathways for its formation in other organisms. The *α-aminoadipate pathway* occurs in a few lower fungi, the higher fungi, and euglenids. The 5-carbon 2-oxoglutarate is the starting compound. Bacteria, other lower fungi, and green plants all use the *diaminopimelate* pathway which originates with the 4-carbon aspartate. The α-aminoadipate pathway parallels that of

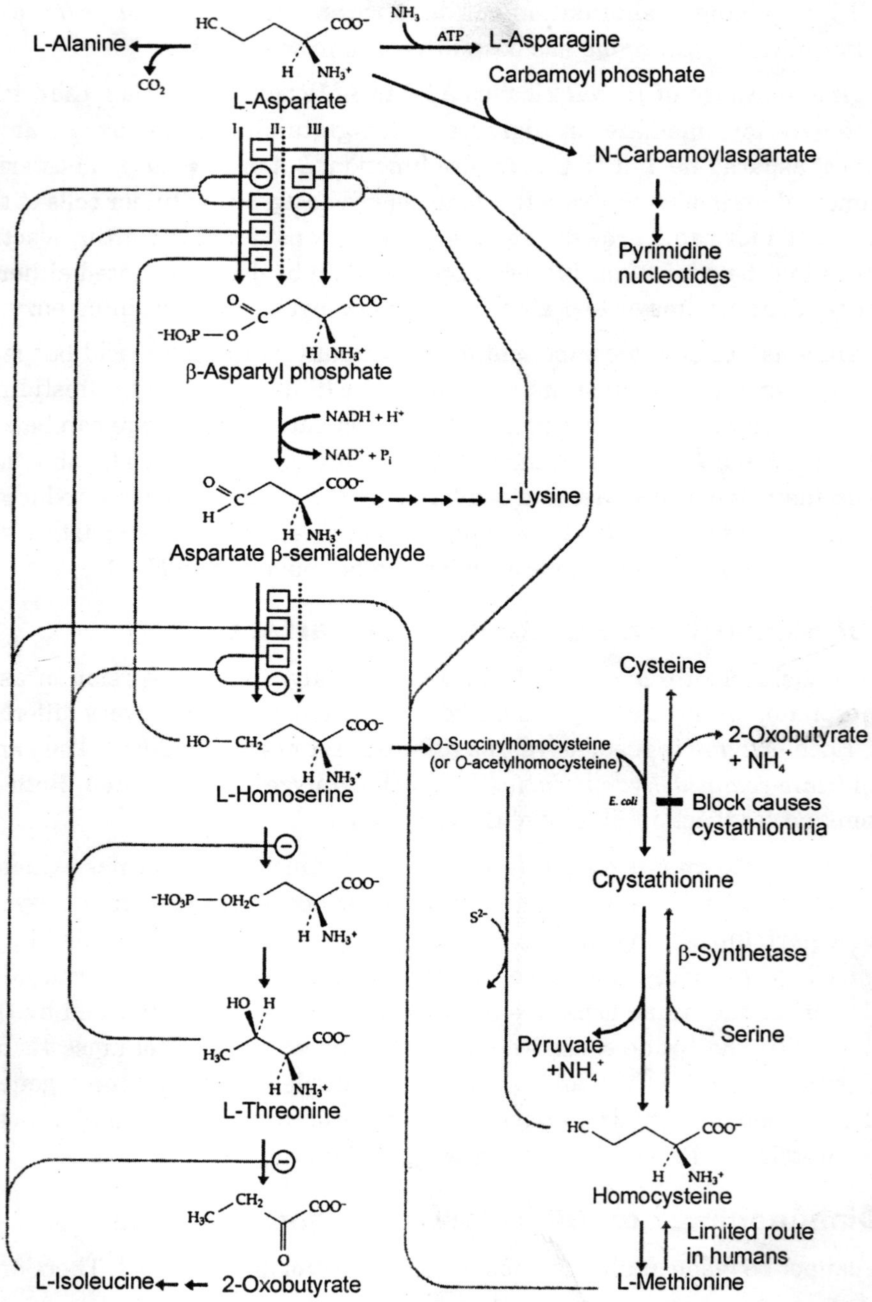

Fig. 6.13 Some biosynthetic reactions of aspartate : ⊖, feedback inhibition and ⊟,feedback repression.

ornithine biosynthesis, 2-oxoglutarate undergoing chain elongation to 2-oxoadipate followed by transamination to α-aminoadipate. This is followed by ATP-dependent reduction to the aldehyde.

The final step of transamination is not accomplished in the usual way (with a PLP-dependent enzyme), but through formation of a Schiff base with glutamate and reduction to *saccharopine.* Oxidation now produces the Schiff base of lysine with 2-oxoglutarate. In the diaminopimelate pathway of lysine synthesis aspartate is converted to aspartate semialdehyde, and a two-carbon unit is added via aldol condensation with pyruvate. Decarboxylation at the end of the sequence yields lysine. A series of cyclic intermediates exist, but it is noteworthy that the initial product of the aldol condensation is converted to diaminopimelic acid by a simple sequence involving α,β elimination of the hydroxy group, reduction with NADPH, and transamination. The process is complicated by the natural tendency for ring closure.

All gram-negative and many gram-positive bacteria use the succinylase pathway shown in fig. 6.14. Succinylation serves to shift the equilibrium back in favour of open-chain compounds. Some species of *Bacillus* use acetylation in the same way, while a few bacteria manage to use a dehydrogenase to reductively aminate tetrahydropimelate to diaminopimelate. The diaminopimelate pathway is of special significance to prokaryotic organisms for the reason that *dipicolinic acid* is formed as an important side product and because of formation of *diaminopimelic acids.* The cyclic dipicolinic acid is a major constituent of bacterial spores but is rarely found elsewhere in nature. Both L, L and meso-diaminopimelic acids are constituents of peptidoglycans of bacterial cell walls. Lysine is not only a constituent of proteins. It can also be trimethylated and converted to *carnitine.* In mammals some specific lysyl side chains of proteins undergo N-trimethylation and proteolytic degradation with release of free trimethyllysine.

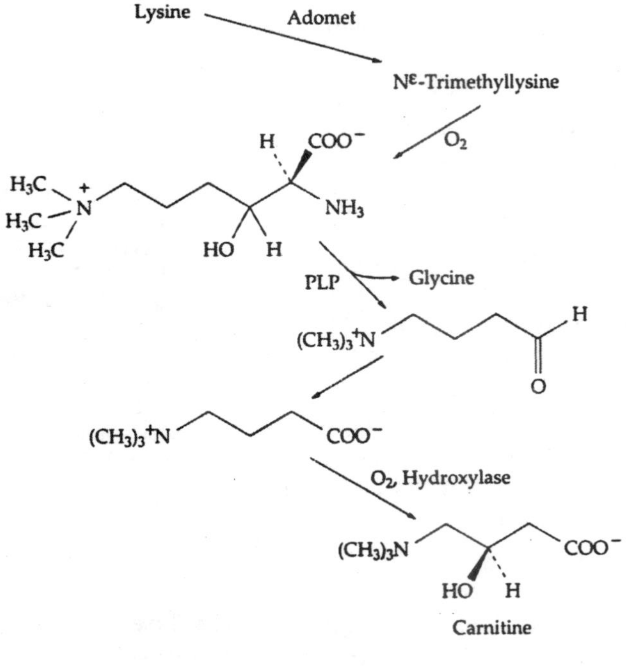

Lysine
Adomet
N^{ε}-Trimethyllysine
O_2
H COO⁻
H_3C + N H_3C H_3C
NH_3
HO H
PLP
Glycine
$(CH_3)_3{}^+N$
H
O
$(CH_3)_3{}^+N$ COO^-
O_2, Hydroxylase
$(CH_3)_3N$ COO^-
HO H
Carnitine

The free trimethyllysine then undergoes hydroxylation by a 2-oxoglutarate-Fe^{2+}-ascorbate-dependent hydroxylase to form β-hydroxytrimethyllysine, which is cleaved by a PLP-dependent enzyme. The resulting aldehyde is oxidized to the carboxylic acid and is converted by a second 2-oxoglutarate-Fe^{2+}-ascorbate-dependent hydroxylase to carnitine *Hypusine* (N^E-(4-amino-2-hydroxybutyl)lysine) occurs in mammalian initiation factor 4D, which is utilized in protein synthesis and is formed by transfer of the 4–carbon butylamine group from spermidine to a lysine side chain followed by hydroxylation. The lupine alkaloid lupinine is formed from two C_5 units of cadaverine which arises by decarboxylation of lysine. Silaffins also contain modified lysines.

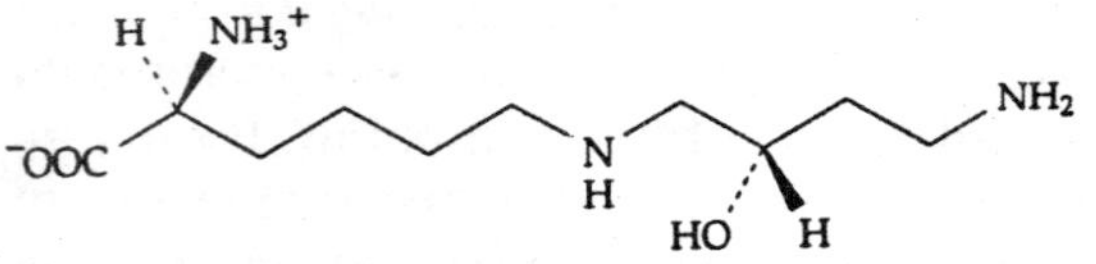

Hypusine: N^{ε}-(4-amino-2-hydroxybutyl) lysine

The Catabolism of Lysine

An unusual feature of lysine metabolism is that the ∝-amino group does not equilibrate with the "nitrogen pool." Catabolism is initiated by deamination and proceeds by β oxidation. At least six variations of the β-oxidation process have been proposed. The evolutionary differences concern the manner in which the two amino groups are moved from the carbon skeleton. In the seemingly simplest pathway, which is used by *Flavobacterium fuscum,* the ε-amino group is removed in a direct (but a typical) transamination. The resulting α-aminoadipate semialdehyde is oxidized to α-aminoadipate, which *is degraded in a sequence characteristic for the catabolism of amino acids.*

Transamination is followed by oxidative decarboxylation of the resulting 2-oxoacid and β oxidation of the coenzyme A derivative. A decarboxylation step by which the terminal

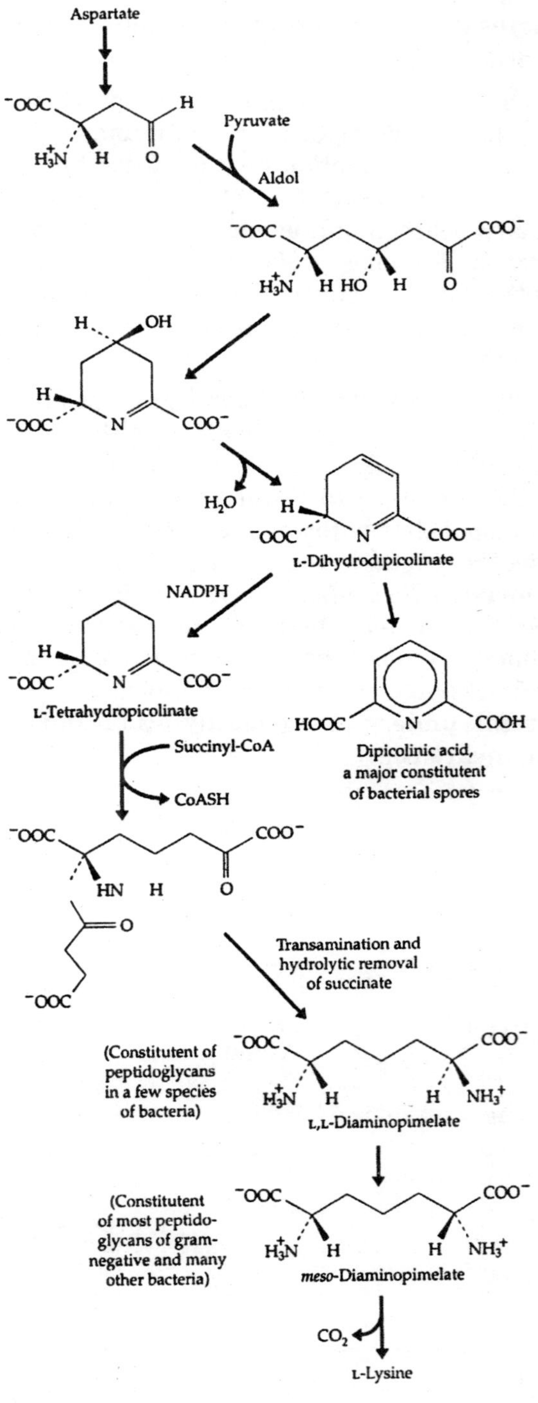

Aspartate → Aldol (Pyruvate) → H_2O → L-Dihydrodipicolinate → NADPH → L-Tetrahydropicolinate → Succinyl-CoA / CoASH → Transamination and hydrolytic removal of succinate → L,L-Diaminopimelate (Constitutent of peptidoglycans in a few species of bacteria) → *meso*-Diaminopimelate (Constitutent of most peptido-glycans of gram-negative and many other bacteria) → CO_2 → L-Lysine

L-Dihydrodipicolinate → Dipicolinic acid, a major constitutent of bacterial spores

Fig. 6.14 The biosynthesis of lysine by the diaminopimelate pathway.

carboxyl group is removed is interposed in the β-oxidation sequence for lysine degradation. Perhaps the initial transamination in pathway A is chemically difficult, for most organisms use more complex sequences to form 2-oxoadipate. In pathway B (which takes place in liver mitochondria and is believed to be the predominant pathway in mammals), the ε-amino group is reductively coupled with 2-oxoglutarate to form saccharopine. The latter is in turn oxidized on the opposite side of the bridge nitrogen to form glutamic acid and α-aminoadipate semialdehyde.

The overall process is the same as direct transamination and just the opposite of that occurring in the aminoadipate pathway of biosynthesis. Absence of one or both of these dehydrogenases causes familial hyperlysinemia. Pathway C has been established for *Pseudomonus putida* and is also followed to some extent in both plants and animals. In most animal tissues it may be used principally for degradation of D-lysine. However, it is the major L-lysine oxidation pathway in brain. In a fungal parasitic species of *Rhizoctonia* L-lysine is converted to saccharopine via pathway B; then using an $NADP^+$-dependent saccharopine oxidase the sequence is shunted to pathway C. L-Pipicolic acid formed in this way also gives rise to various alkaloids including the α-mannosidase inhibitor swainsonine.

Pathway C, like pathway B, makes use of transamination via a reduction-oxidation sequence. It is strictly internal, the oxidizing carbonyl group being formed by transamination of the α-amino group of lysine. Pathway D, apparently used by yeasts, avoids cyclic intermediates by acetylation of the ε-amino group prior to transamination. The 2-oxo group is then effectively blocked by reduction to an alcohol, the blocking group is removed from the ε-amino group, and that end of the molecule is oxidized in a straightforward way to a carboxyl group.

Now the hydroxyl introduced at position 2 is presumably oxidized back to the ketone, which again can be converted to give 2-oxoadipate. Some bacteria, e.g., *Pseudomonas putida* degrade L-lysine with a flavin-dependent oxygenase to δ-aminovaleramide:

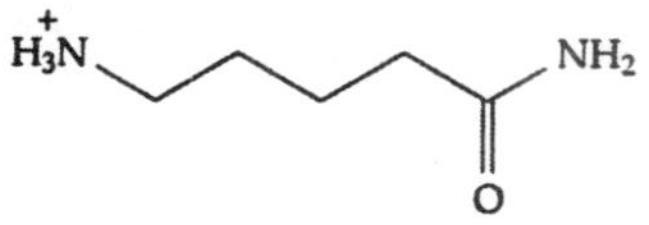

$H_3\overset{+}{N}$ NH$_2$ O

The product is hydrolyzed and oxidized to *glutaryl-CoA*, rejoining the pathways shown in Fig. 6.15 A remarkable and very different approach to lysine breakdown has been developed by clostridia which obtain energy from the fermentation Eq. 6.31 :

$$\text{L-Lysine} + 2\,H_2O \longrightarrow \text{butyrate}^- + \text{acetate}^- + 2\,NH_4^+ \qquad ...(6.31)$$

The reaction is coupled to formation of one molecule of ATP from ADP and P_i. Two pathways have been worked out. In the first lysine is acted upon by a PLP-dependent *L-lysine 2.3-aminomutase* (step *a*) to convert it to β-lysine (3,6-diaminohexanoate). The latter is further isomerized by the vitamin B_{12} and PLP-dependent β-lysine mutase. Oxidative deamination to a 3-oxo compound permits chain cleavage.

The reader can easily propose the remaining reactions of chain cleavage, ATP synthesis, elimination of ammonia, and balancing of the redox steps. An alternative pathway begins with a racemase and isomerization of the resulting D-lysine by another B_{12} and PLP-dependent enzyme. Oxidative deamination presumably occurs, but the mechanism for chain cleavage is not so obvious. It does occur between-C-4 and C-5 as indicated by the dashed line in Eq. 6.32. Another variation is used by *Pseudomonas* β4. Beta-lysine is acetylated on N-6, then undergoes

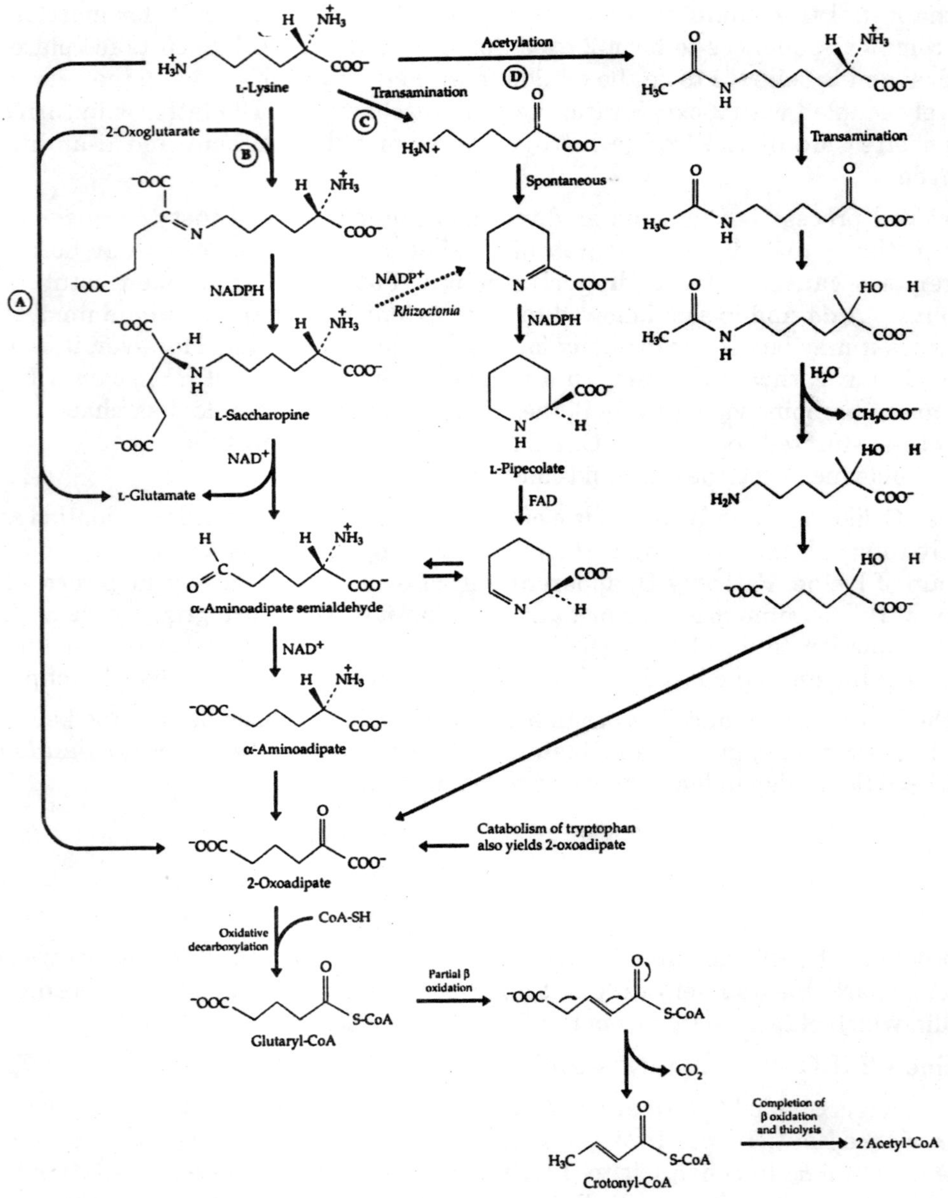

Fig. 6.15. Catabolism of lysine.

transamination to a 2-oxo acid and removal of the first two carbons as acetyl-CoA. The resulting 4-aminobutyrate is then converted to succinate via succinate semialdehyde. Why are there so many pathways of lysine breakdown? The answer is probably related to the ease of spontaneous formation of cyclic intermediates as occurs in the pipecolate pathway. These intermediates may be too stable for efficient metabolism so the indirect pathways evolved. In the fermentation reactions additional constraints are imposed on the pathways by the need for balanced redox processes and a net Gibbs energy decrease.

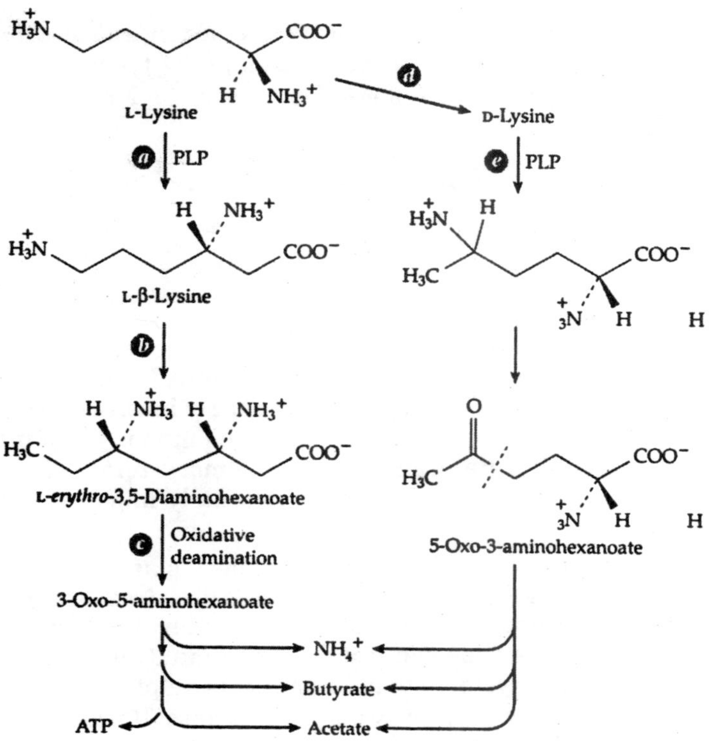

L-Lysine

d

D-Lysine

a PLP

e PLP

L-β-Lysine

b

L-*erythro*-3,5-Diaminohexanoate

c Oxidative deamination

3-Oxo-5-aminohexanoate

5-Oxo-3-aminohexanoate

NH_4^+

Butyrate

ATP

Acetate

Metabolism of Homocysteine and Methionine

Autotrophic organisms synthesize methionine from-aspartate as shown in the lower right side of Fig. 6.13. This involves transfer of a sulfur atom from cysteine into homocysteine, using the carbon skeleton of homoserine, the intermediate *cystathionine,* and two PLP-dependent enzymes, *cystathionine-γ-sinthase* and *cystathionine β-lyase.* This *transsalfuration* sequence is essentially irreversible because of the cleavage to pyruvate and NH_4^+ by the β-lyase.

Nevertheless, this transsulfuration pathway operates in reverse in the animal body, which uses two different PLP enzymes, *cystathionine β-synthase* (which also contains a bound heme) and *cystathionine γ-lyase* in a pathway that metabolizes excess methionine. For human beings methionine is nutritionally essential and comes entirely from the diet. However, the oxoacid analog of methionine can be used as a nutritional supplement.

Dietary homocysteine can also be converted into methionine to a limited extent. Methionine is incorporated into proteins as such and as *N-formylmethionine* at the N-terminal ends of bacterial proteins. In addition to its function in proteins methionine plays a major role in biological methylation reactions in all organisms. It is converted into *S-adenosylmethionine* which is the most widely used methyl group donor for numerous biological methylation reactions.

Adenosylmethionine is also the precursor of the special "wobble base" *queuine.* The product of transmethylation, *S-adenosylhomocysteine* is converted (step *g*) into homocysteine in an unusual NAD-dependent hydrolytic reaction by which adenosine is removed. Homocysteine can be reconverted to methionine, as indicated by the dashed line in Fig. 6.16. This can be accomplished by the vitamin B_{12}-and tetrahydrofolate-dependent *methionine synthase,* which transfers a methyl group from methyl-tetrahydrofolate; by transfer of a methyl group from *betaine,* a trimethylated glycine, or by remethylation with AdoMet.

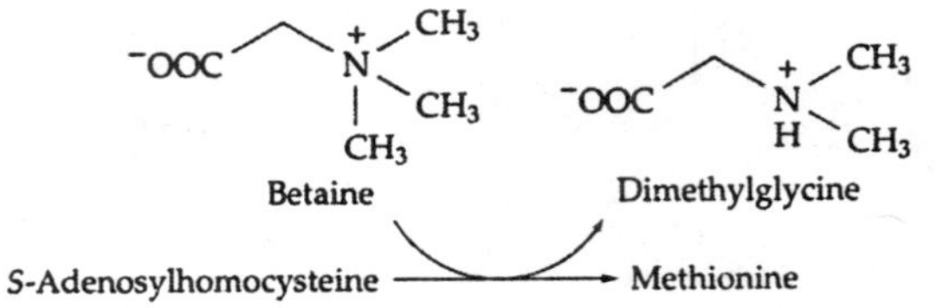

Betaine Dimethylglycine

S-Adenosylhomocysteine → Methionine

When present in excess methionine is toxic and must be removed. Transamination to the corresponding 2-oxoacid occurs in both animals and plants. Oxidative decarboxylation of this oxoacid initiates a major catabolic pathway, which probably involves β oxidation of the resulting acyl-CoA. In bacteria another catabolic reaction of methionine is γ-elimination of methanethiol and deamination to 2-oxobutyrate. Conversion to homocysteine, via the transmethylation pathway, is also a major catabolic route which is especially important because of the toxicity of excess homocysteine.

A hereditary deficiency of cystathionine β-synthase is associated with greatly elevated homocysteine concentrations in blood and urine and often disastrous early cardiovascular disease. About 5–7% of the general population has an increased level of homocysteine and is also at increased risk of artery disease. An adequate intake of vitamin B_6 and especially of folio acid, which is needed for recycling of homocysteine to methionine, is helpful. However, if methionine is in excess it must be removed via the previously discussed transsulfuration pathway.

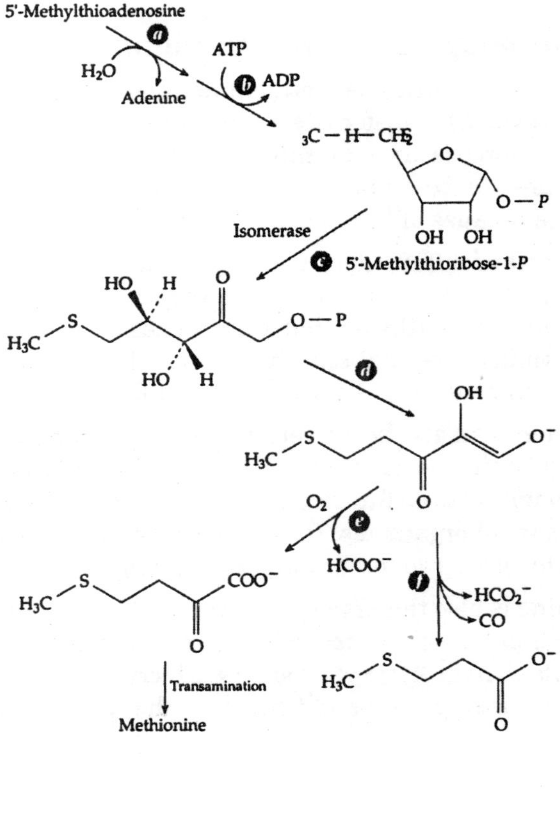

5'-Methylthioadenosine

H_2O Adenine ATP ADP

Isomerase

5'-Methylthioribose-1-P

O_2 $HCOO^-$

HCO_2^- CO

Transamination

Methionine

The products are cysteine and 2-oxobutyrate. The latter can be oxidatively decarboxylated to propionyl-CoA and further metabolized, or it can be converted into leucine and cysteine may be converted to glutathione. Methionine in plants can be converted to the sulfonium compound S-methyl-L-methionine, also called vitamin U. It has strong osmoprotectant activity and accumulates in many marine algae and some flowering plants. Other organisms, including mammals, can use S-methylmethionine to methylate homocysteine, converting both reactants back to methionine enabling animals to meet some of their methionine need from this source.

A salvage pathway. Another product of S-adenosylmethionine is 5′-methylthioadenosine, which can be formed by an internal displacement on the γ-methylene group by the carboxylate group. Methylthioadenosine also arises during formation of the compounds spermidine and ACC. Mammalian tissues convert methylthioadenosine back to methionine by the sequence shown Eq. 6.34. It undergoes phosphorol-ysis to 5′-methylthioribose whose ring is opened and converted to the 2-oxoacid analog. Step *c* of Eq 6.34 may occur by ring opening to an enolmethiouine phosphate which ketonizes to the observed product, but step *e* is a more complex multistep oxidative process. The last step is transamination to methionine with a glutamine-specific aminotransferase. Another enzyme from *Klebsiella* converts the same intermediate anion to methylthiopropionate, formate, and CO.

The plant hormone ethylene. A major reaction of S-adenosylmethionine in plants is the formation of *ethylene.* Ethylene has been recognized since 1858 as causing a thickening of stems of plants and a depression in the rate of elongation. In 1917, it was established that ethylene is formed in fruit and that addition of this gaseous compound hastened ripening. Ethylene is now an established plant hormone having a variety of effects including retardation of mitosis, inhibition of photosynthesis, and stimulation of respiration and of the enzyme phenyalanine ammonialyase.

These effects are indirectly a result of the action of ethylene on transcription of certain genes. In *Arabidopsis,* with which genetic studies are being made, ethylene/binds to the N-terminal part of at least two receptor proteins, which have intracellular histidine kinase domains in the C-terminal parts. DNA-binding proteins specific for *ethylene-responsive elements* (EREs), having the conserved sequence AGCCGCC, are presumably phosphorylated by this kinase and affect transcription of many "genes.

A protein homologous to one ethylene receptor of *Arabidopsis* has been identified in the tomato. A proline to leucine mutation at position 36, near the N terminus, destroys the sensitivity to ethylene and prevents ripening of this tomato. Another component in the ethylene signaling pathway in *Arabidopsis* is a protein serine / threonine kinase that resembles the mammalian raf kinas involved in the signaling cascade shown in Fig. 6.13. The formation of ethylene is often induced by the hormone *auxin,* which stimulates activity of the synthase that forms *1-aminocyclopropane-1-carboxylate (ACC)* from S-adenosyl methionine. 'Although ACC has been known as a minor plant product for over 25 years, it was much more recently identified as the immediate precursor of ethylene. ACC is often produced in response to stresses such as wounding, drought, or waterlogging of roots.

$$\text{(ACC: cyclopropane with } NH_3^+ \text{ and } COO^-) + O_2 + \text{Ascorbate} \rightarrow H_2C = CH_2 + HCN + CO_2 + 2H_2O + \text{Dehydroascorbate} \quad ...(6.35)$$

In the last of these cases the ACC is transferred through the xylem from the roots upward to shoots, which respond in characteristic ways to the ethylene that is released. The conversion

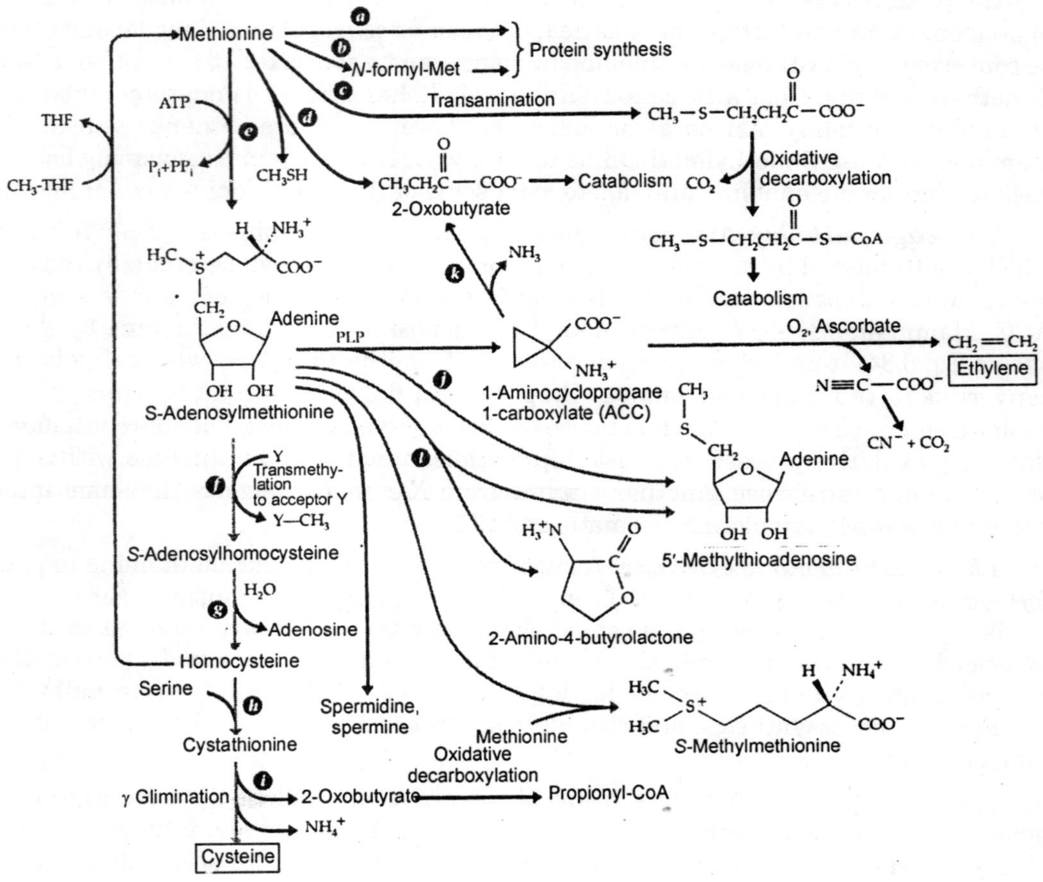

Fig. 6.16. Some metabolic reactions of methionine. Biosynthetic reactions are indicated by green arrows.

of ACC to ethylene, HCN, and CO_2 is catalyzed by ACC oxidase, an Fe_2^+-dependent enzyme of the isopenicillin-*N*-synthase subfamily of oxygenases. However, unlike most of these enzymes,

COO^- NH_3+ $\xrightarrow{e^-,\ H^+}$ COO^- $\overset{+}{N}H_2$

H^+

$\ominus\cdot$ COO^- NH $\xrightarrow{e^-}$ COO^- NH

$H_2C=CH_2$ $\quad N\equiv C-COO^-$

$N\equiv C^- + CO_2$

which utilize 2-oxoglutarate as a co-substrate, ACC oxidase employs ascorbate and forms *HCN* or cyanide ions. It also requires CO_2 or bicarbonate as an activator. A radical mechanism is probable, with two electrons from ACC and two from ascorbate being utilized to reduce O_2 to 2 H_2O. Ethylene is rather inert, but it is metabolized slowly, some of it to ethylene glycol. Plants store N-malonyl-ACC as a metabolically inert pool. Excess ACC can be deaminated in a PLP-dependent reaction to 2-oxobutyrate, a process that also occurs in bacteria able to subsist on ACC. There may also be other mechanisms for ethylene formation, e.g., peroxidation of lipids during scenescence of leaves.

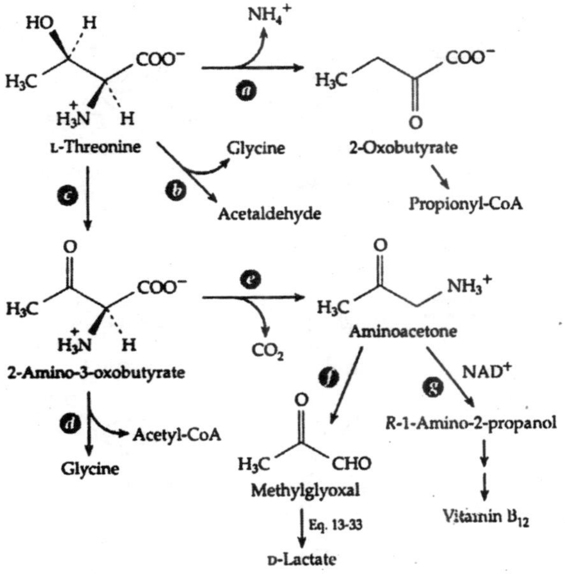

Metabolism of Threonine

Excess threonine is degraded in several ways, one of which is a β elimination reaction catalyzed by L-threonine dehydratase. This PLP-requiring enzyme is produced in high amounts in *E. coli* grown on a medium devoid of glucose and oxygen. Under these circumstances the reaction provides a source of propionyl-CoA, which can be converted to propionate with generation of ATP. This *biodegradative threonine dehydratase* (threonine deaminase) is allosterically activated by AMP, an appronpriate behaviour for a key enzyme in energy metabolism.

A second *biosynthetic threonine dehydratase* is also produced by E. *coli* and is specifically required for production of 2-oxobutyrate needed in the biosynthesis of isoleucine by bacteria, plants, and other autotrophic organisms. In 1956, Umbarger showed that this enzyme is inhibited by isoleucine, the end product of the synthetic pathway. This discovery was instrumental in establishing the concepts of feedback inhibition in metabolic regulation and of allostery. A second catabolic reaction of L-threonine is cleavage to glycine and acetaldehyde. The reaction is catalyzed by serine hydroxymethyl-transferase.

Some bacteria have a very active D-threonine aldolase. A quantitatively more important route of catabolism in most organisms is dehydrogenation to form 2-amino-3-oxobutyrate. This intermediate can be cleaved by another PLP-dependent enzyme to acetyl-CoA plus glycine. It can also be decarboxylated to aminoacetone, a urinary excretion product, or oxidized by amine oxidases to *methylglyoxal*. The latter can be converted to D-lactate through the action of glyox-alase. Aminoacetone is also the source of l-amino-2-propanol for the biosynthesis of vitamin B_{12}.

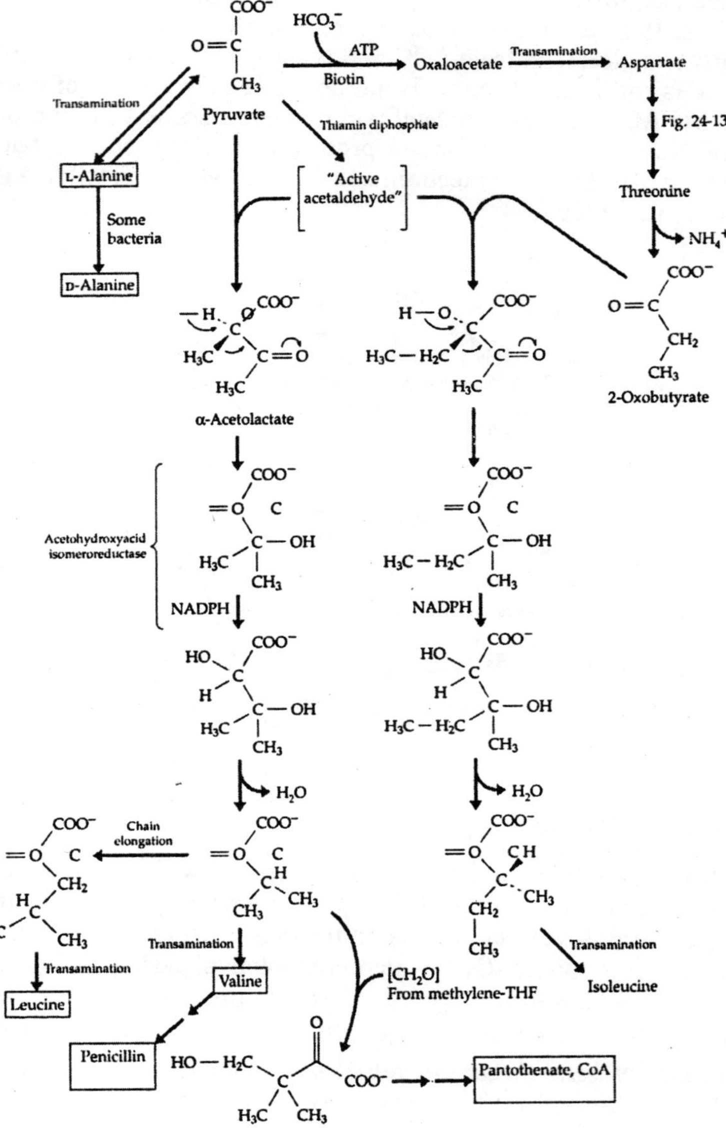

Fig. 6.17. Biosynthesis of leucine, isoleucine, valine, and coenzyme A.

ALANINE AND THE BRANCHED-CHAIN AMINO ACIDS

Pyruvate is the starting material for the formation of both L- and D-alanine and also the branched chain amino acids *valine, leucine,* and *isoleucine.* The chemistry of the reactions has been discussed in the sections indicated in the figure. The first step is catalyzed by the thiamin diphosphate-dependent *acetohydroxyacid synthase* (acetolactate synthase), which joins two molecules of pyruvate or one of pyruvate and one of 2-oxobutyrate. In *E. coli* there are two iso-enzymes encoded by genes *ilv E* and *ilv HI.* Both are regulated by feedback inhibition by valine, probably by an *attenuation* mechanism.

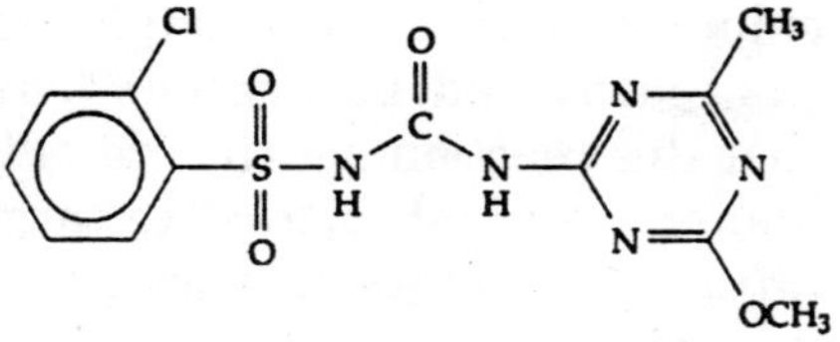

Chlorsulfuron

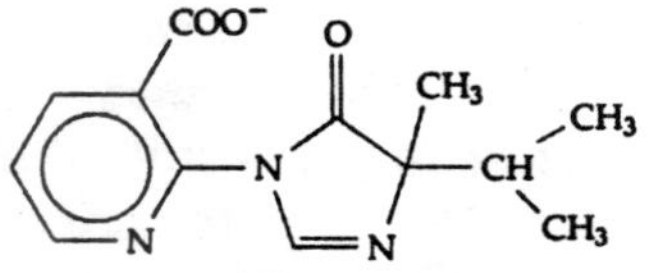

An imidazolinone herbicide

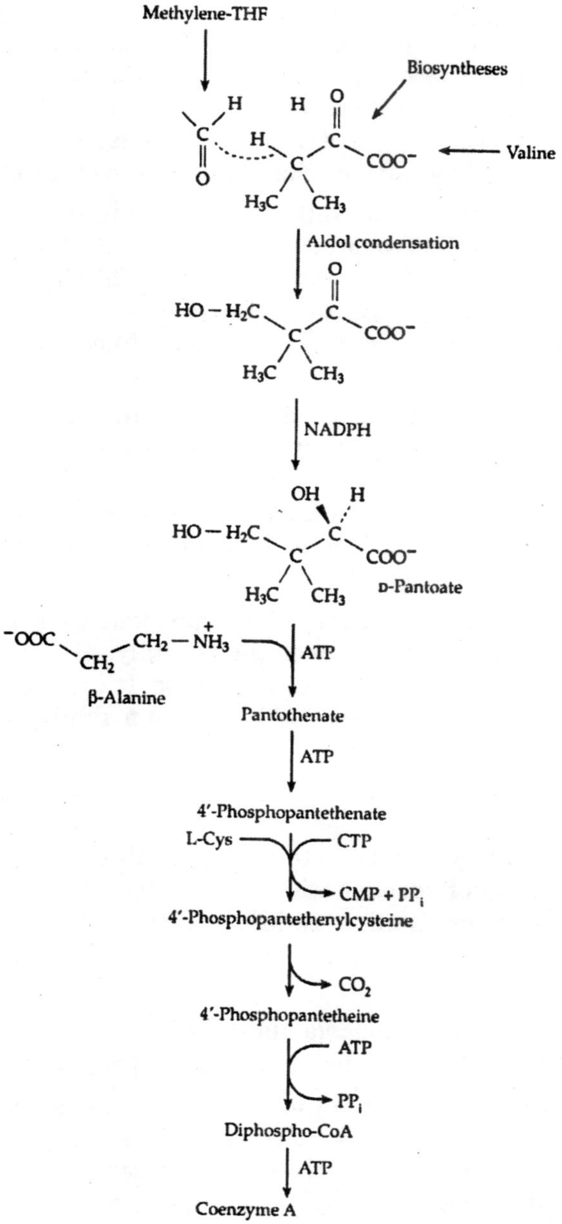

Methylene-THF

Biosyntheses

Valine

Aldol condensation

NADPH

D-Pantoate

β-Alanine

ATP

Pantothenate

ATP

4′-Phosphopantethenate

L-Cys — CTP

$CMP + PP_i$

4′-Phosphopantethenylcysteine

CO_2

4′-Phosphopantetheine

ATP

PP_i

Diphospho-CoA

ATP

Coenzyme A

The enzymes are of some practical interest because they are specifically inhibited by two classes of herbicides, the *sulfonylureas,* of which chlorsulfuron is an example, and the *imidazolinones.* The second step in the synthesis, catalyzed by *acetohydroxyacid isomeroreductase* involves shift of an alkyl group. Neither this reaction nor the preceding one occurs in mammals. For this reason, the enzymes required are both attractive targets for herbicide design. The third enzyme, *dihydroxy acid dehydratase,* catalyzes dehydration followed by tautomerization, resembling 6-phosphogluconate dehydratase.

The dihydroxyacid dehydratase from spinach contains an Fe_2S_2 cluster and may function by an aconitase type mechanism. In *Neurospora* isoleucine and valine are synthesized in the mitochondria. While the 2-oxobutyrate needed for isoleucine formation is shown as originating from threonine in Fig. 6.17. bacteria can often make it in other ways, e.g., from glutamate via β-methylaspartate and transamination to the corresponding 2-oxoacid. It can also be made from pyruvate by chain elongation using acetyl-CoA; citramalate and mesaconate are intermediates. This latter pathway is used by some methanogens as are other alternative routes.

The first step unique to the biosynthetic pathway to leucine is the reaction of the 2-oxo analog of valine with *acetyl-CoA* to form *α-isopropylmalate,* the first step in a chain elongation sequence, leading to the oxoacid precursor of leucine. The third enzyme required in the chain elongation is a decarboxylating dehydrogenase similar to isocitrate dehydrogenase. An additional series of reactions which are shown in Eq. 6.38 leads to *pantoic acid, pantetheine, coenzymne A,* and related cofactors.

The initial reactions of the sequence do not occur in the animal body, explaining our need for pantothenic acid as a vitamin. Alanine also gives rise to a precursor of the vitamin both after a PLP-dependent decarboxylative condensation with the 7-carbon dicarboxylic acid unit of pimeloyl-CoA in a reaction. The resulting alcohol is reduced to 7-oxo-8-amnopelargonic acid which is converted by transamination. with S-adenosylmethionine as the nitrogen donor, to 7,8-diaminopelargonic acid.

This compound undergoes a two-step ATP-dependent cyclization to form *dethiobiotin.* The final step, insertion of sulfur into dethiobiotin, is catalyzed by biotin synthase, a free-radical-dependent enzyme related to pyruvate formate lyase. It transfers the sulfur from cysteine via an Fe–S cluster. Biosynthesis of *lipoic acid* involves a similar insertion of two sulfur atoms into octanoic acid.

Catabolism

Degradation of amino acids most often begins with conversion, either by transamination or by NAD^+-dependent dehydrogenation, to the corresponding 2-oxoacid and oxidative decarboxylation of the latter, Alanine, valine, leucine, and isoleucine are all treated this way in the animal body. Alanine gives pyruvate and acetyl-CoA directly, but the others yield CoA derivatives that undergo

β-oxidation within the mitochondria via the schemes shown in Fig. 6.18. There are some variations from the standard β oxidation sequence for fatty acids shown in Fig. 6.19

In the case of valine the sequence proceeds only to the stage of addition of water to form the β-hydroxy derivative. The latter is converted to free 3-hydroxyisobutyrate, and β oxidation is then completed by oxidation to methylmalonate semialdehyde. The latter is oxidatively decarboxylated to form S-methylmalonyl-CoA. Further metabolism of the latter is indicated in

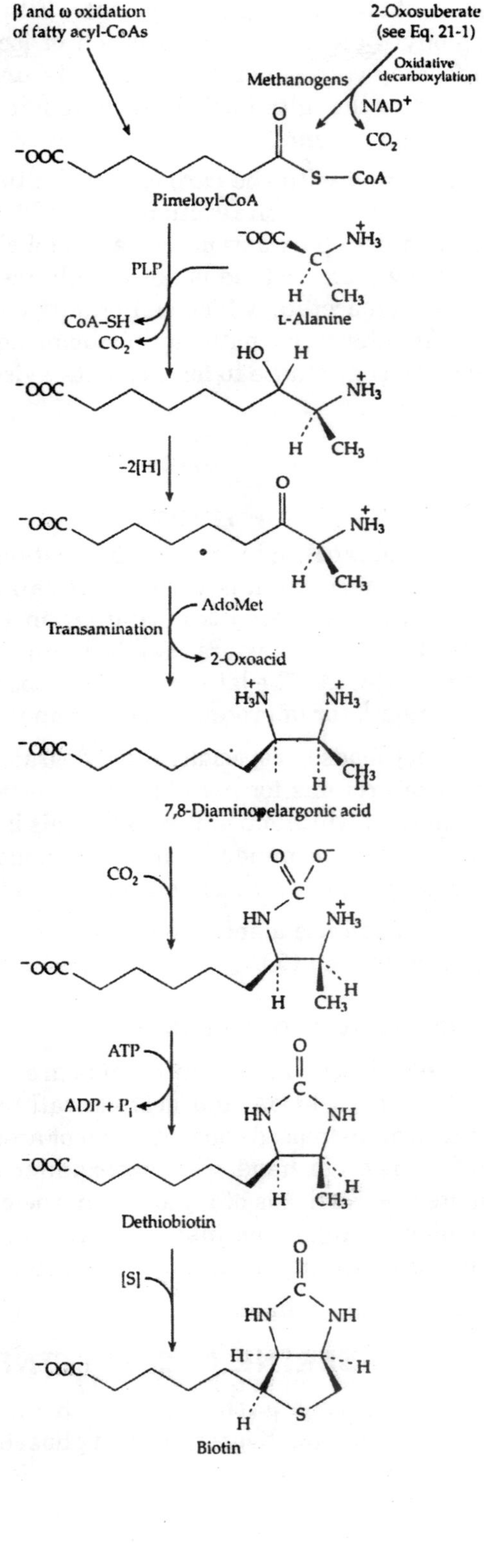

β and ω oxidation of fatty acyl-CoAs

2-Oxosuberate (see Eq. 21-1)

Methanogens

Oxidative decarboxylation

NAD^+

CO_2

^-OOC … S—CoA

Pimeloyl-CoA

PLP

^-OOC, $\overset{+}{N}H_3$, C, H, CH_3

L-Alanine

CoA-SH

CO_2

HO H

$\overset{+}{N}H_3$

H CH_3

-2[H]

O

$\overset{+}{N}H_3$

H CH_3

AdoMet

Transamination

2-Oxoacid

$H_3\overset{+}{N}$ $\overset{+}{N}H_3$

H CH_3 H

7,8-Diaminopelargonic acid

CO_2

O O^-

C

HN $\overset{+}{N}H_3$

H CH_3 H

ATP

ADP + P_i

O

C

HN NH

H CH_3 H

Dethiobiotin

[S]

O

C

HN NH

H H

S

H

Biotin

Fig. 6.20. However, some methylmalonate semialdehyde may be decarboxylated to propionaldehyde, which could be oxidized to propionate. Either of these compounds could then be metabolized to propionyl-CoA. In the degradation of isoleucine, β oxidation proceeds to completion in the normal way with generation of acetyl-CoA and propionyl-CoA. However, in the catabolism of leucine after the initial dehydrogenation in the β-oxidation sequence, carbon dioxide is added using a biotin enzyme.

The double bond conjugated with the carbonyl of the thioester makes this carboxylation analogous to a standard β-carboxylation reaction. Why add the extra CO_2? The methyl group in the β position blocks complete β oxidation, but an aldol cleavage would be possible to give acetyl-CoA and acetone. However, acetone is not readily metabolized further. By addition of CO_2 the product becomes acetoacetate, which can readily be completely metabolized through conversion to acetyl-CoA.An alternative pathway of leucine degradation in the liver is oxidative decarboxylation by a cytosolic oxygenase to form α-hydroxyisovalerate.

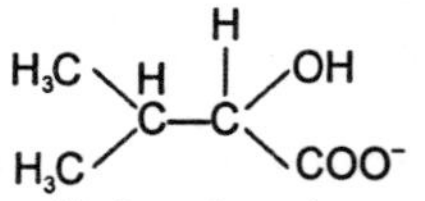

$(H_3C)_2CH{-}CH(OH){-}COO^-$

α-Hydroxyisovalerate

This compound may be metabolized via the valine catabolic pathway of Fig. 6.21. A third pathway, present in some bacteria, begins with the vitamin B_{12}-dependent isomerization of leucine to β-leucine, which can undergo transamination to 3-oxoisocaproate. This can be converted to its CoA ester by a CoA transferase and can undergo β cleavage by free CoA-SH to form acetyl-CoA and isobutyryl-CoA. The latter may enter the valine catabolic pathway. Leucine has long been known as a regulator of protein degradation in muscle.

Dietary protein deficiency leads to especially rapid degradation of the branched-chain amino acids. The daily turnover of proteins for a 70-kg adult ingesting 70 g of protein per day has been estimated as 280 g, most of which must be reused. This large turnover can lead to excessive muscle wasting in disease states. A minor leucine metabolite found in muscle, β-hydroxy β-methylbutyrate has been proposed as a possible endogenous inhibitor of muscle breakdown.

Clostridium propionicum can use alanine as substrate for a balanced fermentation to form ammonium propionate, acetate, and CO_2.

Ketogenic and Glucogenic Amino Acids

According to a long-used classification amino acids are *ketogenic* if (like leucine) they are converted to acetyl-CoA (or acetyl-CoA and acetoacetate). When fed to a starved animal, ketogenic amino acids cause an increased concentration of acetoacetate and other ketone bodies in the blood and urine. On the other hand, *glucogenic* amino acids such as valine, when fed to a starved animal, promote the synthesis of glycogen (in the case of valine via methylmalonyl-CoA, succinate, and oxaloacetate). Examination of Fig. 6.21 hows that isoleucine is both ketogenic and glucogenic, a fact that was known long before the pathway of catabolism was worked out.

SERINE AND GLYCINE

Serine originates in a direct pathway from 3-phosphoglycerate that involves dehydrogenation, transamination, and hydrolysis by a phosphatase. It can also be formed from

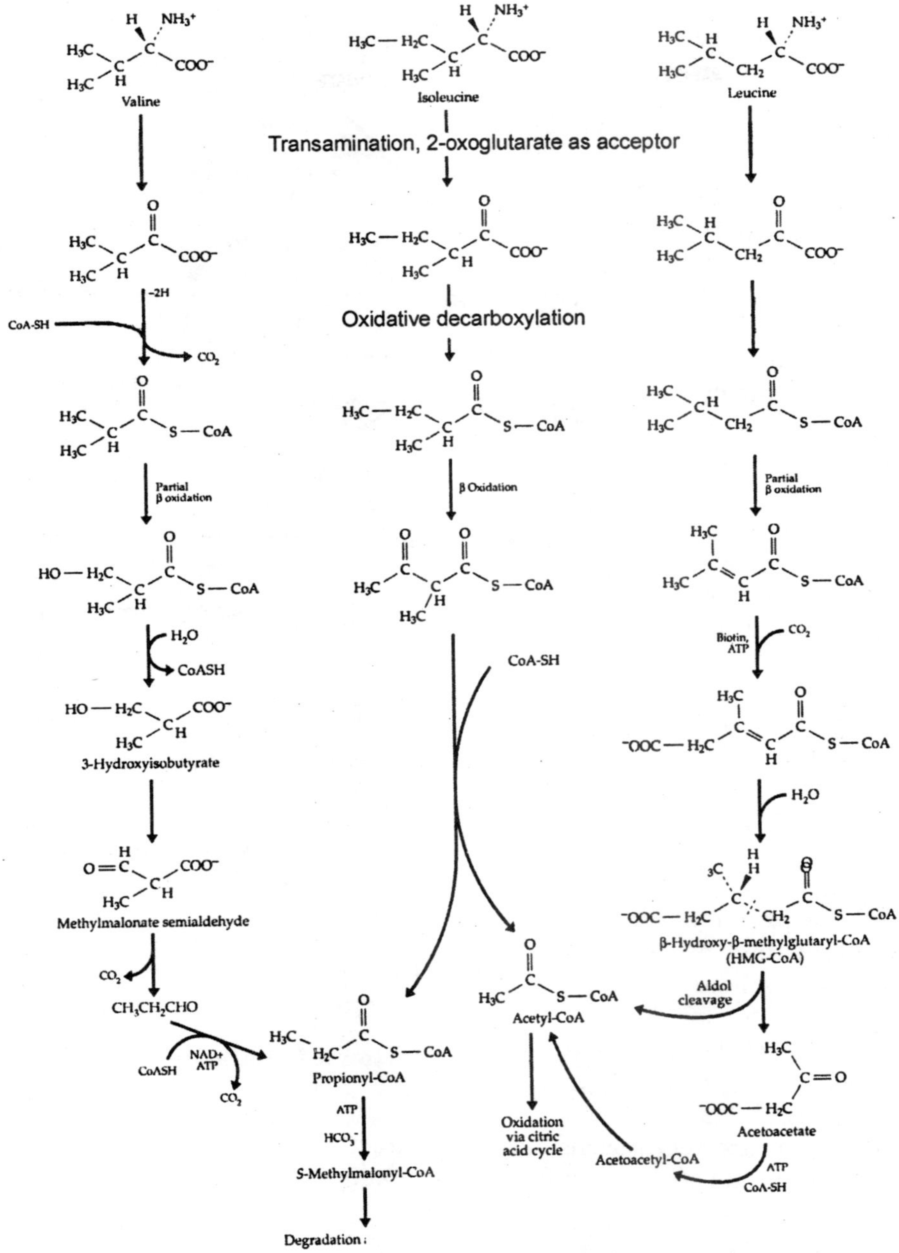

Valine
Isoleucine
Leucine
Transamination, 2-oxoglutarate as acceptor
Oxidative decarboxylation
CoA-SH
CO_2
-2H
Partial β oxidation
β Oxidation
Partial β oxidation
H_2O
CoASH
3-Hydroxyisobutyrate
Biotin, ATP
CO_2
H_2O
Methylmalonate semialdehyde
β-Hydroxy-β-methylglutaryl-CoA (HMG-CoA)
CO_2
CH_3CH_2CHO
Aldol cleavage
Acetyl-CoA
CoASH
NAD+ ATP
CO_2
Propionyl-CoA
ATP
HCO_3^-
S-Methylmalonyl-CoA
Oxidation via citric acid cycle
Acetoacetyl-CoA
Acetoacetate
ATP
CoA-SH
Degradation

Fig. 6.18. Catabolism of valine, leucine, and isoleucine.

glycine by the action of serine hydroxymethyltransferase. This occurs in chloroplasts during photorespiration and also with some methanogens and other autotropic bacteria and methylotrophs. The glycine decarboxylase cycle shown in Fig. 6.22 provides another mechanism available in bacteria, plants, and animal mitochondria for reversible interconversion of glycine and serine. The principal route of catabolism of serine in many microorganisms is deamination to pyruvate. An alternative catabolic pathway is transamination to *hydroxypyruvate,* which as in plants can be reduced to D-glycerate and back to 3-phosphoglycerate. That this pathway is important in human beings is suggested by the occurrence of a rare metabolic defect *L-glyceric aciduria* (or primary hyperoxaluria type II).

The biochemical defect may lie in the lack of reduction of hydroxypyruvate to D-glycerate. When hydroxypyruvate accumulates, lactate dehydrogenase effects its reduction to L-glycerate, which is excreted in large amounts (0.3-0.6 g / 24 h) in the urine. Surprisingly, the defect is accompanied by excessive production of oxalate from glyoxylate. This is apparently an indirect result of the primary defect in utilization of hydroxypyruvate. It has been suggested that oxidation of glyoxylate by NAD$^+$ is coupled to the reduction of hydroxypyruvate by NADH. This and other hyperoxalurias are very serious diseases characterized by the formation of calcium oxalate crystals in tissues and often death from kidney failure before the age of 20.

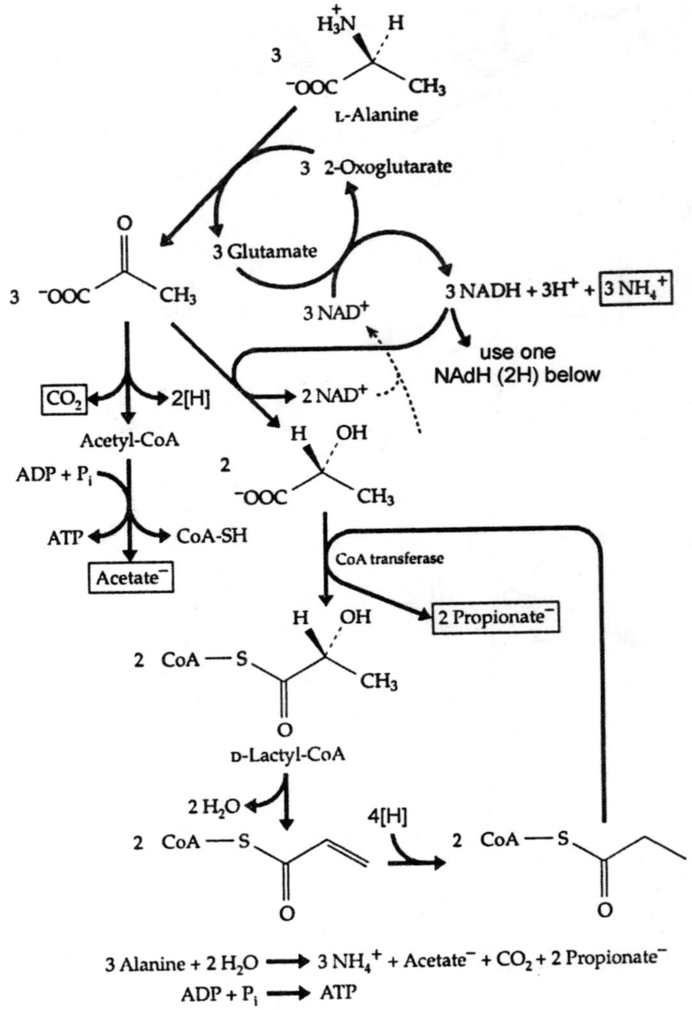

$$3\ \text{Alanine} + 2\ H_2O \longrightarrow 3\ NH_4^+ + \text{Acetate}^- + CO_2 + 2\ \text{Propionate}^-$$

$$\text{ADP} + P_i \longrightarrow \text{ATP}$$

Fig. 6.22. Fermentation of L-alanine *by Clostridium propionicum.*

Biosynthetic Pathways from Serine

L-Serine gives rise to many other substances including *sphingosine* and the *phosphatides.* In many bacteria conversion to O-acetyl-L-serine provides for the formation of *cysteine* by a β-replacement reaction. Serine is also the major source of glycine (step *d)* and of the single-carbon units needed for the synthesis of methyl and formyl groups. The enzyme *serine hvdroxymethyltranferase* also provides the principal route of formation of glycine from serine, but a lesser portion comes via phosphatidylserine, phosphatidylethanolamine, phosphatidylcholine, and free choline. This pathway includes decarboxylation of phosphatidylserine by a pyruvoyl group-dependent enzyme. In contrast, in green plants the major source of ethanolamine is a direct PLP-dependent decarboxylation of serine. Because the body's capacity to generate methyl groups is limited, *choline* under many circumstances is a dietary essential and has been classified as a vitamin.

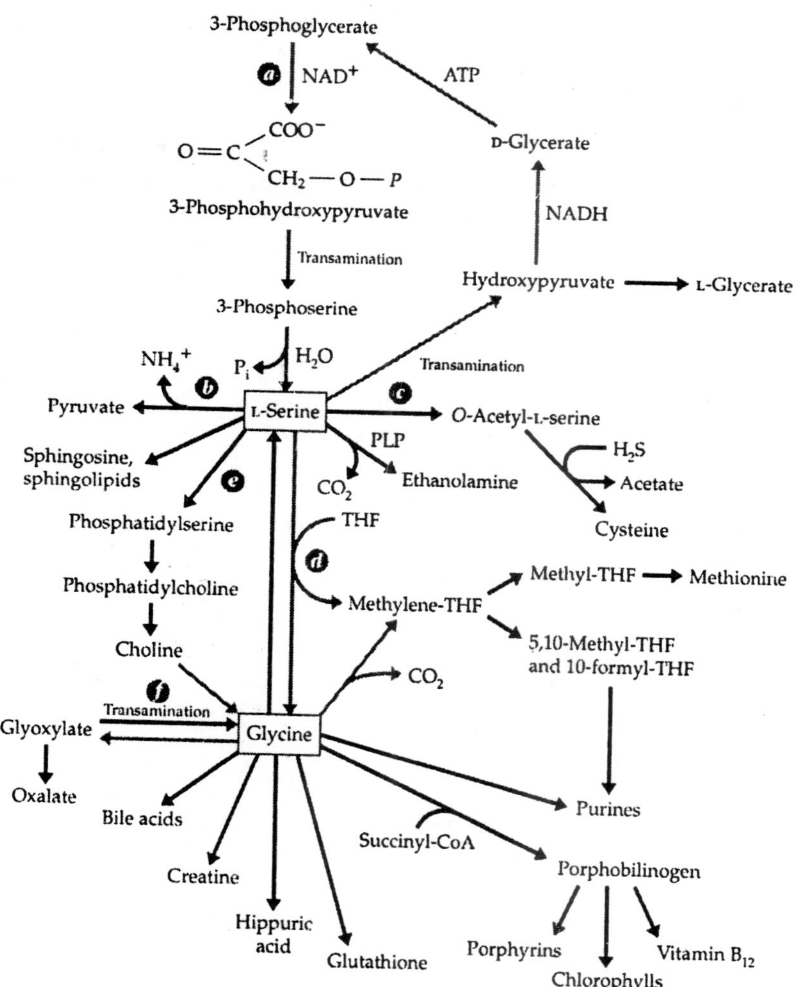

Fig. 6.20. Metabolism of serine and glycine.

However, in the presence of adequate amounts of folic acid and vitamin B_{12}, it is not absolutely required. Choline can be used to reform phosphatidylcholine, while an excess can be dehydrogenated to *glycine betaine,* which is one of the osmoprotectant substances in plants. This quaternary nitrogen compound is one of a small number of substances that, like methionine, are able to donate methyl groups to other compounds and which are also capable of methylating homocys-teine to form methionine. However, the product of transmethylation from betaine, dimethylglycine, is no longer a methylating agent. The two methyl groups are removed oxidatively as formic acid to produce glycine. A third source of glycine is transamination of glyoxylate. The equilibrium constant for the reaction favours glycine strongly for almost any amino donor.

Metabolism of Glycine

While glycine may be formed from glyoxylate by transamination, the oxidation of glycine by an amino acid oxidase permits excess glycine to be converted to glyoxylate. That this pathway, too, is quantitatively important in humans is suggested by the existence of *type 1 hyperoxaluria*. It is thought that a normal pathway for utilization of glyoxylate is blocked in this condition leading to its oxidation to oxalate. The biochemical defect is frequently the absence of a liver specific alanine:glyoxylate aminotransferase that efficiently converts accumulating glyoxylate back to glycine. In some cases the disease arises because, as a result of a mutation in its N-terminal targeting sequence the aminotransferase is targeted to mitochondria, where it functions less efficiently.

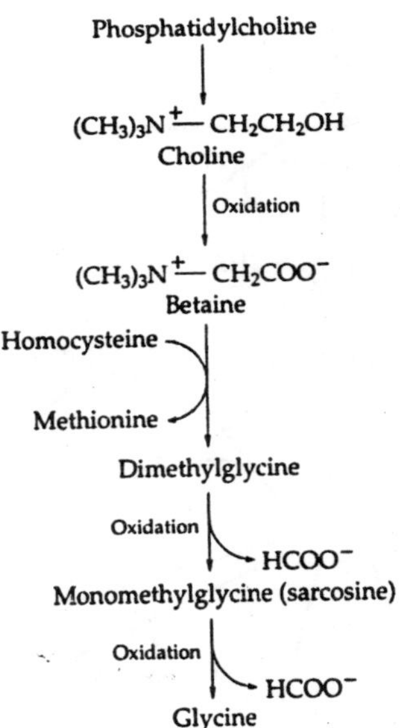

Another possible defect lies in a thiamin-dependent enzyme that condenses glyoxylate with 2-oxoglutarate to form 2-hydroxy-3-oxoadipate. The function of this reaction is uncertain, but the product could undergo decarboxylation and oxidation to regenerate 2-oxoglutarate. This would provide a cyclic pathway (closely paralleling the dicarboxylic acid cycle, for oxidation of glyoxylate without formation of oxalate. Bear in mind that the demonstrated enzymatic condensation reactions of the rather toxic glyoxylate are numerous, and that its metabolism in most organisms is still not well understood. An alternative route of catabolism is used by organisms such as *Diplococcus glycinophilus,* which are able to grow on glycine as a sole source of energy, of carbon, and of nitrogen, and is also used in mitochondria of plants and animals.

The glycine cleavage system, involves decarboxylation, oxidation by NAD^+, release of ammonia, and transfer of the decarboxylated α-carbon of glycine to tetrahydrofolic acid (THF) to form methylene-THF. The C-1 methylene unit of the latter is used primarily for purine biosynthesis but can also be oxidized to CO_2 or can condense with another molecule of glycine to form serine. This can in turn be converted to pyruvate and utilized for biosynthetic processes. Glycine can be reduced to acetate and ammonia by the selenium-dependent clostridial glycine reductase system. A variety of additional products can, be formed from glycine *Hippuric acid* the usual urinary excretion product in the "*detoxication*" of benzoic acid, is formed via benzoyl-CoA :

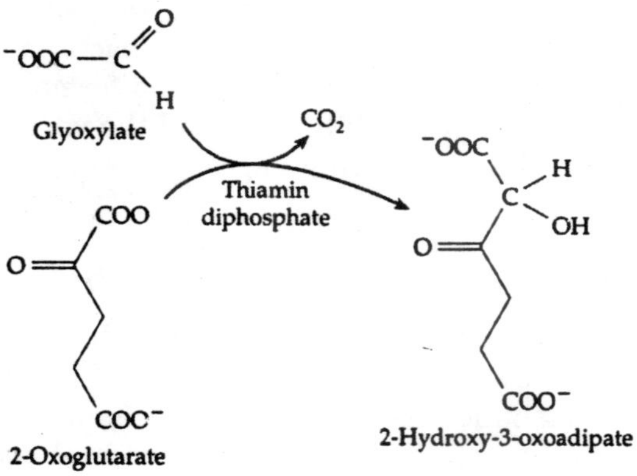

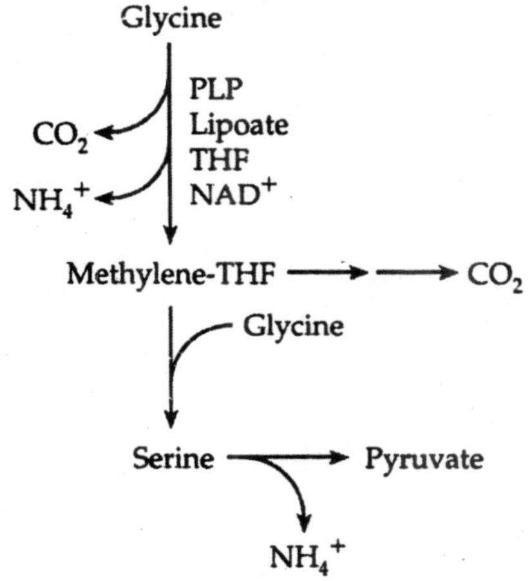

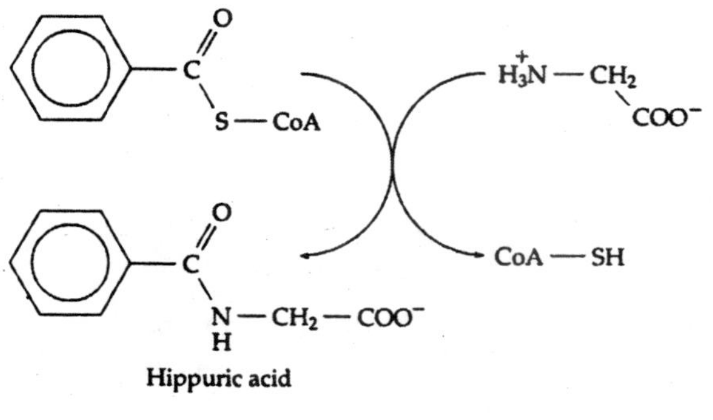

Hippuric acid

N-Methylation yields the monomethyl derivative *sarcosine* and also dimethylglycine, compounds that may function-as osmoprotectants. Many bacteria produce *sarcosine oxidase*, a flavoprotein that oxidizes its substrate back to glycine and formaldehyde, which can react with tetrahydrofolate. The formation of porphobilinogen and the various pyrrole pigments derived from it and the synthesis of the purine ring represent two other major routes for glycine metabolism.

Porphobilinogen, Porphyrins, and Related Substances

In 1946, Shemin and Rittenberg described one of the first successful uses of radiotracers in the study of metabolism. They reported that the atoms of the porphyrin ring in heme have their origins in the simple compounds acetate and glycine. As we now know, acetate is converted to succinyl-CoA in the citric acid cycle. Within the mitochondrial matrix of animal cells succinyl-CoA condenses with glycine to form 5(δ)-*aminolevulinic acid* which is converted to *porphobilinogen*, the immediate precursor to the porphyrins. The same pathways lead also to other tetrapyrroles including chlorophyll, the nickel-containing F, vitamin B_{12}, and other corrins. By degradation of ^{14}C-labeled porphyrins formed from labeled acetate and glycine molecules, Shemin and Rittenberg established the labeling pattern for the pyrrole ring that is indicated for porphobilinogen.

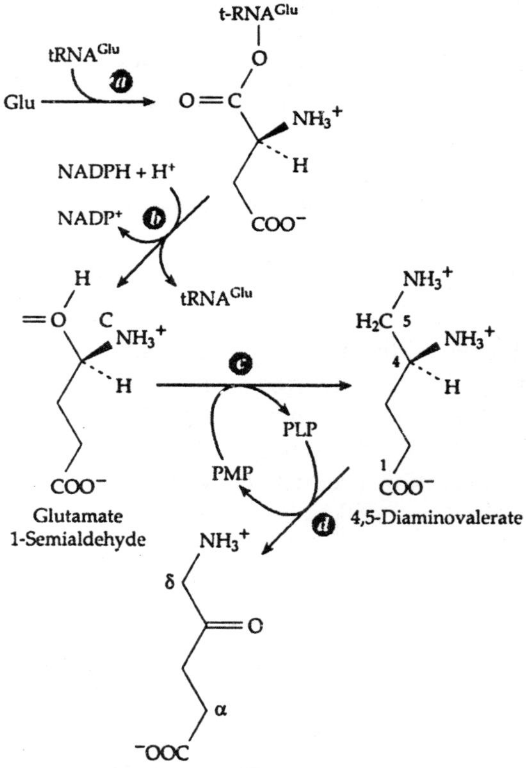

Glutamate 1-Semialdehyde

4,5-Diaminovalerate

5(δ)-Aminolevulinate

The solid circles mark those atoms that were found to be derived from methyl carbon atoms of acetate (bear in mind that acetyl groups of acetyl-CoA pass around the citric acid cycle more than once to introduce label from the methyl group of acetate into both the 2 and 3 positions of succinyl-CoA). Those atoms marked with open circles in Fig. 6.21 were found to be derived mainly from the methyl carbon of acetate and in small part from the carboxyl carbon. Atoms marked with asterisks came from glycine, while unmarked carbon atoms came from the carboxyl carbon of acetate. In cyanobacteria and in chloroplasts the intact 5-carbon skeleton of glutamate enters δ-aminolevulinate.

A surprising finding was that the glutamate becomes coupled to one of the three known glutamate isoacceptor tRNAs that are utilized for protein synthesis. The aminoacyl-tRNA is formed in the usual manner with an ester linkage to the CCA end of the tRNA. This ester linkage can be reductively cleaved by NADPH to form glutamate 1-semialdehyde. Isomerization of the glutamate semialdehyde to δ-aminolevulinate is accomplished by an aminomutase that is structurally and functionally related to aminotransferases. The enzyme utilizes pyridoxamine phosphate (PMP) to transaminate the substrate carbonyl group to form 4,5-diaminovalerate plus bound PLP. A second transmination step yields the product and regenerates the PMP.

Porphyrins. The conversion of two molecules of 5-aminolevulinate into porphobilinogen is a multistep reaction initiated by 5-*aminolevulinate dphydratase* (porphobilinogen synthase). The enzyme binds two molecules of substrate in distinct sites known as the A site and the P site. The substrate in the P site forms a Schiff base with a lysine side chain (K247 in the *E. coli* enzyme), while a bound Zn^{2+} is thought to polarize the carbonyl of the substrate in the A site. An aldol condensation ensues and is followed by dehydration to form a carbon-carbon double bond and ring closure. Tautomerization step leads to porphobilinogen.

The enzyme is a sensitive target for poisoning by lead ions. Condensation to form porphyrins requires two enzymes, *porphobilinogen deaminase* (hydroxymethylbilane synthase) and *uroporphyrinogen III cosynthase.* Porphobilinogen deaminase has a bound coenzyme (prosthetic group) consisting of two linked pyrromethane groups, also derived from porphobilinogen. The first step in assembling the porphyrin ring is condensation of porphobilinogen with this coenzyme.

To initiate this step ammonia is eliminated, probably not by the direct displacements, but by electron flow from the adjacent nitrogen in the same pyrrole ring as indicated in the figure to give an exocyclic double bond. The terminal ring of the coenzyme then adds to the double bond. The condensation process is repeated four times to produce *preuroporphyrinogen* (hydroxymethylbilane). This intermediate is a precursor of the symmetric uroporphyrin I. In the presence of the cosynthase a different ring-closure reaction takes place. The five-membered ring in porphobilinogen has a symmetric arrangement of double bonds. Thus, a condensation reaction can occur at either of the positions ∝ to the ring nitrogen. A sequence of condensation, tautomerization, cleavage, and reformation of the ring leads to the unsymmetric uroporphyrinogen III with its characteristic pattern of the carboxymethyl and carboxyethyl side chains.

A series of decarboxylation and oxidation reactions then leads directly to protoporphyrin IX. The first of these decarboxylations is catalyzed by the cytoplasmic *uroporphyrinogen decarboxylase,* which removes the carboxylate groups of the four acetate side chains sequentially from the D, A, B, and C rings. A possible mechanism, utilizing a tautomerized ring, is illustrated in the accompanying structural formula. The decarboxylated product, coproporphyrinogen, enters the mitochondria and is acted upon by *coproporphyrinogen oxidase,* which oxidatively decarboxylates and oxidizes two of the propyl side chains to vinyl groups. A flavoprotein, *protoporphyrinogen oxidase,* oxidizes the methylene bridges between the pyrrole rings to form protoporphyrin IX. A somewhat different pathway from uroporphyrinogen is followed by sulfate-reducing bacteria. *Ferrochelatase* (protoheme ferrolyase inserts Fe^{2+} into protoporphyrin IX to form heme.

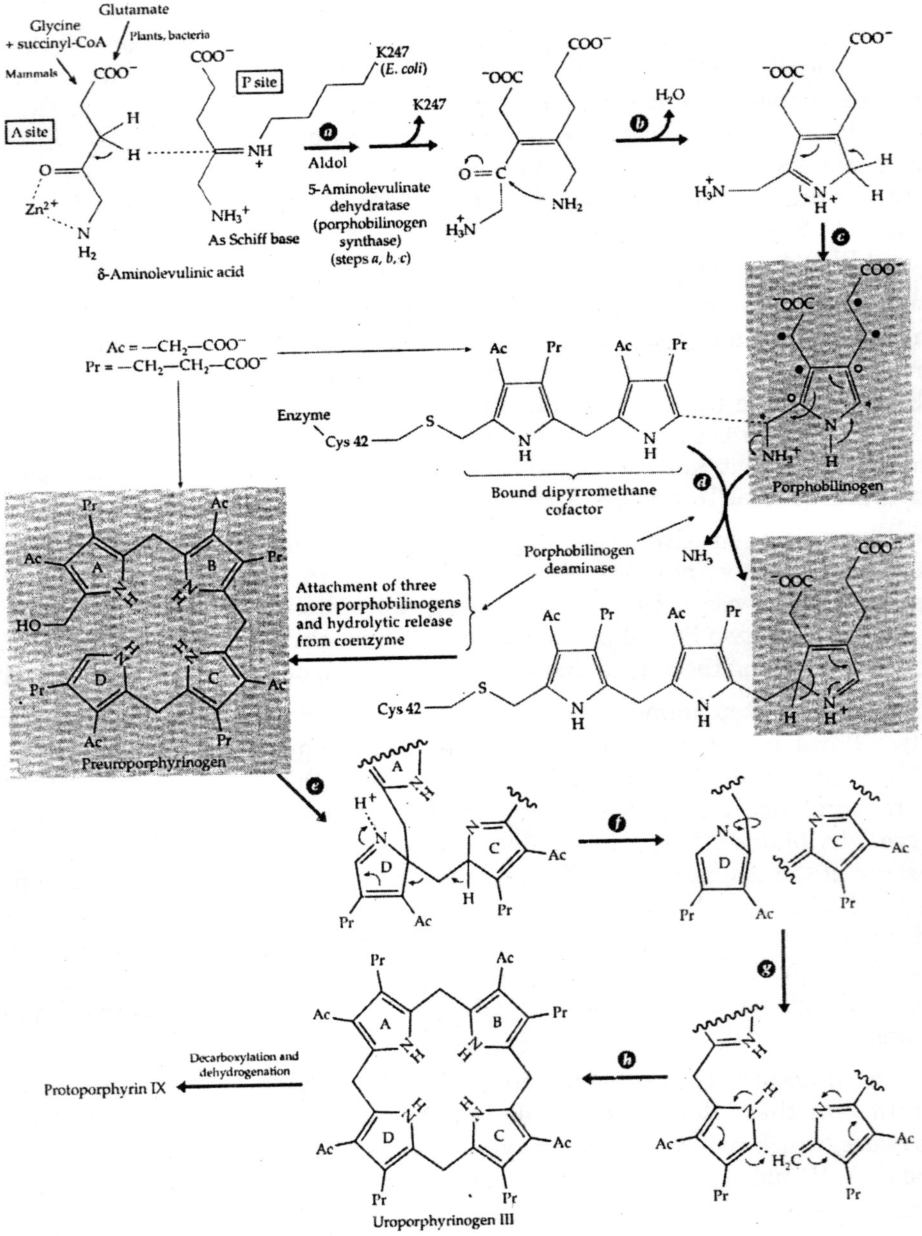

Fig. 6.21. Biosynthesis of porphyrins, chlorins, and related compounds.

Tautomerized pyrrole ring of uroporphyrinogen undergoing decarboxylation

The enzyme is found firmly bound to the inner membrane of mitochondria of animal cells, chloroplasts of plants, and chromatophores of bacteria. While Fe^{2+} is apparently the only metallic ion ordinarily inserted into a porphyrin, the Zn^{2+} protoporphyrin chelate accumulates in substantial amounts in yeast, and Cu^{2+} heme complexes are known. Ferrochelatase, whose activity is stimulated by Ca^{2+}, appears to be inhibited by lead ions, a fact that may account for some of the acute toxicity of lead. Heme *b* is utilized for formation of hemoglobin, myoglobin, and many enzymes. It reacts with appropriate protein precursors to form the cytochromes c. Heme *b* is converted by prenylation to heme *0* and by prenylation and oxidation to heme *a*. The porphyrin biosynthetic pathway also has a number of branches that lead to formation of corrins, chlorins, and chlorophylls.

Corrins. The formation of vitamin B_{12}, other corrins, siroheme, and related chlorin chelates requires a ring contraction with elimination of the methine bridge between rings A and D of the porphyrins. It is natural to assume that the methyl group at C–1 of the corrin ring might arise from the same precursor carbon atom as does the methine bridge in porphyrins, and it is easy to visualize a modified condensation reaction by which ring closure at step *e* in Fig 6.21 occurs by nucleophilic addition to a C = N bond of ring A. However, ^{13}C-NMR data ruled out this possibility.

When vitamin B_{12} was synthesized in the presence of ^{13}C-methyl-containing methionine, it was found that seven methyl groups contained ^{13}C. All of the "*extra*" methyl groups around the periphery of the molecule as well as the one at C-1 were labeled. Other experiments established uroporphyrinogen III as a precursor of vitamin B_{12}. Therefore, it appeared that the ring first closed in a normal way and then reopened between rings A and D with removal of the carbon that forms the methylene bridge. This turned out to be true, but with some surprises. The complex pathways of corrin synthesis have been worked out in detail.

This has been possible because of extensive use of ^{13}C and ^{1}H NMR and because the group

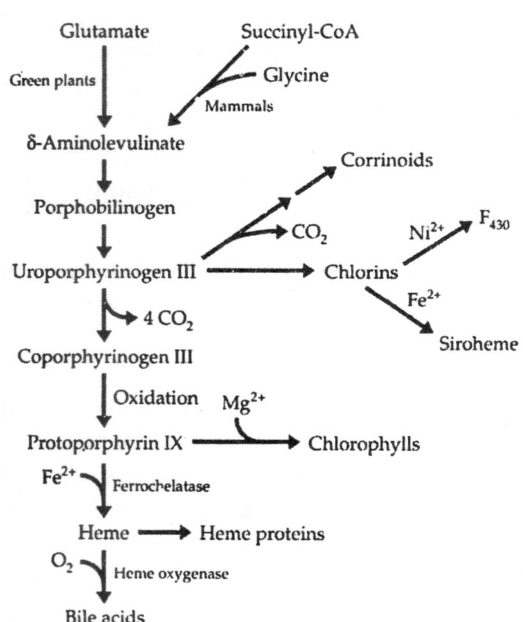

Fig. 6.22 Abbreviated biosynthetic pathways from δ-aminolevuinate to heme proteins, corrins, chlorophylls, and related substances.

of ~20 enzymes required has been produced in the laboratory from genes cloned from *Pseudomonas denitrificans*. The first alterations of uroporphyrinogen are AdoMet-dependent methylations on carbon atoms. Surprisingly, one of these is on the bridge atom that is later removed. The details, including the insertion of Co^{2+} by a *cobaltochelatase,* are described by Battersby and portrayed in Michal's *Biochemical Pathways*.

Chlorophyll. The pathway of chlorophyll synthesis has been elucidated through biochemical genetic studies of *Rhodobacter spheroides* which produces bacteriochlorophyll, from studies of cyanobacteria, and from investigations of green algae and higher plants, which make chlorophyll *a*. The first step in the conversion of protoporphyrin IX into chlorophyll is the insertion of Mg^{2+}. This reaction does not occur readily spontaneously but is catalyzed by an ATP-dependent *magnesium protoporphyrin chelatase*. Subsequently, the carboxyethyl side chain on ring C undergoes methylation and β oxidation.

Oxidative closure of ring E (step *d)* is followed by reduction of the vinyl group of ring B and of the double bond in ring D to form *chlorophyllide a***.** The latter is coupled with phytol, via phytyl diphosphate, to form chlorophyll *a*. Chlorophyll *b* is derived from chlorophyll *a,* evidently by action of an as yet uncharacterized oxygenase, which converts the methyl group on ring B into a formyl group. Bacteriochlorophylls also arise from chlorophyllide *a* and involve reduction of the double bond in ring B. Most photosynthetic bacteria make bacteriochlorophylls esterified with the C_{20} phytol, but some substitute the unsaturated C_{20} geranylgeranyl group and a variety of other isoprenoid alcohols.

The porphyrias. The human body does not use all of the porphobilinogen produced, and a small amount is normally excreted in the urine, principally as coproporpyrins. In a number of hereditary and acquired conditions blood porphyrin levels are elevated and enhanced urinary excretion (porphyria) is observed. Porphyrias may be mild and almost without symptoms, but the intensely fluorescent free porphyrins are sometimes deposited under the skin and cause photosensitivity and ulceration. In extreme cases, in which the excreted porphyrins may colour the urine a wine red, patients may have acute neurological attacks and a variety of other symptoms.

Lucid accounts of such symptoms, experienced by King George III of England, have been written. However, there are doubts about the conclusion that the king suffered from porphyria. Porphyria may result from several different enzyme deficiences in the porphyrin biosynthetic pathway. The condition is often hereditary but may be induced by drugs or other xenobiotic substances and may be continuous or intermittent. In one type of congenital porphyria uroporphyrin I is excreted in large quantities. The biochemical defect appears to be a deficiency of the cosynthase that is required for formation of protoporphyrin IX. Another type of porphyria results from overproduction in the liver of δ-aminolevulinic acid, a compound with neurotoxic properties, possibly as a result of its similarity to the neurotransmitter γ-aminobutyrate. This may account for some of the neurological symptoms of porphyria. Some mild forms of intermittent porphyria may go unrecognized.

However, ingestion of drugs can precipitate an acute attack, probably by inducing excessive synthesis of δ-aminolevulinate synthase. Among compounds having this effect are hexachlorobenzene and tetrachlorodibenzodioxin.

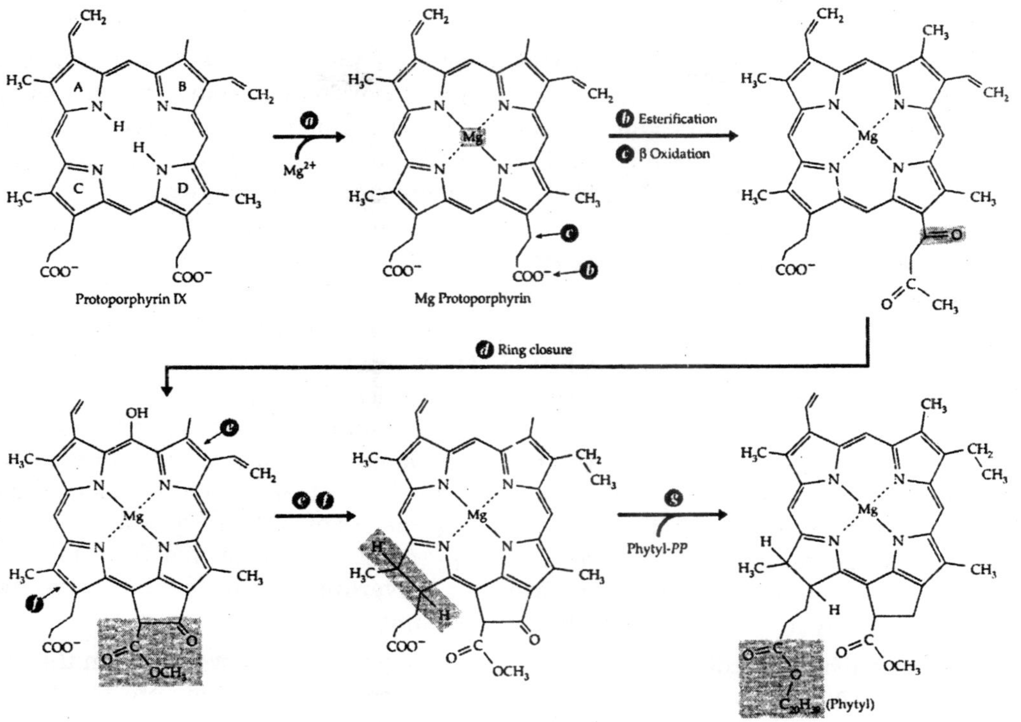

Fig. 6.23. Outline of the biosynthetic pathways for conversion of protoporphyrin IX into the chlorophylls and bacteriochlorophylls.

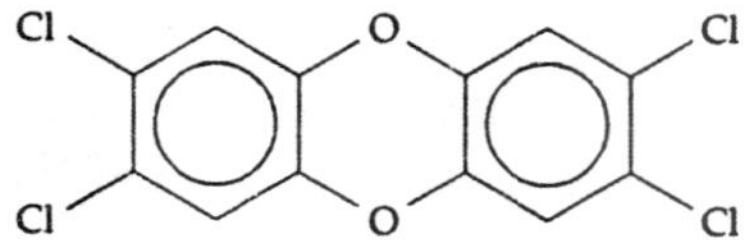

2,3,7,8-Tetrachlorodibenzo-*p*-dioxin

The latter is one of the most potent inducers of the synthase known. The tendency for this dioxin to be present as an impurity in the herbicide 2, 4, 5-trichlorophenoxyacetic acid (2, 4, 5-T) has caused concern. For rodents this dioxin may be the most toxic small molecule known, the oral LD_{50} for guinea pigs being only 1 μg/kg body weight. However, it is over 1000 times less toxic to humans. It is also a potent teratogenic agents

Synthesis of porphyrins in the liver is controlled by δ-aminolevulinate synthase. This key enzyme is sensitive to feedback inhibition by heme, but the increased synthesis of the enzyme induced by drugs can override the inhibition. Several times as much heme is synthesized in the erythroid cells of bone than in liver, but this is not subject to feedback regulation or to stimulation by drugs. Heme is a potent and toxic regulator. Malaria mosquitos, which utilize blood for food but do not have a heme oxygenase detoxify heme by inducing its aggregation into an insoluble hydrogen-bonded solid known as β-hematin.

The bile pigments. The enzymatic degradation of heme is an important metabolic process if only because it releases iron to be reutilized by the body. Some of the pathways are illustrated in Fig. 6.24. The initial oxidative attack is by the microsomal *heme oxygenases* which catalyze the uptake of three molecules of O_2, formation of CO, and release of the chelated Fe.

The electron transport protein NADPH-cytochrome P450 reductase brings electrons from NADPH to the oxygenase. An enzyme-substrate heme complex is formed with the oxygenase. Then the Fe^{3+} is reduced to Fe^{2+} which binds O_2 as in myoglobin or hemoglobin. The complex hydroxylates its own heme α-carbon, the other oxygen atom being reduced to OH^- by the Fe^{2+} and an additional electron from NADPH. The same enzyme catalyzes the next steps in which the ∝ carbon is split out as CO by reaction with two molecules of O_2 to form the open tetrapyrrole dicarbonyl compound *biliverdin,* one of the bile pigments. When $^{18}O_2$'was used, it was found that the biliverdin contains two atoms of ^{18}O, and that the CO contains one.

Heme from the cytochromes *c* appears to be degraded by the same enzymes after proteolytic release from the proteins to which it is bound. There are two human heme oxygenases. The first (HO-1) is synthesized principally in the liver and spleen. Its formation is strongly induced by heme. The second'heme oxygenase (HO-2) is distributed widely among tissues, but it is most abundant in certain neurons in the brain. Its major function may be to generate CO, which is now recognized as a probable neurohormone. Bacteria, such as *Corynebacterium diphtheriae,* employ their own heme oxygenase as *a* means of recovering iron that they need for growth. A large number of other open tetrapyrroles can be formed from biliverdin by reduction or oxidation reactions. Within our bodies biliverdin is reduced to *bilirubin,* which is transported to the liver as a complex with serum albumin.

In the liver bilirubin is converted into glucuronides glycosylation occurring on the propionic acid side chains. A variety of these bilirubin conjugates are excreted into the bile. In the intestine they are hydrolyzed back to free bilirubin, which is reduced by the action of intestinal bacteria to urobilinogen, stercobilinogen, and *meso*-bilirubinogen. These compounds are colourless but are readily oxidized by oxygen *to urobilin* and *stercobilin.* Some of the urobilin and other bile pigments is reabsorbed into the blood and excreted into the urine where it provides the familiar yellow colour. The yellowing ot the skin known as *jaundice* can occur if the heme degradation system is overburdened (e.g., from excessive hemolysis), if the liver fails to conjugate bilirubin, or if there is obstruction of the flow of heme breakdown products into the intestinal tract.

Bilirubin is toxic, and continued exposure to excessive bilirubin levels can cause brain damage. Bilirubin has a low water solubility and tends to form complexes with various proteins, perhaps partly because it assumes folded conformations rather than the linear one. These properties make it difficult to excrete. Thousands of newborn babies are treated for jaundice every year by prolonged irradiation with blue or white light which isornerizes 4Z, 15Z bilirubin to forms that are more readily transported, metabolized, and excreted. A more difficult problem is posed by the fatal deficiency of the glucuronosyl transferase responsible for formation of bilirubin glucuronide.

Efforts are being made to develop a genetic therapy. The open tetrapyrroles of algae and the chrpmophore of phytochrome are all derived from *phycoerythrobilin,* which is formed from biliverdin. The animal bile pigments have not been found in prokaryotes. However, *Clostridium tetanomorphun,* which accumulates uroporphyrinogen III, a precursor to vitamin B_{12}, and does not synthesize protoporphyrin IX, makes a blue bile pigment *bactobilin.* This is a derivative of uroporphyrin rather than of protoporphyrin.

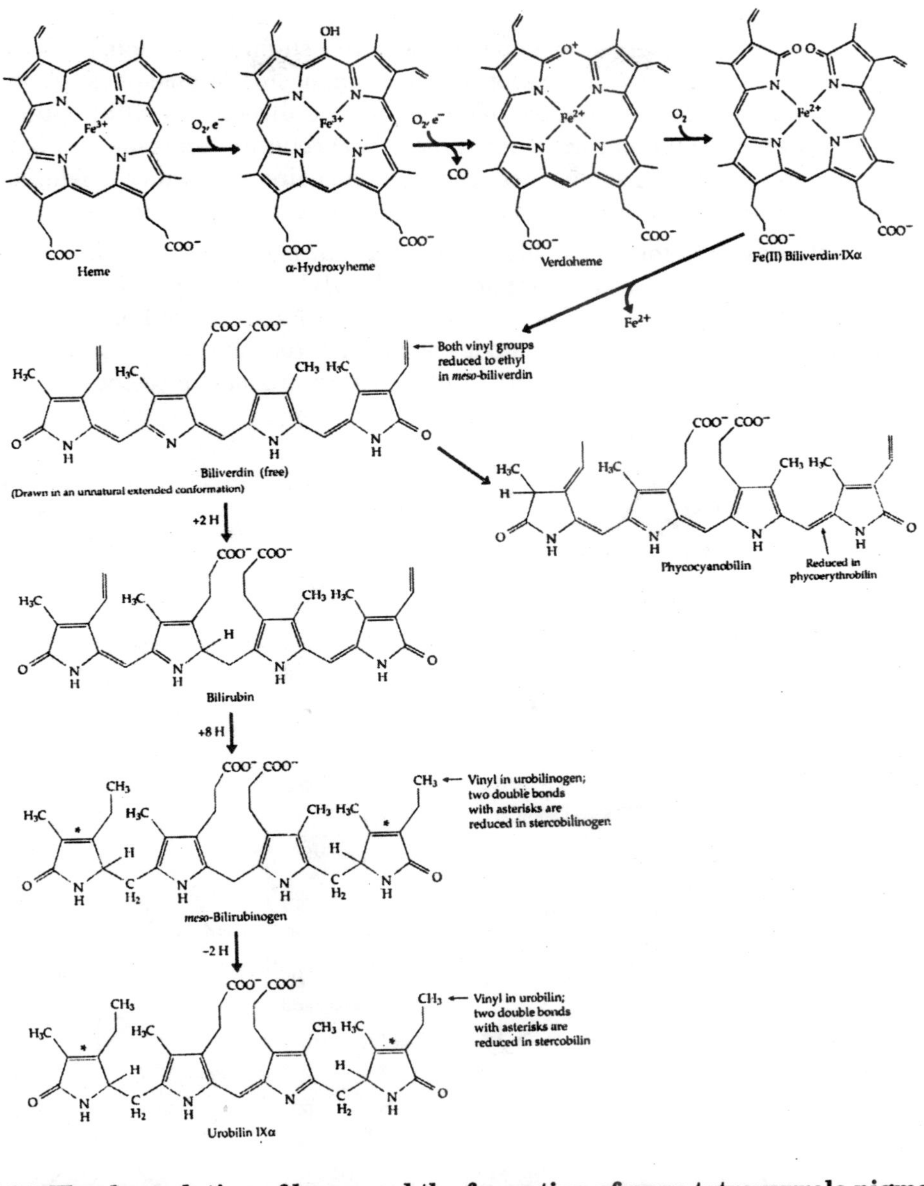

Fig. 6.24. The degradation of heme and the formation of open tetrapyrrole pigments.

G. CYSTEINE AND SULFUR METABOLISM

Cysteine not only is an essential constituent of proteins but also lies on the major route of incorporation of inorganic sulfur into organic compounds. Autotrophic organisms carry out the stepwise reduction of sulfate to sulfite and sulfide (H_2S). These reduced sulfur compounds are the ones that are incorporated into organic substances. Animals make use of the organic sulfur compounds formed by the autotrophs and have an active oxidative metabolism by which the compounds can be decomposed and the sulfur reoxidized to sulfate. Some of the chemistry of inorganic sulfur metabolism has been discussed in earlier chapters. Sulfate is reduced to H_2S by sulfate-reducing bacteria. The initial step in *assimilative* sulfate reduction, used by autotrophs including green plants and *E. coli,* is the formation of adenosine 5′-phosphosulfate (APS) (step *a*). The sulfate reducing bacteria reduce adenylyl sulfate directly to sulfite step *b*), but the assimilative pathway of reduction in *E. coli* proceeds through 3′-phospho-5′-adenylyl sulfate (PAPS), a compound whose function as "*active sulfate*" has alreardy been considered. Reduction of PAPS to sulfite, step *d*) is accomplished by an NADPH-dependent enzyme.

The same pathway is found in the alga *Chlorella,* but a second route of sulfate reduction occuring in green plants may be more important. Adenylyl sulfate transfers its sulfo group to a thiol group of a carrier, step *a*). The resulting thiosulfonate is reduced by a ferredoxin-dependent reductase. Finally, a sulfide group is transferred from the –S—S⁻ group of the reduced carrier directly into cysteine in a β-substitution reaction analogous to that described in the next paragraph.

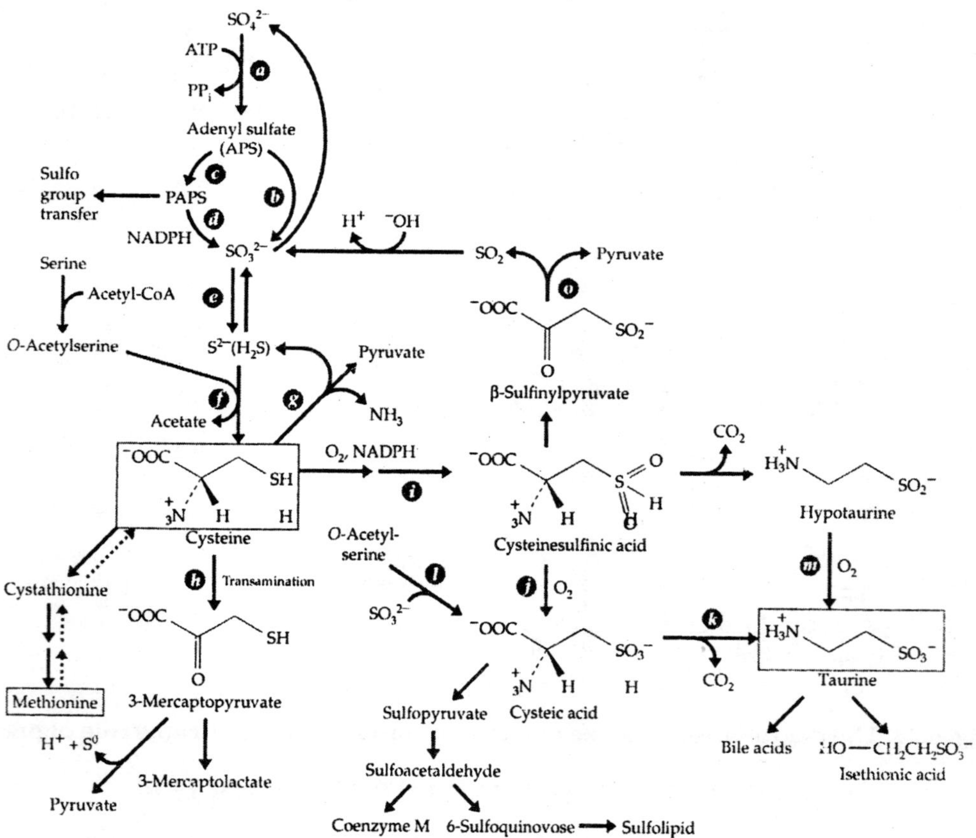

Fig. 6.25 Pathways of biosynthesis (green arrows) and catabolism of cysteine as well as other aspects of sulfur metabolism. Solid arrows are major biosynthetic pathways. The dashed arrows represent more specialized pathways; they also show processes occurring in the animal body to convert methionine to cysteine and to degrade the latter.

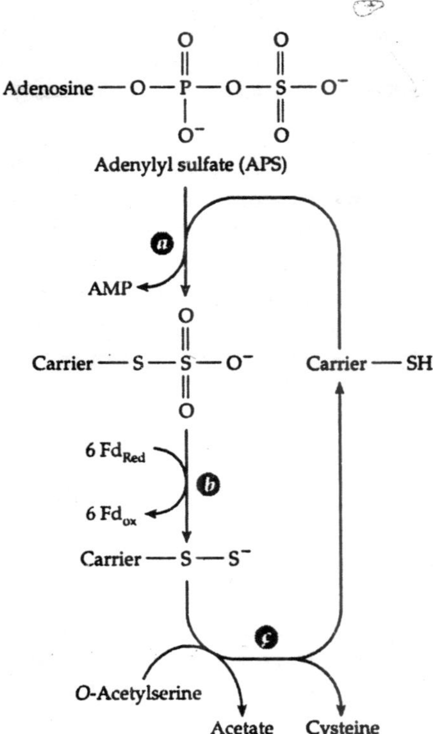

Synthesis and Catabolism of Cysteine

Cysteine is formed in plants and in bacteria from sulfide and serine after the latter has been acetylated by transfer of an acetyl group from acetyl-CoA (step *f*). This standard PLP-dependent β replacement is catalyzed by *cysteine synthase* (O-acetylserine sulfhydrase). A similar enzyme is used by some cells to introduce sulfide ion directly into homocysteine, via either O-succinyl homtoserine or O-acetyl homoserine. In *E. coli* cysteine can be converted to methionine.

In animals the converse process, the conversion of methionine to cysteine is important. Animals are unable to incorporate sulfide directly into cysteine, and this amino acid must be either provided in the diet or formed from dietary methionine. The latter process is limited, and cysteine is an essential dietary constituent for infants. The formation of cysteine from methionine occurs via the same transsulfuration pathway as in methionine synthesis in autotrophic organisms. However, the latter use cystathionine γ–synthase and (β-lyase while cysteine synthesis in animals uses cystathionine β-synthase and γ-lyase. Some bacteria degrade L-cysteine-or D-cysteine via the PLP-dependent α, β elimination to form H_2S, pyruvate, and ammonia (reaction g,).

Another catabolic pathway is transamination (rection *h*) to *3-mercaptopyruvate.* The latter compound can be reductively cleaved to pyruvate and sulfide. Cysteine can also be oxidized by NAD^+ and lactate dehydrogenase to 3-mercaptopyruvate. An interesting PLP-dependent β-replacement reaction of cysteine leads to *β-cyanoalanine,* the lathyritic factor present in some plants.[449] This reaction also detoxifies the HCN produced during the biosynthesis of ethylene from ACC. Cysteine and cystine are relatively insoluble and are toxic in excess. Excretion is usually controlled carefully. However, in *cystinuria,* a disease recognized in the medical literature since 1810, there is a greatly increased excretion of cystine and also of the dibasic amino acids.

As a consequence, stones of cystine develop in the kidneys and bladder. Patients may excrete more than 1 g of cystine in 24 h compared to a normal of 0.05 g, as well as excessive amounts of lysine, arginine, and ornithine. The defect can be fatal, but some persons with the condition remain healthy indefinitely. Cystinuria is one of several human diseases with altered membrane transport and faulty reabsorption of materials from kidney tubules or from the small intestine. Substances are taken up on one side of a cell and discharged into the bloodstream from the other side of the cell. In another rare hereditary condition, *cystinosis,* free cystine accumulates within lysosomes.

Cysteine Sulfinate and Taurine.

A quantitatively important pathway of cysteine catabolism in animals is oxidation to *cysteine sulfinate* (reaction *i*) a two-step hydroxylation requiring O_2, NADPH or NADH; and Fe^{2+}. Cysteine sulfinic acid can be further oxidized to *cysteic acid* (cysteine sulfonate), which can be decarboxylated to *taurine.* The latter is a component of bile salts and is one of the most abundant free amino acids in human tissues. Its concentration is high in excitable tissues, and it may be a neurotransmitter. Taurine may have a special function in retinal photoreceptor cells. It is an essential dietary amino acid for cats, who may die of heart failure in its absence, and under some conditions for humans. In many marine invertebrates, teleosts, anu amphibians taurine serves as a regulator of osmotic pressure, its concentration decreasing in fresh water and increasing in salt water.

A similar role has been suggested for taurine in mammalian hearts. A chronically low concentration of Na^+ leads to increased taurine. Taurine can be reduced to *isethionic acid,* another component of nervous tissue. Cysteic acid can arise in an alternative way from O-acetylserine and sulfite, and taurine can also be formed by decarboxylation of cysteine sulfinic acid to *hypotaurine* and oxidation of the latter (reaction *m).* Cysteic acid can be converted to the sulfolipid of chloroplasts. Another route ot metabolism for cysteine sulfinic acid is transamination to 3-sulfinylpyruvate, a compound that undergoes ready loss of SO_2 in a reaction analogous to the decarboxylation of oxaloacetate (reaction *o,*). This probably represents one of the major routes by which sulfur is removed from organic compounds in the animal body.

However, before being excreted the sulfite must be oxidized to sulfate by the Mo-containing sulfite oxidase. The essentiality of sulfite oxidase is evidenced by the severe neurological defect observed in its absence. Most of the sulfate generated in the body is excreted unchanged in the urine, but a significant fraction is esterified with oligosaccharides and phenolic compounds. These sulfate esters are formed by sulfo transfer from PAPS.

Mercaptopyruvate, Thiosulfate, and Assembly of Iron - Sulfur Centers

An important property of 3-mercaptopyruvate arises from electron withdrawal by the carbonyl group. This makes the SH group electrophilic and able to be transferred as SH^+, S°, to a variety of nucleophiles. Thus sulfite yields thiosulfate ($S_2O_3^{2-}$ + H^+, *a),* cyanide yields thiocyanate, and cysteine sulfinate yields alanine thiosulfonate. The reactions are catalyzed by *mercaptopyruvate sulfurtransferase,* an enzyme very similar to *thiosulfate sulfurtransferase.* The latter is a liver enzyme often called by the traditional name *rhodanese.* It acts by a double displacement mechanism, a thiolate anion in the active site of the enzyme serving as a carrier for the S atom being transferred. Equation 24-46 illustrates the transfer of S^0 from thiosulfate to CN^-, converting that ion to the less toxic thiocyanate. Crystallographic studies show that the negative charge on the thiol and dithiol anions of rhodanese is balanced by the partial. Positive charges at the N termini of two helices and by hydrogen bonds to protons of several side chains. This evidently explains how the negative thiosulfate anion can react with another anion, E-S⁻. Another interesting feature of this enzyme is that the monomer has a nearly perfect twofold axis of symmetry with respect to the protein folding pattern.

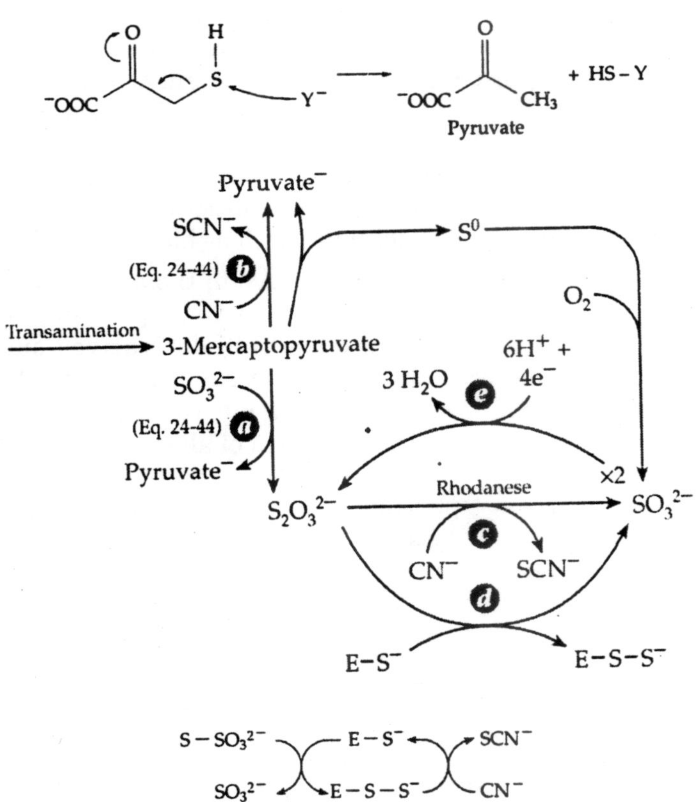

However, the symmetry is lacking in the sequence and only one-half of the molecule contains an active site. Yet another enzyme able to release or transfer sulfur in the S^0 oxidation state is the PLP-dependent *cysteine desulfurase* that is encoded by the *nifS* gene of the nitrogenase gene cluster shown in Fig 6.4. This enzyme releases S^0 from cysteine with formation of alanine for release of Se°from selenocysteine. As with rhodanese an active site cysteine accepts the departing S^0 of cysteine to form an enzyme-bound persulfide. This protein may in turn transfer the sulfur into the forming Fe–S or Fe–S–Mo clusters. Three PLP-dependent persulfide-forming sulfurtransferases related to the NifS protein have been found in *E. coli.* Similar enzymes are present in other organisms.

A sulfur atom may be transferred from the bound persulfide anion to acceptor proteins involved in metal cluster formation. Some members of the nifS-like family act on cystine to release free thiocysteine (cysteine peroxide), which may also serve as a sulfur atom donor.

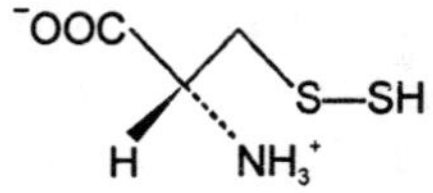

^{-}OOC — S—SH, H, NH_3^+

Thiocysteine (cysteine persulfide)

Thiocysteine can also arise in a similar manner through action of cystathionine β lyase on cystine. Thiocysteine is eliminated with production of pyruvate and ammonia from the rest of the cystine molecule. One of the nifS-like proteins of *E. coli* is thought to transfer a selenium atom from selenocysteine into *selenophosphate.* The latter can be formed by transfer of a phospho group from ATP to selenide HSe^-. The other products of ATP cleavage are AMP and P_i. Reduction of Se^0 to HSe^- is presumably necessary. Several additional proteins identified as necessary for metal-sulfide cluster formation are present in bacteria and in eukaryotes, both in the cytosol and in mitochondria. They may serve as intermediate sulfur carriers, as scaffolds or templates for cluster formatioa, or for insertion of intact Fe–S, Fe–S–Mo, or other types of clusters into proteins and into 2-selenouridine. Sulfurtransferases are also thought to be involved in insertion of sulfur atoms into organic molecules such as biotin, lipoic acicjor methanopterin.

A reaction that is ordinarily of minor consequence in the animal body but which may be enhanced by a deficiency of sulfite oxidase is the reductive coupling of two molecu'es of sulfite to form thiosulfate. Several organic hydrodisulfide derivatives such as thiocysteine, thioglutathione, and thiotaurine occur in animals in small amounts. Another biosyn-thetic pathway, outlined in Eq 6.47 converts sulfite and PEP into coenzyme M. This cofactor is needed not only for methane formation but also for utilization of alkenes by soil bacteria.

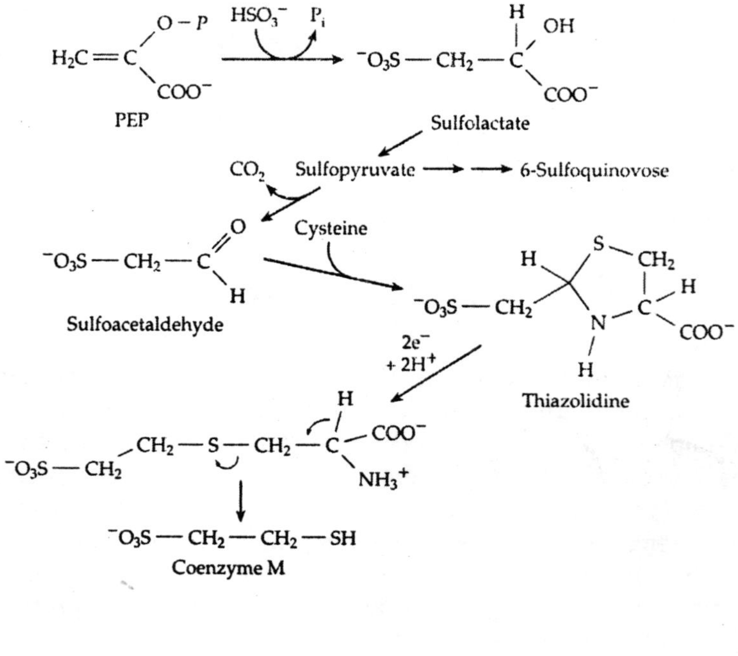

$H_2C{=}C(O{-}P)COO^-$ (PEP) $\xrightarrow{HSO_3^-,\ P_i}$ $^-O_3S{-}CH_2{-}C(H)(OH)COO^-$ (Sulfolactate)

Sulfolactate → Sulfopyruvate → → 6-Sulfoquinovose

Sulfopyruvate $\xrightarrow{-CO_2}$ $^-O_3S{-}CH_2{-}C(=O)H$ (Sulfoacetaldehyde)

Sulfoacetaldehyde $\xrightarrow{\text{Cysteine}}$ Thiazolidine ($^-O_3S{-}CH_2{-}$, H, S, CH_2, H, N–H, COO^-)

Thiazolidine $\xrightarrow{2e^- + 2H^+}$ $^-O_3S{-}CH_2{-}CH_2{-}S{-}CH_2{-}C(H)(COO^-)NH_3^+$

↓

$^-O_3S{-}CH_2{-}CH_2{-}SH$

Coenzyme M

IMMUNOLOGICAL RESPONSES

Considerable efforts have been made to determine the mechanistic changes that are responsible for immune deficiency and autoimmunity in the chronic alcoholic. Alterations that could reflect regulatory disturbances involving some aspect of T-cell function include: 1) increased serum levels of IgA, IgG, and IgM; 2) reduced cell-mediated immunity such as a loss of DTH reactions; 3) reduced lymphocyte numbers, altered subset distributions, persistent T-cell activation, and increased CD4/CD8 ratio; 4) changes in cytokine levels, which can be acute or chronic. In work to characterize the T cells of chronic alcoholics, it was established that alcoholics have an increased percentage of activated CD8+ T cells, measured as HLADR surface expression.

Subsequent work by three- and four-colour flow cytometry showed that these patients also have a significant shift from a naive cell expression of the leukocyte common antigen (CD45RA+) toward a memory cell phenotype (CD45RO+). This change is present in both CD8+ and CD4+ T cells and is accompanied by a significant reduction of L-selectin and an increase in CD 1 Ib on the surface of CD8+ cells, and by a loss of L-selectin in some but not all patients' CD4+ T cells. Most alcoholics also have stably increased expression of the carbohydrate-rich marker, CD57. The CD57+ T cell subsets of both patients and controls respond to polyclonal stimulation through the TCR with a rapid burst of IFN-γ and TNF-∝ production, but not IL-4.

In addition, the CD57+ subset does not require a second signal for this production, whereas the CD57- subset does. These findings are consistent with the concept that this subset is a differentiated effector cell with cytotoxic potential and a TH1 immediate response cytokine profile. It has been demonstrated also that there is a substantial increase in rapid cytoplasmic IFNγ production in the T cells of chronic ethanol consuming mice, along with an increase in the memory phenotype in the activated T cells of these mice. Because there are also increases in IL-4 production, the findings do not exclude TH2 excess in the chronic ethanol mice, but it has not been possible to demonstrate any TH1/TH2 skewing in antigen-specific T cells of chronic ethanol mice.

Recently, it has been observed in our laboratory that in spite of the activated phenotype of the CD8+ T cells in these mice, the production of antigen-specific CD8+ cells during the primary response to the intracellular bacterium *Listeria monocytogenes,* is significantly reduced. The adequacy of both polyclonal and antigen-specific T-cell responses is of increasing interest as we attempt to understand the mechanisms of the 19 Polyclonal and Antigen-specific Responses of T Cells and T Cell Subsets immunodeficiency of chronic alcohol abuse. This chapter describes some of the basic techniques available for characterization of these responses.

MATERIALS

Isolation of T cells From Spleen, Thymus, or Lymph Node

1. Dulbecco's phosphate buffered saline (DPBS, Invitrogen, Carlsbad, CA).
2. RPMI-1640 media (Invitrogen, Carlsbad, CA).
3. Lysis buffer (lyses red blood cells in cell suspension): 155*mM* NH_4Cl, 1 *mM* Tris-HCl, pH 7.20-7.5. Mix 4.15 *g* of NH_4Cl with 0.061 g of Trizma. Base (Sigma, St. Louis, MO.) to make 1L with reagent grade water. Adjust pH to 7.2-7.5. Filter to sterilize and store at room temperature (RT).
4. Trypan blue stain 0.4% (Invitrogen, Carlsbad, CA). Stain used to discriminate live cells from dead cells when counting cells in a hemocytometer.

Column Fractionation of Spleen Cell Suspension

1. Magnetic Cell Sorting (MACS) Kit (Miltenyi Biotech, Auburn, CA). Kits contain biotinylated-antibody cocktail and streptavidin microbeads for cell separation. Negative-selection kits are available for: CD4+ cells (cat. no. 130-090-860), CD8+ cells (cat. no. 130-090-859), and B cells (cat. no. 130-090-862).
2. Column buffer (for the MACS separation system): 2mM ethylene diamine tetraacetic acid (EDTA), 0.5% bovine serum albumin (BSA) in DPBS. Place 1 liter of DPBS on stir plate with stir bar. Add 0.75 g of EDTA (Sigma, St Louis, MO) and stir until dissolved. Then add 5 g of biotechnology grade BSA (Amresco, Solon, OH) and stir until dissolved. Store at 4°C.
3. Columns and magnetic holders for cell separation (Miltenyi Biotech, Auburn, CA).
4. Biotinylated-anti CD45R/B220 antibody (clone # RA3-6B2) for B-cell depletion of splenocytes.
5. Streptavidin microbeads (Miltenyi Biotech, Auburn, CA) for B-cell depletion of splenocytes.
6. Lipopolysaccharide (LPS, Sigma, St Louis, MO). Purified from bacteria, lyophilized and sterilized by γ-irradiation.

Flow Cytometry Staining

1. Flow cytometry wash buffer: PBS-0.1% azide (used for washing flow cytometry tubes). Add 1.Og of sodium azide (Fisher Scientific, Pittsburgh, PA) to 1 liter of DPBS. Store at RT.
2. Flow cytometry fixative (used to fix cells before flow cytometry): Mix 1 part 10% buffered formalin with nine parts flow cytometry wash buffer. Make fresh each week. Store at RT.

Stimulation and Cytokine Response of Cells

1. Conjugation buffer (for immobilizing antibody on culture plates): Mix 6.055 g of Tris-HCl (Sigma, St. Louis, MO) with 0.9 mL of *2N* HCl (Sigma, St. Louis, MO) in 1 liter of reagent grade water. Adjust pH to 9.25-9.50. Filter to sterilize and store at RT.

2. Purified anti-CD3 antibody (clone #1445-2C11), (Becton Dickinson, Sparks, MD). Dilute antibody in conjugation buffer (1 μg/mL). Dilute antibody just before use. Allow antibody to conjugate to plate for 4 h at RT or overnight at 4°C.
3. Purified, azide-free, low endotoxin, anti-CD28 antibody (clone # 37.51; Becton Dickinson, Sparks, MD). Add 1 μg/mL to media in anti-CD3 positive control well after washing and before adding cells.
4. Culture medium - RPMI-10, Mix 500 mL of RPMI with L-glutamine, 50 mL of heat-inactivated fetal calf serum (HI-FCS, Invitrogen, Carlsbad, CA), 100 μL of 50 mM beta-mercaptoethanol (Sigma, St. Louis, MO), 100 μL of 50mg/ml Gentamicin (Invitrogen, Carlsbad, CA). Filter to sterilize and store at 4°C. Use 1.0mL/well for a 24-well tissue culture plate.
5. BD Cytofix/Cytoperm™ Plus Kit (with BD GolgiPlug™Protein Transport Inhibitor; Becton Dickinson, Sparks, MD).
6. Micro-porous tape (3 M, Minneapolis, MN).

Antigen-Specific T Cells, Using Attenuated, L. Monocytogenes as a Model Organism

1. ACT A- *Listeria monocytogenes,* DPL 1942. The stock we used, was kindly provided by John Harty, University of Iowa for production of frozen stocks of attenuated Listeria.
2. Tryptic Soy Broth (TSB) Media for culture of, *L. monocytogenes:* Mix 30.0 g of BD Bacto™ Tryptic Soy Broth powder (Becton Dickinson, Sparks, MD) in 1 liter of reagent grade water. Warm to dissolve powder and autoclave. For plates, add 15.0 g of Bacto™ Agar (Becton Dickinson, Sparks, MD) per liter. Add streptomycin sulfate (Sigma, St. Louis, MO) to a final concentration of 50μg/mL.
3. *Listeria monocytogenes* Listeriolysin O molecule (LLO)-specific peptide (Bio-Synthesis, Inc., Lewisville, TX). The peptide sequence for the C57B1/6 CD4+ epitope used in our lab is NEKYAQAYPNVS (LLO AA# 190-201). For Balb/c the CD8+ epitope is GYKDGNEYI (LLO AA# 91-99).
4. Intracellular staining buffer for cytoplasmic staining. Flow cytometry wash buffer with 1 % HI-FCS.
5. Sterile saline solution for injection of Listeria. 0.9% Sodium chloride, inj., USP, (Hospira, Inc., Lake Forest, IL).
6. Igepal for homogenization of Listeria-containing mouse organs. Igepal C630 (o ctylphenoxy)polyethoxyethanol (Sigma, St. Louis, MO). Igepal is a surfactant to help release Listeria from spleen/liver cells.
7. Variable speed tissue homogenizer to homogenize organs. Tissue Tearor Model 985-370 (Biospec Products, Inc., Bartlesville, OK).
8. P815 cell line - (ATCC # TIB-64™, cultured in RPMI-10).

Adoptive Transfer

1. Pentamer wash buffer: 0.1% sodium azide, 0.1% biotechnology grade BSA in DPBS.

2. Pentamer fixation buffer: 1% HI-FCS, 2.5% formaldehyde (Fisher Scientific, Pittsburgh, PA) in DPBS.

METHODS

Isolation of T Cells From Spleen, Thymus, or Lymph Node

1. Organ Harvest

Place animals in a CO_2 chamber until respiration ceases. Remove animals and perform cervical dislocation. Excise desired organs and place them in 5 mL of either buffer (DPBS) or culture medium (RPMI-10).

2. Mechanical Disruption

Several alternative methods can be used :

(a) Transfer organs and buffer to a 60 × 15-mm Petri dish. Take two sterilized glass microscope slides with frosted ends. Place the organ on the frosted side of one slide and crush the tissue with the frosted end of the other. Spleens can be cut into two pieces to facilitate crushing.

(b) Place a cell dissociation sieve with a 60-mesh screen in a 100 × 15-mm Petri dish and transfer the organs and buffer to the sieve. Press the tissue through the screen using either a tissue grinder pestle or the plunger of a 10-mL syringe. The screen helps to sift out pieces of connective tissue. This method is especially effective when a number of organs are being pooled.

(c) Place a small volume of buffer in a glass tissue homogenizer. Grind tissue with the pestle to make a single-cell suspension.

2. Removal of Red Blood Cells

Transfer the cell suspension to a 15-mL culture tube, rinsing with enough buffer to fill the tube. Pellet the cells for 5min at 250 RCF. Aspirate the media from the pellet. Resuspend the cell pellet in lysis buffer, using 1 ml of lysis buffer/organ and incubate at RT for 1-2 min. Rinse away the lysis buffer by adding approx 10 volumes of buffer or media.

3. Quick Spin

Pull the chunks of connective tissue to the bottom of the tubes by centrifuging them until the force just reaches 70 RCF. Remove the tubes from the centrifuge and decant the supernatant into clean tubes. Pellet the cells in the new tube as before. Alternatively, cell suspensions .may be filtered through nylon mesh cell strainers (Becton Dickinson, Sparks, MD).

4. Count

Resuspend the cells in a small volume (~1mL) of buffer or media. Determine the cell density by counting an aliquot of the cells diluted 1:100 with Trypan blue stain in a hemocytometer. The cells are now ready for culture as whole splenocytes or for further fractionation of the population. Save cells for lineage marker analysis if using flow cytometry (10^6 cells per tube).

Column Fractionation of Spleen Cell Suspension

Cell suspensions may be fractionated to allow studies of purified populations or to enrich for rare cell types. The refinem^nt of magnetic bead separations allows easy fractionation of T and B cells with 90%+ purity. Cells are isolated by either positive or negative selection. The positive selection methods can result in activation of the selected cells, which is often an undesirable effect in culture.

Briefly, cells are incubated with a biotinylated antibody cocktail followed by antibiotin microbeads that react with the antibodies, attached to the cells. Elution of the cell/bead suspension over a column in a magnetic field retains the labeled cells on the column allowing the un-labeled (negatively selected) cells to pass unimpeded. The column is then removed from the magnetic field and the labeled (positively selected) cells are eluted, if desired.

Column Purification of CD4+ T Cells by Negative Selection

1. Antibody Incubation

Prepare single-cell suspension as discussed in Subheading 3.1. Suspend the cells in column buffer with 40mL of buffer for every 10^7 total cells. Add 10mL of biotinylated-antibody cocktail from kit (Miltenyi cat. no. 130-090-860) for every 10^7 total cells and incubate at 4°C for 10 min.

2. Microbead incubation

Add an additional 30µL of column buffer and 20 µL of anti-biotin microbeads (provided in kit) for every 10^7 cells. Mix and incubate 15min at 4°C.

3. Wash cells in approx 10x volume of column buffer and pellet by centrifuging for 5 min at 250 RCF. Resuspend them in de-gassed column buffer with 500 µL of column buffer for every 10^8 cells.

4. Magnetic column elution

Use one Miltenyi LS size column for every 10^8 cells. Place the columns in the proper magnetic holder and pre-wet columns with 3 mL of de-gassed column buffer, discarding the eluate. Load 10^8 cells (0.5 mL) on each column and let them run through, collecting the eluate. Wash the column four times using 3 mL of de-gassed column buffer in each wash. Pellet and resuspend the cells in the desired buffer, pooling tubes as needed. The negatively selected cells are then ready for culture or further treatment. Take aliquots of 10^6 cells for lineage marker analysis if using flow cytometry to determine the purity of the cell fractionation. If positively selected cells are to be collected, remove the column from the magnet holder and elute the retained cells into a clean tube with four 3 mL.washes of column buffer. Otherwise, discard the used column.

Column Purification of CD8+ T Cells by Negative Selection

Follow the same protocol as the CD4+ purification given above (Miltenyi Biotech Inc., cat. no. 130-090-859).

Column Purification of B Cells with Miltenyi B-Cell Kit; Overnight Activation

1. Antibody incubation: prepare single-cell suspensions of splenocytes from two or three normal mice. Resuspend cells in column buffer with 40 µL of buffer and 10µL of antibody

cocktail from kit (Miltenyi Biotech Inc., cat. no. 130-090-862) for every 10^7 cells. Incubate at 4°C for 10 min.

2. Microbead incubation: Add an additional 30 μL of column buffer and 20 μL of anti-biotin microbeads for every 10^7 cells. Incubate 15 min at 4°C. Wash once with approx 10x volumes of column buffer. Resuspend cells in de-gassed column buffer at 500μL/10^8 cells.
3. Magnetic column elution: prepare column and elute cells.
4. Resuspend purified cells in RPMI-10 at 2×10^6 cells/mL and add 20 μg/mL LPS. Incubate overnight at 37°C, 5% CO_2.
5. The next day, wash cells twice with DPBS or RPMI to remove LPS. Resuspend in RPMI-10. Cells are now ready for peptide loading.

B-Cel! Depletion of Splenocytes

1. Antibody incubation

Prepare a single-cell splenocyte suspension as discussed in Subheading 3.1. Resuspend the cells in RPMI or DPBS without serum. Add biotinylated-anti-CD45R/B220 antibody at a concentration of 1 μg antibody/106 spleen cells. Incubate 20 minutes at 4°C.

2. Microbead incubation

Wash away excess antibody with approx 10x volumes of RPMI or DPBS by pelleting cells once. Add streptavidin microbeads (Miltenyi Biotech Inc., cat. no. 130-048-101) with 10 μL of beads for every 107 cells and incubate 15 min at 4°C to allow the attachment of labeled cells to the beads. Wash once with approx 10x volumes of column buffer. Resuspend cells in de-gassed column buffer at 500μL/108 cells. 3. Magnetic column elution : Prepare column and elute cells.

Flow Cytometry Staining

Staining lineage markers on the cells is used to demonstrate both that the starting cell population was as expected and that any.fractionation of the cells was successful. The markers used may vary from experiment to experiment. Markers to document the makeup of the donor animals would include markers for the various cell line-ages: T cells (e.g., CD3, CD4, CD8), B cells (e.g., CD19), macrophages and DCs (e.g., CD11b, CD11c), etc., as determined by the aim of the experiment. The same panel of stains is used on any purified cr enriched cell suspension to demonstrate that the fractionation was successful.

1. Stain cells

Place 10^6 cells in a 12 × 75-mm polystyrene tube in a volume of approx 100μL. Add the antibodies to the cells. Antibodies either can be from commercial sources or can be locally produced from hybridoma lines. Commercially produced antibodies may have a recommended working concen-tration; however, the optimum concentration of antibody should be determined by titrating the response of the cells used in any particular laboratory.

2. Incubate Stain

Hcubate cells with antibodies in the dark at either RT or 4°C for 15-20min.

3. Wash

wash to remove excess antibody with 1 ml of flow cytometry wash buffer. Pellet cells for 5 min at 250 RCF.

4. Secondary Antibody

If a biotin-labeled primary antibody is used, add the fluores-cent-labeled-streptavidin conjugate in wash buffer at 1 μg/10^6 cells and incubate for 20 min. Wash as above.

5. Fix

Resuspend cells in 0.3 mL of flow fixative. Store at 4°C in dark until ready to analyze.

Stimulation and Cytokine Response of Cells After Polyclonal Stimulation

The response of cells is measured by exposing, a controlled population of cells to defined stimuli in culture. Positive and negative control conditions are included to aid in determination of the cells' response to the test stimuli. The volume of cells to be used in culture is determined by the parameters being measured and by the number of cells available. Cytokine production provides one means to quantify the functional response of the cells being tested. Cytokine production can be quantified by measuring either the total (soluble) cytokines secreted or the intracellular cytokines.

Intracellular staining allows the identification of the cell type producing the cytokines within a given 4-6 hour period of stimulation. Brefeldin A prevents the release of cytokines from cells by blocking protein transport from the endoplasmic reticulum (ER) to the Golgi apparatus. The number and subset of cells producing cytokines can be identified using combined cell surface marker and cytoplasmic staining.

1. Preparation of the Culture Plate

Volumes (1.0mL/well) and concentrations given are for a 24-well culture plate. Prepare the plates by adding a polyclonal stimulator, such as anti-CD3 antibody, to the positive-control wells. Immobilized monoclonal anti-CD3 antibody (clone #145-2C11) activates T cells in culture in a polyclonal fashion providing a good positive control. Dilute the antibody to a working concentration of 1 μg/mL with conjugation buffer and pipet into (positive control) culture wells (1.0mL/well). Seal with Parafilm® (Pechiney Plastic Packaging, Menasha, WI) and place plate at 4°C overnight or keep at RT for 4h. Wash the wells.

Do not allow wells to dry out. After the final wash of the anti-CD3 wells, add 0.5 mL of culture medium to *all* wells that will have cells. If intracellular cytokine levels will be measured, use media containing 2μl/mL BD GolgiPlug™(BD Cytofix/Cytoperm™Plus kit, Becton Dickinson, Sparks, MD) and place 0.5 mL in the plates/wells for those samples. The final Brefeldin A concentration in the wells after the addition of cells will be 1 μg/mL. The cells being cultured can be either from whole organ preps (e.g., whole splenocytes), B-depleted or purified T cells (CD4+ or CD8+).

2. Adjust the concentration of cells to be cultured to 4.0 × 10^6/mL and add 0.5 mL of the cell suspension to the wells (final concentration of 2 × 10^6 cells/mL). Cover and seal lid

to plate with microporous tape (3M, Minneapolis, MN.). Incubate for the desired time period. Cells and/or culture supernatants may be harvested for assay at the end of the incubation period.

3. For supernatant collection, pipet the liquid contents of each well into a centrifuge tube. Centrifuge the tubes to pellet any cells and then transfer the supernatants to tubes. Freeze at -80°C for cytokine ELISA or Cytokine Bead Array (CBA) analysis.
4. To collect cells resuspend them with a Pasteur pipet. Transfer the suspended cells to a centrifuge tube. Wash the well twice with the same volume of media to increase cell recovery. Centrifuge to pellet the cells and discard the supernatant. The cells are then ready for cell surface marker staining, cytoplasmic staining, or RNA/DNA isolation, etc.

(Antigen-Specific T Cells, Using Attenuated, L. Monocytogenes as a Model Organism

Antigen-specific T cells can be induced in the experimental animal model for study of adaptive immunity in the experimental animal encompassing primary, secondary and memory responses. Inoculation of the experimental animal with an infectious organism (either virulent or attenuated) or with an isolated immunogenic molecule produces an antigen-specific response that can be measured later by stimulating cells in culture with the infectious organism, a specific molecule or a peptide epitope. Any cell preparation method discussed above may be used, depending on the aim of the experiment.

The cell preparation may be from either individual mice or pooled populations. Both the economics and logistics of cell population purification discourage processing a large number of samples. Listeriolysin O (LLO) T-cell peptide epitopes have been identified that can be used to measure T-cell specific responses. The peptides used in our lab are targeted to CD4+ or CD8+ T cells in C57B1/6 or Bal/6 mice, respectively. The specific peptide chosen must be titered with the cell system being studied to determine the optimum concentration for culture.

Inoculation

ACT A - *Listeria monocytogenes* (DPL 1942), may be used to infect mice and elicit a measurable immune response to known specific peptide epitopes. ACT A - Listeria has been engineered so that the intracellular bacterium is unable to nucleate host cellular actin and therefore is incapable of moving from cell to cell to amplify the infection; it is thus less likely to kill the host.

1. Preheat a shaking incubator to 37°C. Prepare a 50-mL culture tube with 10 mL of TSB media containing 50 µg/mL streptomycin. Remove 1 mL of media and save to use as the blank in a spectrophotometer. Thaw a 1-mL aliquot of ACT A- Listeria and add it to the 9 mL of media in the culture tube. Incubate in the pre-warmed shaking incubator for approx 1 h. After 30 min, the OD_{600} should be at least 0.01. Continue monitoring OD_{600} until it is 0.06-0.1 (log phase). Remove the tube from the incubator and hold it at RT. An OD_{600} of 1.0 = 10^9 CFU/mL.
2. Calculate the amount of inoculum to be prepared. Example: Inoculation of 10 mice with 0.3-mL doses of 5×10^5 CFU Listeria requires 3,OmL, but extra should be prepared. Prepare 6.0 mL (20 doses):

20 doses × (5 × 10^5 CFU/dose) = 1 x 10^7 CPU needed.

A culture with an OD_{600} = 0.06 contains 6 × 10^7 CFU/mL.

(1 × 10^7 CFU needed) / (6 × 10^7 CFU/mL) = 0.167mL culture needed. Mix the 0.167mL of Listeria culture with 5.833 mL of sterile saline and inject mice intraperitoneally. For intravenous infection, the number of CFU's of Listeria given the mice is reduced to 10^5 CFU and the injected volume is decreased to 0.1 mL.

3. Prepare four or five 10-fold serial dilutions of the inoculum. Plate 20 μL of each dilution on separate TSB-strep agar plates to determine the actual dose of bacteria that was given to the experimental animals. Incubate plates overnight at 37°C and count the colonies the next day. (# colonies) × (dilution factor) × 50 = # CFU/mL.

Cell Preparation

Harvest organs from Listeria-infected mice. Prepare singel-cell suspensions for culture from mice as discussed above. Prepare postitive and negative control wells, as well as the LLO-specific peptide wells, and add cells. Positive control wells contain anti-CD3 +/– anti-CD28 as needed. If measuring intracellular cytokines, add Brefeldin A (GolgiPlug™in kit) to the wells 4–6 h before harvesting the cells.

Culture

Culture the cells in a 37°C, 5% CO_2 incubator for the time period determined to provide the desired response for the cells being tested (usually 6h;). A complete set of test/control wells is required for each time period used. After the time has elapsed, supernatants can be collected to measure soluble cytokines or cells may be collected from Brefeldin A-treated wells for measurement of intracellular cytokines. Freeze supernatants for soluble cytokine measurement, at –80°C until assayed. Stain cells for intracellular cytokine expression.

Antigen Presenting Cells (APCs)

When culturing purified T cells (CD4+, CD8+ cells), APCs may be pre-loaded with specific peptide antigen. Either activated purified B cells or cell lines (e.g., P815 cells) may be used.

1. Whole Splenocytes

In a whole spleen preparation, the native APCs are present and functional and therefore no additional APCs are required. This is the starting point for determining optimum antigen concentrations for culture. Prepare whole spleen, single-cell suspensions from either pooled or individual animals. Add prepared cells to the test wells (negative control, positive control and peptide) with a final concentration of 2 × 10^6 cells/mL. The concentration of peptide must be titrated in the cell system being studied. Start with a range of serial dilutions (e.g. 10^{-4} *M*, 10^{-5} *M*, etc.) and determine the concentration that induces the optimum response. Culture the cells for the specified time and collect cells and/or supernatants for analysis.

2. Loaded B Cells

Prepare activated B cells, beginning.the day before, as discussed in Subheading 3.2.3., culturing (activating) them overnight with LPS. Wash two or three times with approx 10x volumes of.DPBS or RPMI to remove LPS. Adjust B-cell concentration to 2 × 10^6 cells/mL in RPMI-10 and add the specific peptide at the concentration found to be optimum when culturing whole splenocytes with the peptide. Incubate for 1 h at 37°C. Wash twice with 10x volumes of RPMI. Add loaded B cells to culture wells along with the purified T cells in a range of several T cell:B cell ratios such as 100:1 down to 1:1. No additional peptide is added to the wells.

3. P815 Cell line as APCs

P815 (ATCC # TIB-64™) is a mouse lymphoblast-like mastocytoma cell line that can be used as antigen presentation cells. They can be loaded with specific peptides in the same manner as loading purified B cells P815 cells, grown in suspension, are washed once with RPMI-10. The cell concentration is adjusted to 5 × 10^6 cells/mL and peptide added to the optimal concentration as determined by-titration in the cell system being used. Incubate for 1 h at 37°C. Wash once or twice with RPMI-10. The cells are then ready to add to the prepared culture plate. It is best to initially test a range of cell ratios as with the B cells. A negative control well is required in which nonpeptide-loaded P815 cells are added to the purified T cells. This is used to determine non-specific binding during analysis of the flow cytometry data.

Intracellular Staining for Cytokines after Brefeldin A Treatment

This staining is a multistep process. *The cell* surface markers (CD4, CDS, etc.) must be stained first, then the cells are permeabilized to allow the antibodies against the cytokines to reach the cytoplasmic constituents of the cells. Figure 7.1 shows a plot of purified CD4+ cells stained for cytoplasmic interferon-γ (IFN-γ) and CD44+ expression. CD4+ T cells were isolated from C57B1/6 mice 7d after inoculation with. ACT A-Listeria, and incubated with LLO peptide for 6 h.

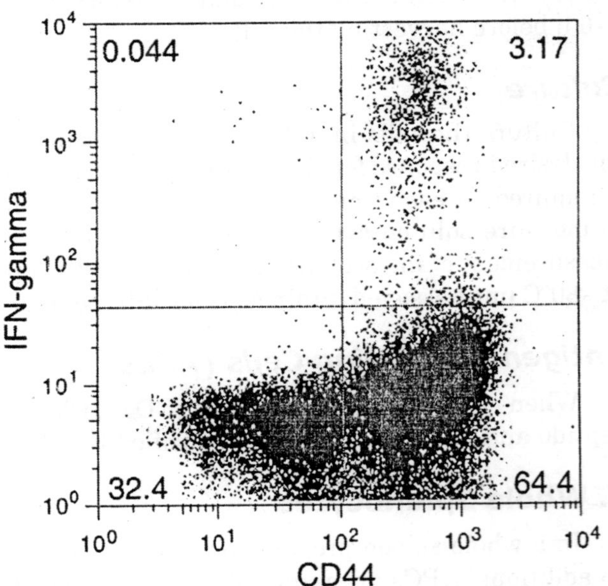

Fig. 7.1. Pattern of cytoplasmic IFN-γ and surface CD44+ expression in a culture of purified CD4+ cells. CD4+ cells were isolated from C57B1/6 mice 7d after inoculation with ACT A- *Listeria monocytogenes* and cultured for 6h with B cells loaded with listeriolysin peptide (AA# 190–201) at a ratio of 1 : 1 with added GolgiPlug™, then permeabilized, stained and analyzed on a BD™FACS Calibur. The IFN-γ+ cells represent those responding specifically to the'peptide during the peak of the primary response. The numbers represent the percentage of CD4+ cells in each quadrant.

1. Stain the surface markers after the final wash step, omit the fixation buffer. Instead, the cells are washed twice with 1.0 mL of intracellular staining buffer.
2. Resuspend the cell pellet in 250 μL of the Cytofix/Cytoperm solution provided in the BD

Cytofix/Cytoperm™Plus kit (Becton Dickinson, Sparks, MD.). Vortex the cells as the buffer is added to decrease clumping of the cells. Incubate 20min at 4°C.

3. Prepare the Perm/Wash buffer (approx 4.2mL/sample will be needed) by diluting the 10 x stock provided in the kit with reagent grade water. Wash the cells twice with 1.0 mL of diluted Perm/Wash buffer, vortexing the tube as the buffer is added.
4. During the second wash of the cells, prepare fluorescent-labeled anti-cytokine solution by diluting the antibody in Perm/Wash buffer (50 μL of diluted antibody/sample). Vortex the cell pellet while adding the antibody and incubate in the dark at 4°C for 30 min.
5. Wash twice with 1.0 mL of Perm/Wash buffer. Vortex while adding buffer.
6. Resuspend in 250 μL of staining buffer and vortex well. Keep at 4°C in the dark until the samples can be run on a flow cytometer. They must be run within 24 h or signal could be decreased.

Time Course of Response

Establish a response curve by measuring the cytokine response over time. A typical response curve for *Listeria monocytogenes* infection is shown in Fig. 7.2. Culture cells from Listeria-infected mice at various times after inoculation (e. g. daily, weekly, monthly) to measure immune response over time depending on the parameters being studied. Soluble (total) cytokines may be measured by either cytokine ELISA or by CBA assay (e.g. Becton Dickinson's BD™ CBA or Bio-Rad's Bio-Plex™).

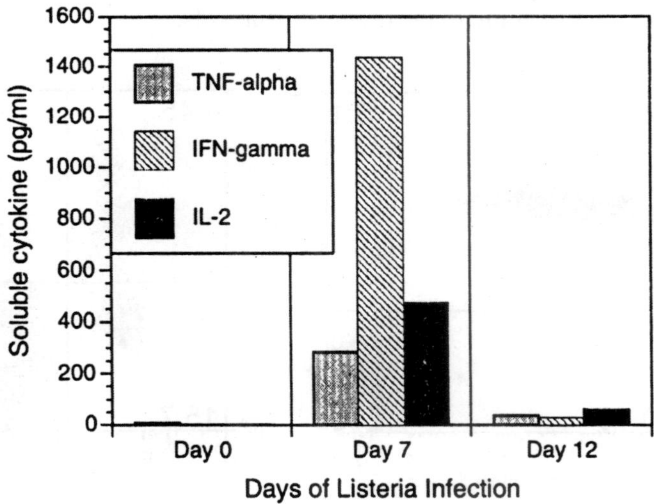

Fig. 7.2. Time course of soluble cytokine production after ACT A- *Listeria monocytogenes* inoculation Unfractionated spleen cells cultured for 8h with listeriolysin peptide (190–201) at a concen-tration of 10^{-4} *M*. Time course with uninfected (Day 0), seven and twelve day Listeria inoculated C57B1/6 mice

Adoptive Transfer

Adoptive transfer of immunity can be used to study primary and secondary immune responses. Congenic strains of mice (e.g., C57 B1/6J [CD45.2+] and B6.SJL-Ptprc^a Pep^b/BoyJ

[CD45.1+]) allow one to discriminate between the responding cell types. Monoclonal antibodies are available to separate CD45.1+ cells (clone # A20) and CD45.2+ cells (clone # 104).

An immune response is induced in the donor mice. Then T cells are transferred to congenic recipient mice. Inoculation of the recipient animals produces cytokine responses from both donor (secondary response) and recipient primary response) T cells. Donor mice are inoculated with Listeria. T cells are isolated, usually after 28+ days, and transferred intravenously to congenic recipients. The recipient mice are then inoculated immediately with Listeria.

At the peak primary time for Listeria infection, 6–7 d later, spleen cells are cultured with and without Listeria-specific peptide and then the cytokine response is measured. Figure 7.3A illustrates the resolution of congenic cells in a whole spleen culture one day after adoptive transfer of CD4+ T cells. Staining for intracellular cytokine expression and gating on CD4+ cells allows the discrimination of cytokine-producing donor cells from cytokine-producing recipient cells as shown in Fig. 7.3B.

Adoptive Transfer Utilizing Congenic Strains of Mice

1. Inoculate the donor mice (e.g., the CD45.1+ strain, B6.SJL-Ptprc^{a} Pep^{b}/BoyJ) with Listeria.
2. Isolate T cells from donor mice at the desired interval by column purification or B220-depletion. Resuspend cells in DPBS and inject intravenously or via retro-orbital sinus into recipient mice (e.g., the CD45.2+ strain, C57B1/6J). Adoptive transfers have been performed in our lab using up to a 1:1 ratio of donor to recipient splenic lymphocytes as starting material. Average cell recoveries after purification of donor cells allow the routine transfer of 1×10^7 CD4+ cells and 8×10^6 CD8+ cells. Fewer transferred cells may be preferred in some experimental protocols.

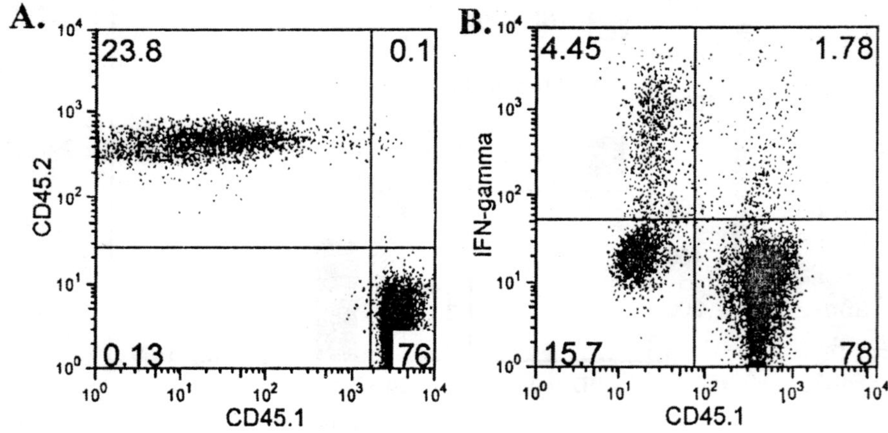

Fig. 7.3. Patterns of congenic CD4+ cells after adoptive transfer. Donor (CD45.2+) CD4+ cells were transferred to recipient (CD45.1+) mice. Recipients were inoculated with Listeria 1 day later and organs harvested 6d after Listeria inoculation. Analyzed on BD™ FACS Calibur. (A) Whole splenocytes gated for total CD4+ cells. (B) Whole splenocytes cultured 6 hours with CD4+ LLO-specific peptide and GolgiPlug™. Stained for cytoplasmic IFN-γ. In this case the donor cells were obtained 58 d after Listeria inoculation. Thus, the donor haplotype (CD45.1$^-$) response is the response of memory cells at the peak time of memory response (6d), whereas the recipient response (CD45.1+) is a primary response, and is one day before the usual peak primary response at 7 d after inoculation.

3. Infect recipient mice. This can either be done the same day or the day following the cell transfer.
4. After 6d (peak memory response time), harvest organs and culture cells, either as whole spleen or purified T-cell fractions as described previously. Intracellular staining for cytokines combined with haplotype staining discriminate the response of donor cells from those of the recipient. Analysis requires gating for the transferred T cells (CD4+ or CD8+) followed by displaying anticytokine-labeled cells against either the donor or recipient haplotype.

Fluorescent-labeled MHC Class I Pentamers and MHC Class II Tetramers

These are commercially available to identify antigen-specific T cells (CD8+ and CD4+ respectively) without the need for congenic strains of mice. Antigen-specific T cells are generated and cultured as discussed previously.

1. Collect 10^6 cells to stain for flow cytometry. Wash with 1.0 mL and resuspend in 50 μL of pentamer wash buffer.
2. Add 10μL of fluorescent-labeled pentamer and incubate 10-15 min at RT in the dark.
3. Wash with 1.0 mL of pentamer wash buffer and add 50μL of pentamer wash buffer containing fluorescent-labeled anti-T-cell antibody.
4. Incubate on ice 20-30 min in the dark.
5. Wash twice with 1.0 mL of pentamer wash buffer and resuspend in 0.3 mL of pentamer fix buffer. Store the .samples in the dark at 4°C until, they can be analyzed by flow cytometry.

NOTES

1. Column fractionation buffer: The EDTA must be completely dissolved in the DPBS before the BSA is added or the EDTA will not dissolve well. This buffer must be de-gassed at RT for at least 1 h before applying it to a column. If the buffer is not de-gassed, bubbles may form inside the column, greatly slowing the flow rate and decreasing the cell recovery. De-gassing is easily accomplished: place column buffer in a sterile Erlenmeyer flask and apply a vacuum for 1 h (ordinary house vacuum, if available, is usually sufficient).
2. LPS preparations from several different bacterial strains are available. It is necessary to screen a selection of preparations with any cell system to find the optimal combination of LPS source and concentration. All LPS preparations used should be purified to eliminate cell stimulation by TLR ligands other than LPS.
3. Wash the anti-CD3-conjugated wells of the culture plate before adding cells to remove all traces of the conjugation buffer. Wash twice with 2mL of DPBS or RPMI. Wash once more with 1 mL of culture media. Do not allow wells to dry out between wash steps. Purified antibody that is suitable for use in co-stimulation, immunoprecipitation, flow cytometry, and western blotting is desired.

   The concentration used here (1 (μg/mL) has been titered in our lab and has proven to provide maximum response in T-cell cultures. If a less intense response is desired, the antibody concentration may be diluted. Soluble anti-CD3 antibody may be used in some culture applications in place of immobilized antibody and is simply added to the culture well. The antibody preparation used for this must be azide-free and low endotoxin (NA/LE).

4. Consistency in the plasticware (Culture plates and flasks) used for tissue culture is important. Slight differences in manufacturing processes can have an effect on how cells will fare in culture. Different sources of plasticware should be tested with the cell system under investigation. Cell volumes in culture can be scaled proportionately up or down (from the 1.0 mL/well used for a 24-well tissue culture plate) if a different plate configuration is preferred.
5. Production of frozen stock of attenuated *Listeria monocytogenes*. Listeria must be cycled through mice periodically to maintain the ability to enter cells.
   a. Inoculate a 10-mL TSB-strep broth culture with Listeria and grow to log phase (OD_{600} = 0.06).
   b. Inject 10^7 CPU (in 0.3 mL of sterile saline) into two or three Balb/c mice (intraperitoneally). Do this as late in the day as possible.
   c. As early as possible the next day, harvest and weigh livers and spleens from all infected mice. Attenuated Listeria is cleared from the body quickly so keep the in vivo time to no more than 16-18 h.
   d. Homogenize individual organs in 0.2% Igepal in reagent grade water, 10 mL for liver and 5mL for spleen, using the high speed on a Tissue Tearor Homogenize 10–15 s until tissue is liquefied. Prepare three 10-fold serial dilutions of the organ homogenates with 0.2% Igepal and plate 20 µL of each sample, including the whole organ homogenates, on individual TSB-strep agar plates and incubate overnight at 37°C. Listeria infection is typically reported as CFU/gram in liver and total CFU in spleen.
   e. Late the next day, pick one, isolated, colony to inoculate a 10 mL TSB-strep overnight culture. Incubate in a shaking incubator at 37°C.
   f. Use the entire 10 mL overnight culture to inoculate 0.5 – 1.0 liter of TSB-strep broth. Incubate in a shaking incubator at 37°C until the OD_{600} = 0.1 (log phase).
   g. Freeze 1.0 mL aliquots of bacteria in TSB-strep broth and store at –80°C. No glycerol is required.
6. Cells that have been stained for flow cytometry should be analyzed as quickly as possible. Storing them for more than 1-2 d can result in auto-fluorescence that may interfere with the analysis.
7. The time period for stimulation depends on the functional response being measured. Memory cells (CD44+) will respond rapidly and are typically measured after 6 hours of stimulation. If the responsiveness of the native subset (CD44–) is sought, soluble cytokine is measured after 24-72 hours, in the presence of anti-CD3 antibody with anti-CD28 antibody for the second signal.
8. For the primary Listeria response, seven days after inoculation is typically used. A secondary exposure (boost) is best measured at 6 d after the booster inoculation. The contraction phase of the T-cell response is estimated at 11 or 12 ds after inoculation and memory levels after 28+ days.
9. Phycoerythrin works quite well as the fluor for intracellular staining. Use other fluorescent tags for the surface markers.

8 THE DEFENSIVE SYSTEM

We are constantly exposed to an incredible diversity of bacteria, viruses, and parasites, many of which would flourish in our cells or extracellular fluids were it not for our immune system. Remarkably, we are often even able to defend ourselves against organisms that we have never before encountered. How does the immune system protect us? The key is our ability to produce more than 10^8 distinct *antibodies* and more than 10^{12} *T-cell receptors,* each of which presents a different surface for specifically binding a molecule from a foreign organism and initiating the destruction of the invader. The presence of this remarkable repertoire of defensive molecules poses a challenge.

What prevents the immune system from attacking cells that express molecules normally present in our bodies; that is, *how does the immune system distinguish between nonself and self*? We shall examine these questions, focusing first on the structures of the proteins participating in the molecular recognition processes and then on the mechanisms for selecting cells that express molecules useful for protecting us from a specific pathogen. Emphasis will be on the modular construction of the proteins of the immune system — identifying structural motifs and considering how spectacular diversity can arise from modular construction.

PRINCIPLES OF EVOLUTION

The immune system comprises two parallel but interrelated systems. In the *humoral immune response,* soluble proteins called *antibodies (immunoglobulins)* function as recognition elements that bind to foreign molecules and serve as markers signaling foreign invasion. Antibodies are secreted *by plasma cells,* which are derived from *B lymphocytes (B cells).* A foreign macromolecule that binds selectively to an antibody is called an *antigen.* In a physiological context, if the binding of the foreign molecule stimulates an immune response, that molecule is called an *immunogen.* The specific affinity of an antibody is not for the entire macromolecular antigen but for a particular site on the antigen called the *epitope* or *antigenic determinant.* In the *cellular immune response,* cells called *cytotoxic T lymphocytes* (also commonly called *killer T cells)* kill cells that display foreign motifs on their surfaces.

Another class of T cells called *helper T lymphocytes* contributes to both the humoral and the cellular immune responses by stimulating the differentiation and proliferation of appropriate B cells and cytotoxic T cells. The celluar immune response is mediated by specific receptors that are expressed on the surfaces of the T cells. The remarkable ability of the immune system to adapt to an essentially limitless set of potential pathogens requires a powerful system for transforming, the immune cells and molecules present in our systems in response to the presence of pathogens. *This adaptive system operates through the principles of evolution, including reproduction with variation followed by selection of the most well suited members of a population.* If the human genome contains, by the latest estimates, only 40,000 genes, how can the immune system generate more than 10^8 different antibody proteins and 10^{12} T-cell receptors? The answer is found in a novel mechanism for generating a highly diverse set of genes from a limited set of genetic building blocks.

Linking different sets of DNA regions in a combinatorial manner produces many distinct protein-encoding genes that are not present in the genome. A rigorous selection process then leaves for proliferation only cells that synthesize proteins determined to be useful in the immune response. The subsequent reproduction of these cells without additional recombination serves to enrich the cell population with members expressing a particular protein species. Critical to the development of the immune response is the selection process, which determines which cells will reproduce.

The process comprises several stages. In the early stages of the development of an immune response, cells expressing piolecules that bind tightly to self-molecules are destroyed or silenced, whereas cells expressing molecules that do not bind strongly to self-molecules and that have the potential for binding strongly to foreign molecules are preserved. The appearance of an immunogenic invader at a later time will stimulate cells expressing immunoglobulins or T-cell receptors that bind specifically to elements of that pathogen to reproduce — in evolutionary terms, such cells are selected for. Thus, the immune response is based on the selection of cells expressing molecules that are specifically effective against a particular invader; the response evolves from a population with wide-ranging specificities to a more-focused collection of cells and molecules that are well suited to defend the host when confronted with that particular challenge.

Just as Medieval defenders used their weapons and the castle walls to defend their city, the immune system constantly battles against foregn invaders such as viruses, bacteria, and parasites to defend the organism.

ANTIGEN-BINDING UNITS

Antibodies are central molecular players in the immune response, and we examine them first. A fruitful approach in studying proteins as large as antibodies is to split the protein into fragments that retain activity. In 1959, Rodney Porter showed that *immunoglobulin G (IgG),* the major antibody in serum, can be cleaved into three 50-kd fragments by the limited proteolytic action of papain. Two of these fragments bind antigen. They are called F_{ab} (F stands for fragment, *ab* for antigen binding). The other fragment, called F_c because it crystallizes readily, does not bind antigen, but it has other important biological activities, including the mediation of responses termed *effector functions.* These functions include the initiation of the *complement cascade,* a process that leads to the lysis of target cells.

Although such effector functions are crucial to the functioning of the immune system, they will not be discussed further here. How do these fragments relate to the three-dimensional structure of whole IgG molecules? Immunoglobulin G consists of two kinds of polypeptide chains, a 25-kd *light (L) chain* and a 50-kd *heavy (H) chain* Figure 8.2). The subunit composition is L_2H_2. Each L chain is linked to an H chain by a disulfide bond, and the H chains are linked to each other by at least one disulfide bond. Examination of the amino acid sequences and three-dimensional structures of IgG molecules reveals that each L chain comprises two homologous domains, termed *immunoglobulin domains.* Each H chain has four immunoglobulin domains.

Overall, the molecule adopts a conformation that resembles the letter Y, in which the stem, corresponding to the F_c fragment obtained by cleavage with papain, consists of the two carboxyl-terminal immunoglobulin domains of each H chain and in which the two arms of the Y, corresponding to the two F_{ab} fragments, are formed by the two amino-terminal domains of each H chain and the two amino-terminal domains of each L chain. The linkers between the stem and the two arms consist of relatively extended polypeptide regions within the H chains and are quite flexible. Papain cleaves the H chains on the carboxyl-terminal side of the disulfide bond that links each L and H chain

Thus, each F_{ab} consists of an entire L chain and the amino-terminal half of an H chain, whereas F_c consists of the carboxyl-terminal halves of both H chains. Each F_{ab} contains a single antigen-binding site: Because an intact IgG molecule contains two F_{ab} components and therefore has two binding sites, it can cross-link multiple antigens. Furthermore, the F_c and the two F_{ab} units of the intact IgG are joined by flexible polypeptide regions that allow facile variation in the angle between the F_{ab} units through a wide range. This kind of mobility, called *segmental flexibility,* can enhance the formation of an antibody-antigen complex by enabling both combining sites on an antibody to bind an antigen that possesses multiple binding sites, such as a viral coat composed of repeating identical monomers or a bacterial cell surface.

The combining sites at the tips of the F_{ab} units simply move to match the distance between specific determinants on the antigen. Immunoglobulin G is the antibody present in highest concentration in the serum, but other classes of immunoglobulin also are present. Each class includes an L chain (either μ, or ν) and a distinct H chain. The heavy chains in IgG are called i chains, whereas those in immunoglobulins A, M, D, and E are called χ, *o*, ϕ, and γ, respectively. *Immunoglobulin M (IgM)* is the first class of antibody to appear in the serum after exposure to an antigen.

The presence of 10 combining sites enables IgM to bind especially tightly to antigens containing multiple identical epitopes. The strength of an interaction comprising multiple independent binding interactions between partners is termed *avidity* rather than affinity, which denotes the binding strength of a single combining site. The presence of 10 combining sites in IgM compared with 2 sites in IgG enables IgM to bind many mukivalent antigens that would slip away from IgG. *Immunoglobulin A (IgA)* is the major class of antibody in external secretions, such as saliva, tears, bronchial mucus, and intestinal mucus. Thus, IgA serves as a first line of defense against bacterial and viral antigens.

The role of *immunoglobulin D (IgD)* is not yet known. *Immunoglobulin E (IgE)* is important in conferring protection against parasites, but IgE also causes allergic reactions. IgE-antigen complexes form cross-links with receptors on the surfaces of mast cells to trigger a cascade that leads to the release of granules containing pharmacologically active molecules. Histamine, one of the agents released, induces smooth muscle contraction and stimulates the secretion of mucus. A comparison of the amino acid sequences of different IgG antibodies from human beings or mice shows that the carboxyl-terminal half of the L chains and the carboxyl-terminal three-quarters of the H chains are very similar in all of the antibodies.

Importantly, the amino-terminal domain of each chain is more variable, including three stretches of approximately 7 to 12 amino acids within each chain that are hyperyariable. The amino-terminal immunoglobulin domain of each chain is thus referred to as the *variable region,* whereas the remaining immunoglobulin domains are much more similar in all antibodies and are referred to as *constant regions.*

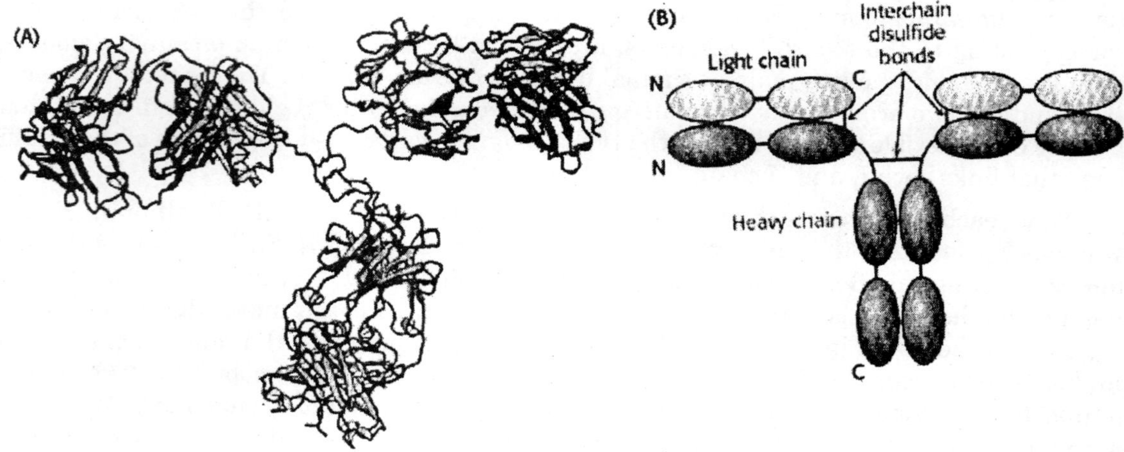

Fig. 8.1. Immunoglobulin G Structure. (A) The three-dimensional structure of an IgG molecule showing the light chains the heavy chains. (B) A schematic view of an IgG molecule indicating the positions of the interchain disulfide bonds. N, amino terminus; C, carboxyl terminus.

Table 8.1 Properties of immunoglobulin classes

Class	Serum concentration (mg/ml)	Mass(kd)	Sedimentation coefficients(s)	Light chains	Heavy chains	Chain structure
IgG	12	150	7	μ or ν	$\iota \times 3$	$\mu_2\iota_2$ or $\nu_2\, \iota_2$
IgA	3	180–500	7, 10, 13	μ or ν	χ	$(\mu_2\chi_2)_n$ or $(\nu_2\, \chi_2)_n$
IgM	1	950	18 – 20	μ or ν	ο	$(\mu_2 o_2)_5$ or $(\nu_2\, o_2)_5$
IgD	0.1	175	7	μ or ν	ϕ	$\mu_2\, \phi_2$ or $\nu_2\, \phi_2$
IgE	0.001	200	8	μ or ν	γ	$\mu_2\gamma_2$ or $\nu_2\, \gamma_2$

Note: n = 1, 2, or 3. IgM and oligomers of IgA also contain J chains that connect immunoglobulin molecules. IgA in secretions has an additional component.

HYPERVARIABLE LOOPS

An IgG molecule consists of a total of 12 immunoglobulin domains. These domains have many sequence features in common and adopt a common structure, the *immunoglobulin fold.* Remarkably, this same structural domain is found in many other proteins that play key roles in the immune system. The immunoglobulin fold consists of a pair of δ sheets, each built of antiparallel δ strands, that surround a central hydrophobic core. A single disulfide bond bridges the two sheets. Two aspects of this structure are particularly important for its function. First, three loops present at one end of the structure form a potential binding surface. Thes loops contain the hypervariable sequences present in antibodies and in T-cell receptors.

Variation of the amino acid sequences of these loops provides the major mechanism for the generation of the vastly diverse set of antibodies and T-cell receptors expressed by the immune system. These loops are referred to as *hypervariable loops* or *complementaritydetermining regions (CDRs)*. Second, the amino terminus and the carboxyl terminus are at opposite ends of the structure, which allows structural domains to be strung together to form chains, as in the L and H chains of antibodies. Such chains are present in several other key molecules in the immune system. The immunoglobulin fold is one of the most prevalent domains encoded by the human genome — more than 750 genes encode proteins with at least one immunoglobulin fold recognizable at the level of amino acid sequence. Such domains are also common in other multicellular animals such as flies and nemotodes.

However, from inspection of amino acid sequence alone, immunoglobulin-fold domains do not appear to be present in yeast or plants. However, structurally similar domains are present in these organisms, including the key photosynthetic electron-transport protein plastocyanin in plants. Thus, the immunoglobulin-fold family appears to have expanded greatly along evolutionary branches leading to animals — particularly, vertebrates.

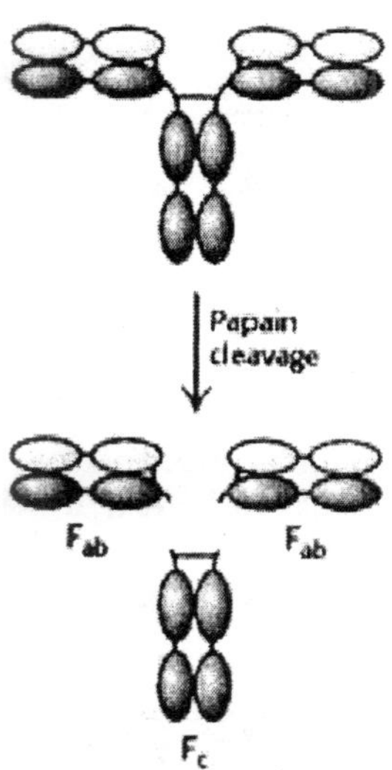

Fig. 8.2 Immunoglobulin G Cleavage. Treatment of intact IgG molecules with the protease papain results in the formation of three large fragments: F_{ab} fragments that retain antigen-binding capability and one F_c fragment that does not.

Fig. 8.3. Antigen Cross-Linking. Because IgG molecules include two antigen-binding sites, antibodies can crosslink multivalent antigens such as viral surfaces.

SPECIFIC MOLECULES

For each class of antibody, the amino-terminal immunoglobulin domains of the L and H chains (the variable domains, designated V_L and V_H) come together at the ends of the arms extending from the structure. The positions of the complementarity-determining regions are striking. These hypervariable sequences, present in three loops of each domain, come together so that all six loops form a single surface at the end of each arm. Because virtually any V_L can pair with any V_H, *a very large number of different binding sites can be constructed by their combinatorial association.*

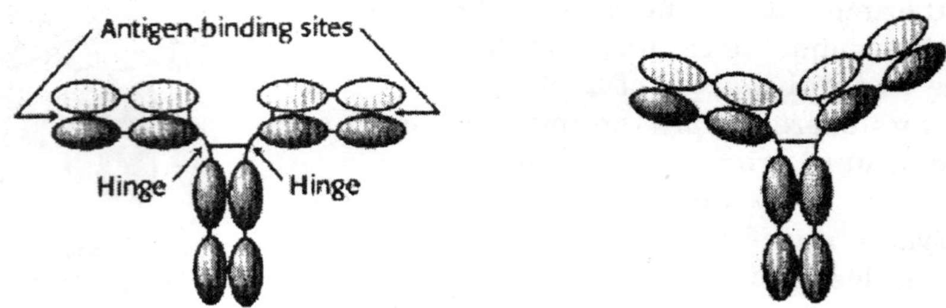

Fig. 8.4. Segmental Flexibility. The linkages between the F_{ab} and the F_c regions of an IgG molecule are flexible, allowing-the two antigen-binding sites to adopt a range of orientations with respect to one another. This flexibility allows effective interactions with a multivalent antigen without requiring that the epitopes on the target be a precise distance apart.

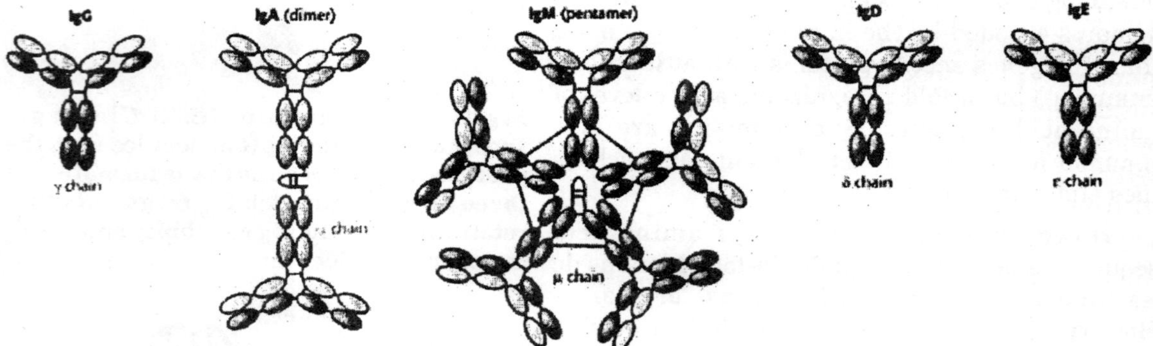

Fig. 8.5. Classes of ImmunoGlobulin. Each of five classes of immunoglobulin has the same light chain (shown in yellowj-eefribined with a different heavy chain (τ, χ, o, ϕ, *or* μ). Disulfide bonds are indicated by green lines. The IgA dimer and the IgM pentamer have a small polypeptide chain in addition to the light and heavy chains.

X-Ray Analyses

The results of x-ray crystallographic studies of many large and small antigens bound to F_{ab} molecules have been sources of much insight into the structural basis of antibody specificity. The binding of antigens to antibodies is governed by the same principles that govern the binding of substrates to enzymes. The apposition of complementary shapes results in numerous contacts between amino acids at the binding surfaces of both molecules.

Numerous hydrogen bonds, electrostatic interactions, and van der Waals interactions, reinforced by hydrophobic interactions, combine to give specific and strong binding. A few aspects of antibody binding merit specific attention, inasmuch as they relate directly to the structure of immunoglobulins. The binding site on the antibody has been found to incorporate some or all of the CDRs in the variable domains of the antibody. Small molecules (e.g., octapeptides) are likely to make contact with fewer CDRs, with perhaps 15 residues of the antibody participating in the binding interaction.

Macromolecules often make more extensive contact, interacting with all six CDRs and.20 or more residues of the antibody. Small molecules often bind in a cleft of the antigen-binding region. Macromolecules such as globular proteins tend to interact across larger, fairly flat apposed surfaces bearing complementary protrusions and depressions. A well-studied case of small-molecule binding is seen in an example of phosphorylcholine bound to F_{ab}. Crystallographic analysis revealed phosphorylcholine bound to a cavity lined by residues from five CDRs — two from the L chain and three from the H chain.

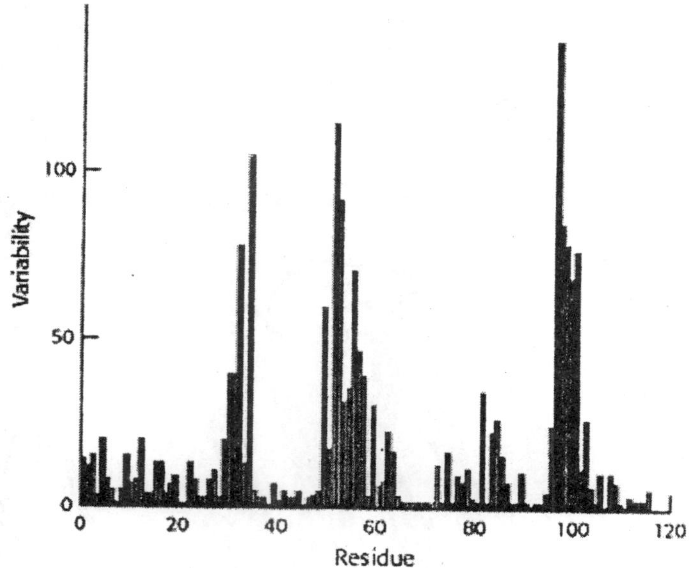

Fig. 8.6. Immunoglobulin Sequence Diversity. A plot of sequence variability as a function of position along the sequence of the amino-terminal immunoglobulin domain of the H chain of human IgG molecules. Three regions (in red) show remarkably high levels of variability. These hypervariable regions correspond to three loops in the immunoglobulin domain structure.

The positively charged trimethylammonium group of phosphorylcholine is buried inside the wedge-shaped cavity, where it interacts electrostatically with two negatively charged glutamate residues. The negatively charged phosphate group of phosphorylcholine binds to the positively charged guanidinium group of an arginine residue at the mouth of the crevice and to a nearby lysine residue. The phosphate group is also hydrogen bonded to the hydroxyl group of a tyrosine residue and to the guanidinium group of the arginine side chain.

Numerous van der Waals interactions, such as those made by a tryptophan side chain, also stabilize this complex. The binding of phosphorylcholine does not significantly change the structure of the antibody, yet induced fit plays a role in the formation of many antibody-antigen complexes. A malleable binding site can accommodate many more kinds of ligands than can a rigid one. Thus, induced fit increases the repertoire of antibody specificities.

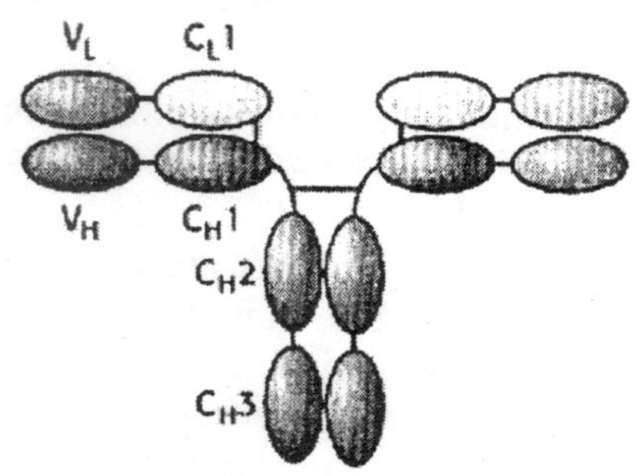

Fig. 8.7. Variable and Constant Regions. Each L and H chain includes one immunoglobulin domain at its amino terminus that is quite variable from one antibody to another. These domains are referred to as V_L and V_H. The remaining domains are more constant from one antibody to another and are referred to as constant domains (C_L1, C_H1, C_H2, and C_H3).

Antigens and Antibodies with Numerous Interactions

How do large antigens interact with antibodies? A large collection of antibodies raised against hen egg-white lysozyme has been structurally characterized in great detail. Each different antibody binds to a distinct surface of lysozyme. Let us examine the interactions present in one of these complexes in detail. This antibody binds two polypeptide segments that are widely separated in the primary structure, residues 18 to 27 and-116 to 129. All six CDRs of the antibody make contact with this epitope.

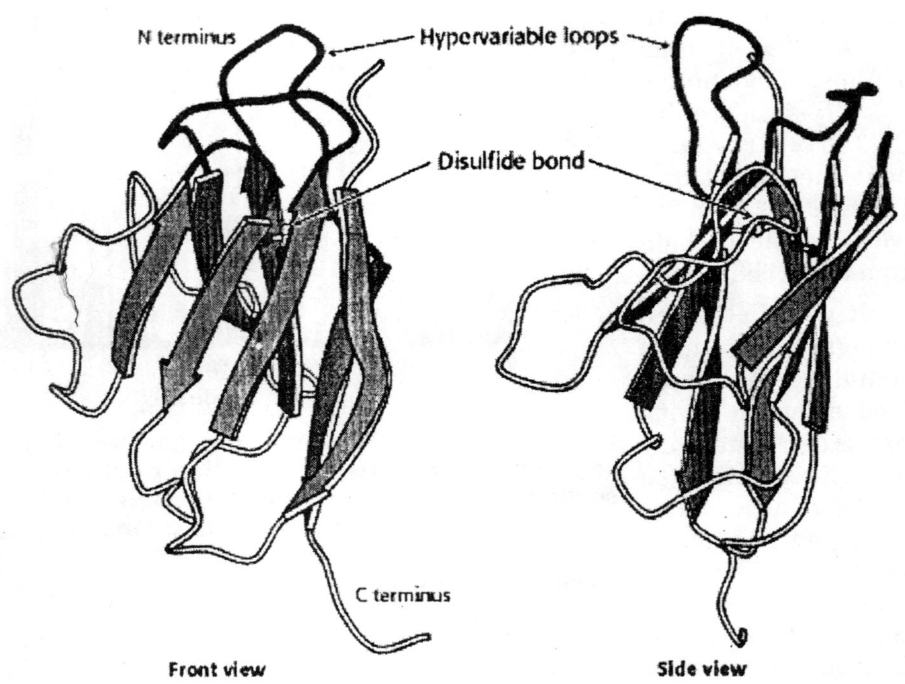

Fig. 8.8. Immunoglobulin Fold. An immunoglobulin domain consists of a pair of δ-sheets linked by a disulfide bond and hydrophobic interactions. Three hypervariable loops lie at one end of the structure.

The region of contact is quite extensive (about 30 × 20 Å). The apposed surfaces are rather flat. The only exception is the side chain of glutamine 121 of lysozyme, which penetrates deeply into the antibody binding site, where it forms a hydrogen bond with a main-chain carbonyl oxygen atom and is surrounded by three aromatic side chains. The formation of 12 hydrogen bonds and numerous van der Waals interactions contributes to the high affinity (K_d = 20 nM) of this antibody-antigen interaction. Examination of the F_{ab} molecule without bound protein reveals that the structures of the V_L and V_H domains change little on binding, although they slide 1 Å apart to allow more intimate contact with lysozyme.

GENE REARRANGEMENTS

A mammal such as a mouse or a human being can synthesize large amounts of specific antibody against virtually any foreign determinant within a matter of days of being exposed to it. We have seen that antibody specificity is determined by the amino acid sequences of the variable regions of both light and heavy chains, which brings us to the key question: How are different variable-region sequences generated? The discovery of distinct variable, and constant regions in the L and H chains raised the possibility that the genes that encode immunoglobulins have an unusual architecture that facilitates the generation of a diverse set of polypeptide 'products.

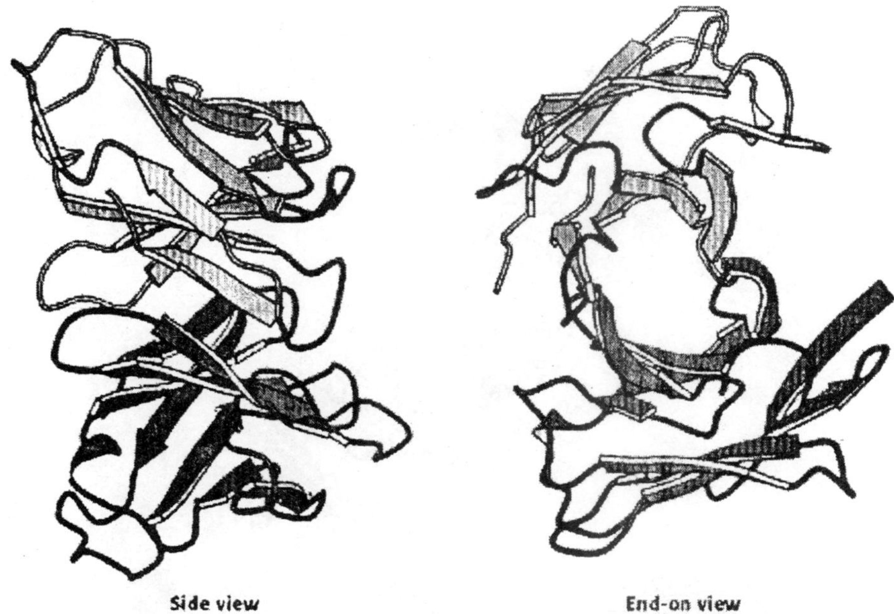

Fig. 8.9. Variable Domains. Two views of the variable domains of the L chain and the H chain (blue); the complementarity-determining regions (CDRs) are shown in the figure. The six CDRs come together to form a binding surface. The specificity of the surface is determined by the sequences and structures of the CDRs.

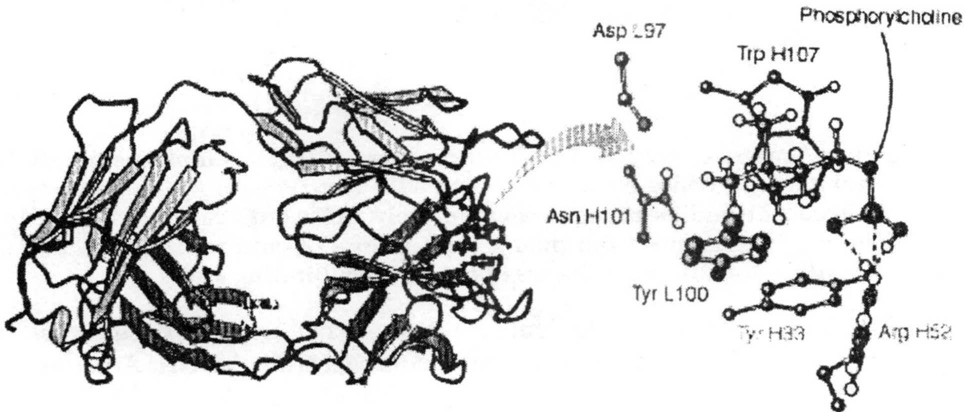

Fig. 8.10. Binding of a Small Antigen. The structure of a complex between an F_{ab} fragment of an antibody and its target in this case, phosphoryl-choline. Residues from the antibody interact with phosphorylcholine through hydrogen bonding and electrostatic and van der Waals interactions.

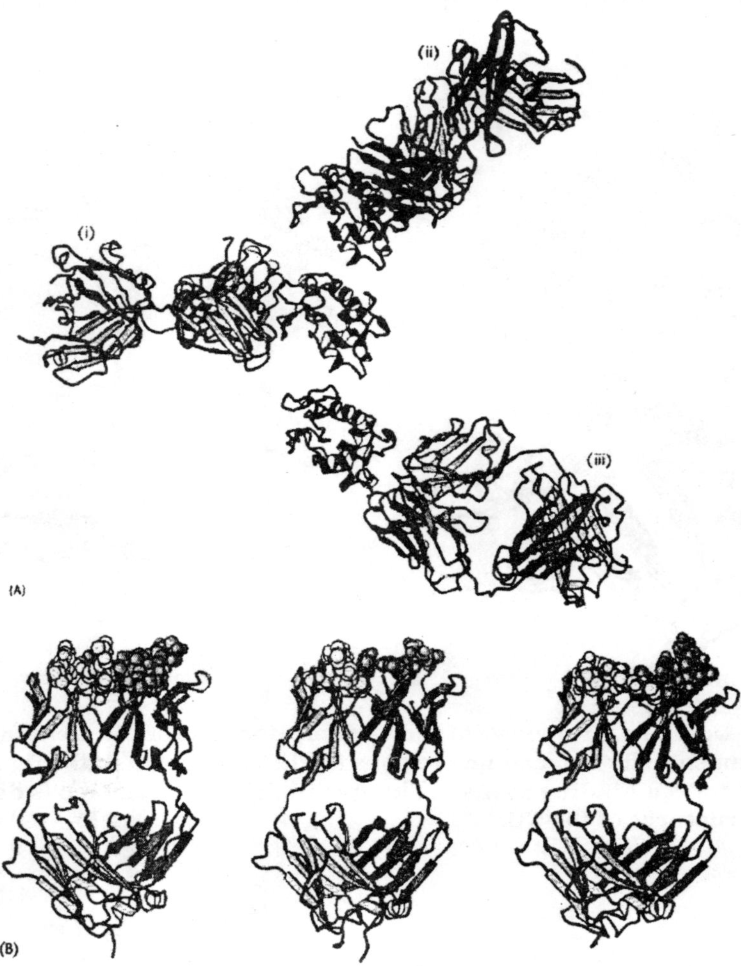

Fig. 8.11. Antibodies Against Lysozyme. (A) The structures of three complexes (i,ii,iii) between F_{ab} fragments and hen egg-white lysozyme shown with lysozyme in the same orientation in each case. The three antibodies recognize completely different epitopes on the lysozyme molecule. (B) The F_{ab} fragments from part A with points of contact highlighted as space-filling models, revealing the different shapes of the antigen-binding sites.

In 1965, William Dreyer and Claude Bennett proposed that multiple *V (variable) genes* are separate from a single *C (constant) gene* in embryonic (germ-line) DNA. According to their model, one of these V genes becomes joined to the C gene in the course of differentiation of the antibody-producing cell. A critical test of this novel hypothesis had to await the isolation of pure immunoglobulin mRNA and the development of techniques for analyzing mammalian genomes. Twenty years later, Susumu Tonegawa found that V and C genes are indeed far apart in embryonic DNA but are closely associated in the DNA of antibody-producing cells. Thus, immunoglobulin genes are rearranged in the differentiation of lymphocytes.

Antibody Diversity

Sequencing studies carried out by Susumu Tonegawa, Philip Leder, and Leroy Hood revealed that V genes in embryonic cells do not encode the entire variable region of L and H chains. Consider, for example, the region that encodes the p. light-chain family. A tandem array of 40 segments, each of which encodes approximately the first 97 residues of the variable domain of the L chain, is present on human chromosome 2. However, the variable region of the L chain extends to residue 110. Where is the DNA that encodes the last 13 residues of the V region?

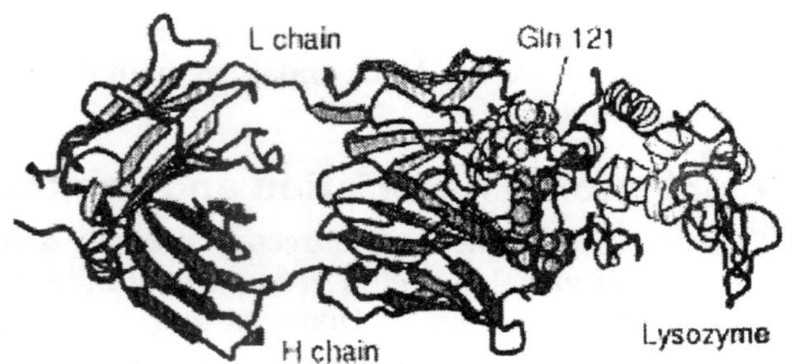

Fig. 8.12. Antibody - Protein Interactions. The structure of a complex between an F_{ab} fragment and lysozyme reveals that the binding surfaces are complementary in shape over a large area. A single residue of lysozyme, glutamine 121, penetrates more deeply into the antibody combining site.

For L chains in undifferentiated cells, this stretch of DNA is located in an unexpected place: near the C gene. It is called the *J gene* because it *j*oins the V and C genes in a differentiated cell. In fact, a tandem array of five J genes is located near the C gene in embryonic cells. In the differentiation of an antibody-producing cell, a V gene becomes spliced to a J gene to form a complete gene for the variable. RNA splicing generates an mRNA molecule for the complete L chain by linking the coding regions for the rearranged VJ unit with that for the C unit J genes are important contributors to antibody diversity because they encode part of the last hypervariable segment (CDR3).

In forming a continuous variable-region gene, any of the 40 V genes can become linked to any of five J genes. Thus, somatic recombination of these gene segments amplifies the diversity already present in the germ line. The linkage between V and J is not precisely controlled. Recombination between these genes can take place at one of several bases near the codon for residue 95, generating additional diversity. A similar array of V and J genes encoding the v light chain is present on human chromosome 22. This region includes 30 V_{υ} gene segments and four J_{υ} segments. In addition, this region includes four distinct C genes, in contrast with the single C gene in the μ locus. In human beings, the genes encoding the heavy chain are present on chromosome 14. Remarkably, the variable domain of heavy chains is assembled from *three* rather than two segments. In addition to V_H genes that encode residues 1 to 94 and J_H segments that encode residues 98 to 113, this chromosomal region includes a distinct set of segments that encode residues 95 to 97.

These gene segments are called D for diversity. Some 27 D segments lie between 51 V_H and 6 J_H segments.

The recombination process first joins a D segment to a J_H segment; a V_H segment is then joined to DJ_H. A greater variety of antigen-binding patches and clefts can be formed by the H chain than by the L chain because the H chain is encoded by three rather than two gene segments. Moreover, CDR3 of the H chain is diversified by the action of terminal deoxyribonucleotidyl transferase, a special DNA polymerase that requires no template. This enzyme inserts extra nucleotides between V_H and D. The *V(D)J recombination* of both the L and the H chains is executed by specific enzymes present in immune cells. These proteins, called *RAG-1* and *RAG-2,* recognize specific DNA sequences called *recombination signal sequences (RSSs)* adjacent to the V, D, and J segments and facilitate the cleavage and religation of the DNA segments.

Combinatorial Association and Somatic Mutation

Let us recapitulate the sources of antibody diversity. The germ line contains a rather large repertoire of variable-region genes. For μ light chains, there are about 40 V-segment genes and five J-segment genes. Hence, a total of 40 × 5 = 200 kinds of complete V_{μ} genes can be formed by the combinations of V and J. A similar analysis suggests that at least 120 different v light chains can be generated. A larger number of heavy-chain genes can be formed because of the role of the D segments. For 51 V, 27 D, and 6 J gene segments, the number of complete V_H genes that can be formed is 8262. The association of 320 kinds of L chains with 8262 kinds of H chains would yield 2.6×10^6 different antibodies. Variability in the exact points of segment joining and other mechanisms increases this value by at least two orders of magnitude. Even more diversity is introduced into antibody chains by *somatic mutation* — that is, the introduction of mutations into the recombined genes.

In fact, a 1000-fold increase in binding affinity is seen in the course of a typical humoral immune response, arising from somatic mutation, a process called *affinity maturation.* The generation of an expanded repertoire leads to the selection of antibodies that more precisely fit the antigen. Thus, nature draws on each of three sources of diversity — a germ-line repertoire, somatic recombination, and somatic mutation — to form the rich variety of antibodies that protect an organism from foreign incursions..

Oligomerization of Antibodies

The processes heretofore described generate a highly diverse set of antibody molecules — a key first step in the generation of an immune response. The next stage is the selection of a particular set of antibodies directed against a specific invader. How does this selection occur? Each immature B cell, produced in the bone marrow, expresses a monomeric form of IgM attached to its surface. Each cell expresses approximately 10^5 IgM molecules, but *all of these molecules are identical in amino acid sequence and, hence, in antigen-binding specificity.* Thus, the selection of a particular immature B cell for growth will lead to the amplification of an antibody with a unique specificity.

The selection process begins with the binding of an antigen to the membrane-bound antibody. Associated with each membrane-linked IgM molecule are two molecules of a heterodimeric membrane protein called Ig χ-Ig-δ. Examination of the amino acid sequences of Ig-χ and Ig-δ is highly instructive. The amino of each protein lies outside the cell and corresponds to a single immunoglobulin, and the carboxyl terminus, which lies inside the cell, includes a sequence of 18 amino acids called an *immunoreceptor tyrosine-basedactivation motif (ITAM).* As its name suggests, each ITAM includes key tyrosine residues, which are subject to phosphorylation by particular protein kinases present in immune-system cells. A fundamental observation with regard to the mechanism by which the binding of antigen to membrane-bound antibody triggers the subsequent steps of the immune response is that *oligomerization or clustering of the antibody molecules is required.* The requirement for oligomerization is reminiscent of the dimerization of receptors triggered by growth hormone and epidermal growth factor indeed, the associated signaling mechanisms appear to be quite similar.

The oligomerization of the membrane-bound antibodies results in the phosphorylation of the tyrosine residues within the ITAMs by protein tyrosine kinases including Lyn, a homolog of Src. The phosphorylated ITAMs serve as docking sites for a protein kinase termed spleen tyrosine kinase (Syk), which has two SH2 domains that interact with the pair of phosphorylated tyrosines in each ITAM. Syk, when activated by phosphorylation, proceeds to phosphorylate other signal-transduction proteins' including an inhibitory subunit of a transcription factor called NF-μB and an isofor of phpspholipase C. The signaling processes continue downstream to activate gene expression, leading to the stimulation of cell growth and initiating further B-cell differentiation. Drugs that modulate the immune system have served as sources, of insight into immune-system signaling pathways.

For example, *cyclosporin,* a powerful suppressor of the immune system, acts by blocking a phosphatase called *calcineurin,* which normally activates a transcription factor called NF-AT by dephosphorylating it.

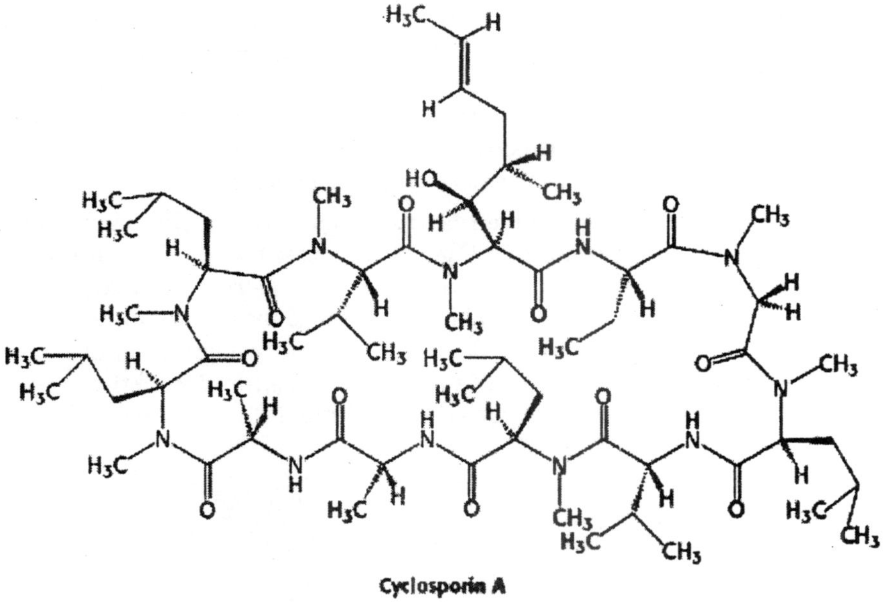

Cyclosporin A

The potent immune supression that results reveals how crucial the activity of this transcription factor is to the development of an immune response. Without drugs such as cyclosporin, organ transplantation would be extremely difficult because transplanted tissue expresses a wide range of foreign antigens, which causes the immune system to reject the new tissue. The role of oligomerization in the B-cell signaling pathway is illuminated when we consider the nature of many antigens presented by pathogens.

The surfaces of many viruses, bacteria, and parasites are characterized by arrays of identical membrane proteins or membrane-linked carbohydrates. Thus, most pathogens present multiple binding surfaces that will naturally cause membrane-associated antibodies to oligomerize as they bind adjacent epitopes. In addition, the mechanism accounts for the observation that most small molecules do not induce an immune response; however, coupling multiple copies of the small molecule to a large oligomeric protein such as keyhole limpet hemocyanin (KLH), which has a molecular mass of close to 1 million daltons or more, promotes antibody oligomerization and, hence, the production of antibodies against the small-molecule epitope. The large protein is called the *carrier* of the attached chemical group, which is called a *haptenic determinant.* The small foreign molecule by itself is called a *hapten.* Antibodies elicited by attached haptens will bind unattached haptens as well.

Different Classes of Antibodies

The development of an effective antibody-based immune response .depends on the secretion into the blood of antibodies that have appropriate effector functions. At the beginning of this response, an alternative mRNA splicing pathway is activated so that the production of membrane-linked IgM is supplanted by the synthesis of secreted IgM secreted IgM is pentameric and has a relatively high avidity for multivalent antigens. Later, the antibody-producing cell makes either IgG, IgA, IgD, or IgE of the same specificity as the intially secreted IgM. In this switch, the light chain is unchanged, as is the variable region of the heavy chain.

Only the constant region of the heavy chain changes. This step in the differentiation of an antibody-producing.cell is called *class switching.* In undifferentiated cells, the genes for the constant region of each class of heavy chain, called C_o, C_ϕ, C_τ, C_γ, and C_χ, are next to each other. There are eight in all, including four genes for the constant regions of ι chains. A complete gene for the heavy chains of IgM antibody is formed by the translocation of a V_H gene segment to a DJ_H gene segment. How are other heavy chains formed?

Class switching is mediated by a gene-rearrangement process that moves a VDJ gene from a site near one C gene to a site near another C gene. Importantly, *the antigen-binding specificity is conserved in class switching because the entire $V_HDJ_H$gene is- translocated in an intact form.* For example, the antigen-combining specificity of IgA produced by a particular cell is the same as that of IgM synthesized at an earlier stage of its development. The biological significance of C_H switching is that a whole recognition domain (the variable domain) is shifted from the early constant region (C_O) to one of several other constant regions that mediate different effector functions.

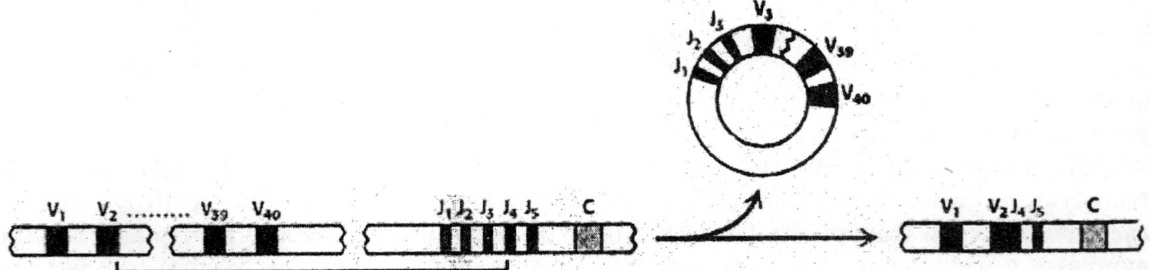

Fig. 8.13. VJ Recombination. A single V gene (in this case, V_2) is linked to a J gene (here, J_4) to form an intact VJ region. The intervening DNA is released in a circular form. Because the V and J regions are selected at random and the joint between them is not always in exactly the same place, many VJ combinations can be generated by this process.

MAJOR-HISTOCOMPATIBILITY-COMPLEX PROTEINS

Soluble antibodies are highly effective against extracellular pathogens, but they confer little protection against microorganisms that are predominantly intracellular, such as viruses and mycobacteria (which cause tuberculosis and leprosy). These pathogens are shielded from antibodies by the host-cell membrane. A different and more subtle strategy, *cell-mediated immunity,* evolved to cope with intracellular pathogens. *T cells* continually scan the surfaces of all cells and kill those that exhibit foreign markings. The task is not simple; intracellular microorganisms are not so obliging as to intentionally leave telltale traces on the surface of their host. Quite the contrary, successful pathogens are masters of the art of camouflage.

Vertebrates have devised an ingenious mechanism—cut and display—to reveal the presence of stealthy intruders. Nearly all vertebrate cells exhibit on their surfaces a sample of peptides derived from the digestion of proteins in their cytosol. These peptides are displayed by integral membrane proteins that are encoded by the *major histocompatibility complex (MHC).* Specifically, peptides derived from cytosolic proteins are bound to *class I MHC proteins.* How are these peptides generated arid delivered to the plasma membrane? The process starts in the cytosol with the degradation of proteins, self proteins as well as those of pathogens. Digestion is carried out by proteasomes.

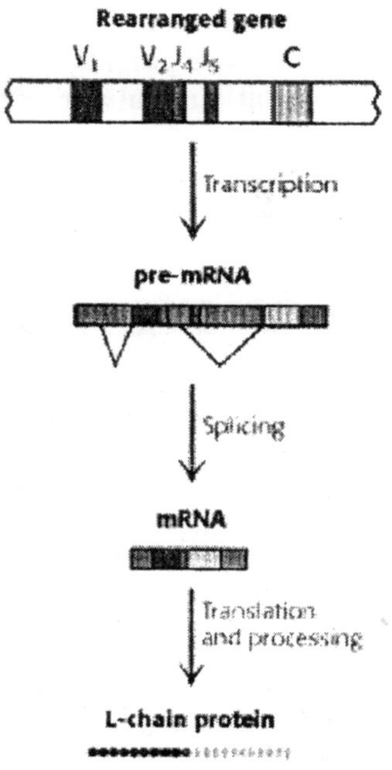

Fig. 8.14. Light-Chain Expression. The light-chain protein is expressed by transcription of the rearranged gene to produce a pte-RNA molecule with the VJ and C regions separated. RNA splicing removes the intervening sequences to produce an mRNA molecule with the VJ and C regions linked. Translation of the mRNA and processing of the initial protein product produces the light chain.

The resulting peptide fragments are transported from the cytosol into the lumen of the endoplasmic reticulum by an ATP-driven pump. In the ER, peptides combine with nascent class I MHC proteins; these complexes are then targeted to the plasma membrane. MHC proteins embedded in the plasma membrane tenaciously grip their bound peptides so that they can be touched and scrutinized by T-cell receptors on the surface of a killer cell. Foreign peptides bound to class I MHC proteins signal that a cell is infected and mark it for destruction by cytotoxic T cells. An assembly consisting of the foreign peptide-MHC complex, the T-cell receptor, and numerous accessory proteins triggers a cascade that induces apoptosis in the infected cell. Strictly speaking, infected cells are not killed but, instead, are triggered to commit suicide to aid the organism.

Peptides Presented by MHC Proteins

The three-dimensional structure of a large fragment of a human MHC class I protein, *human leukocyte antigen A2 (HLA-A2),* was solved in 1987 by Don Wiley and Pamela Bjorkman. Class I MHC proteins consist of a 44-kd χ chain noncovalently bound to a 12-kd polypeptide called δ_2 - *microglobulin.* The χ chain has three extracellular domains (χ_1, χ_2, and χ_3), a transmembrane segment, and a tail that extends into the cvtosol. Cleavage by papain of the HLA χ chain several residues-before the transmembrane segment yielded a soluble heterodimeric fragment. The δ_2-microglobulin subunit and the χ_3 domains have immunoglobulin folds, although the pairing of the two domains differs from that in antibodies. The χ_1 and χ_2 domains exhibit a novel and remarkable architecture.

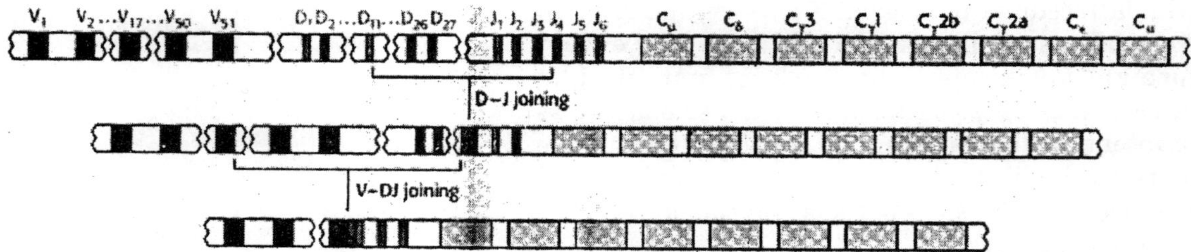

Fig. 8.15. V(D) J Recombination. The heavy-chain locus includes an array of 51 V segments, 27 D segments, and 6 J segments. Gene rearrangement begins with D-J joining, followed by further rearrangement to link the V segment to the DJ segment.

They associate intimately to form a deep groove that serves as the peptide-binding site. The floor of the groove, which is about 25 A long and 10 A wide, is formed by eight 8 strands, four from each domain. A long helix contributed by the χ_1 domain forms one side, and a helix contributed by the χ_2 domain forms the other side. *This groove is the binding site for the presentation of peptides.* The groove can be filled by a peptide from 8 to 10 residues long in an extended conformation. As we shall see, MHC proteins are remarkably diverse in the human population; each person expresses as many as six distinct class I MHC proteins and many different forms are present in different people. The first structure determined, HLA-A2, binds peptides that almost always have leucine in the second position and valine in the last position.

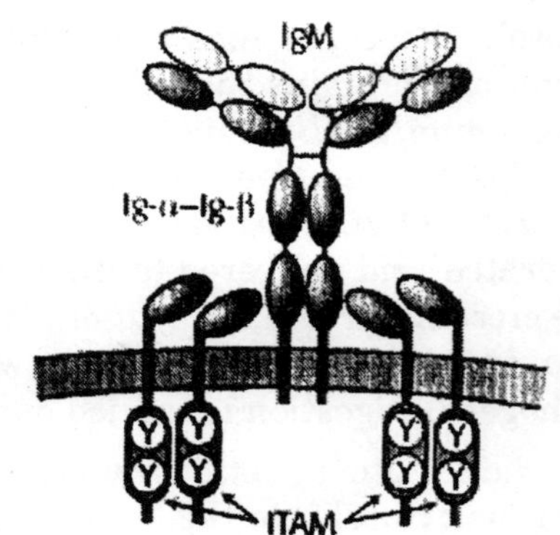

Fig. 8.16. B–Cell Receptor, This complex consists of a membrane-bound IgM molecule noncovalently bound to two Ig-χ-Ig-δ heterodimers. The intracellular domains of each of the Ig-χ and Ig-δ chains include an immunoreceptor tyrosine-based activation motif (ITAM).

Side chains from the MHC molecule interact with the arnino and carboxyl termini and with the side chains in these two key positions. These residues are often referred to as the *anchor residues.* The other residues are highly variable. Thus, many millions of different peptides can be presented by this particular class I MHC protein; the identities of only two of the nine residues are crucial for binding. Each class of MHC molecules requires a unique set of anchor residues. Thus, a tremendous range of peptides can'be presented by these molecules. Note that *one face of the bound peptide is exposed to solution where it can be examined by other molecules, particularly T-cell receptors.* An additional remarkable feature of MHC-peptide complexes is their kinetic stability; once bound, a peptide is not released, even over a period of days.

Antibody-like Proteins

We are now ready to consider the receptor that recognizes peptides displayed by MHC proteins on target cells. The *T-cell receptor* consists of a 43-kd χ chain (T_χ) joined by adisulfide bond to a43-kd δ chain T_δ . Each chain spans the plasma membrane and has a short carboxyl-terminal region on the cytosolic side. A small proportion of T cells express a receptor consisting of τ and ϕ chains in place of χ and δ. T_χ and T_δ , like immunoglobulin L and H chains, consist *of variable* and *constant* regions. Indeed, *these domains of the T-cell receptor are homologous to the V and C domains of immunoglobulins.* Furthermore, hypervariable sequences present in the V regions of T_χ and T_δ form the binding site for the epitope. The genetic architecture of these proteins is similar to that of immunoglobulins. The variable region of T is encoded by about 50 Vsegment genes and 70 J-segment genes. T_δ is encoded by two D-segment genes in addition to 57 V and 13 J-segment genes.

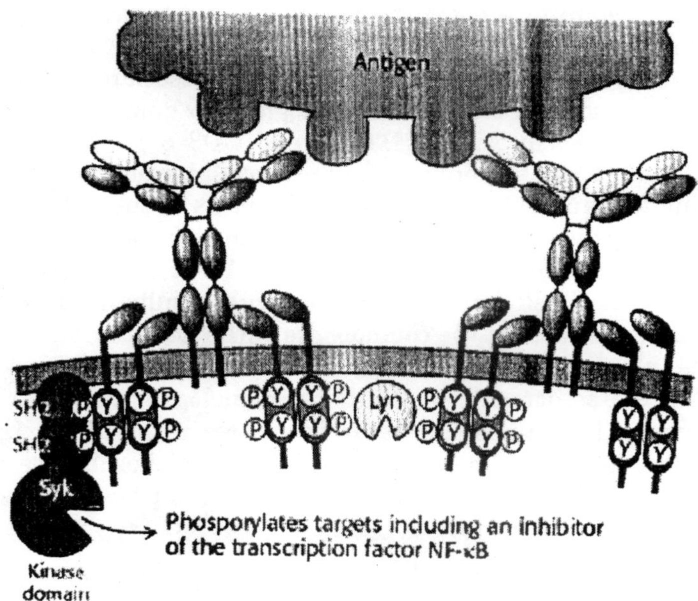

Fig. 8.17. B-Cell Activation. The binding of multivalent antigen such as bacterial or viral surfaces links membrane-bound IgM molecules. This oligomerization triggers the phosphorylation of tyrosine residues in the ITAM sequences by protein tyrosine kinases such as Lyn. After phosphorylation, the ITAMs serve as docking sites for Syk, a protein kinase that phosphorylates a number of targets, including transcription factors.

Again, the diversity of component genes and the use of slightly imprecise modes of joining them increase the number of distinct proteins formed. *At least 10^{12} different specificities could arise from combinations of this repertoire of genes.* Thus, T-cell receptors, like immunoglobulins, can recognize a very large number of different epitopes. All the receptors on a particular T cell have the same specificity. How do T cells recognize their targets? The variable regions of the χ and δ chains of the T-cell receptor form a binding site that recognizes a combined epitope-foreign peptide bound to an MHC prntein. Neither the foreign peptide alone nor the MHC protein alone forms a complex with the T-cell receptor. Thus, fragments of an intracellular pathogen are presented in a context that allows them to be detected, leading to the initiation of an appropriate response.

Cytotoxic T Cells

The T-cell receptor does not act alone in recognizing and mediating the fate of target cells. Cytotoxic T cells also express a protein termed *CDS* on their surfaces that is crucial for the recognition of the class I MHC-peptide complex. The abbreviation CD stands for *cluster of differentiation,* referring to a cell-surface marker that is used to identify a lineage or stage of differentiation. Antibodies specific for particular CD proteins have been invaluable in following the development of leukocytes and in discovering new interactions between specific cell types. Each chain in the CD8 dimer contains a domain that resembles an.immunoglobulin variable domain CD8 interacts primarily with the relatively constant χ3 domain of class I MHC proteins. This interaction further stabilizes the interactions between the T cell and its target. The cytosolic tail of CDS contains a docking site for Lck, a cytosolic tyrosine kinase akin to Src. The T-cell receptor itself is associated with six polypeptides that form the CD3 complex. The ι, ϕ, and γ chains of CD3 are homologous to Ig-χ and Ig-δ associated with the B-cell receptor 53; each chain consists of an extracellular immunoglobulin domain and an intracellular ITAM region. These chains associate into CD3 ι, ϕ and CD3 ϕ γ 7 heterodimers. An additional component, the CD3 l chain, has only a small extracellular domain and a larger intracellular domain containing three ITAM sequences. On the basis of these components, a model for T-cell activation can be envisaged that is closely parallel to the pathway for B-cell activation. The binding of the T-cell receptor with the class I MHC-peptide complex and the concomitant binding of CDS from the T-cell with the MHC molecule results-in the association of the kinase Lck with the ITAM substrates of the components of the CD3 complex. Phosphorylation of the tyrosine residues in the ITAM sequences generates docking sites for a protein kinase called ZAP-70 (for 70-kd zeta-associatedprotein) that is homologous to Syk in B cells. Docked by its two SH2 domains, ZAP-70 phosphorylates downstream targets in the signaling cascade. Additional molecules, including a membrane-bound protein phosphatase called CD45 and a cell-surface protein called CD28, play ancillary roles in this process. T-cell activation has two important consequences.

First, the activation of cytotoxic T cells results in the secretion of *perform.* This 70-kd protein makes the cell membrane of the target cell permeable by polymerizing to form transmembrane pores 10 nm wide. The cytotoxic T cell then secretes proteases called *granzymes* into the target cell. These enzymes initiate the pathway of apoptosis, leading to the death of the target cell and the fragmentation of its DNA, including any viral DNA that may be present. Second, after it has stimulated its target cell to commit suicide, the activated T cell disengages and is stimulated to reproduce. Thus, additional T cells that express the same T-cell receptor are generated to continue the. battle against the invader after these T cells have been identified as a suitable weapon.

Class II MHC Proteins

Not all T cells are cytotoxic. *Helper T cells, a different class, stimulate the pro-liferation of specific B lymphocytes and cytotoxic T cells and thereby serve as partners in determining the immune responses that are produced.* The importance of helper T cells is graphically revealed by the devastation wrought by AIDS, a condition that destroys these cells. Helper T cells, like cytotoxic T cells, detect foreign peptides that are presented on cell surfaces by MHC proteins. However, the source of the peptides, the MHC proteins that bind them, and the transport pathway are different. Helper T cells recognize peptides bound to MHC molecules referred to as class II. Their helping action is focused on B cells, macrophages, and dendritic cells. *Class II MHC proteins* are expressed only by these *antigen-presenting cells,* unlike class I MHC proteins, which are expressed on nearly all cells.

The peptides presented by class II MHC proteins do not come from the cytosol. Rather, *they arise from the degradation of proteins~that have been internalized by endocytosis.* Consider, for example, a virus particle that is captured by membrane-bound immunoglobulins on the surface of a B cell. This complex is delivered to an endosome, a membrane enclosed acidic compartment, where it is digested. The resulting peptides become associated with class II MHC proteins, which move to the cell surface. Peptides from the cytosol cannot reach class II proteins, whereas peptides from endosomal compartments cannot reach class I proteins. This segregation of displayed peptides is biologically critical. The association of a foreign peptide with a class II MHC protein signals that a cell has *encountered* a pathogen and serves as a call for *help.* In contrast, association with a class I MHC protein signals that a cell has *succumbed to* a pathogen and is a call for *destruction.*

Foreign Peptides on Antigen-Presenting Cells

The overall structure of a class II MHC molecule is remarkably similar to that of a class I molecule. Class II molecules consist of a 33-kd *Z* chain and a noncovalently bound 30-kd δ chain. Each contains two extracellular domains, a transmembrane segment, and a short cytosolic tail. The peptide-binding site is formed by the χ_1 and δ_1 domains, each of which contributes a long helix and part of a δ sheet. Thus, the same structural elements are present in class 1 and class II MHC molecules, but they are combined into polypeptide chains in different ways. Class II MHC molecules appear to form stable dimers, unlike class I molecules, which are monomeric. The peptide-binding site of a class II molecule is open at both ends, and so this groove can accommodate longer peptides than can be bound by class I molecules; typically, peptides between 13 and 18 residues long are bound. The peptide-binding specificity of each class II molecule depends on binding pockets that recognize particular amino acids in specific positions along the sequence. Helper T cells express T-cell receptors that are produced from the same genes as those on cytotoxic T cells. These T-cell receptors interact with class II MHC molecules in a manner that is analogous to T-cell-receptor interaction with class I MHC molecules. Nonetheless, helper T cells and cytotoxic T cells are distinguished by other proteins that they express on their surfaces. In particular, helper T cells express a protein called CD4 instead of expressing CD8. *CD4* consists of four immunoglobulin domains that extend from the T-cell surface, as well as a small cytoplasmic region.

The amino-terminal immunoglobulin domains of CD4 interact with the base on the class II MHC molecule. Thus, helper T cells bind cells expressing class II MHC specifically because of the interactions with CD4. When a helper T cell binds to an antigen-presenting cell expressing an appropriate class II MHC-peptide complex, signaling pathways analogous to those in cytotoxic T cells are initiated by the action of the kinase Lck on ITAMs in the CD3 molecules associated with the T-cell receptor. However, rather than triggering events leading to the death of the attached cell, *these signaling pathways result in the secretion of cytokines from the helper cell.* Cytokines are a family of molecules that include, among others, interleukin-2 and interferon.

Cytokines bind to specific receptors on the antigenpresenting cell and stimulate growth, differentiation, and in regard to plasma cells, which are derived from B cells, antibody secretion. Thus, the internalization and presentation of parts of a foreign pathogen help to generate a local environment in which cells taking part in the defense against this pathogen can flourish through the action of helper T cells.

Diverse MHC Proteins

MHC class I and II proteins, the presentears of peptides to T cells, were discovered because of their role in *transplantation rejection.* A tissue transplanted from one person to another or from one mouse to another is usually rejected by the immune system. In contrast, tissues transplanted from one identical twin to another or between mice of an inbred strain are acceptead. Genetic analyses revealed that rejection occurs'when tissues are transplanted between individuals having different genes in the major histocompatibility complex, a cluster of more than 75 genes playing key roles in immunity.

The 3500-kb span of the MHC is nearly the length of the entire *E. coli* chromosome. The MHC encodes class I proteins (presenters to cytotoxic *T* cells) and class II proteins (presenters to helper T cells), as well as class III proteins (components of the complement cascade) and many other proteins that play key roles in immunity. Human beings express six different class I genes (three from each parent) and six different class II genes. The three loci for class I genes are called HLA–A, - B, and –C; those for class II genes are called HLA–DP, –DQ, and –DR. These loci are *highly polymorphic:* many alleles of each are present in the population. For example, more than 50 each of HLA–A, –B, and –C alleles are known; the numbers discovered increase each year.

Hence, the likelihood that two unrelated persons have identical class I and II proteins is very small ($<10^{-4}$), accounting for transplantation rejection unless the genotypes of donor and acceptor are closely matched in advance, Differences between class I proteins are located mainly in the χ_1 and χ_2 domains, which form the peptide-binding site. The χ_3 domain, which interacts with a constant δ_2-microglobulin is largely conserved. Similarly, the differences between class II proteins cluster near the peptide-binding groove. Why are MHC proteins so highly variable? *Their diversity makes possible the presentation of a very wide range of peptides to T'cells.* A *particular* class I or class II molecule may not be able to bind any of the peptide fragments of a viral protein.

The likelihood of a fit is markedly increased by having several kinds (usually six) of each class of presenters in eacn individual. If all members of a species had identical class I or class II molecules, the population would be much more vulnerable to devastation by a pathogen that had evolved to evade presentation. The evolution of the diverse human MHC repertoire has been driven by the selection for individual members of the species who resist infections to which otler members of the population may be susceptible.

Human Immunodeficiency Viruses

In 1981, the first cases of a new disease now called *acquired immune deficiency syndrome (AIDS)* were recognized. The victims died of rare infections because their immune systems were crippled. The cause was identified two years later by Luc Montagnier and coworkers. AIDS is produced by *human immunodeficiency virus (HIV),* of which two major classes are known: HIV-1 and the much Less common HIV-2. Like other *retroviruses,* HIV contains a single-stranded RNA genome that is replicated through a double-stranded DNA intermediate. This viral DNA becomes integrated into the genome of the host cell. In fact, viral genes are transcribed only when they are integrated into the host DNA. The HIV virion is enveloped by a lipid bilayer membrane containing two glycoproteins:gp41 spans the membrane and is associated with gp120, which is located on the external face.

The core of the virus contains two copies of the RNA genome and associated transfer RNAs, and several molecules of reverse transcriptase. They are surrounded by many copies of two proteins called p18 and p24. *The host cell for HIV is the helper T cell.* The gp120 molecules on the membrane of HIV bind to CD4 molecules on the surface of the helper T cell. This interaction allows the associated viral gp41 to insert its amino-terminal head into the host-cell membrane. The viral membrane and the helper-cell membrane fuse, and the viral core is released directly into the cytoso. Infection by HIV leads to the destruction of helper T cells because the permeability of the host plasma membrane is markedly increased by the insertion of viral glycoproteins and the budding of virus particles.

The influx of ions and water disrupts the ionic balance, causing osmotic lysis. The development of an effective AIDS vaccine is difficult owing to the antigenic diversity of HIV strains. Because its mechanism for replication is quite error prone, a population of HIV presents an everchanging array of coat proteins. Indeed, the mutation rate of HIV is more than 65 times as high as that of influenza virus. A major aim now is to define relatively conserved sequences in these HIV proteins and use them as immunogens.

RESPONSES AGAINST SELF-ANTIGENS

The primary function of the immune system is to protect the host from invasion by foreign organisms. But how does the immune system avoid mounting attacks against the host organism? In other words, how does the immune system distinguish between self and nonself? Clearly, proteins from the organism itself do not bear some special tag identifying them. Instead, selection processes early in the developmental pathways for immune cells kill or suppress those immune cells that react strongly with self-antigens. The evolutionary paradigm still applies; immune cells that recognize self-antigens are generated, but selective mechanisms eliminate such cells in the course of development.

Positive and Negative Selection in the Thymus

T cells derive their name from the location of their production — the thymus, a small organ situated just above the heart. Examination of the developmental pathways leading to the production of mature cytotoxic and helper T cells reveals the selection mechanisms that are crucial for distinguishing self from nonself. These selection criteria are quite stringent; approximately 98% of the thymocytes, the precursors of T cells, die before the completion of the maturation process. Thymocytes produced in the bone marrow do not express the T-cellreceptor complex, CD4, or CD8. On relocation to the thymus and rearrangement of the T-cell-receptor genes, the immature thymocyte expresses all of these molecules.

These cells are first subjected *to positive selection*. Cells for which the T-cell receptor can bind with reasonable affinity to either class I or class II MHC molecules survive this selection; those for which the T-cell receptor does not participate in such an interaction undergo apoptosis and die. The affinities of interaction required to pass this selection are relatively modest, and so contacts between the T-cell receptor and the MHC molecules themselves are sufficient without any significant contribution from the bound peptides (which will be derived from proteins in the thymus). *The role of the positive selection step is to prevent the production of T cells that will not bind to any MHC complex present, regardless of the peptide bound.* The cell population that survives positive selection is subjected to a second step, *negative selection.*

Here, T cells that bind with high affinity to MHC complexes bound to self-peptides expressed on the surfaces of antigen-presenting cells in the thymus undergo apoptosis or are otherwise suppressed. Those that do not bind too avidly to any such MHC complex complete development and become mature cytotoxic T cells (which express only CD8) or helper T cells (which express only CD4). The negative selection step leads to *self tolerance;* cells that bind an MHC-self-peptide complex are removed from the T-cell population. Similar mechanisms apply to developing B cells, suppressing B cells that express antibodies that interact strongly with self-antigens.

Autoimmune Diseases

Although thymic selection is remarkably efficient in suppressing the immune response to self-antigens, failures do occur. Such failures results in *autoimmune diseases.* These diseases include relatively common illnesses such as insulin-dependent diabetes mellitus, multiple sclerosis, and rheumatoid arthritis. In these illnesses, immune responses against self-antigens result in damage to selective tissues that express the antigen. In many cases, the cause of the generation of self-reactive antibodies or T cells in unclear.

However, in other cases, infectious organisms such as bacteria or viruses may play a role. Infection leads to the generation of antibodies and T cells that react with many different epitopes from the infectious organism. If one of these antigens closely resembles a self-antigen, an autoimmune response can result. For example, *Streptococcus* infections sometimes lead to rheumatic fever owing to the production of antibodies to streptococcal antigens that cross-react w/ith exposed molecules in heart muscle.

The Immune System and Cancer

The development of immune responses against proteins encoded by our own genomes can be beneficial under some circumstances. Cancer cells have undergone significant changes that often result in the expression of proteins that are not normally expressed. For example, the mutation of genes can generate proteins that do not correspond in amino acid sequence to any normal protein. Such proteins may be recognized as foreign, and an immune response will be generated specifically against the cancer cell.

Alternatively, cancer cells often produce proteins that are expressed during embryonic development but are not expressed or are expressed at very low levels after birth. For example, a membrane glycoprotein protein called *carcinoembryonic antigen (CEA)* appears in the gasterointestinal cells of developing fetuses but is not normally expressed at significant levels after birth. More than 50% of patients with colorectal cancer have elevated serum levels of CEA. Immune cells recognizing epitopes from such proteins will not be subject to negative selection and, hence, will be present in the adult immune repertoire. These cells may play a cancer surveillance role, killing cells that overexpress antigens such as CEA and preventing genetically damaged cells from developing into tumors.

BOX : MAPLE SYRUP URINE DISEASE AND JAMAICAN VOMITING SICKNESS

In a rare autosomal recessive condition (discovered in 1954) the urine and perspiration has a maple syrup odor. High concentrations of the branched-chain 2-oxoacids formed by transamination of valine, leucine, and isoleucine are present, and the odor arises from decomposition products of these acids. The branched-chain amino acids as well as the related alcohols also accumulate in the blood and are found in the urine. The biochemical defect lies in the enzyme catalyzing oxidative decarboxylation of the oxoacids. Insertions, deletions, and substitutions may be present in any of the subunits

The disease which may affect one person in ~ 200,000, is usually fatal in early childhood if untreated. Children suffer seizures, mental retardation, and coma. They may survive on a low-protein (gelatin) diet supplemented with essential amino acids, but treatment is difficult and a sudden relapse is apt to prove fatal. Some patients respond to administration of thiamin at 20 times the normal daily requirement. The branched-chain oxoacid dehydrogenase from some of these children shows a reduced affinity for the essential coenzyme thiamin diphosphate. Polled hereford calves in Australia develop maple syrup urine disease relatively of often. One cause was established as a mutation that introduces a stop codon that causes premature termination within the leader peptide during synthesis of the thiamin diphosphate-dependent El suburut.

A similar biochemical defect in Trnuta of *'Bacillus subtilis* causes difficulties for this bacterium, which requires branched-chain fatty acids in its membranes. Branched acyl-CoA derivatives are needed as starter pieces for their synthesis. With the oxidative decarboxylation of the necessary oxoacids blocked, the mutant is unable to grow unless supplemented with branched-chain fatty acids. Because persons may be born with defects in almost any gene, a variety of other problems leading to accumulation of organic acids are also known. *Methylmalonic aciduria* and propionic acidemia and propionic acidemia. *Lactic acidemia* often results from a defect in pyruvate dehydrogenase.

A rare defect of catabolism of leucine is *isovaleric acidemia,* a failure in oxidation of isovaleiy 1-CbA. The symptoms of this disease are also present in the Jamaican vomiting sickness, caused by eating unripe ackee fruit. Although the ripe fruit is safe to eat, unripe fruit contains a toxin *hypoglycin A* with the following structure. It is metabolized to an acyl-CoA derivative as shown. This is an enzyme-activated inhibitor of the medium-chain fatty acyl-CoA dehydrogenase required for β oxidation of fatty acids.

The compound also inhibits isovaleryl-CoA dehydrogenase, causing an accumulation of isovaleric acid in the blood. Depression of the central nervous system by isovaleric acid in the blood could be responsible for some symptoms. However, death from the highly fatal Jamaican vomiting sickness comes from the hypoglycemic effect. Blood glucose levels may fall as low as 0.5 mM, one-tenth the normal concentration and patients must be treated by infusion of glucose.

SULFUR COMPOUNDS OF GARLIC, ONIONS, SKUNKS, ETC.

Many familiar odors and tastes come from sulfur-containing com-potinds. Crushing onions or garlic releases the pyridoxal phosphate-dependent enzyme *alliinase.* In garlic it acts upon the amino acid *alliin* (accompanying scheme) releasing, by β elimination, a sulfenic acid that dimerizes to form *allicin,* a chemically unstable molecule that accounts for the odor of garlic.

Among the breakdown products of allicin is the nonvolatile ajoene, a compound with anticoagulant activity and perhaps accounting for one aspect of the purported medical benefits of garlic. Another is an antibacterial activity.

Onions contain an amino acid that is a positional isomer of alliin. When acted upon by alliinase it produces 2-propene sulfenic acid, which isomerizes to the *lachrimatory factor* that brings tears to the eyes of onion cutters.c This, too, decomposes to form many other compounds.

The defensive secretion of the striped skunk has intrigued chemists for over 100 years. The components were shown to contain sulfur, and one was incorrectly identified and was long accepted as being butyl mercaptan. Modern capillary gas chromatography by Wood as revealed the presence of seven major components, with the indicated structures. Two are simple volatile mercaptans, but three are thioacetates which hydrolyze in water only slowly, releasing smell for days or weeks from sprayed animals. Washing with mildly basic soap hastens the hydrolysis.

Many readers (~ 40%) may be aware that after eating asparagus a strong odor appears in their urine. These genetic "stinkers" secrete S-methyl thioacrylate, and related compounds, derived from a plant constituent.

The sulfur compound sulfuraphane, extracted from fresh broccoli, has received attention in recent years because of its strong action in inducing synthesis of quinone reductase and glutathione *S*-transferases that help detoxify xenobiotics and may have significant anticancer activity[f].

9 BIOCHEMISTRY OF NUCLEIC ACID

Both the replication and transcription of DNA are complex processes. Although the basic chemistry is relatively simple many enzymes and other proteins are required. In part this reflects organizational and topological problems associated with the huge amount of DNA present as a single molecule within a chromosome.

THE TOPOLOGY AND ENVIRONMENT OF DNA

Although we can isolate DNA in the form of simple double helical fragments, the topology of natural DNA is always more complex. Covalently closed circular DNA such as that in plasmids, mitochondria, and bacterial chromosomes is supercoiled and bound to proteins.

The DNA of chromosomes and bacteriophage particles is folded further into more compact forms. For example, the chromosome of *E. coli* contains DNA about 1.5 mm in length folded within a cell that is only 2 μm long. The diploid length of DNA in a 20 μm cell of a human is about 1.5 meters. At the time of cell division human DNA must all be replicated and packaged into chromosomes, 23 pairs in each cell. The density of the compacted DNA varies. A bacterial nucleoid may contain 10–30 mg/ml of DNA.[2] In chromatin of a eukaryotic nucleus there may be 200 mg / ml of DNA and in nucleosomes 330–400 mg / ml. The tightly compacted head of the T4 bacteriophage (Box 7-C) contains 520 mg/ml.

DNA in Viruses

In the simplest filamentous DNA viruses such as M13 the DNA is coated by à helical protein sheath, as it is extruded from a cell. The sheath is peeled off as the virus enters another cell. However, in the large tailed phage, which contains ~ 160 kb of polynucleotide chains, the DNA is closely packed within the heads. In a model that seems to accommodate most experimental results, the DNA rod bends sharply into a series of folds, which are laid down around the long

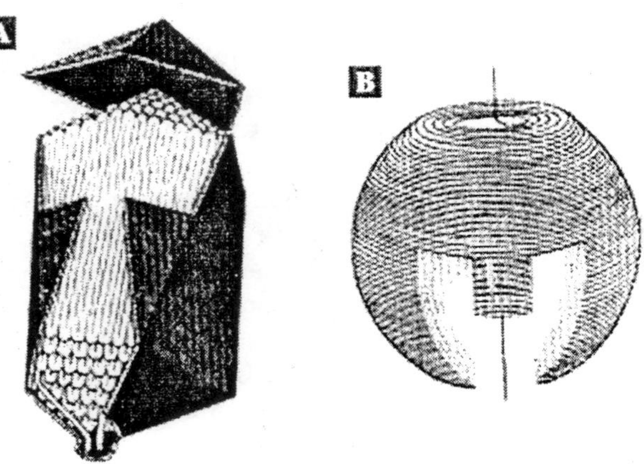

Fig. 9.1 Possible ways of packing DNA into the heads of bacteriophage particles. (A) Spiral-fold. (B) Concentric shell model.

axis of the head in spirally arranged shells. The end of the DNA that enters the phage head first appears to be located in the center with successive shells of DNA around it. In the large bacteriophage G the DNA appears to be folded to form 12 icosahedrally arranged pear-shaped rings in the corners of the capsid. The 2.0 nm diameter double helical segments of DNA lie roughly parallel and are separated by only 0.5 – 1.0 nm of solvent,[6] which contains cations such as the polycation of spermidine. Another possibility is that the DNA may be wound as on a spool of thread. The DNA chains have an external diameter of ~2 nm with 0.6 nm additional for a hydration layer. In a phage head the adjacent parallel chains are 2.6 – 2.7 nm apart. Thus, the packing is very tight. Even so, capsids tend to be only about half filled with DNA. An exception is provided by the tobacco mosaic virus in which the RNA genome is held precisely by protein subunits, which dissociate to release the RNA during infection.

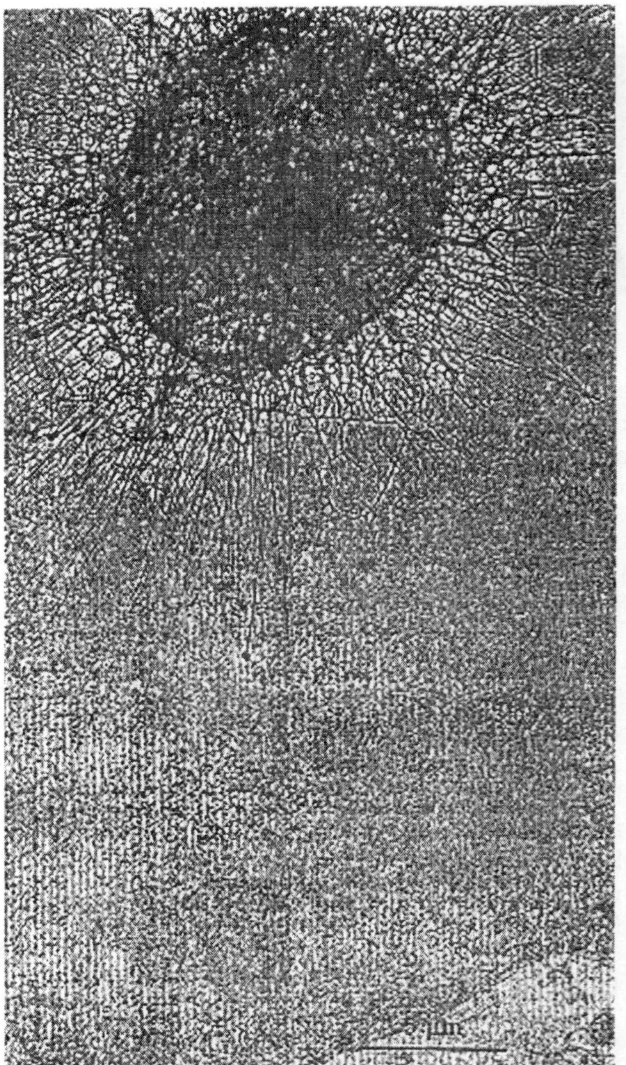

Fig. 9.2. Electron micrograph of a bacterial nucleoid. The DNA is usually contained within the 'cage', but has been spread, using Kleinschmidt's procedure, to yield a surrounding skirt. The cage contains a protein network which includes elements of the cytoskeleton, enclosing the residual nuclear substructures. The denser fibrils radiating from this cage disappear to nuclease digestion and are probably aggregates of DNA fibers which merge with individual DNA strands at the extremities of the skirt. These strands are highly supercoiled, indicative of intact DNA.

Bacterial Chromosomes and Plasmids

Most DNA in living organisms, whether bacteria or eukaryotes, is underwound. That is, the superhelix density is about –0.05 or one supercoil per 200 base pairs. In eukaryotes this negative supercoiling can be accounted for by the winding of DNA around the histones within the nucleosomes. The situation in bacteria is not as clear. There are many bacterial DNA-binding proteins but one of them known as HU is particularly abundant: In E. *coli* it exists as mixed dimers and tetramers of 9.5 (α) and 9.2 (β) kDa subunits. Each HU tetramer can bind~ 60 bp of DNA. There are about 60,000 HU monomers per cell, enough to coat ~ 20% of the genome.

Possible modes of interaction with DNA have been proposed on the basis of the X-ray structure of HU. Binding to HU causes the DNA to be more tightly wound, introducing additional negative supercoils. The resulting unreleased torsional stress may be important in the functioning of the DNA. Binding is strongest to four-way junctions and to DNA with nicks and gaps or to structures induced by supercoiling. Other basic histonelike proteins may also bind to the DNA. However, there are no structures that resemble eukaryotic nucleosomes. If bacterial cells are lysed under certain conditions, e.g., in 1 M NaCl or in the presence of a "physiological" 5 mM spermidine, the entire bacterial chromo-some can be isolated.

The DNA in these isolated chromosomes retains some torsional tension that, however, can be relaxed by nicking with nucleases or by γ-irradiation. However, a single nick relaxes the DNA very little. The explanation appears to be that the DNA is held by proteins of the nucleoid matrix in a series of loops. A single nick relaxes just one loop. On this basis there are 43 ± 10 loops per genome with ~ 100 kb of DNA per loop. A 136-residue protein H-NS is involved in condensation of bacterial DNA. It may act as a scaffolding protein, but it also functions in controlling transcription.

Protamines, Histones, and Nucleosomes

Within bacterial cells the negatively charged phosphate groups of the DNA are neutralized to a large extent by the positively charged polyamines, by cations such as K^+ and Mg^{2+}, and by basic proteins such as HU. Within the mature heads of sperm cells of fish, the tightly packaged DNA is neutralized by the *protaniinea,* small ~5-kDa proteins rich in arginine. Similar basic proteins are found in mammalian sperm. However, within most eukaryotic cells, the charges on DNA are balanced principally by a group of basic proteins, first isolated and named *histon* by Kossel in 1884.

There are five classes of histones, which range in molecular mass from ~ 11 to 21.5 kDa.

H_1 (including $H_1°$, H_5; lysine-rich "linker" histone)

H_2A, H_2B (moderately lysine-rich) ⎤
⎬ "Core" histones
H_3, H_4 (arginine-rich) ⎦

All of the core histones share a conserved 65-residue *histone fold.* The arginine-rich histones have a strongly conserved amino acid sequence, histone H_4 from pea seedlings differing from that of the bovine thymus by only two amino acids. On the other hand, the lysine-rich H_1 is almost species-specific in its sequence. Differentiated tissues contain at least seven variant forms of histone H_1 including proteins designated $H_1°$, Hit, and H_5. The N-terminal 25-40 amino acids of the core histones are positively charged and highly conserved. The 135-residue histone H_3 of calf thymus carries a net charge of +18 within the first 53 residues. This is probably the portion that binds to DNA. On the other hand, the carboxyl terminal end is hydrophobic and only slightly basic.

Histones undergo substantial amounts of *micromodification* including phosphorylation, acetylation, and methylation. Mono-, di-, and tri-methyllysine residues may be present. The core histones all undergo acetylation on specific lysyl side chains. Nuclear histone acetyltrans-ferases transfer acetyl groups from acetylCoA and hydrolytic deacetylases may remove them. The amount of acetylation varies during different stages of the cell cycle, suggesting a regulatory role. Acetylation sites in H_3 and H_4 are highly conserved in all eukaryotes. A small fraction of histone H_2A undergoes phosphorylation and dephosphorylation continuously, but H_1 and H_3 are phosphorylated and dephosphorylated at specific stages of the cell cycle.

Phosphorylation of H_1 has been thought essential for "condensation" of chromatin, the folding into the tightly packed chromosome structures. However, more recent experiments point to the N terminus of histone H_2B as the required site of phosphorylation for chromosome condensation. In addition, histone H_1 may interact with membrane lipids. Histone H_2A exhibits the greatest heterogeneity and appears to function in regulation of transcription, of gene silencing, and of repair of double-strand breaks in DNA. Each of the histones appears to be regulated separately. In animal chromatin ~10% of histone H_2A and small fractions of H_2B and of tissue-specific histones are covalently linked to ubiquitin. However, this monoubiquitination may not be related to proteolytic degradation. Archea contain histones that dimerize and bind DNA to form nucleosomes.

Nucleosomes. An early idea of the function of histones was that they serve as gene repressers. To some extent this view is still valid. However, the large quantity of histone and uniform distribution over the DNA suggested some other role. This was. clarified when electron micrographs showed that chromatin fibers form *nucleosomes,* regular repeating structures resembling beads on a string. The same structure is seen in the "minichromosomes" formed from virus SV40.

Two molecules each of histones H_2A, H_2B, H_3, and H_4 form the core of the nucleosome around which ~146 bp of dsDNA is coiled into approximately two negative, left-handed toroidal superhelical turns. Digestion of chromatin by nucleases causes rapid cleavage into ~200-bp fragments and slower cleavage to 146 ± 20-bp fragments. This suggested that ~ 200-bp segments of DNA are folded around a histone octomer, contracting the 68 nm extended length of relaxed B-DNA into a 10-nm nucleosome. A short linker region of variable length, up to 80 bp, lies between the nucleosomes. The fifth histone, H_1 (or H_5 in some species), may bind to this linker DNA.

A nucleosome with bound H_1 is sometimes called a *chromatosome***.** The superhelix density of ~0.05 observed for DNA extracted from eukaryotic cells is just equal to one negative superhelix turn per nucleosome. For example, the number of nucleosomes seen in the minichromosome of Fig. 9.3 matches the numbers of supercoils in the SV40DNA. If there are two negative supercoils per nucleosome, the DNA in the nucleosome must be wound more tightly than in relaxed DNA (10.0 bp per turn instead of the 10.6 of relaxed DNA).

NMR data suggest that within the nucleosome tne regular base pairing in the DNA may be partially disrupted and that someparts of the histone have a high degree of mobility. Although nucleosomes are distributed rather evenly along the DNA of a cell, there are some DNA sequences that favour nucleosome formation. The resulting *positioned nuclzosomes* are often found in the vicinity of gene promoters, enhancers, and other control sequences. It seems likely that positioning of nucleosomes is related to control of transcription or of other activities. The same basic structure for chromatin has been found in animals, fungi, and green plants.

The linker histones. Anucleosome is pictured in Fig 9. 3A as if held in a compact configuration by the binding ~ of histone H_1 at a position that marks a pseudo twofold axis that lies in the plane of the nucleosome. However, this is only one of several possible locations for linker histones of the H_1, H family. The structure of the linker histones is somewhat different from that of core histones. They have an 80-residue globular domain with long N-terminal *and* C-terminal chains, both of which are rich in basic residues and evidently available for binding to DNA.

Perhaps they bind both to the DNA entering the nucleosome and to that leaving the nucleosome, reducing electrostatic repulsion of those two parts of the DNA superhelix. Another possible location for histone H_1 or H_5 is above the histone surface and *inside* the DNA loop. A third suggested location for the globular linker core is *between* the two turns of the DNA strand. While one function of the linker histones may be to stabilize mononucleosomes, they may also play a role in compaction of the DNA into the 30-nm fibers universally seen in nuclei of cell Histone H_1 can also be regarded as a general represser, holding chromatin tightly folded and preventing transcription.

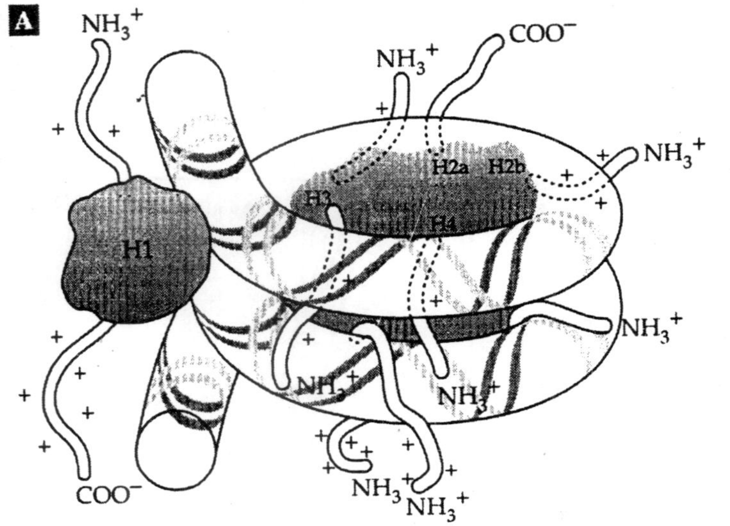

Fig. 9. 3(A) A nucleosome is formed when dsDNA, shown schematically as a tube, wraps roughly twice around a histone octomer. This complex. or nucleosomal core particle, includes at its center two copies of histone H_3, two of histone H_4, and two H_2A–H_2B dimeric pairs, one of which is not visible, Ends of each histone molecule are thought to protrude like tails from the core, ready to interact with other molecules. In many organisms, histone H_1, portrayed at left in one possible position, helps to anchor DNA to the core and promotes further compaction of the DNA into a 30-nanometer fiber.

The possible roles of acetylation, phosphorylation, methylation, ubiquitination, and other modifications of histones in controlling transcription, replication, and DNA repair are receiving increasing attention. *Active chromatin,* where transcription is occurring, has an altered nucleosome structure and increased susceptibility to nuclease action. It appears to be less tightly packed than inactive chromatin and to contain regions called *hypersensitive sites* that are accessible to nucleases or chemical modification reagents.

There seems to be a direct link between increased acetylation of histones and enhanced initiation of transcription by RNA polymerase II. Conversely, deacetylation is associated with repression of transcription. Both histone acetylase and deacetylase activities have been found in transcriptional regulators. The observation that H_1 becomes phosphorylated during the initiation step of mitosis suggests another control mechanism for its represser functions. Several multiprotein complexes that "remodel" chromatin have been identified. These complexes contain ATP-dependent *helicase* activities that open up DNA for transcription.

Folding of nucleosome chains; chromosomes. Electron microscopy shows that chromatin is packed into the nucleus largely as fibers of ~30- to 36-nm diameter. Thoma *et al.* proposed that the fibers could be formed by winding the string of nucleosomes into a simple one-start helix with six nucleosomes per turn and a pitch of ~11 nm. The H_1 molecules would be close together in the center. Its zigzag pattern of adjacent nucleosomes would generate a two-start helix. Another alternative model envisions larger solenoids with interdigitated nucleosomes. A somewhat similar structure appears to be present in the specialized eukaryotic *lampbrush chromosomes* which are observed during the meiotic prophase of oocytes. They have been studied intensively in amphibians such as *Xenopus*.

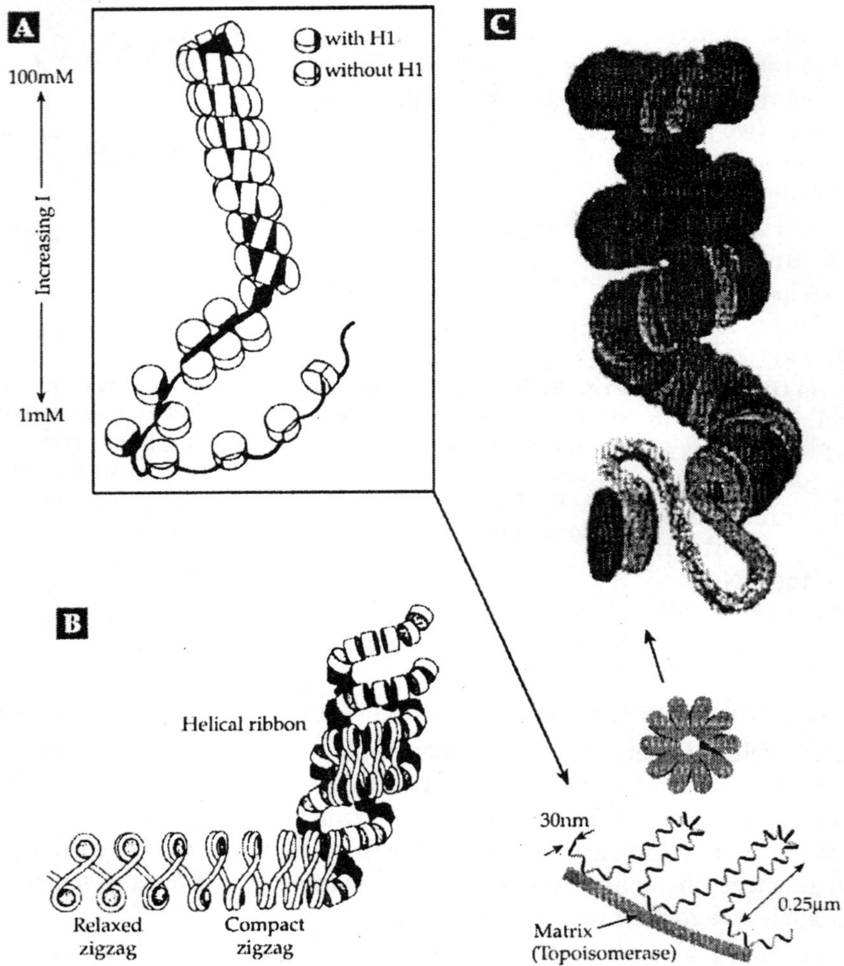

Fig. 9.4 (A, B) Two possible models of the 30-nm chromatin fiber. (A) Thoma *et al.* (B) Woodcock *et al.* The fully compacted structure is seen at the top of each figure. The bottom parts of the figures illustrate proposed intermediate steps in the ionic strength-induced compaction. (C) Possible organization of the DNA within a metaphase chromosome. Six nucleosomes form each turn of a solenoid in the 30-nm filament as in (A). The 30-nm filament forms -30 kb-loop domains of DNA and some of these attach at the base to the nuclear matrix that contains topoisomerase II. About ten of the loops form a helical radial array of 250-nm diameter around the core of the chromosome. Further winding of this helix into a tight coil ~ 700 nm in diameter, as at the top in (C), forms a metaphase chromatid.

A lampbrush chromosome is actually a homologous pair of chromosomes, each one in turn consisting of two closely associated chromatids. The chromosomes are highly expanded, and about 5% of the DNA is extended in the form of ~4000 perfectly paired loops visible with an electron microscope. Each loop consists of ~50 μm or ~ 150 kb of extended DNA. No evidence of any breaks in the DNA is seen, a fact that supports the belief that a single DNA molecule extends from one end of the chromosome to the other through all of the loops. Like the puffs of polytene chromosomes, which may have a similar structure, lampbrush chromosomes appear to be actively engaged in transcription.

Approximately 3% of the DNA may be functional in producing mRNA that is accumulated within the oocyte and is used as a template for protein synthesis during early embryonic development. A different arrangement is present in metaphase chromosomes, which appear as two dense parallel sister chromatids of ~ 700-nm diameter. The DNA must be highly folded.

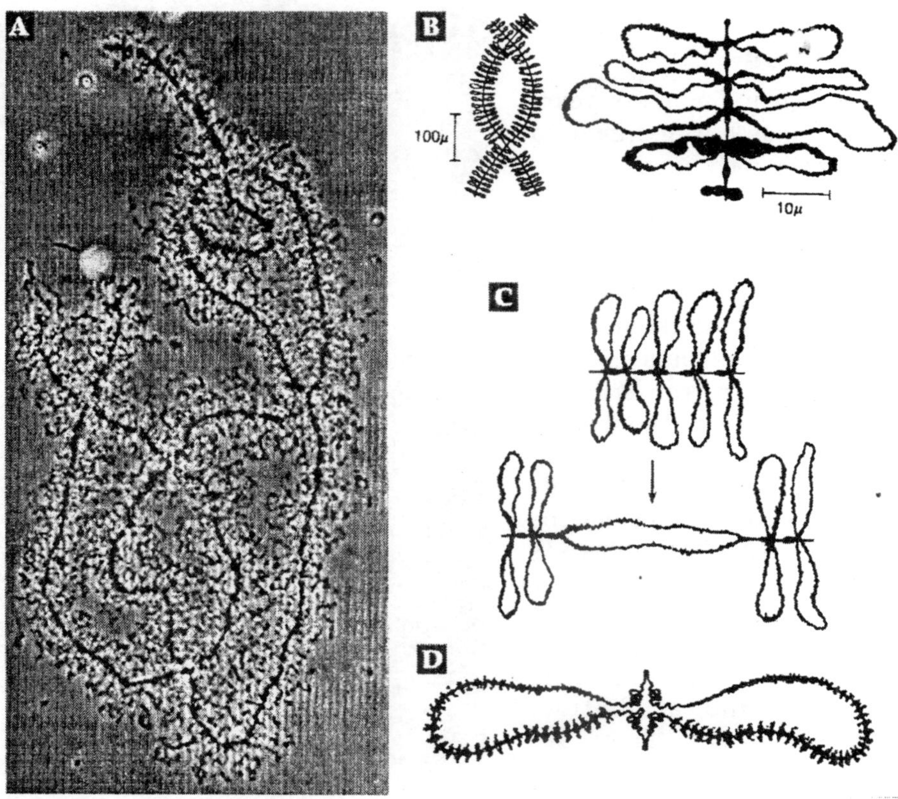

Fig. 9.5. (A) Photomicrograph of a lampbrush chromosome from the nucleus of an oocyte of the newt *Triturus*. (BD) Diagrammatic views of lampbrush chromosomes. (B) The two homologous chromosomes (left) are held together by two chiasmata. A portion of the central chromosome axis (right) shows that two loops with identical morphology emerge at a given point, evidence that each chromosome has already split into two chromatids. (C) Accidental stretching of a chromosome reveals the continuity of the loop axis with the central axis. (D) A single loop pair, showing the single DNA molecules on which RNA chains (indicated by fuzzy shading) are being transcribed.

In the model shown in Fig. 9. 5C the 30-nm fiber is folded into ~ 30-kb loops, each one formed from ~25 turns of the 30-nm helix. The loops then form a helical array 250 nm in diameter with ~ ten loops (300 kb) per turn. This helix is further wound into a tight helix of ~700-nm diameter. A single turn of this helix may contain as much as 9 Mb of DNA. About four hundred coils (an average of 18 coils per human haploid chromosome) could accommodate the entire genome.

A group of five proteins, some of which are designated *SMC* (structural maintenance of chromosomes) proteins, form a complex called *cohesin*. SMC proteins, large multidomain proteins found in all eukaryotes, are also present in bacteria. Interphase chromatin must be much less tightly packed and contains regions in which large loops, e.g., of 20–120-kb size, are uncoiled enough to allow transcription factors and other proteins to locate their target sequences. Many models of interphase chromatin have been proposed.

The Cell Nucleus

It has been clear for many years that the nucleus has a well-organized structure. However, techniques such as fluorescent labeling have only recently been used to provide important details. These studies show that individual interphase chromosomes occupy discrete territories within the nucleus. Parts of the chromosomes may be unfolded and active in transcription, while others are more tightly coiled. Some parts, known as *heterochromatin,* are very, tightly coiled and metabolically almost inert. These regions include the highly repetitive DNA of telomeres and centromeres as well as other regions and complete inactivated female X chromosomes. Replication, transcription, and RNA splicing complexes are found at distinct locations within cells. They are apparently fixed, perhaps attached to the inner nuclear membrane, while the DNA passes through the complexes.

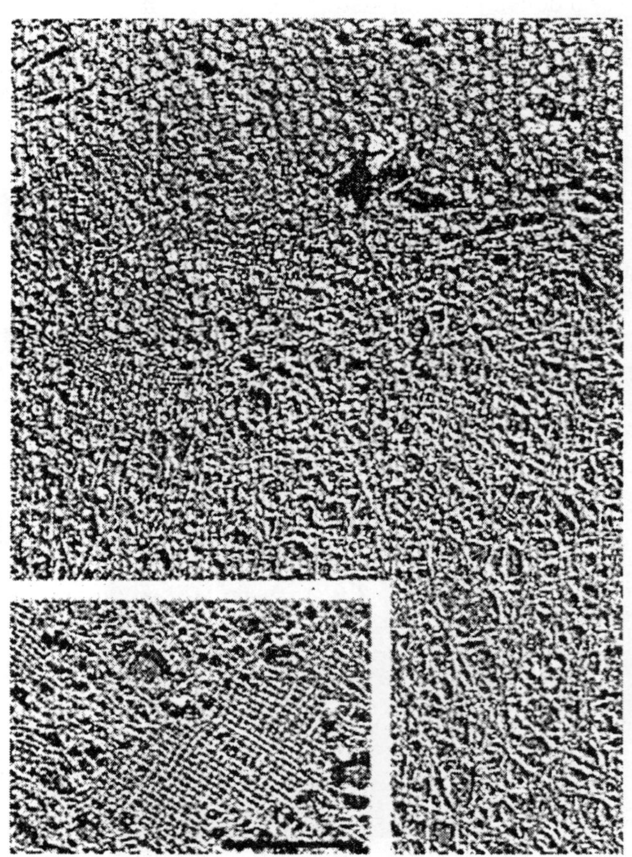

Fig. 9.6. Native nuclear lamina of *Xenopus* oocytes. Freeze-dried metal-shadowed nuclear envelope extracted with Triton X-100, revealing the nuclear lamina meshwork partially covered with arrays of nuclear pore complexes. Inset, relatively well-preserved area of the meshwork of nearly orthogonal filaments from which pore complexes have been mechanically removed.

The nuclear matrix. The lipid bilayers, the histones and other soluble proteins, and the DAN can all be removed from nuclei by extraction and enzymatic digestion. An in soluble residue, **the nuclear** matrix, is left. Largely protein in nature, this matrix is spread throughout the nucleus. Remnants of the membranes remain in the form of proteins that were in or along the bilayer. The nucleolus is clearly defined. The DNA appears to be bound to the nuclear matrix proteins. A specific 320-kb piece of *a Drosophila* chromosome has been mapped and used to locate nontranscribed scaffold (or matrix) attachment regions of DNA bound to matrix proteins. These were found at intervals of 26-112 kb, the intervening loops containing up to five or more genes. A 120-kDa protein together with topoisomerase II may be components of a *nuclear scaffold* that constrains the loops of DNA. The scaffold may also provide locations for the complexes of proteins involved in replication and other processes.

The *matrix attachment regions* (MARs) may also act as *insulators* that shield promoters for transcription within certain loops from control elements such as *enhancers* that may be present in adjacent loops. At least one nuclear matrix component becomes phosphorylated and moves to the nuclear poles during mitosis.

Other nonhistone nuclear proteins. Polyacrylamide gel electrophoresis revealed more than 450 components in HeLa cell nuclei. Most are present in small amounts of <10,000 molecules per cell and are not detectable in cytoplasm. Among the more acidic proteins are many enzymes including RNA polymerases. There are also gene repressors, hormone-binding proteins, protein kinases, and topoisomerases.

Among the six most abundant nonhistone nuclear proteins in the rat are the cytoskeletal proteins myosin, actin, tubulin, and tropomyosin. A group or' small (<30 kDa) proteins, the *high mobility group (HMG)* proteins, can be extracted from chromatin with 0.35 M NaCl. Two pairs, HMG–1 + HMG–2 R (renamed *–HMGB*) and HMG–14 + HMG–17 (renamed *HMGN*) are present in nuclei of all mammals and birds. HMG–14 and HMG-17 appear to be concentrated where active transcription is occurring. HMG-17 undergoes a variety of modifications including acetylation, methylation, phosphorylation, ubiquitination, glycosylation, and ADP-ribosylation, suggesting that it may assist in regulation of transcription. Other members of the HMG family bind more specifically to certain DNA sequences or structures.

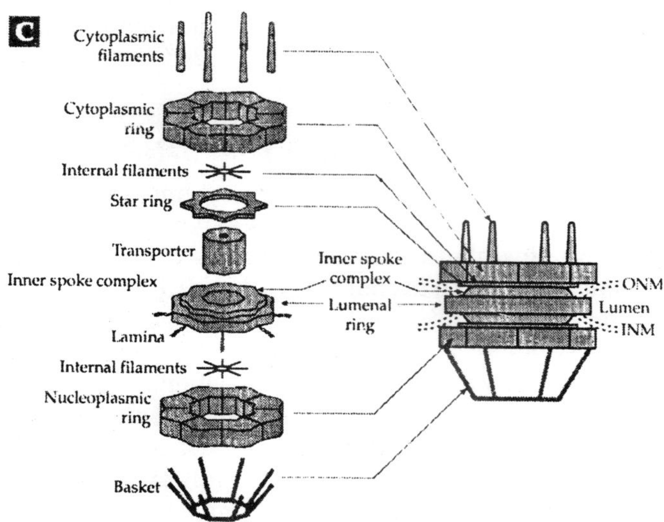

Fig. 9.7. Electron micrographs of nuclear pores from a Xenopus oocyte.

Nuclear membrane, pores, and lamina. The nuclear membrane consists of two bilayer membranes separated by a 40– to 60–nm perinuclear space. Both of the membranes and the intervening space are penetrated by large proteinaceous *nuclear pores.* The two membranes are fused together around" the pores. On the inside of the innermost nuclear membrane is a relatively insoluble meshwork of intermediate f ilaments, the *nuclear lamina.* It acts as a scaffold for tfie pores and may also interact with chromatin. Three proteins, the *lamins* A, B, and C, are the major components of the nuclear lamina. The A and C lamins, which differ only at their C termini, are homologous to cytoplasmic intermediate filament proteins and may exist as a network of coiled-coil polymers. Lamins are phosphorylated, and the lamina is disassembled during the prophase of mitosis, apparently under the influence of the cdc2-cyclin B complex and reappears in telophase at the end of mitosis. Nuclear pores consist of octameric rings of protein subunits with a complex structure and an outside diameter of 120 nm and an inside diameter of –80 nm.

A pore may consist of as many as 80 to 100 different proteins and have a mass of ~125 MDa. The pores are sometimes seen to be blocked by -35 nm granules, perhaps pre-ribosomes. Transport through nuclear pores occurs in both directions. Numerous proteins enter the nucleus where many bind, to RNA and are then exported as ribosomal subunits. Transfer RNAs and messenger RNAs must also be exported. A family of nuclear transporters known as *importing* and *exporting* mediate the movement of protein and RNA through the pores. They depend upon the G protein Ran, which also functions in spindle formation, and the hydrolysis of GTP. Large conformational changes may be linked to the transport cycle and also have been observed in response to changes in the Ca^{2+} level.

The nucleolus. This organelle is a dynamic structure, which breaks down and reforms in each mitotic cycle. It is organized around clusters of genes for the 28S, 18S, and 5.8S ribosomal RNA subunits. Both chromatin and ribonucleoproteins that transport the rRNA out to the cytoplasm are present. The major nucleolar protein *nucleolin* is characterized by its glycine-rich C terminus, which contains seven repeats of GGRGG and also contains $N^{\gamma}N^{\gamma}$-dimethylarginine. Nucleolin may control transcription of the DNA that carries the rRNA genes. It appears to be required for ribosome synthesis and for attachment to the nuclear matrix. A variety of other proteins in smaller amounts are also needed in nucleoli. In addition to the true nucleoli other regions described as *speckles* and *coiled bodies* may develop around RNA splicing centers.

ORGANIZATION OF DNA

A vast amount of new information on DNA sequences and structures is available for yeast, *Arabidopsis, Drosophila,* the nematode *Caenorhabditis elegans,* human beings, and other species. Nevertheless, it may be worthwhile to consider older discoveries, some dating back 20 years or more. When DNA is cut into ~10-kb fragments by shearing and the fragments are denatured by heat, the renaturation of the resulting single-stranded fragments upon cooling takes place in two or more steps. Some material reforms double helices rapidly, whereas other material is slow to renature. At least four kinetically distinct fractions have been recognized: (1) About 70% of mammalian DNA appears to exist largely as single copies, i.e., with unique sequences. (2) About 20% of the total DNA is moderately repetitious, containing sequences that may be present 10^3–10^5 times. (3) About 5% reassociates very rapidly and is identified as highly repetitive *satellite DNA* of which there may be ~10^6 copies. (4) A smaller fraction contains long palindromic sequences.

As a consequence, the single-stranded pieces can fold back almost instantaneously to form hairpin structures. These differences also show up in density gradient centrifugation in CsCl, whether on large pieces of DNA in the presence of ligands such as Ag^+ or on smaller restriction fragments which have often been studied by polyacrylamide gel electrophoresis. If there are ~35,000 human genes, protein or RNA coding sequences must occupy only 2–3% of the genome. The pufferfish *Fuga rubripes* probably has almost as many genes as we but only –13% as much DNA. In further contrast the newt *Triturus cristatus* has six times as much DNA as a human.

The compact genome of the green plant *Arabidopsis thaliana* occupies only 120 Mb. In contrast are the 415–, 2500–, and 5300-Mb genomes of rice, maize, and barley, respectively. What is the function of all of the apparently noncoding DNA present in some organisms? It is often viewed as *"junk,"* whose only function may be to facilitate evolutionary changes in the genome. However, there is doubtless important undiscovered information in these regions. It has been very hard to determine sequences, in part because of the large amount of highly repetitive DNA. Some regions have been *"unclonable"* in prokaryotic systems because of the presence of transposon-like sequences or *"kinkable"* elements (TG•CA steps), palindromes, etc.

Repetitive DNA

Rapidly renaturing DNA fragments often have a different base composition than the bulk of the DNA and, consequently, often separate as small satellite bands upon centrifugation in a CsCl gradient. Satellite DNA is usually associated with regions of the chromosome that do not unravel in telophase as does the bulk of the DNA. Satellite DNA usually consists of short highly repetitive sequences, which occur in large clusters of up to 100 Mb of DNA, often near centromeres or telomeres or on a Y chromosome. The DNA of a satellite band from the kangaroo rat contains the sequence 5'-GGACACAGCG-3' repeated so often that it accounts for 11% of the entire DNA of the cell. Longer repeating sequences -170 bp are also often present as are *microsatellites* of 2– to 5-base-pair repeats. At least 30,000 microsatellite loci are present in the human genome.

Centromeres. The attachment of spindle fibers to chromosomes depends upon the segments of DNA known as centromeres to organize the attached kinetochores. In the yeast *Saccharomyces cerevisiae* a 120-bp region containing three short conserved sequences is present in the centrosomes of all ten chromosomes. This may fold into a distinctive looped structure. Human centrosomes are large and complex, but the DNA is highly repetitive, giving rise to *α-satellite DNA*. Sequences such as $(TGGAA)_n$ are repeated many times. Such sequences can form self-complementary looped structures containing some unpaired guanines that intercalate and stack between sheared G • A pairs.

Complex regional centromeres involving kilobases or megabases of DNA have also been identified in fission yeasts, *Drosophila,* and green plants. The great variability indicates that centromeric sequences undergo rapid evolution. This may be related to the fact that of the four cells formed by female meiosis only one becomes an egg. A series of unique *centromeric proteins* (CENP-A to CENP-G) bind to the DNA sequences of centromeres and direct the formation of the kinetochores. Even for the simpler centromere of budding yeasts, kinetochores have a complex structure. The CENP proteins were first identified as autoantigens in sera of patients with the autoimmune disorder *scleroderma.*

Telomeres. The DNA sequences at the chromosome ends have a TG-rich strand, such as the *(TTGGGG) (Tetrahymena)* and the $(TTAGGG)_n$ of both human and trypanosome chromosomes. The complementary DNA strand is CA-rich. The *S. cerevisiae* telomers have ~350 base pairs containing the sequences ($TG_{1-3}/C_{1-3}A$) as well as one or more copies of a 6.7-kb nonrepetitive sequence and other elements. In many species the repetitive telomeric sequences have 3' poly(G) tails at the ends of the DNA molecules. These tails are able to form G quartet structures. A variety of telomere-binding proteins have been isolated. Some of these bind to G quartet structures and some, such as RAP1 of yeast, to double-stranded telomere repeat regions. Special problems associated with replication of telomeres are discussed in Section C,8.

Short interspersed sequences *(SINES).* Much of the reiterated DNA is present in repeated segments 100–500 bp in length that lie between 1– to 2–kb segments of unique DNA. The best known example is the human ***Alu*** family, socalled because it contains a site for cleavage by restriction enzyme *Alu*I. Sequences of the *Alu* family also exists in other primates and in rodents. The ~300-bp *Alu* sequence is reiterated over 500,000 times in the human genome with various sequence alterations, but an 80–90% homology. This sequence consists of two similar ~130-bp segments called the *"left monomer"* and *"right monomer."* The right monomer contains a 31-bp insertion and the left end carries a poly(dA) sequence.

In addition there is a short 7– to 20-bp sequence, which is variable between different *Alu* sequences but is directly repeated at each end of a given *Alu* sequence. The *Alu* sequence has strong homology with the 7S RNA that is part of the signal recognition particle involved in transport of newly synthesized peptide chains across the membranes of the ER. *Alu* sequences are transcribed into hnRNA, the precursors to mRNA. Some *Alu* sequences are present in intervening sequences (introns) within genes anizf others are in noncoding sequences between genes.

Sharp suggested that specific proteins in the nucleus may bind to the *Alu* sequences, preventing hnRNA from leaving the nucleus before it has been processed to remove introns and other sequences absent in mRNA. However, the presence of the poly(dA) regions and the direct terminal repeats suggests that the whole *Alu* sequence is *pseudogene* derived from a *retroposon,* a type of transposon that originated from an RNA molecule. *Alu* and other SINES contain an RNA polymerase III promoter and are transcribed. Active retroposons form reverse transcripts (cDNAs) that can be integrated at various points in the genome. One theory is that retroposons have no biological function but have invaded the genome at random locations. Many SINES and other families of repetitive sequences have been characterized by the presence of a restriction enzyme cleavage site in each copy of the sequence. *Eco*RI cuts the previously mentioned α-satellite DNA.

A 319-bp reiterated sequence in the human genome surrounds a *Hint* site, and Sau3A cleaves a 849-bp sequence with ~1000 copies per haploid genome. In the hermit crab 30% of the genome consists of repeated sequences, one 156-bp unit occuring ~7 million times. Many identical 14-bp GC-rich inverted repeats are present. *Neurospora* and yeast both contain many copies of the GC-rich palindromic sequence 5'-CC<u>CTGCAG</u>TA<u>CTGCAG</u>G-3', which contains the two underlined *Pstl* sites. Some repetitive sequences such as the **CAT** family and the *homeodomain sequence* are found in the control regions preceding the 5' ends of groups of genes that are regulated coordinately. Some repetitive sequences seem to be unstable in the genome and may be excised and lost or may increase in number during aging.

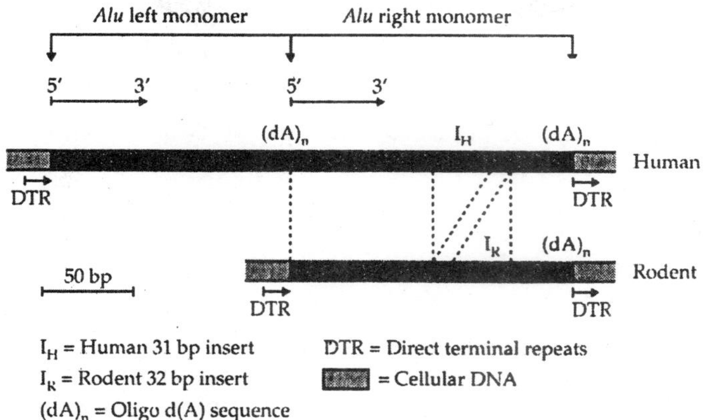

Fig. Structure of human and rodent *Alu* sequences in DNA

Repeated DNA sequences have apparently originated not only with 7S RNA but also from mRNA or tRNA molecules. A characteristic SINE in rodent DNA is known as the ID (identifier) sequence because it was once thought to be a marker for genes transcribed in neural tissues. Rat DNA contains ~130,000 copies of the ID sequence. Other *clustered repeats* are frequently found between genes that are present in large numbers. These include the genes for ribosomal RNA, tRNA, small nuclear RNAs, and histones.

Long interspersed repeat sequences (LINES). These moderately repetitive sequences may be several kb in length. Just one type seems to be abundant in each mammalian species. The human LI sequence is over 6 kb in length. Like the *Alu* sequence it has a poly(dA) sequence at the 3'-end, but it does not contain short terminal direct repeats. These interspersed repeat sequences usually contain genes which may be functional, but many of the copies contain pseudogenes or genes that are randomly truncated, having lost a segment from the 5' end. These sequences, too, seem to have been dispersed throughout the genome by retroposons.

Human chromosome 22, sequenced in 1999, contains within its 33.4 Mb of euchromatin at least 545 genes, 134 pseudogenes, and 8043 L1 sequences (9.7% of the DNA), as well as 20,188 *Alu* sequences (16.8% of the DNA) and many other interspersed repeats. Repetitive DNA sequences are less common in prokaryotes, but they do exist. For example, many dispersed and clustered repeating units are present in *Halobacterium* DNA.

Genes for Ribosomal RNA and Small RNA Molecules

Most genes are present as one copy each per hap-loid genome. However, there are many copies of the genes for ribosomal RNA and tRNA. In *Xenopus* DNA there are ~450 repeats of the 28 S and 18 S rRNA genes on one chromosome and ~24,000 copies of the 5 S RNA genes at the ends of the long arms of most of the chromosomes. Nontranscribed spacer regions lie between the repeats of the 28S and 18S gene pairs, as can be seen with the electron microscope in Fig. 9.9. The gene for 5 S RNA has a high GC content. A-denaturation map of the DNA shows easily melted regions separated by shorter 120-bp sequences, apparently of high GC content and presumably coding of the 5 S RNA. The easily denatured AT-rich spacers are ~630 bases long.

Using restriction enzymes much of this DNA can be cut into segments that contain repeats within repeats. One 15-unit polynucleotide contains the sequence $A_4CUCA_3CU_3G$ repeated about

30 times. About 200 copies of rRNA genes per haploid genome are located at the constrictions in the short arms of human chromosomes 13,14,15, 21, and 22. As in *Xenopus,* clusters of tandemly repeated 5S RNA genes are found at the ends of the long arms of most chromosomes.

A similar organization of rRNA genes is found in the rat. The ten different chromosome ends carrying rRNA genes in the human diploid nucleus come together to form the nucleolus the site of synthesis of ribosomal subunits. At first there are ten small nucleoli, but fuse these to form the single highly structured but mem-braneless nucleolus. The rRNA genes of the macronucleus of *Tetrahymena* have been *"amplified"* and are found on linear 21-kb palindromic molecules, about 10^4 copies being present per cell.

Initiation of transcription begins near the center and proceeds outward in both directions. These short chromosomes contain typical telomeric ends. Genes for the tRNAs are spread throughout the genome of bacteria, mitochondria chloroplasts, and eukaryotic nuclei. They sometimes occur in clusters but more often are far apart. In *Drosophila* there are probably at least 600 tRNA genes, many occurring in pairs of opposite polarity, i.e., as inverted repeat sequences. The genes for small nuclear RNAs U1-U6 are organized in a variety of ways. The ~ 100 human U1 genes occur on a single chromosome, perhaps organized in a tandem array and interspersed with as many as 10,000 defective pseudogenes.

Other Gene Clusters and Pseudogenes

Closely related structural genes often occur in clusters. Among these are clusters of *immunoglobulin* genes and clustered genes for the α and β *globins*, which encode the protein sequences for the hemoglobins. The human α globin gene cluster occupies about 30 kb on chromosome 16 and the β globin genes 60 kb on chromosome 11. The α1 and α2 genes encode identical peptides, while the related ξ gene encodes the corresponding subunit of embryonic hemoglobin. The β cluster includes, in addition to the adult gene, a pair of fetal globin genes (γ^G and γ^A) differing by only one amino acid (Gly vs Ala) at position 136, the embryonic ε chain, and the minor adult δ chain. Besides the functional genes the globin cluster contains *pseudogenes,* which are given the prefix ψ in fig 9.10. These are nonfunctional genes, which appear to encode peptides homologous to the known globins.

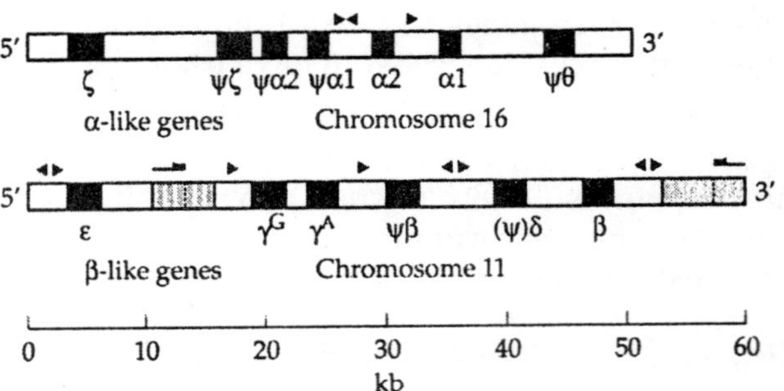

Fig. 9.10 Organization of the globin genes on human chromosomes 11 and 16. The composition of the various embryonic, fetal, and adult hemoglobins is also indicated. Closed boxes indicate active genes and open boxes pseudogenes. The triangles (⇀) indicate *Alu* repetitive sequences and their orientation. The shaded boxes indicate Kpn repeat sequences and the half-arrows (↽) their respective orientation. The Kpn sequence between the ε and γ^G genes in fact consists of two tandemly linked Kpn repeats.

However, they contain mutations that prevent expression. For example, deletion of a single nucleotide near the beginning of the pseudogene will scrambl: the genetic message by changing the reading frame. Globin pseudogene ξψ has a nonsense mutation at codon 6, ψα1 contains a whole set of mutations that prevent transcription and translation, while ψα2 is mutated almost to the point of being unrecognizable as a globin relative. Do pseudogenes have a physiological function?

The presence of pseudogenes in close proximity to functional genes suggests that there may be an as yet unrecognized controlling function. On the other hand, they may be relics of evolution. If a gene family arises by duplication and mutation of duplicated copies, we may expect to find gene copies that have not been selected as useful but are still present in the genome. There are regions of the genome that seem to be largely single copy DNA with few reiterated sequences or gene clusters. While some gene families exist as clusters others are dispersed throughout the genome.

Introns, Exons, and Overlapping Genes

A major difference between prokaryotes and eukaryotes is the presence of intervening sequences (*introns*) between the coding sequences (*exons*) in eukaryotes. Introns are especially numerous in higher organisms. For example, the gene for the myosin heavy chain present in rat embryonic skeletal muscle encodes a 1939-residue peptide but occupies a length of 22 kb of DNA. The gene is split into 41 exons, whose transcribed RNA must be cut and spliced at 40 places to form the mRNA. By comparison the corresponding gene from the nematode *Caenorhabdatis elegans* is far less fragmented.

It is not clear why some genes have so many introns and others so few. The very compact chromosomes of viruses and most bacteria do not contain introns. However, they are sometimes present in mitochondrial and chloroplast genes. A surprise, which was first recognized in viral RNA and DNA, is that genes sometimes overlap. For example, two proteins, one long and one shorter, are synthesized starting at the same point in the RNA genome of phage Qβ.

In the DNA of phage ϕX174 the third nucleotide of the stop signals for some genes are also the first nucleotides for the start signal for translation of the next gene.' Pairs of genes have been found in which one of the genes of the pair is found completely inside the other gene but is translated in a different reading frame. A fourbase overlap between dihydrofolate reductase and thymidylate synthase has been found in the DNA of phage T4. A transposable DNA insertion sequence in *E. coli* encodes two genes, one of which is contained within the other and which is transcribed from the opposite strand of DNA.

The double-stranded RNA of a reovirus produces two peptides from the same sequence using two different AUG initiation codons in different reading frames. Overlapping obviously limits severely the mutational alterations that are allowable. Perhaps it is for this reason that eukaryotic genes seem to be dispersed in widely separated locations and overlap is rare. However, the human type IV collagen genes for the αl and α2 chains are encoded on opposite strands with their 5' flanking regions overlapping. Some introns contain genes. Chlorarachniophyte algae contain a multimembraned chloroplast thought to be a vestigial remnant of an endosymbiont. Its 380-kb genome consists of three short chromosomes that encode overlapping genes and contain the shortest introns (18–20 base pairs) known.

DNA of Organelles

Mitochondrial DNA (mtDNA), discussed in Chapter 18, varies in size from <6 kb to 367 kb. The small circular mtDNA of animals is extremely efficiently packed with genes for tRNAs, rRNA, and a small number of protein subunits. However, the 78-kb yeast mtDNA contains many long AT-rich spacers as well as long introns, some of which contain genes for splicing enzymes. Otherwise, it is similar to animal mitochondrial DNA. Mitochondrial genomes of higher plants are much larger; that of maize is a 570-kb circle, which contains both direct and inverted repeat sequences. Recombination between these sequences is apparently the origin of smaller incomplete circular genomes that are also present.

Plant mtDNA appears to contain the genes present in animal mitochondria plus additional genes. For example, one encodes a 5S RNA. The most unusual mtDNA is found in hemoflagellates such as *Trypanosorna* and *Leishmania.* The single large mitochondrion or *kinetoplast* located near the base of the flagellum contains a network of catenated circular DNA molecules (Fig. 9 .16). The genetic makeup is similar to that of other mtDNA, but the rRNA genes are unusually short. Chloroplast DNA varies in size from 120 to 160 kb

Methylation of DNA

As mentioned in already a significant fraction of the pyrimidine and purine bases in DNA is methylated. One function of such methylation in bacteria, to protect against the action of restriction endonucleases. For example, the gene for the well-known *EcoRl* endonuclease is carried in *E. coli* cells by an R factor. This plasmid also carries (just 29 base pairs away) the gene for a 326-residue *N^6-adenine methylase.* This enzyme uses S-adenosylmethionine (AdoMet) to methylate the two adenines (marked by asterisks) in the six-base-pair recognition sequence 5'-G A A* T T C / 3'-C T T A* AG converting them to N^6-methyladenines (m^6A). Other DNA methyltransferases place methyl groups on N-4 of cytosine or on C-5 of cytosine.

The latter utilizes a mechanism illustrated in Eq. 9.4. Such enzymes are components of both type II and type I restriction-methylation systems. However, most of the m^6A in the E. *coli* chromosome arises from action of a different methylase, one that recognizes the palindromic sequence 5'-GATC and methylates adenines in both chains. This *DNA adenine methylase,* a product of the *dam* gene, plays an important role in mismatch repair, transposition, regulation of transcription, and initiation of DNA replication. The same methylase regulates at least 20 genes induced during infection by *Salmonella typhimurium.* Some of these genes are essential to virulence. A similar methyl-transferase appears to control differentiation of the stalked *Caulobacteria* cells.

CG doublets. The only modified base commonly found in eukaryotes is 5-methylcytosine, which upon deamination becomes thymine. Most methylation occurs when C is followed by G. Usually 60–90% of all 5'–CG sequences (CpG sequences) in eukaryotic DNA are methylated. However, the fraction of methylated cytosine varies from almost zero for *Drosophilia, Caenorhabditis,* and *Saccharomyces* to as much as 30% in higher plants. When CG pairs are methylated on both cytosines, an interesting result arises upon replication of the DNA. The methyl groups don't prevent replication, but only one of the DNA strands in each daughter duplex is methylated.

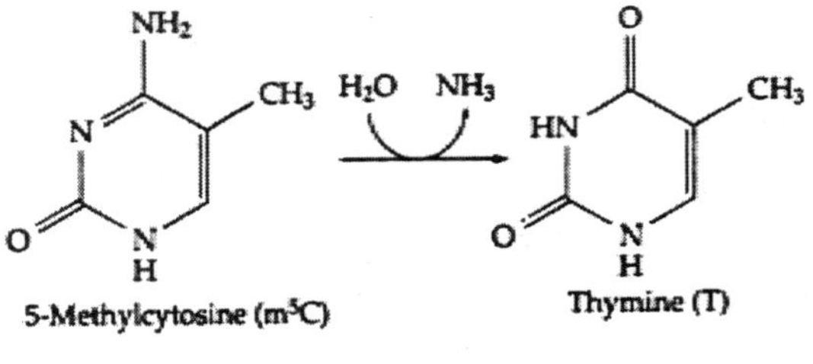

NH_2 CH_3 N O N H — H_2O → NH_3 — O CH_3 HN O N H

5-Methylcytosine (m^5C) Thymine (T)

However, additional methylation of DNA occurs within three hours of DNA replication by a *maintenance methyltrans-ferase* that recognizes methylated CpG sequences in the old DNA strand and methylates the cytosine in the 3′-GpC of the newly synthesized strand. Mammalian cells contain at least three different DNA (cytosine-5) methyltransferases. Enzymatic *demethylation* which converts m^5C residues back to cytosine, may also occur. It has been difficult to demonstrate enzymatic demethylation of 5-methylcytosine has been reported.

The observed loss of methylation during embryonic development if may be a result of loss of the maintenance methylase. Most CpG doublets are found in large *"islands"* of several hundred bases to about *2* kb in length, which are unusually rich in G + C. These islands lie near the 5' ends of ~ 60% of all human genes and near origins of replication. They are also found in regions of the DNA that are compacted into heterochromatin. The genes in methylated regions tend to be *"silent"* i.e., they are not actively transcribed. Demethylation may permit transcription. The degree of methylation of the CG doublets is variable both in position within a chromosome and with stage of development. Cytosine methylation is essential to embryonic development, and mice lacking the maintenance DNA methytransferase are developmentally retarded and die at mid-gestation.

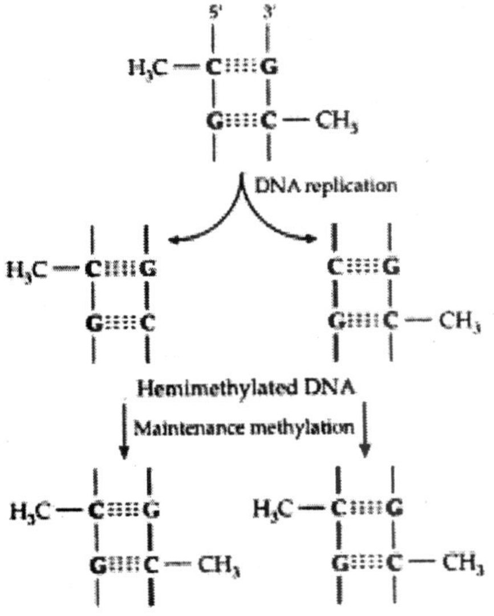

DNA replication

Hemimethylated DNA

Maintenance methylation

Normal methylation pattern retained

Not all CG sequences are methylated and various patterns of DNA methylation are generated at different stages of development by rounds of methylation and by the action of demethylases. The maintenance methylase ensures that a stable methylation pattern persists until altered by new rounds of methylase or demethylase action. A parallel is found in the prokaryotic *Canlobacter* in which three chromosomal sites are fully methylated in swarmer cells, become hemimethylated in stalked cells, and are fully methylated again just prior to cell division. DNA methylation affects transcription, either directly by preventing the binding of transcription factors or indirectly via a series of binding proteins specific for methylated CG doublets.

In early stages of embryonic development there is very little methylation, but some genes are quickly silenced as methylation takes place. Heterochromatic regions including inactivated

X chromosomes are heavily methylated. However, additional alteration in chromatin is required for complete silencing of genes. Recent studies indicate that an abundant mammalian protein binds to the methylated DNA along with a *historic deacetylase.* The latter acts on acetylated mstones to free lysine side chains, which may interact in an inhibitory manner with the DNA. In female mammalian cells most of the genes on one of the two X-chromosomes are completely inactivated.

DNA methylation plays a major role in this process. A perfect correlation has been observed between 5'-methylation of cytosines in CpG islands and inactivation of X-chromosome genes. Methylation may also play a role in recombination and repair. Methylation of DNA decreases with increasing age. It increases as a result of oncogenic transformation of cells.

Imprinting. With the exception of X-linked genes each person has two copies of each gene, one of maternal and one of paternal origin. Both copies of most of these genes are expressed. However, a few of the genes receive from one parent or the other an *imprint,* a mark that distinguishes the parental origin. Such imprints are maintained in cells through embryonic development but are erased in embryonic gonads to allow for a new imprint in the germ cells. Imprinting depends upon DNA methylation, and all imprinted genes show the presence of differentially methylated regions.

REPLICATION

Following the discovery of the double helix and the enthusiasm that it engendered many people thought that the synthesis of DNA was simple. The nucleptide precursors would align themselves along separated DNA template strands and perhaps spontaneously snap onto the growing chains. In fact, replication is a complex process that requires the cooperative action of many different gene products and perhaps an association with membrane sites. The matter is made more complex by the fact that some of the enzymes involved in replication are also required in the processes of genetic recombination, in repair of damaged DNA molecules, and in defensive systems of cells.

Early Studies

That the DNA content doubles prior to cell division was established by microspectrophotometry. It was clear that both daughter cells must receive one or more identical molecules of DNA. However, it was not known whether the original double-stranded DNA molecule was copied in such a way that an entirely new double-stranded DNA was formed or whether, as we now know to be the case, the two chains of the Original molecule separated. The latter is called *semiconservative* replication, each of the separated strands having a new complementary strand synthesized along it to form the two identical double-stranded molecules. The first definitive evidence for semiconservative replication was reported by Meselson and Stahl in 1958.

Cells of *E. coli* were grown on a medium containing isotopically pure $^{15}NH_4^+$ ions as the sole source of nitrogen. After a few generations of growth in this medium the DNA contained exclusively ^{15}N. Then the cells were transferred abruptly to a medium containing $^{14}NH_4^+$ and were allowed to grow and to double and quadruple in number. At various stages DNA was isolated and subjected to ultracentrifugation in a CsCl gradient. Small but easily detectable differences in density led to separation of dsDNA molecules containing only ^{15}N from those containing partly ^{15}N and from those containing only ^{14}N. At the beginning of the experiment only DNA containing entirely ^{15}N was present. However, after one generation of growth in the ^{14}N-containing medium, the density of *all* the DNA was such as to indicate a content of one-half ^{14}N and one-half ^{14}N.

After a second generation of growth half of the DNA atill contained both nitrogen isotoped in equal quantity, whereas half contained only ^{14}N, exactly the result expected for semiconservative replication.

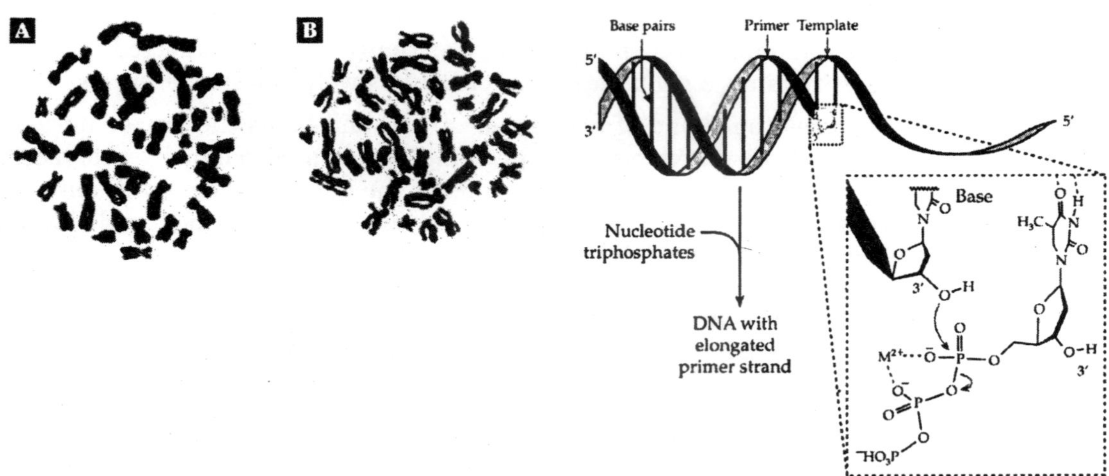

Fig. 9.11 (A) Human chromosomes after one replication in the presence of 5-bromodeoxyuridine (BrdU). Both chromatids of each chromosome contain BrdU in one strand of the DNA duplex and normal thymidine in the other. (B) After two replications in the presence of BrdU one chromatid of each chromosome contains BrdU in both strands of the duplex and stains strongly with a special differential staining procedure. The other chromatid contains only normal thymidine in one strand of the duplex and is not stained.

Autoradiography. Later a technique of direct autoradiography of DNA using ^{3}H-labeled thymidine was applied by Cairns. Cells of *E. coli* were grown on a medium containing the radioactive thymidine for various times but typically for 1 h (~2 generations). The cells were then ruptured, the DNA was spread on a thin membrane filter, and autoradiograms were prepared. When the DNA contained [^{3}H]thymidine the exposed trace in the autoradiogram could be followed around the entire 1.1–1.4 mm circumference of the spread DNA molecule.

Molecules partially labeled and in the process of replication could also be identified. After two hours of growth in the presence of ^{3}H-labeled thymidine about half of the bacterial DNA was fully labeled, but half contained regions that were labeled only half as heavily. They presumably contained ^{3}H label in a single strand and, therefore, represented unreplicated regions. All of the molecules had undergone one round of replication with tritiated thymidine to yield lightly labeled molecules; parts of the molecules had not completed the second round. The more heavily labeled regions were interpreted as fully replicated.

The shapes of the *"replication figures"* suggested that DNA is synthesized in a continuous manner starting from one point and continuing around the circular molecule at a constant rate. Although subsequent experiments (considered in Section 4) show that replication is usually *bidirectional,* the experiments of Cairns were important because they introduced a technique for direct visualization of replication *in vivo.*

The chemistry of DNA polymerization. What are precursors of DNA? Early experiments showed that the nucleoside [^{3}H]thymidine was efficiently incorporated into DNA, but for energetic reasons it seemed unlikely that thymidine was an immediate precursor. Evidence favouring the nucleoside triphosphates was provided in 1958 when Arthur Kornberg identified a DNA polymerase from *E. coli.* Kornberg's enzyme, now known as *DNA polymerase I,* was isolated in the amount of 600 mg from 90 kg of bacterial cells (over 400 molecules of enzyme per cell). The 928-residue enzyme displayed many of the properties expected of a DNA-synthesizing enzyme. It requires a *template strand* of DNA as well as a shorter *primer strand.*

The enzyme recognizes the 3' end of the primer strand and binds the proper nucleoside triphosphate to pair with the next base in the template strand. Then it catalyzes the displacement of pyrophosphate, at the same time linking the new nucleotide unit onto the 3' end of the primer strand. Continuing in this way, the enzyme is able to turn a single-stranded template DNA into a double-stranded DNA in which the newly synthesized strand contains, at each point, the base complementary to the one in the template strand. Although the action of the DNA polymerase I, according to Eq. 9.3 provided a straightforward way to form a complementary strand of DNA, it did not explain how double-stranded DNA could be copied.

One problem is that the two strands must be separated and unwound. If unwinding and replication occured at a single *replication fork* in the DNA, as indicated by Cairns' experiment, the entire molecule would have to spin at a speed of 300 revolutions per second to permit replication of the *E. coli* chromosome in 20 min. It also required that some kind of a *swivel,* or at least a nick in one chain, be present in the chromosome as indicated in the following drawing. Another problem was posed by the fact that the two chains in DNA have opposite orientations. Thus, at the replication fork one of the new chains might be expected to grow by addition of a new nucleotide at the 3' end, while the other chain would grow at the 5' end. *If so, there should be two DNA polymerases,* one specific for polymerization at each of the two kinds of chain end. Nevertheless, despite intensive search the only DNA polymerases found added new units only at the 3' end.

Discontinuous replication and RNA primers. In 1968, Okazaki reported that during the time that replication of DNA is taking place bacterial cells contain short fragments of DNA. These are now called *Okazaki fragments* or *replication fragments*. A second development was the discovery of the enzyme *DNA ligase* which is able to join two pieces of DNA to form a continuous chain. These two discoveries provided an explanation for the lack of a second kind of DNA polymerase. One strand, the *leading strand,* of the replicating DNA could be synthesized continuously in the 5' to 3' direction while the other strand, the *lagging strand,* would have to be synthesized in segments (the *Okazaki fragments*), which could then be joined by the DNA ligase.

In 1971 Brutlag *et al.* reported that initiation of synthesis of the DNA of phage M13 in E. *coli* required formation of a short segment of RNA as a primer. It was subsequently shown that *with few exceptions priming by RNA synthesis is always required for replication.* Newer studies have revealed great complexity in the mechanisms of replication. As for synthesis of any polymer there are distinct steps of *initiation, elongation,* and *termination.*

Topological problems associated with the unwinding and rewinding of the double helices and with disconnection of catenated circles and untying of knots are solved with the aid of special enzymes, the *helicases* and *topoisomerases*. Replication requires both DNA and RNA polymerases, a ligase, and ancillary proteins, some of whose functions aren't yet clear. Many of these proteins associate to form large multiprotein complexes, which are given names such as *primosome* (for priming) and *replisome.* Many bacteriophages and plasmids also replicate within cells of *E. coli* utilizing bacterial proteins as well as proteins encoded in the viral or plasmid genome.

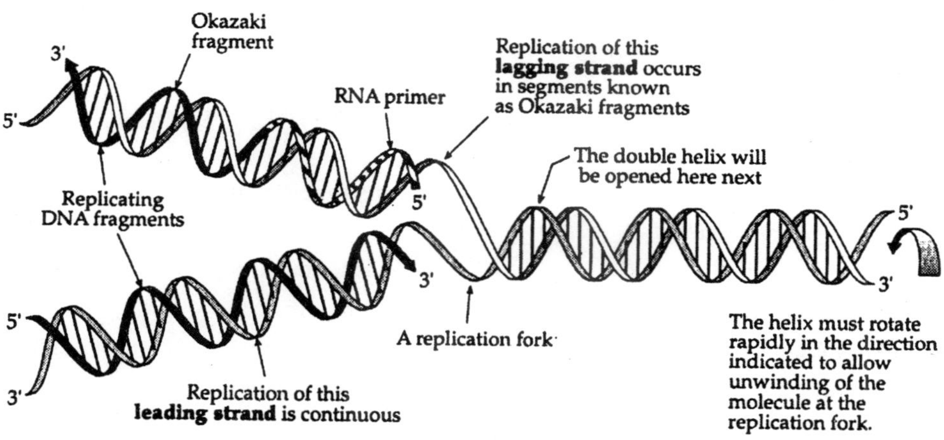

DNA Polymerases

Immediately after its identification DNA polymerase I was generally assumed to be the principal chain elongation enzyme. However, an *amber* mutant of *E. coli deficient* in DNA polymerase I (gene *polA.;* Fig. 9.4A) synthesized DNA normally. This finding stimulated' an intensive search for new polymerases. Two were found: DNA polymerases II (gene *pol* B) and III. Both are present in amounts less than 25% of that of DNA polymerase I. Both have properties similar to those of polymerase I, but there are important differences.

By now DNA polymerases have been isolated from many organisms, many genes have been cloned and many sequences, both of bacterial and eukaryotic polymerases are known. Comparisons of both sequences and three-dimensional structures, suggest that the polymerases belong to at least six families. These include the RNA-dependent DNA polymerases known as *reverse transcriptases* as well as some *RNA polymerases*. Some of the polymerases exist as single polypeptide chains, while others function only as large complexes. In every case a two-metal ion catalytic mechanism with in line nucleotidyl transfer, appears to be used by the enzymes. Two-metal ion catalysis is also observed for phosphatases and ribozymes.

Exonuclease activities, proofreading, and editing, DNA polymerase I not only catalyzes the growth of DNA chains at the 3' end of a primer strand but also, at about a 10-fold slower rate, the hydrolytic removal of nucleotides from the 3' end (*3'- 5' exonuclease activity*). The same enzyme also catalyzes hydrolytic removal of nucleotides from the 5' end of DNA chains. This latter 5'- 3' *exonuclease activity,* the DNA polymerase activity, and the 3'-5' exonuclease activity all arise from separate active sites in the protein. DNA polymerases II and III do not catalyze hydrolysis from the 5' end. Treatment with proteolytic enzymes cuts a 323-residue piece containing the 5'-3' exonuclease from DNA polymerase I leaving a larger C-terminal piece known as the *Klenow fragment.* This fragment retains the polymerase activity as well as the 3'-5' exonuclease activity and is widely used in genetic engineering.

Table 9.1. Families of DNA Polymerases.

Class	*Name*	*Function*	*Molecular mass (kDa)*
A	E. Coli polymerase I (pol I)	DNA excision repair	103
	Klenow fragment		68
	Bacillus subtilis Pol I[b]	DNA excision repair	
	Thermus aquaticus DNA polymerase (Teq)[c.d]	DNA excision repair	
	T7 DNA polymerase[e]	Virus replication	80
	T7 RNA polymerase[f.g.]		99
	Eukaryotic poly γ (gamma)	Mitochondirial replication	
	Eukaryotic Pol θ (theta)	DNA repair	
B	Eukaryotic poly α (alpha)	DNA replaction	180 (core)
	Eukaryotic poly δ (delta)	DNA replaction	
	Eukaryotic poly ε (epsilon)	DNA replaction	
	Eukaryotic poly ζ (zeta)	Bypass synthesis	
	Bacteriophage T4 DNA Pol + accessory proteins[h,i]	DNA replaction	43
	E. coli Poll II[j]		90
C	Bactrial DNA Pol III[p] + accessory proteins	DNA replication	~ 900 (holoenzyme)
D	Euryarchaeotic Pol II		
X	Eukaryotic DNA Pol β[n,o]	DNA repair	39 × 2
	Eukaryotic Pol λ (lambda)	Base excision repair	
	Eukaryotic Pol μ (mu)	Non-homologous end-joining	
	Eukaryotic Pol σ (sigma)	Sister chromated cohesion	
Y	E.coli UmuC protein	SOS response	
	Eukaryotic Pol η (eta, XP-V, RAD 30)	Bypass synthesis	
	Eukaryotic Pol ι (iota)	Bypass synthesis	
	Eukaryotic Pol κ(kappa)	Bypass synthesis	
Reverse transcriptase family			
	HIV reverse transcriptase[k,1]		
	Telomerase[m]		
	RNA-dependent RNA polymerses		

In the Klenow fragment the large C-terminal domain contains the polymerase. The N-terminal domain contains the 3',5'-exonuclease activity, which is thought to fulfill a *proofreading* and *editing* function. The polymerase acts at the 3' end of the growing DNA chain. Before moving on to the next position, the enzyme verifies that the correct base pair has been formed in the preceding polymerization event. If it has not, the exonuclease action removes the incorrect nucleotide and allows the polymerase to add the correct one. Thus, each base pair is checked twice, first before polymerization and then after polymerizations. A puzzle was the fact that structural studies indicate that the editing center is over 3 nm away from the catalytic center in Pol I.

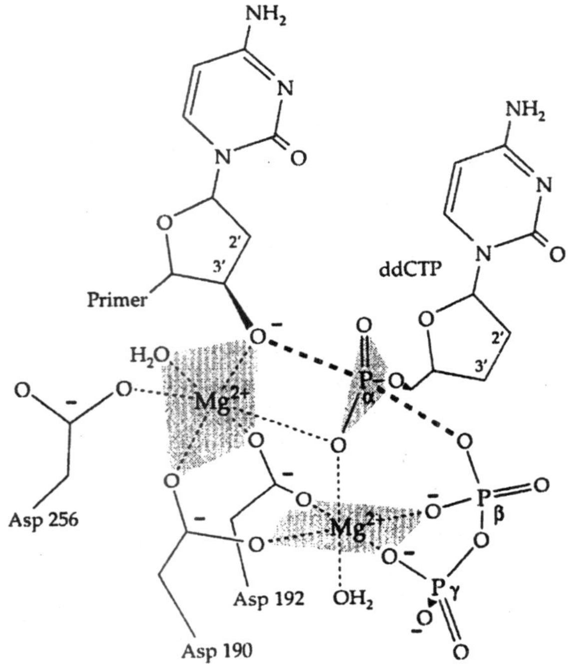

When the catalytic center *"identifies"* a nucleotide triphosphate as able to form a proper Watson-Crick nucleotide pair, it catalyzes the formation of the new nucleotide linkage. Then it releases the 3'-chain end, which sometimes *"melts"* and over a 10- to 100-ms time interval is able to reach over into the exonuclease site where the newly added nucleotide may be hydrolytically removed. However, if the newly formed nucleotide is properly paired, it will be less likely to melt, and the new nucleotide will be more likely to be retained. DNA polymerase I and other related polymerases utilize *processive mechanisms,* moving from one site to the next without diffusing away from the DNA. The schematic picture shown in fig. 9.14 also indicates how the 5'-3' exonuclease activity can come into play, when the polymerase reaches the end of a single-stranded gap.

Fig. 9.13 Proposed mechanism and transition state structure for the synthetic nucleotidyl transfer activity of DNA polymerase β (and other DNA polymerases). The chain-terminating inhibitor dideoxy CTP is reacting with the 3'-OH group of a growing polynucleotide primer chain. This -OH group (as $-O^-$) makes an in-line nucleophilic attack on P_α of the dideoxy-CTP. Notice the two metal ions, which interact with the phospho groups and which are held by three aspartate side chains. Two of the latter, Asp 190. and Asp 256, are present in similar positions in all of the polymerases. The active centers for the hydrolytic 3'-5' and 5'-3' exonuclease activities of some of the polymerases also appear to involve two-metal catalysis and in-line displacement.

Other Class A polymerases. The *Thermus aquaticus (Taq)* polymerase is best known for its widespread use in the polymerase chain reaction. Like *E. coli* I the enzyme is a large multidomain protein. The structure of the catalytic domains of the two enzymes are nearly identical, but the *Taq* polymerase has poor 3'-5' editing activity. The enzyme has been carefully engineered to improve its characteristics for use in the PCR reaction. Some bacteriophage encodetheir own DNA polymerases. However, they usually rely on the host cell to provide accessory proteins. The sequence of the DNA polymerase from phage T7 is closely homolo-gous to that of the Klenow fragment and the 3D structurs are similar. The 80-kDa T7 polymerase requires the 12-kDa thioredoxin from the host cell as an additional subunit.

It has been genetically engineered to improve its usefulness in DNA sequencing. About 45% of the sequence of the *RNA polymerase* Encoded by phage T7, which transcribes RNA from the phage DNA, is also similar to that of the Klenow fragment. Sequences of these DNA polymerases are distantly related to those of reverse transcriptases. The 136-kDa polymerase y functions in mitochondria but is encoded in a nuclear gene. It is the only DNA polymerase that is inhibited by antiviral nucleotide analogs such as AZT.

Polymerases of Class B. Although E. *coli* polymerase II is a member of this family, relatively little is known about its function. It may participate in DNA repair in the "SOS" response. The catalytic subunits of the major *eukaryotic* DNA polymerases as well as of archaeal DNA polymerases are members of the B family. Eukaryotic cells contain at least 13 DNA polymerases which are desig-nated by Greek letters. Polymerases α, β, and ε are essential for nuclear DNA synthesis. They function together with accessory proteins in pri-mase and replisome complexes. Others participate in DNA repair. Phage T4 encodes a DNA polymerase much used in the laboratory because of its ability to polymerize using a long single-stranded template. It also depends upon accessory factors provided by the bacterial host.

Other DNA polymerases. *Reverse transcriptases* synthesize DNA using an RNA template strancr. Tney are best known for their function in retroviruses. The HIV reverse transcriptase is a heterodimer of 51- and 66-kDa subunits. The larger subunit contains a *ribonuclese H* domain. The enzyme is a prime target for drugs such as AZT and others. A different reverse transcriptase is found in all eukaryotic cells *telomerase,* an enzyme essential for replication of chromosome ends. Reverse transcriptases have also been found in rare L1 sequences that are functioning retrotransposons. The ~335-residue catalytic subunit of *eukaryotic polymerase* β , which has a DNA repair function, is the simplest known DNA polymerase. The active enzyme is as small as 38 kDa. It lacks proofreading and is less accurate than the eukaryotic replicative polymerases. Structural analysis revealed a folding pattern related to that of a *nucleotidyltransferase superfamily* that includes enzymes such as terminal deoxynucleotidytransferase and the glutamine synthase adenylytransferase. However, its active site is similar to that of other DNA polymerases.

DNA polymerase III. This Class C enzyme is the major bacterial polymerase for DNA replication. In its complete holoenzyme form it can synthesize new DNA strands at rates as high as ~1 kilobase s^{-1} without dissociation from a template. A genetic approach has provided important information about DNA replication. A series of temperature-sensitive mutants of E. *coli* unable to carry out DNA synthesis were obtained. From these mutations, genes *dnaA, F, G, I, J, K, L, P, I, X,* and Z were identified and located at various points on the chromosome map. Genes C and D map close together at 89 min, and it now appears that they are one gene. Gene F encodes a ribonucle-otide reductase.

The functions of genes *A-E, G, X,* and Z are discussed in the following sections. None of these genes code for DNA polymerase I, but gene *'dnaE* was identified as that of DNA polymerase III, which is now known to be the major polymerase in bacterial DNA replication. To obtain rapid error free synthesis it must be combined with a number of other subunits to form *polymerase III holoenzyme* (or replisome). Ten different subunits, some as two or more copies, form the holoenzyme.

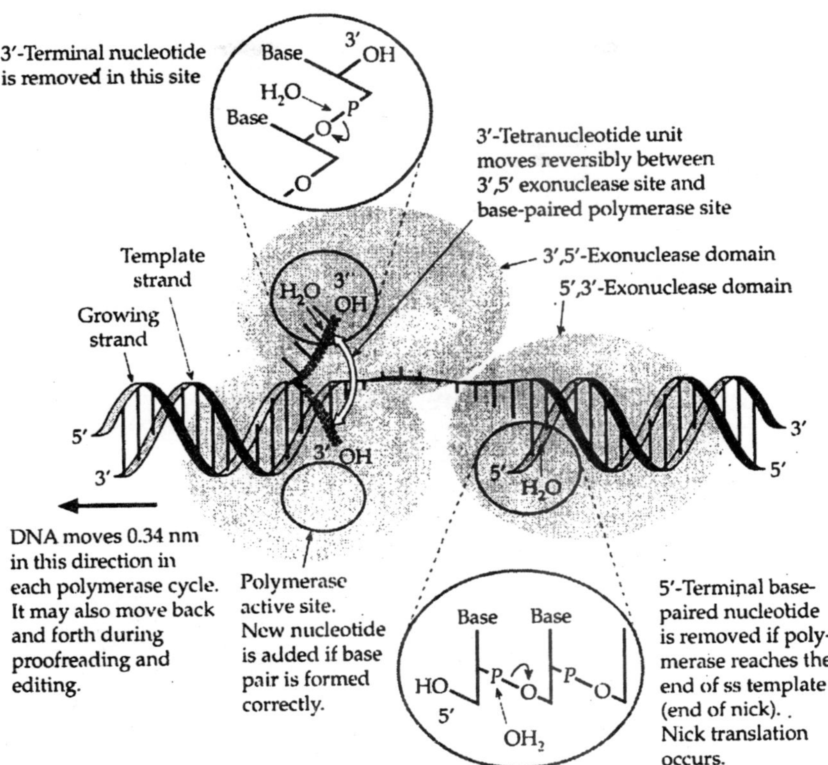

Fig. 9.14. Schematic representation of DNA polymerase action on a nicked strand of DNA in which the nick has been enlarged. At the catalytic center new nucleotide units are added at the 3' end of a growing strand. At the 3'-5' exonuclease site the 3' terminal nucleotide may be removed hydrolytically. This will happen to the greatest extent if the nucleotide is poorly paired in the duplex. At the 5'-3' exonuclease site nucleotides are hydrolytically removed from the 5' end of a strand in the chain.

The polymerase catalytic center is in the 132-kDa a subunit. The 27.5-kDa ε subunit contains the 3'-5' exonuclease editing activiry. Mutation in its gene *(dnaQ)* leads to a high spontaneous mutation rate in bacteria. Together with the θ subunit, α and ε form the polymerase III core. This complex has polymerase activity and improved proofreading ability but is still unable to act rapidly, accurately, and processively. Full catalytic activity requires at least the additional β, γ, δ, and τ subunits. The presence of the i subunit causes the core complex to dimerize to form Pol III' can add only about ten nucleotides to a growing DNA chain before it dissociates. The presence of the β_2 dimer, known as the *processivity factor* or *sliding clamp,* is essential For highly processive polymerization. The β protein forms a ring around the duplex DNA and interacts with the polymerase clamping it tightly to the DNA. Putting the β_2 clamp onto the DNA is an ATP-dependent process that first involves binding of the *γ-complex* or *clamp loader* ATP-dependent opening of the β_2 ring, and insertion of the DNA duplex (step *c*). This complex may form a replisome structure that acts simultaneously on the leading and lagging strands. The *Bacillus subtilis* replisome appears to contain two different catalytic (a) subunits, perhaps one for each strand.

Other Replication Proteins

DNA ligases. These enzymes, which are essential to replication, have a specific function of repairing "nicked" DNA. Such DNA, as indicated in Eq. 9.5, has a break in one strand and contains a 3'-hydroxyl group and a 5'-phosphate group, which must be rejoined. The ligase from *E. coli* activates the phosphate group in an unusual way by transfer of an adenylyl group from NAD^+, with displacement of nicotinamide mononucleotide (Eq. 9.5, step *a*). The reaction is completed by displacement of AMP as indicated in Eq. 9.5, step *b*. Cells infected by bacteriophage T4 synthesize a virally encoded ligase, which utilizes ATP rather than NAD^+ as the activating re-agent. The ~ 190-kDa mammalian DNA ligase I has been found deficient in some patients with the Bloom syndrome, a condition associated with poor DNA repair and a high incidence of cancer.

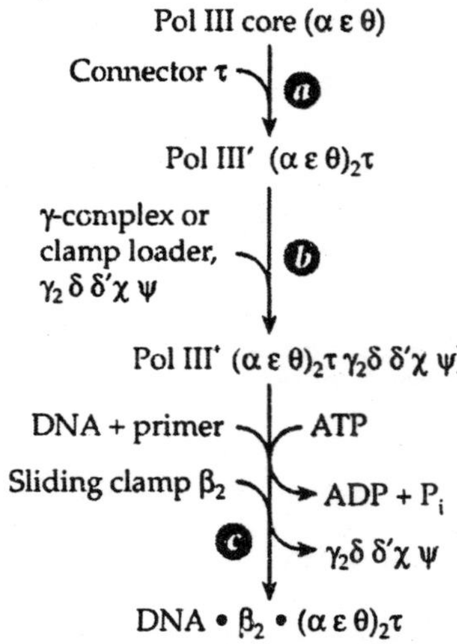

Single-strand binding proteins. Genetic analysis of replication of the DNA of phage T4 within cells of *E. coli* revealed that at least five genes of the virus are required. One of these, gene 43 specifies the T4 DNA polymerase, while gene 32 codes for a single-strand binding protein, also known as the DNA unwinding, melting, or helix-destabilizing protein. It has a greater affinity for ssDNA than for dsDNA and binds to a length of ssDNA causing unwinding of the double helix and exposure of the purine and pyrimidine bases of the template strand. The protein is required for replication, genetic recombination, and repair of DNA. Similar proteins are encoded in the genomes of many viruses. The 87-residue single-strand binding pro-tein encoded by gene 5 of phage M13 forms a dimer, which completely coats newly synthesized ssDNA preventing the DNA polymerase system of the host bacteria from converting it into dsDNA. The polynucleotide chain binds into a groove in the protein with one tyrosine intercalated between the DNA bases.

The *E. coli* single-strand binding protein, another helix-destabilizing protein that is usually called simply **SSB,** is a tetramer of 18.5-kDa subunits. It is essential to DNA replication. About 35 nucleotides may bind to each tetramer. The situation is not as clear in eukaryotes where DNA is largely coiled around histones in the nucleosomes. Several single-strand binding proteins have been identified, but the need for SSB proteins in eukaryotic nuclear replication is uncertain. A human mitochondrial SSB resembles that of *E. coli.*

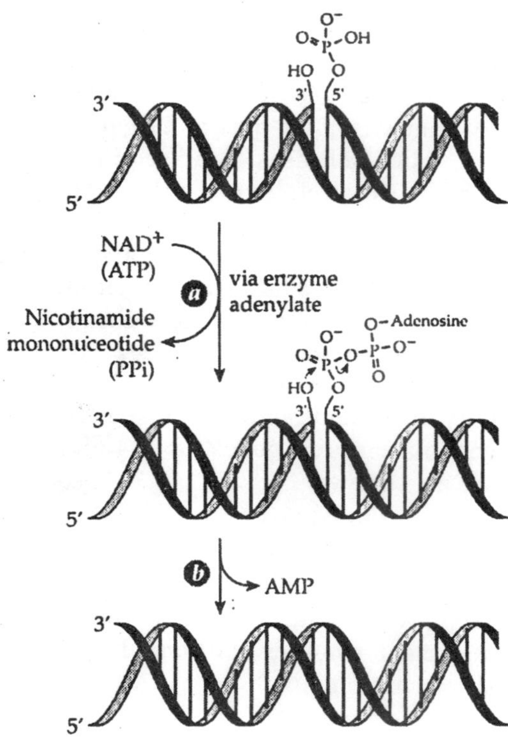

Helicases and tapoisomerases. Cells of *E. coli* contain at least 12 DNA-dependent ATPases that cause unwinding of DNA at the expense of hydrolysis of ATP. The activity of these **helicases** is essential to replication, repair, and recombination of DNA in all organisms. The primary replicative helicase of *E. coli,* which unwinds DNA ahead of the replication fork, is encoded by gene *dnaB*. The active form is a hexamer, which exists in at least two conformational states. A segment of ~ 20 nucleotide units of DNA binds to one hexamer. Helicases have ATPase activity, and the dnaB hexamer contains six ATP-binding sites. However, only three of them may be occupied. Many helicases have a hexameric ringlike structure; that of *Bacillus subtilis* is seen clearly in electron micrographs. Three-dimensional structures are known for some. Although these enzymes may bind to duplex DNA, they also bind to and move along single-stranded DNA in either the 5'→3' or 3'→5' direction. The directionality of a helicase can be determined by annealing two small pieces of ssDNA to the 5' and 3' ends of a longer strand of ssDNA. A 5'→3' helicase will translocate along the long ssDNA and displace the oligonucleotide annealed at the 3' end of the strand. while a 3'→5' helicase will displace the oligonucleotide annealed at the 5' end.

The dnaB protein is a 5'→3' helicase. However, the first helicase identified, the product of the *E. coli rep* gene, is a 3'→5' helicase. It is one of the host proteins needed for the propagation of phages such as φX174 and M13. It catalyzes the unwinding of the double-stranded replication forms of these viruses. It binds to a stretch of ~ 20 nucleotides in a single-stranded region of nicked DNA. The hydrolysis of ATP moves the enzyme along the bound strand in a 3′ to 5′ direction opening up the DNA at *a* replication fork. Another E. *coli* 3′→5′ helicase is protein *priA* (also called n'), a component of some primasome structures involved in replication of viruses.

The bacteriophage T4 gene 41 protein, a 5′→3′ helicase, functions together with the gene 61 primase in replication of that virus. The phage T7 gene 4 protein and virus SV40 large T antigen are also hexameric ringlike helicases. The *E. coli* protein **RecQ** is required for various aspects of recombination and is the prototype of a large group of helicases present in both prokaryotes and eukaryotes. The bacterial ruvB protein is a hexameric helicase that propels branch migration in Holliday junctions (Fig. 9.28) during genetic recombination, while helicase rho is required for termination of RNA synthesis. Numerous eukaryotic helicases have been identified and purified. Helicase DNA2 is needed for DNA replication in nuclear extracts from yeast. Human 3′→5′ DNA helicases, members of the RecQ family, are defective in some patients with **Bloom syndrome** and **Werner Syndrome.** Bloom syndrome causes growth retardation, immunodeficiency, sensitivity to sunlight, and a predisposition to skin cancers and leukemias. The yeast *(S. cerevisiae)* genome contains genes for 134 different proteins that are prob-ably helicases. RNA helicases are also known.

A characteristic of all helicases is their ATPase activity, which apparently provides energy for "melting" the DNA. The mechanisms are not clear, but rapid separation of the stacked and hydrogen-bonded base pairs may be impossible without some assistance from an ATP-dependent process. In the case of the rep protein hydrolysis of two molecules of ATP seems to be required to melt one base pair. It isn't clear whether one strand of DNA passes through the center of the oligomeric ring, as shown in Fig. 9.15, or whether both strands pass through. Helicases vary in their amino acid sequences, but they all possess several characteristic **signature sequences** including the Walker A and B motifs found in other ATPases, in sequences related to the *E. coli* **recA** proteins and in synthases.

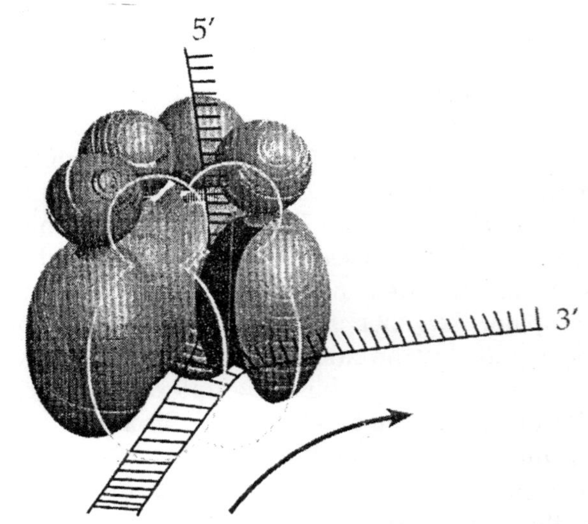

Fig. 9.15 Model of DnaB hexamer proposed by Jezewska *et al.* The arrow indicates the direction of movement of the DNA relative to the position of the helicase.

While helicases can cause the unwinding of linear DNA duplexes, they cannot alter the linking number of covalently closed circular double-stranded DNA. The latter is the function of **topoisomerases,** which have been found in organisms from bacteria to higher plants and animals and may also be encoded by viruses. There are two basic types of DNA topoisomerase. Those of type I change the linking number in steps of 1. One way that this might be accomplished is for an enzyme to nick one strand in the DNA allowing one of the cut ends to swivel around the unbroken strand, then to reseal the chain. However, it was found that topoisomerases can also cause catenation or decatenation of circular duplex DNA as long as at least one of the reacting DNA molecules is nicked. This observation suggested that a topoisomerase binds to a single-stranded region at a nick and cuts the chain but does not release the ends. This permits either a single strand or a duplex to pass through the broken strand, which is then resealed.

Topoisomerases of type I usually act most rapidly on negatively supercoiled DNA. They relax it by decreasing the number of negatively supercoiled turns one at a time. Negative supercoiling presumably facilitates binding of the enzyme to a single-stranded region by unwinding of the duplex. No ATP or other obvious source of energy is needed by type I topoi-somerases. The chain cleavage involves a simple nucleophilic displacement by an –OH group of a tyrosine side chain (Tyr 319 of *E. coli* topoisomerase I), which attacks a phosphorus atom in the DNA chain (Eq. 9.6). The result is covalent attachment of the enzyme to the

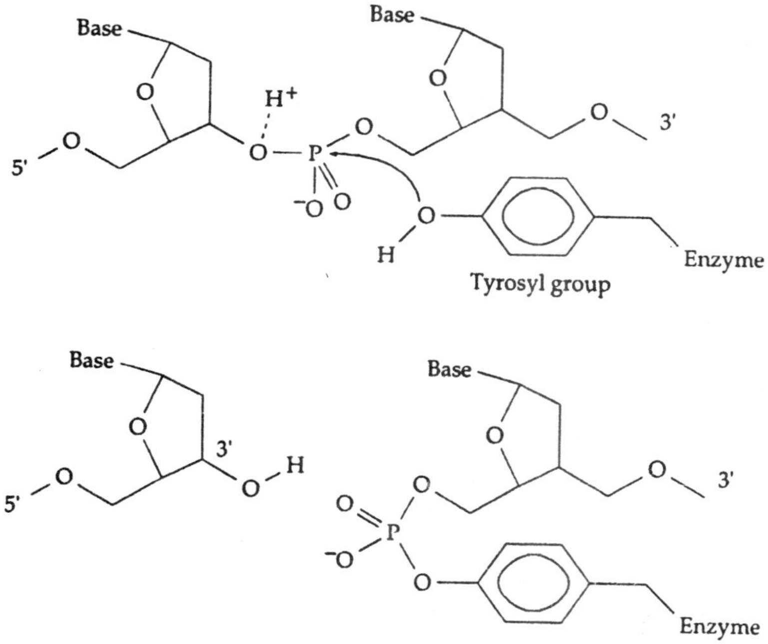

Base
O
H^+
O
5'
O—P
$^-$O O
Base
O
O
3'
O
H
Enzyme
Tyrosyl group

Base
O
3'
5'
O
O
H
Base
O
O
3'
O
O
P
$^-$O
O
Enzyme

5'-end of the cut strand in type IA topoisomerases and at the 3'-end in type IB topoisomerases. After the passage of the other strand through the gap in the cut strand of a type IB topoisomerase (Fig. 9.16) the free 3'-OH oxygen atom (or 5'-OH) attacks the phosphorus atom in the phos-photyrosine diester to reform the chain and release the enzyme. X-ray diffraction studies show that both the Type IA E. *coli* topoisomerase I and the human type IB topoisomerase are large proteins with holes of appropriate diameter for a DNA double helix. As illustrated in Fig. 9.16B, the protein may open to allow a double helix to enter and occupy a suitable position for cleavage of one chain by the active-site Tyr 319. Topoisomerases are metalloenzymes, usually functioning best with Mg^{2+}. *E. coli* topoisomerase I also contains 3–4 tightly bound Zn^{2+} ions.

Table 9.2. Some Proteins of DNA Replication E. coli

Name	*Gene*	*Mass of monomer (kDa)*	*Map location (min) Fig. 9.4*
Ploymerase I (also has 3′–5′ and 5′–3′ exonulcease, and RNase activities	*polA*	103	87
Polymerase III			
Core			
α subunit (polymerase)	*dnaE*	130	4
ε subunit (3′–5′ exonuclease)	*dnaQ*	27.5	5
θ subunit	*hol E*	10	
Sliding clamp, β_2	*dnaN*	37 × 2	83
Connector τ (ATPase)	*dnaX*	71	
Clamp loader (*Y* complex)			
γ subunit	*dnaX*	47.5 × 2	11
σ subunit		35	
σ′ subunit		33	
Z subunit		15	
ψ subunit		12	
DNA binding proteins			
Single-strand, SSB	*ssb*	18.5 × 4	92
Double-strand, HU			
α subunit,		9.5 × 2	
β subunit		9.5 × 2	
Helicases (ATP-dependent)			
Primary replicative	*dnaB*	52 × 6	92
Dna C protein	*dnaC*		99
PriA(n′), primosome	*priA*	76	88
Rep		76.4	85
Initiation and priming proteins			
Dna A protein	*dna A*	52	83
Primase (an RNA polymerase)	*dnaG*	60	67
PriB (n) primosome	*priB*		96
PriC (n″) primosons	*priC*		
DnaT (primosome assembly)	*dnaT*		99
Ribonuclease HI	*rnhA*		
DNA ligase	*lig*	75	52
Topoisomerases			
Type I	*top A*		28
Type II, DNA gyrase ($\alpha_2\beta_2$)			
Subunit α	*gyrA*		97
Subunit β	*gyrB*		90

Topoisomerases of type II change the linking number by 2 in either the positive or negative direc-tion and hydrolyze ATP in the process. The best known of these is the *E. coli* **DNA gyrase,** an $\alpha_2\beta_2$ dimer of 97-kDa (α) and 90-kDa (β) subunits. The enzyme catalyzes the ATP-dependent introduction of negative supercoils into DNA. It also relaxes negatively supercoiled DNA slowly in the absence of ATP. Type II topoisomerases are found in all organisms. They are encoded by some bacteriophage such as T4 and by plasmids. However, most differ from bacterial DNA gyrase in not coupling DNA supercoiling to ATP hydrolysis. They require ATP but like topoisomerase I cause a relaxation of the supercoiling. Strands of one segment of DNA (called the "gate" or **G-segment)** are cleaved by the enzyme with staggered cuts four base pairs apart. Another segment of DNA (the "transport" or **T-segment)** is men passed through the gate and is thought to be released from a second gate in the complex (Fig. 9.16C). The enzyme subunits bind through phosphotyrosine linkages as in Eq. 9.5 to the 5' phospho groups of the two cleaved chains, while the subunits bind ATP and may like tiny muscles twist the DNA. Topoisomerases II are large dimeric proteins. The subunits of yeast topoisomerase II (Fig. 9.16C) are 1200-residue multidomain proteins. Mechanisms of DNA cleavage by types I and II topoisomerases appear to be related. However, the ATP-dependent conformational changes involved in a two-gate mechanism are unique to topoisomerases H. The bacterium *Sulfolobus* contains a type I topoisomerase that is called **reverse gyrase** because it utilizes ATP to introduce *positive* supercoils into DNA. This is in contrast to gyrase, a type II topoisomerase that introduces negative supercoils.

Type II topoisomerases are essential and func-tion in replication, DNA repair, transcription, and chromosome segregation at mitosis. Yeast with a *top2* mutation dies during mitosis with hopelessly entangled daughter chromosomes. A fluorescent antibody to eukaryotic topoisomerase II binds to chromosomes, probably at the bases of the radial loops present during mitosis, and also to centrosomes Higher organisms contain more than one topoisomerase II. Their specific functions are uncertain, but one appears to be to unknot entangled chromosomal DNA. In the crowded conditions of a cell nucleus topoisomerase can also cause inadvertent *formation* of knots.

The functional role of topoisomerases of type I is less clear. Staining with fluorescent antibodies to the enzyme has revealed its presence in the transcriptionally active "puffs" of polytene chromosomes and in centromeres of mitotic cells. A current hypothesis is that in *E. coli* class I topoisomerases act to relax negatively supercoiled strands of DNA behind transcription complexes, while gyrase acts to generate superhelical twists, which favour opening of the duplex ahead of transcription complexes. Transcription of a supercoiled rRNA gene *in vitro* is diminished by the selective topoisomerase I inhibitor **camptothecin,** one of a group of antitumor drugs directed against topoisomerase of both types I and II.

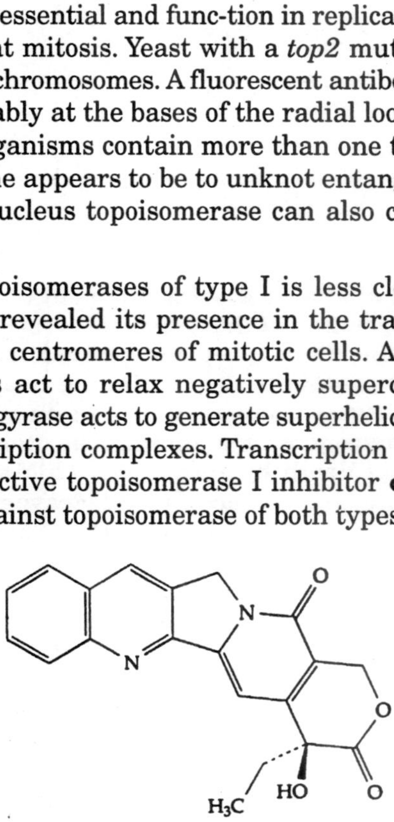

O N N O HO O H_3C

Camptothecin, a topoisomerase I inhibitor

In the autoimmune diseases scleroderma and systematic lupus erythematosus antigens to nuclear proteins or nucleic acids ate present in the blood. Many patients with severe scleroderma have an antibody against topoisomerase I.

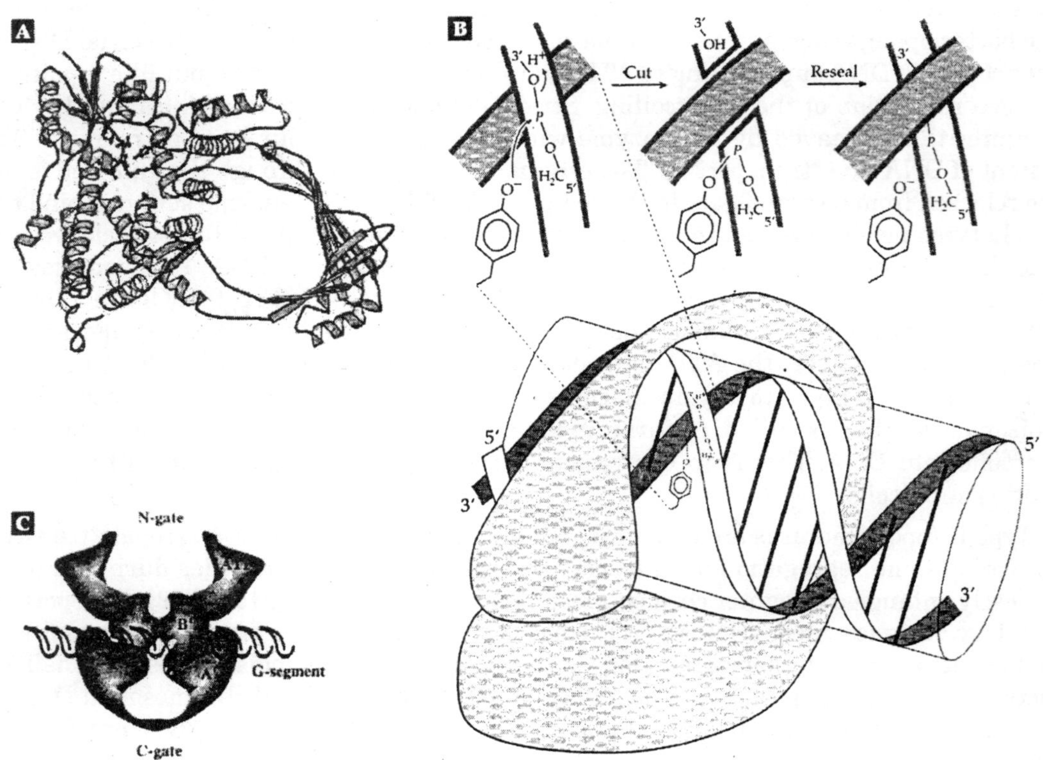

Fig. 9.16 (A) Ribbon drawing of a large 67-kDa fragment of the 97-kDa (864-residue) *E. coli* topoisomerase I showing the position of the activesite tyrosine 319 and an adjacent arginine. (B) Schematic diagram indicating a way in which topoisomerases of type 1 may pass one strand of DNA through another. The protein is shown binding to a single strand of a DNA duplex. This binding is facilitated by negative supercoiling. The enzyme then cuts the same strand by means of a nucleophilic displacement on a phosphorus atom using a tyrosinate side chain. The other cut end is held noncovalently by the enzyme, while the second strand passes through the gap. Then the gap is resealed by a reversal of the cleavage reaction. (C) Schematic model of a type II topoisomerase bound to a G-segment of DNA. This double helix is cut and another double helical strand, the T-strand, enters the N-gate. The gate then closes, and the central gate opens to allow the T-strand to pass through and exit through the C-gate. The shapes of the three domains are based on crystallographic data for the yeast enzyme. The ATPase, B', and A' domains consist of residues 1 to 409, 410 to 660, and 660 to 1200, approximately.

Primuses, initiator proteins, and ribonucleases. The priming segment needed for initiation of DNA replication is either a short segment of RNA or an oligonucleotide containing a mixture of ribonucleotide and deoxyribonucleotide units. The enzyme forming the primer is an RNA polymerase called **primase.** In *E. coli* it is encoded by gene *dna*G. Under some circumstances other RNA polymerases can act as the primase. Bacteriophages and plasmids may also encode primases (Table 9.1). For example, gene 61 of phage T4 (Fig. 9.2) encodes a primase, which together with the T4 helicase forms the priming particle or **primosome.** The phage T7 gene 4 encodes a 63-kDa mutlifunctional protein that is both a primase and a helicase. The primase active site is on the out-side of the hexameric ring. Additional proteins representing products of genes *dnaA* and C are also required for initiation of replication in *E. coli.* Several molecules of the dnaA **initiator protein** bind to a specific DNA **origin** sequence and participate in assembling a primosome that also contains the hexameric dnaB helicase and, transiently, protein dnaC. Replication of some single-stranded phages, such as ϕX174, in E. *coli* also require the *E. coli* priA, priB, and priC proteins (Table 9.2). For successful completion of replication chaperonins the products of genes *dnaJ* and *grp* E are needed as is a ribonuclease that digests the primer segments after they have been used as replication initiators.

Replication of Bacterial DNA

The basic mechanisms of replication implied in Eq. 9.3 seems to be universal, but several questions had to be asked. "Is replication initiated at a fixed point or points in a chromosome?" and "Does replication occur in one direction only or do two forks form at the point of origin and travel in opposite directions?" To answer these questions both genetic methods and electron microscopy have been employed.

Directions of replication. One technique for establishing the direction of replication in *E. coli* was to insert a λ prophage at the *att* site (Fig. 9.4,17 min) and DNA from phage Mu–1 at a variety of other sites around the chromosome. Phage Mu–1 was especially useful because it can be integrated at many different locations within the well-mapped genome. Integration within a gene inactivates that gene (an addition mutation) and allows the localization of the Mu prophage. Bacteria were prepared containing both λ and Mu-1 prophage, the latter at various sites. The bacteria were also auxotrophic for certain amino acids. Because of this, replication could be stopped by amino acid starvation. The bacteria usually completed any replication cycle in progress and then stopped. When the missing amino acids were added, replication began, again starting from the replication origin. Bromouracil, which enters DNA in place of thymine, was added at the same time. Consequently, the newly synthesized DNA strands were denser than the parent strands. After various times of replica-tion the newly formed strands were separated by centrifugation in a CsCl gradient and were tested for hybridization with both A, and Mu-1 DNA. Since the cells did not all begin replication at the same time after addition of amino acids, a variety of lengths of newly replicated DNA were present. Nevertheless, from the observed ratios of Mu-1 DNA to *K* DNA for the various strains it was possible to map the progress of replication beginning at an **origin** or/C near gene *ih* at 74 min. Replication was found to progress bidirectionally around the chromosome arid to terminate between genes *trp* and *his* at ~25 min.

The use of autoradiographic methods confirmed bidirectional replication. Strains of amino acid auxotrophs with small nucleoside triphosphate pools were used. The addition of amino acids after starvation led to initiation of replication with only a 6-min lag. The cells were labeled with [^{3}H]thymidine, and after the replication forks had moved a short distance from the origin of replication the cells were given a pulse of "superhot" [^{3}H]thymidine. Using autoradiography it was possible to observe the clearly bidirectional replication forks. Replication in other bacteria is also bidirectional.

Origins of replication. Replication of the *E. coli* chromosome begins and proceeds bidirectionally from its defined origin *oriC*. Replication of linear phage T7 is also bidirectional and begins at a point 17% of the way from one end. In mammalian mitochondrial DNA the origin of replication for the H-strand is in the D-Ioop but that for the L-strand is 2 / 3 of the way around the circular chromosome within a cluster of tRNA genes. The single-stranded circular DNA genomes of Ff, 0X174, and G4 phage also have distinct origins for initiation of replication to give RF circles.

Most origins have quasipalindromic nucleotide sequences, perhaps so that DNA can be looped out from the main duplex as is shown in Fig. 9.18A and B. The lengths of on sequences vary, as does the com-plexity of their possible folding patterns. Plasmids have been constructed which not only contain the *E. coli* origin, but are dependent upon that origin for their own replication. Study of those plasmids indicate that a 245-bp *oriC* sequence is essential. This sequence, which is shown in Fig. 9.18, contains several repeated oligonucleotides including 11 GATC sequences, which are sites of adenine N^6-methylation, and four "9-mers" (commonly known as dnaA "boxes") with the consensus sequence

$$\mathrm{TTAT}^{\mathrm{A}}_{\mathrm{C}}\mathrm{CA}^{\mathrm{A}}_{\mathrm{C}}\mathrm{C}$$

These appear in both 5' to 3' and 3' to 5' orientations, allowing them to form two base paired "stems." .In addition, there are three direct repeats of a 13-residue consensus sequence 5'-GATCTNTTNTTTT, which form an AT rich duplex. Other bacterial replication origins often follow a pattern similar to that in Fig. 9.18. However, the origin for *Mycoplasma genitalium* has been hard to detect. Replication origins of archaea have character-istics similar to those of bacteria and of organelles.

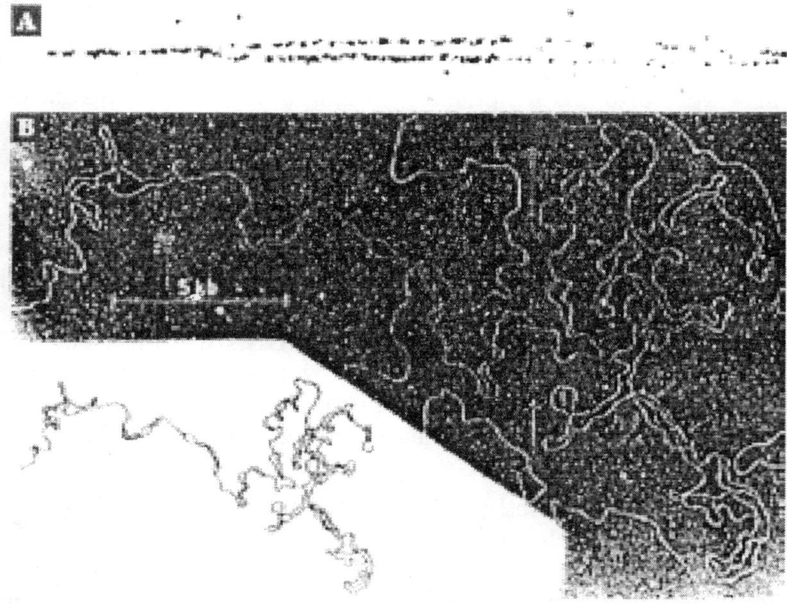

Fig. 9.17. Bidirectional replication. (A) Replication forks in the *E. coli* chromosome. The autoradiographic pattern was produced by a chromosome that initiated replication with [^{3}H]thymine (5 Ci/mmol) and was subsequently labeled with [^{3}H]thymidine (5 ci/mmol) for 6 min. The total length of the grain track is 370 µn. From Kuempel *et al.* (B) Fragment of replicating chromosomal DNA from cleavage nuclei of *Drosophila melanogaster*. The DNA, which was spread in the presence of formamide, contains several "eyes" formed where the DNA has been replicated. From Kriegstein and Hogness.

Priming and initiation of DNA synthesis. The first step in initiating a new round of replication and a new cell cycle in *E. coli* appears to be the binding of the initiator protein, dnaA, to the 9-mers in the negatively supercoiled origin. The resulting complex is visible in an electron microscope and may consist of a core of up to 20–40 molecules of dnaA protein with the DNA wrapped around them. ATP is also required and is bound to the protein where it hydrolyzes slowly. Similar initiator proteins are found in many bacteria and are encoded by some viruses and plasmids. In archaea, as well as in humans, the initiator proteins seem to combine functions of the *E. coli* dnaA protein with those of SSB.

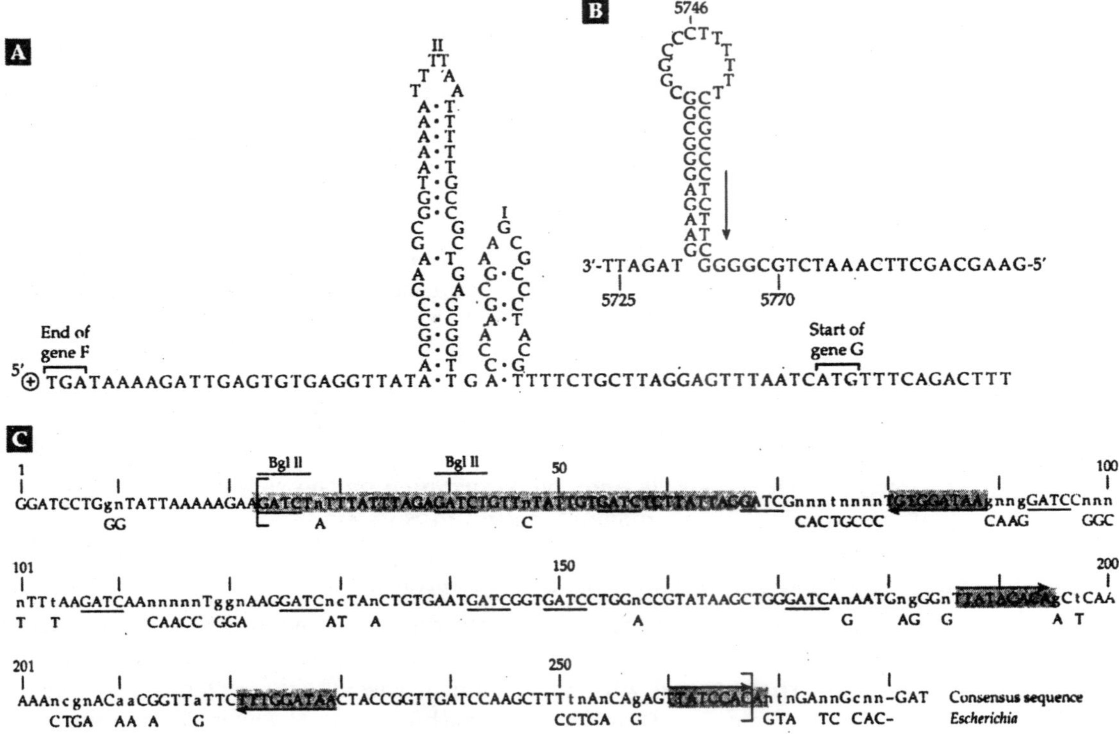

Fig. 9.18 Nucleotide sequences and proposed secondary structures for some origins of replication. (A) Bacteriophage ϕX174. The ends of the neighbouring genes F and G are shown. (B) Human mitochondrial DNA; origin of replication of the L strand. The sequence shown is for the H strand. The arrow indicates the direction of replication. (C) Bacterial *ori* region. The consensus sequence of origins of replication that will function in *E. coli.* Derived from sequences from *E. coli, Salmonella typhimurium, Enterobacter aerogenes, Klebsiella pneumoniae,* and *Erwinia carotovora.* In the consensus sequence a large capital letter means that the same nucleotide is present in all six origins; a lowercase letter means that the same nucleotide is present in three of the five origins; n means that any of the four nucleotides may be present; represents a deletion, G being present here in the *Enterobacter* origin. Underlined are *dam* methylation sites. Green shaded sequences, three 13-mers in a region that undergoes easy opening of the duplex for insertion of replication proteins; gray boxes with arrows, four 9-mers that are specific sites for binding of replication protein A. Brackets enclose a 245-bp minimum origin.

Following the binding of the dnaA protein the hexameric helicase dnaB is loaded onto the adjacent DNA in the region of the 13-residue repeated sequences, which are labeled 1, 2, and 3. As shown in this figure introduction of the helicase is assisted by protein dnaC, which forms a complex with the helicase. Additional binding of protein HU and a temperature of >30°C are essential for tight binding of the proteins. Although the dnaC protein is needed for formation of the preprinting complex, it dissociates after the dnaB protein becomes firmly bound. One dnaB hexamer binds to each single strand of the DNA duplex with opposite orientations. With the dnaB helicase in place on each strand, this ATP-driven enzyme processes along the DNA, unwinding the duplex in both directions. (Perhaps it may be more accurate to say that the *DNA moves throtigh the helicase.)* The resulting "bubble" is held open by binding of SSB tetramers. The primase (dnaG protein) then binds adjacent to the helicase and synthesizes the RNA primer along each strand of DNA. As the helicase processes to the right in the fork, the complex with the dnaA protein dissociates, permitting primer synthesis into the origin region. Alternatively, RNA polymerase can prime replication by initiating transcription on both strands of the DNA. Suitable promoters are present and oriented in opposite directions on the two strands.

It has long been postulated that the bacterial chromosome is attached to the plasma membrane. At least one such attachment site may be at or near the origin of replication. Furthermore, the exchange of ADP for ATP in the dnaA protein is catalyzed specifically by cardiolipin and phosphatidylglycerol containing the unsaturated oleic acid. Inositol polyphosphates may also play a role.

Elongation of DNA chains. DNA polymerase III in its holoen-zyme form is the major polymerase for DNA replication. It elongates the primer chains rapidly and processively leaving only very small gaps at the ends of single-stranded regions. The rate of elongation, which is ~ 3 nucleotides / s for 8 kb *oriC* plasmids, may be determined by the rate of action of the dnaB helicase. A completely unwound oriC plasmid, bound to SSB, undergoes primer elongation ~ 10 times faster. However, the rate for the intact *E. coll* replisome is nearly 1000 nucleotide s^{-1} with a rate of misincorporation of only one in 10^9 nucleotides.

Small *oriC* plasmids need to be primed at only one location, but the large bacterial chromosome must undergo priming at many sites on the lagging strand to permit DNA polymerase III to act on that strand with formation of the Okazaki fragments. The DNA polymerase complex may be a dimer that works on both strands at once. The lagging strand may be looped out to allow it to lie parallel with the leading strand. The appearance of the electron micrograph in Fig. 9.20B supports this suggestion. However, the manner in which the lagging strand can be shifted to bring the next primed initiation site to the replication complex is not clear. DNA polymerase III holoenzyme itself may be organized as an asymmetric oligomer that operates on both strands in a complex such. The primase, either the dnaG protein or the primosome used by phage ϕX174 (Section 5), may synthesize the RNA primers on the lagging strand.

Termination of replication. *As* each Okazaki fragment is completed along the template for the lagging strand, the RNA primer piece is digested out, replaced by DNA, and the nick sealed by action of DNA ligase. Ribonuclease H, which is found in both bacteria and eukaryotic cells, specifically degrades the RNA component in these RNA-DNA hybrid regions. In bacteria another mechanism for primer removal is available. The 5' – 3' exonuclease activity of DNA polymerase I will cut out the RNA segment, while the 5' – 3' polymerase activity of the same enzyme will fill the gap.

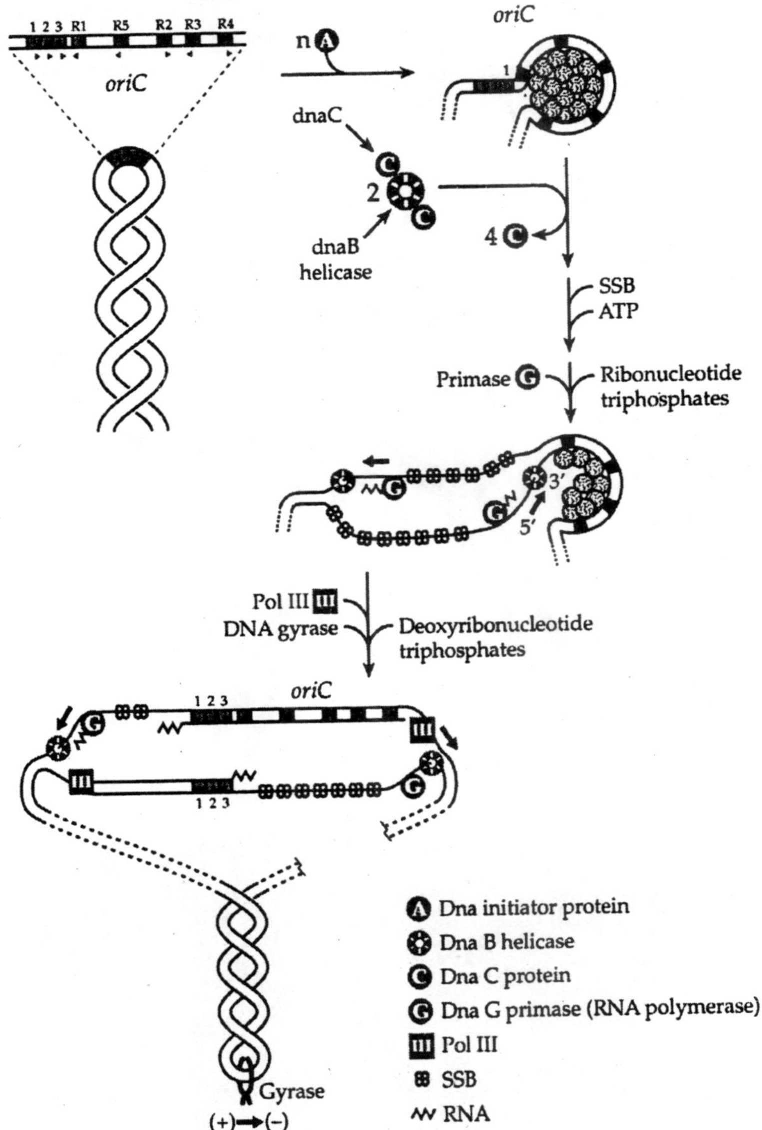

Fig. 9.19. Hypothetical scheme for initiation of bidirectional replication in *E. coli*. The closed boxes R1 through R5 represent the 9-residue recognition sequences for the *E. coli* dnaA protein. The open boxes 1, 2, and 3 represent the three 13-residue repeats, possible sites for binding of the dnaB-dnaC protein complex.

Replication of *oriC* plasmids may occur by simply allowing replication of the leading strands at both replication forks to continue all the way around the circle. However, in *E. coli* bidirectional replication continues only until the two replication forks converge. This can occur anywhere between two **terminators,** Tl and T2, located at 28.1 min and 35.6 min. The terminators slow replication in the counterclock-wise and clockwise directions, respectively. A gene *(tus)* near Tl encodes a **terminator utilization substance,** a DNA-binding protein that associates with Tl and T2 and causes termination. Another problem may be the separating of catenated DNA circles by action of a topoisomerase. Finally, it is essential to **partition** the original chromosome and its replica, one to each daughter cell. This requires at least three other gene products including one large 170-kDa protein.

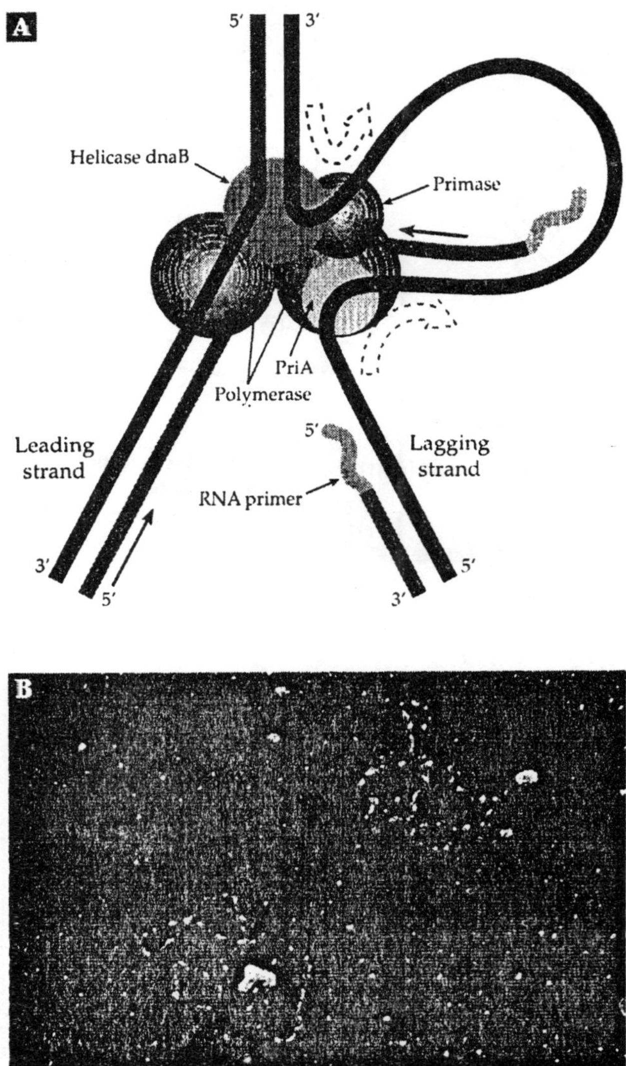

Fig. 9.20. (A) Hypothetical replisome for concurrent replication of leading and lagging strands by a dimeric polymerase associated with helicase dnaB and a primosome. Open arrows indicate directions of movement of DNA, which is forming a loop as the polymerase fills a gap to complete an Okazaki fragment. The primase will then form a new primer and a new loop. (B) Electron micrograph of the primosome bound to covalently closed ϕX174 duplex replicative form. These enzymatically synthesized duplexes invariably contain a single primosome with one or two associated small DNA loops.

The Replication of Viral DNA

The replication of viral DNA usually depends upon the genes of both the .host organism and the virus. For example, *ts* mutations in the *E. coli* genes *dnaB, D, E, F,* and G lead to a loss in ability to support growth of phage λ as well as loss of ability to reproduce under conditions where the *ts* gene products are inactivated. However, the phage can replicate in E. *coli* with mutated genes *dnaA* and C because phage λ encodes its own initiator proteins by genes *O* and P (marked on the gene map in Fig. 9.4). In addition to these two proteins, seven *E. coli* proteins are required to initiate replication at the lambda origin *ori* A, and to complete replication. The *E. coli dna*B helicase and the dnaC protein are needed. The chaperonins dnaj, dnaK, and GrpE are also necessary for replication of phage A, and other viruses. As we have seen (Section 2) many viruses contain genes specifying their own DNA polymerases and primases, which function in cooperation with host proteins. For many dsDNA viruses the origins of replication, priming reactions, and chain elongation processes closely resemble those of *E. coli*.

In contrast, the first step in replication of the filamentous Ff viruses (fl, fd, and M13) or the small icosahedral ϕX174 or G4 is conversion of the single-stranded closed circular DNA molecules of the infecting virus particles into circular double-stranded **replicative forms** (RF). This occurs as the DNA enters the bacterial cell and is accomplished entirely by the enzymes of the host cell. Phage G4 DNA, whose replication has the simplest known requirements, contains a tight hairpin region at its origin. The rest of the DNA must be coated by SSB for replication to occur. The hairpin resists melting and serves as a binding site for *E. coli* primase. This is the only known case in which no other priming proteins are needed. Primase synthesizes up to a 29-ribonu-cleotide primer after which DNA polymerase III holoenzyme copies the rest of the chain.

Replication of the closely related ϕX174 is more complex. It requires assembly of a **primosome** made up of at least seven host proteins : dnaB, dnaC, primase (dnaG), and proteins priB, A, C (n, n′, n″), and dnaT (i), Table 9.2. The 76-kDa helicase priA (n′) may locate the **primosome assembly sequences,** which are ~70 nucleotides in length, and displace SSB from them. These sequences can adopt secondary structures with a pair of hairpin loops. Kornberg and associates suggested that the same kind of primo-some formed at these sites in the ϕX DNA may participate in replication of the lagging strand of the chromosomal DNA. If so, helicase priA presumably functions in the replisome on one strand and dnaB helicase on the other as depicted.

In the second and third stages of replication ϕX174 RF molecules are themselves replicated and are then used for synthesis of new viral (+) strands. At both stages a virally encoded **gene A protein,** which has endonuclease activity, nicks the duplex. Cutting the (+) strand it leaves a free 3″-OH on DNA residue 4305, while the 5′-phospho group of residue 4306 becomes covalently attached to a tyrosyl residue in the A protein. The free 3′-OH serves as the primer for a **rolling-circle synthesis** (Eq. 9.7). As a new viral strand is synthesized along the complementary (–) strand as a template, the original viral DNA (+) strand is displaced (Eq. 9.7) as a single-stranded tail.

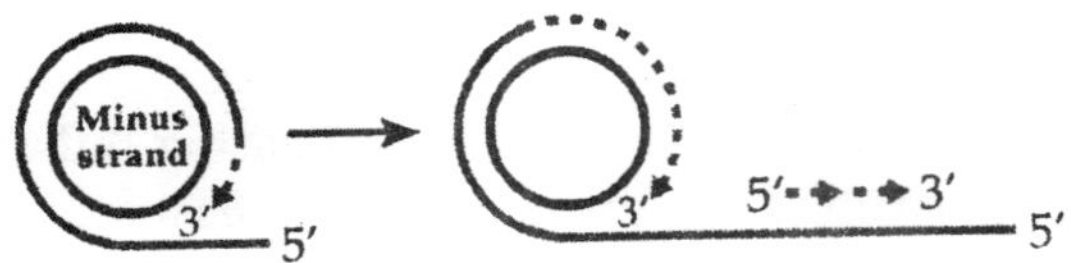

A strand complementary to the single-stranded tail is then formed in segments. A complete turn of the circle produces a viral strand twice the normal length. Cleavage by the endonuclease activity of the gene A protein and closure of the circle completes the replication. The displaced (+) strand can be cut off and either incorporated into a progeny phage or converted into another RF circle. The A protein, attached to the 5'-terminus of the (+) strand, is involved in either case. It can participate in repeated sequences of initiation and termination of viral (+) strand synthesis. Once double-stranded circles are formed, they undergo several replications to give additional RF circles, which serve as templates for the synthesis of many single strands of viral (+) DNA, which are incorporated into the mature viruses. This synthesis of additional RF circles requires transcription of some viral RF genes.

In the final stage of replication the single-stranded (+) chains formed by the rolling-circle mechanism are packaged into phage particles. The gene 5 single-strand binding protein of M13 coats the DNA chains as they are formed, evidently preventing their conversion to RF circles. In the case of ϕX174 the new single-stranded DNA circles are packaged as they are synthesized to form complete icosahedral virus particles.

Replication of the larger tailed viruses, which have many genes, is complex and varied. The lytic phage A. resembles the smaller viruses in using the host replica-tion enzymes. In the final stages a rolling-circle mechanism is utilized to form **concatemers** consisting of linear DNA duplexes with numerous successive copies of the viral DNA. The ssDNA that is formed in the rolling circle is converted to dsDNA as it is formed. Finally a **terminase** cleaves the DNA at specific cos sites, using staggered cuts, to form cohesive ends. However, there are uncertainties. The linear dsDNA enters an empty preformed procapsid, apparently pumped in an ATP-dependent fashion, perhaps by a rotating portal ring.

Packaging or Viral Genomes

The construction of intact virus particles from the genomic DNA and protein subunits is often a complex process. It is simplest for the small filamentous ssDNA viruses (Fig. 9.7). The subunits are synthesized as soluble proteins, which enter the cell membrane, then lose their leader sequences. As the viral DNA coated by the viral gene 5 ssDNA-binding protein enters the membrane, the binding protein is replaced by the coat subunits.

The process is somewhat more complex for the icosahedral viruses. In the ϕX, G4, α3 family the icosa-hedral procapsid is constructed with the aid of both internal and external **scaffolding proteins** as is illustrated in Fig. 9.28. In phage that replicate via concatameric dsDNA the terminase that cleaves the DNA also interacts in a precise way with the packaging apparatus of the prohead. For the tailed phage the ringshaped oligomeric headtail connector, together with an ATPase, may function as a rotatory pump to feed the DNA into the prohead. This has been demonstrated for phage ϕ29. In some cases the terminase produces new phage DNA of unit size, but in other cases, e.g., with phage T4, the DNA may be cut more randomly when the head is full or when another piece is needed to fill it. After the DNA is packaged the virus capsids usually expand and become stronger.

Plasmids

Most bacteria contain plasmids which are self-replicating but stably maintained at well-defined numbers of copies per cell. They are usually not essential to the cell but may carry traits such as anti-biotic resistance or toxin formation that benefit the. bacterial host. A plasmid always carries in its DNA an orign (on) of replication and a gene, usually designated *rep,* for an initiator protein. It usually encodes other proteins as well but may depend largely on host pro-teins for replication. Plasmids may use the *oriC* copied from the bacterial host's DNA, the origin from phage λ, yeast autonomously replicating sequences (ARSs), or other origins. Replication of the small 6.6-kb plasmid ColEl, which is present at ~20 copies per cell, depends entirely on the host-cell replication machinery· However, the control of copy number depends upon synthesis of **antisense RNA** and its reaction with the plasmid DNA. Similar copy number control is used by the larger ~ 100-kb resistance factor **R plasmids.** Some plasmids use replication systems very similar to those of viruses such as ϕX174, often using rolling-circle replication. However, the plasmids lack the proteins for virus coat formation and maturation.

The F factor plasmids are large 100-kb circular DNA molecules containing ~ 60 genes, about 20 of which encode proteins involving transfer of DNA into another bacterial cell F plasmids display strict copy number control with only 1-2 copies per host chromosome. The controls lie in a region known as the partition locus, which resembles regions of the host chromosomes that are involved in partition of the bacterial genome. They have repetitive sequences suggesting a similarity to centromeres of eukaryotic chromosomes.

Chromosome Ends

The T-odd bacteriophages Tl, T3, T5, and T7 are medium-sized phage with linear duplex DNA genomes. Replication of linear DNA in these and in many other genomes presents a problem. Even if the RNA primer segment is made at the very 3' end of the template strand, there will be a gap in the final replicated strand when the primer is digested out. Since there is no known enzyme that will add to the 3' end of a chain, this gap will remain unfilled. The problem is solved by **terminal redundancy,** the presence of a common 260-nucleotide sequence at both ends. Several daughter DNA molecules with gaps at the 5' ends can be joined by their cohesive ends to form a long **concatamer.** DNA polymerase I fills any gaps, and the chains can then be ligated and cleaved at different points to generate complete 5' ends.

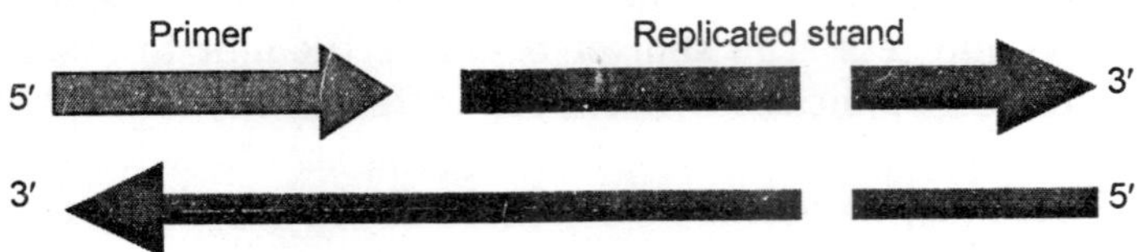

Another mechanism, which is utilized by some single-stranded parvoviruses, obviates the need for an RNA primer by use of a palindromic sequence to form a hairpin loop (Eq. 9.8).

Yet another solution to this problem is used by some viruses. **Phage ϕ29** of *Bacillus subtilis* primes the replication of its 19,285 bp dsDNA at both ends by a **terminal protein,** which is linked covalently through its Ser 232 –OH group to dAMP. The 3'–OH of the deoxyadenosyl group primes the DNA replication. In a similar fashion replication of the eukaryotic **adenoviruses,** whose genome is a 35– to 36-kb linear DNA duplex, starts at the ends and is primed by one residue of dCMP covalently attached through a 5'-phosphodiester linkage to a serine side chain in a 80-kDa **preterminal protein.** It substitutes for the RNA oligonucleotides that prime most DNA synthesis. The dCMP pairs with guanine at the 3' terminus of the template strand and provides the initiating 3'–OH group. During the replication the preterminal protein is cleaved to the 55-kDa **terminal protein,** which remains covalently attached, one molecule at the 5' end of each strand. The genome can be replicated *in vitro by* five proteins: the virally encoded preterminal protein, DNA polymerase, DNA-binding protein, and two cellular transcription factors that bind in the adenovirus origin region.

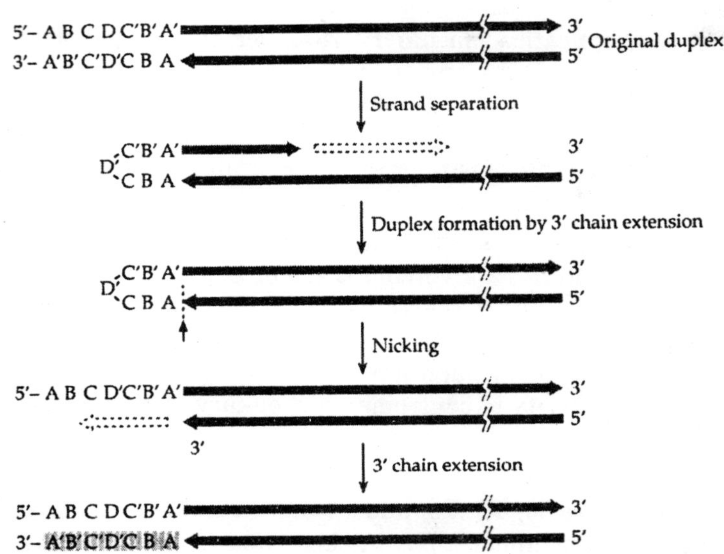

The chromosome end problem is solved in an-other way in eukaryotes. As discussed in Section B,1, **telomeres** (chromosome ends) contain repeated sequences of variable length. One DNA strand is always G-rich. For example, in human cells the sequence 5'–(TTAGGG),, –3', where *n* may be ~20, oc- curs at the 3' ends of the G-rich strands. The other strand, whose 5' end is at the telomere end, is C-rich and has the complementary sequence 3'–$(AACTCCC)_H$ – 5'. The 3' end of the G-rich strand is always longer by 12–16 nucleotides than the end of the C-rich strand. This 3' extension may fold back to form non-Watson-Crick structures that apparently involve G-quartets. The shorter 5' end is thought to result from the need for a short RNA primer during replication. When replicated in the normal fashion the full G-rich leading strand will be formed, but the C-rich lagging strand will be 8-12 nucleotides short, when the RNA primer is digested away. A result of this is that human somatic cells gradually lose telomeric repeats. However, in tumor cells, germline cells, and unicellular organisms the enzyme **telomerase** prevents this telomere loss.

Telomerase is a reverse transcriptase that copies the DNA sequence of the telomeric repeats from a small **guide RNA** that is part of the enzyme. The first telomerase studied was the relatively abundant enzyme from *Tetrahymena.* It contains a 159-nucleotide RNA with the sequence 5'–C AACCCC AA–3' at positions 43–51. This sequence is complementary to the 5'–TTGGGG–3' repeat sequence of the *Tetrahy-mena* telomeres. A 127-kDa human protein contains a similar guide RNA with the sequence 5'-CUAACCCUACC–3', which is complementary to the human telomere repeat sequence. Telomerases evidently allow the cell to elongate the telomere 5'-ends using the guide sequence and the reverse transcriptase activity of the telomerase. Any number of additional repeats may be added to the 5' ends. The shortened 3' ends can also be lengthened in the next round of replication. The control of telomerase must be important. The enzyme is active in early embryonic cells and some stem cells. However, most normal cells have little or no telomerase activity and lose telomere length throughout their lifespan with eventual growth arrest. On the other hand, excessive telomerase activity may induce cancer. Certain mutations in the telomerase guide RNA can cause greatly increased telomerase activity. The control of telom-erase is still poorly understood but involves specific telomere binding proteins.

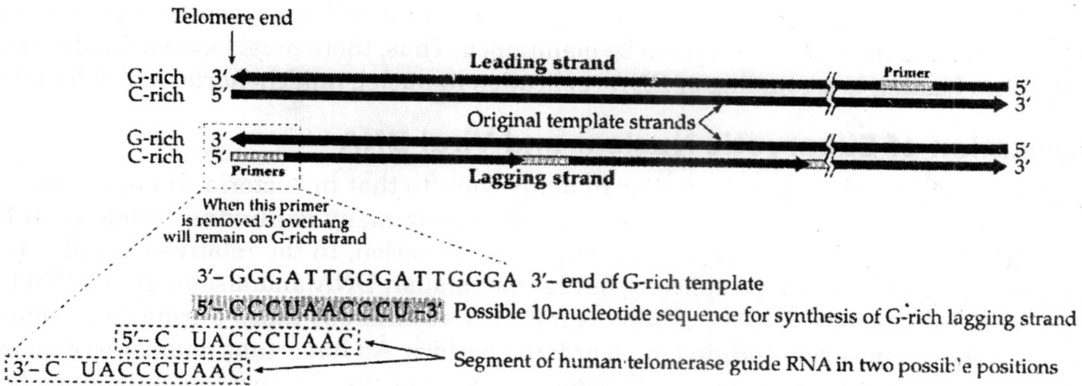

Fig. 9.21. Aspects of telomere synthesis. The end of the chromosome and the 5' end of the C-rich strand is at the left. This is the template for replication of the leading G-rich strand (green). The primer lies far back in the chromosome. The C-rich strand is replicated in segments from several RNA oligonucleotide primers, one of which lies at the 5'–terminus. This first primer is removed by RNase activity leaving a 12–16 nucleotide 3'–overhang. The telomerase guide RNA can hybridize with the 3'–end of the G-rich strands providing a template that allows additional growth of the G-rich strand and extension of the C-rich strand also in the next replication.

Mitochondrial and Chloroplast DNA

Replication of the ~ 16-kb mammalian mtDNA begins with RNA priming within a small **displacement loop** or D-loop. One daughter strand, the heavier or H strand, starts to grow on the primer. As it does, the parental H strand is displaced and the D-loop is enlarged. The H-strand grows until -70% of the parental H strand has been displaced and the L-strand origin is uncovered. Then a new light L strand is laid down to form the second daughter duplex. The rate of formation of the new L strand is only 10 nucleotides / s, an hour being required to complete the process. The DNA formed is initially relaxed, another 40 min being needed to introduce the 100 superhelical turns present in the finished chromosome.

The kinetoplast DNA of trypanosomes consists of thousands of catenated circular DNA molecules. Among these the smaller minicircles always contain the sequence GGGGTTGGTGTA at their origins of replication. The minicircles are individually removed from the mass prior to replication. The two progeny circles are then both recatenated into the mass.

Chloroplasts contain large 120– to 169-kb circular genomes encoding about 100 proteins. A characteristic feature of most chloroplast DNA is the presence of long inverted repeat sequences (10,058 bp in the liverwort, 25,339 bp in tobacco). These are separated by 19,813 and 81,095 bp single copy regions in the liverwort and by similar sized regions in tobacco. Plastid DNA exists as a mixture of monomeric molecules with smaller amounts of dimers, trimers, and tetramers.

Ethidium bromide inhibits the replication of chloroplast DNA and causes partial degradation of existing DNA in chloroplasts without interfering with replication of DNA in the nucleus. The effect is similar to that of the same drug on mitochondrial DNA.

However, cells of *Chlnmydomonas* treated with ethidi-um bromide are able later to regenerate their chloro-plast DNA. This result has been interpreted to mean that there may be one or a few "master copies" of chloroplast DNA in specially protected locations. The result should also be considered in relationship to the following observation. Although nuclear and organelle DNA molecules replicate at different times in the cell cycle, constant proportions of the organelle and nuclear DNA tend to be maintained. Thus, there must be some kind of control mechanism leading to a coupling of DNA replication in nuclei, mitochondria, and chloroplasts.

Replication of Eukaryotic Nuclear and Viral DNA

Replication in eukaryotes is similar in many ways to that in bacteria. However, the ~ 10^6 kb of DNA in a typical eukaryotic genome is divided into many **replicons,** segments of DNA 30-150 kb in length, each having its own origin of replication. In the relatively small ~ 14-Mb yeast genome there are –400 replicons, but in mammalian DNA and also in plant DNA there are probably thousands. DNA synthesis is initiated at different times during the S-phase of the cell cycle at the various origins in an ordered pattern. An important unanswered question is how the cell is able to replicate all segments of all of the chromosomes just once before entering mitosis.

Many of the proteins of eukaryotic replication are closely related in sequences and functions to those of bacteria. There are initiator proteins analogous to *E. coli* dnaA. Eukaryotic polymerases a and 5 and possibly *E* are essential for replication. Polymerase a, which is inhibited specifically by the fungal metabolite **aphidocolin,** is a complex of a ~170-kDa DNA polymerase core, an RNA-synthesizing primase consisting of 58- and 49-kDa subunits, and a 70-kDa subunit

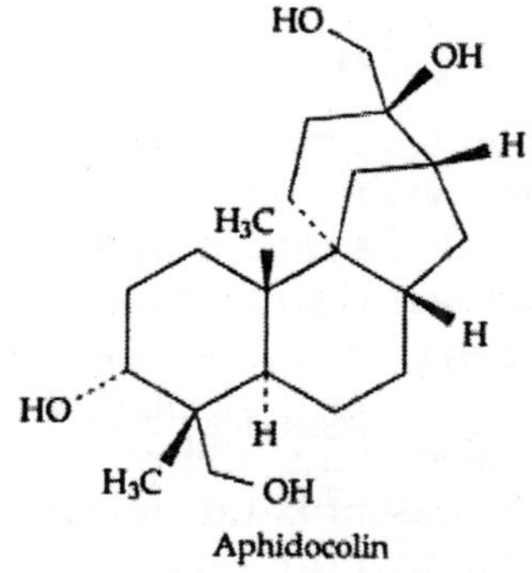

Aphidocolin

of uncertain function. The complex makes an RNA-DNA primer consisting of ~10 nucleotides of RNA and ~30 of DNA. This pol α primer is replaced early in replication by the highly processive polymerase δ and, perhaps, under some conditions by polymerase ε. The ringlike processivity factor or "clamp" that is provided in *E. coli* by protein dnaB is called the **proliferating cell nuclease antigen** (PCNA) in eukaryotes. It is loaded on to the DNA by a **clamp loader,** the 5-subunit replication factor C (RFC). As in *E. coli* an SSB type protein known as **replication protein A (RPA)** is also essential. PCNA is not only essential to eukaryotic replication but is also required for recombination and repair. The poloα/primer primes leading strand synthesis initially but then switches to replication origins on the lagging strand where, together with other proteins, it primes the formation of the Okazaki fragments. Polα/primer is also a logical participant in the control of the initiation of the S-phase of the cell cycle.

Eukaryotic Viruses. Investigation of viruses provided the first insights into eukaryotic DNA replication. Most of the factors needed for replication of the DNA of adenoviruses, simian virus 40 (SV40), and polyomavirus within animal cells are supplied by the host. Replication of the 5-kb SV40 DNA, whose DNA forms typical nucleosomes, appears to be an excellent model for eukaryotic replication in general. The single SV40 origin of replication is a 64-bp sequence containing the 5–bp sequence GAGG C four times as pairs of inverted repeats. These are recognized by the 95-kDa virally encoded initiation protein which also has helicase activity and is known as the T antigen. A nearby 17-bp sequence containing only AT pairs is presumably the region of entry of the host cell's polymerase a/ primer. Single-stranded DNA regions are coated with the replication protein A. After the primer is formed, the RFC complex loads the sliding clamp PCNA, and polymerase a is replaced by polymerase 8 on both leading and lagging DNA strands permitting highly processive bidirectional chain elongation. Topoisomerase activity is required to decatenate the replicated chromosomes. Since SV40 DNA forms typical nucleosome, its replication is thought to mimic chromosomal replication quite closely. The more complex herpes simplex virus HSV–1 has a 153-kb genome, a linear DNA duplex. It has –75 genes and encodes its own DNA polymerase, origin-binding protein, SSB, and other proteins needed for replication within eukaryotic cells.

Artificial chromosomes. Another approach to understanding eukaryotic replication, similar to the use of *ori*C plasmids in *E. coli,* is to study **autonomously replicating sequences** (ARSs) and plasmids and **artificial chromosomes** made from them. ARS sequences were first found in the budding yeast *S. cerevisiae.* Plasmids containing an ARS, whose core consensus sequence is 5'-(A/ T)TT TAT(A/G)TTT(A/T), replicate autonomously during S-phase. Such plasmids have been genetically engi-neered, providing them with telomeres and some kind of functional centromere, to form artificial chromo-somes. **Yeast artificial chromosomes** (YACs) have become extremely important as cloning vehicles and they also serve as important tools for studying eukaryotic replication and its control. They can be cultured in yeast cells or can be transferred into animal cells, etc.

As mentioned human centromeres are rich in the repetitive a-satellite DNA. By joining a-satellite DNA-containing fragments of the X-chrotnosome to cloned telomeric DNA, human **minichromosomes** have been created. These have been developed into **human artificial chromosomes,** which may be practical vehicles for gene transfer in human therapy.

Replication of nuclear DNA. The budding yeast *Saccharomyces cerevisiae* has permitted the most de-tailed picture of DNA replication in a eukaryote. The complete genome sequences are known and the ARSs have been physically mapped. For example, in chromosome VI there are nine origins that differ in frequency of initiation and which replicate sequentially during the S-phase of the cell cycle. The initiation **(replicator)** regions surround the 11-bp consensus sequence of the ARSs, each occupying at most –150 bp. However, in metazoa and even in the fission yeast *Schizosaccharomyces potiibe* the ARSs range from 500 to 1500 bp in length. These origin regions frequently overlap the **promoter** sequences, which control initia-tion of transcription. This association with transcription origins has also been observed in metazoan cells, where replication origins are often clustered. However, there is no sequence homology between the ARSs of S. *cerevisiae* and replication origins in other species, even those of S. *pombe.*

The study of replication in yeast ARSs and artificial chromosomes has revealed that initiation of replication requires not only an initiator protein but a complex of six proteins that form an **origin recogni-tion complex (ORC)**. This complex, which is essential to initiation of replication, may be joined by additional proteins in a **prereplication complex.** At least some of the ORC proteins have their homologous counterparts in metazoa, suggesting a highly conserved initiation machinery.

One or more of the proteins that bind to the ORC may constitute a **license** to replicate. The licensing concept states that when replication occurs the license is destroyed and the origin involved cannot initiate replication again without a new license. The **replication licensing factor** (LRF) is postulated to be unable to pass through the nuclear membrane. It can only reach the replication origin after the S-phase has concluded and mitosis has taken place. At this time the membrane has been disrupted. A second signal, the **S-phase promoting factor** (SPF), cannot act without an intact nucleus and a license in place. This system ensures that DNA is replicated only once per cell cycle. Among the proteins involved in the licensing is a group of **minichromosome mainte-nance (MCM) proteins,** so-named because of their importance to replication of ARSs and artificial chromosomes. Six of these proteins (MCM2–MCM7) can form a hexameric complex with one subunit of each type as well as other complexes, e.g., $(MCM4,6,7)_2$. The latter acts as an ATP-dependent helicase. A somewhat simpler MCM complex is found in archaea. Some of the cell cycle proteins including Cdc6 and the protein kinases Cdc7 and Cdc28 as well as other proteins, are also required for regulation of replication origins. Proteins homologous to those of yeast have been identified in humans and other eukaryotes. Licensing of replication involves association of the MCM helicases with each ORC during the G1 phase of the cell cycle. Binding of the initiation factor Cdc6/18 and of a recently discovered loading factor Cdt-1 apparently completes the licensing. Once licensing has occurred both cdc6/18 and Cdt-1 can dissociate from the DNA. Removal of Cdt is facilitated by its binding to another protein **Geminen,** found first in the frog *X. laevis*. The ORC complexes may remain at the origins. It has been estimated that a yeast cell contains –400–600 molecules of the very stable ORC, about one ORC per replication origin. However, a large excess of MCM proteins may be present. Their concentrations may regulate the number of ORC molecules that associate with DNA. Replication of DNA is not the only aspect of cell growth. For example, as DNA is replicated histones must be synthesized and assembled. This synthesis occurs during the S-phase and is tightly coupled to replication.

Initiation of replication in metazoans is still confusing. Almost any piece of DNA will be replicated if introduced into a *Xenopus* egg, where initiation appears to occur just once at a random position. However, in differentiated tissue the origins of replication seem to be fewer in number and more specifically located. A possible explanation is that high concentrations of ORC and MCM proteins in the embryo may lead to many relatively nonspecific origins and a replicon size of ~7 kb. The lower concentrations of these auxiliary factors in somatic cells may lead to fewer but more specific origins with a replicon size of ~170 kb. Three distinct mammalian origins have been studied in detail. One is in the β globin locus (Fig. 9.10). A second is near the dihydrofolate reductase gene. A third, which is activated early in S-phase, is at the 3' end of the lamin B2 gene. The latter has been localized to a 500-bp region. These findings suggest that the replicon concept, as developed for yeast, may be generally applicable.

Replication of the intact genome of *Drosophila* has been studied in rapidly dividing nuclei by electron microscopy. The replication rate in these nuclei is ~300,000 bases/s, but it has been estimated that replication forks in animal chromosomes move no faster than ~50 bases/s. Thus, we would anticipate at least 6000 forks, or one fork per 10 kK bases. Indeed, this number of forks has been observed. They oc.ur in pairs with many short regions containing single-stranded DNA as if one strand at the fork is replicated more rapidly than the other as in mitochondrial DNA. The arrangement of the ssDNA regions at the two forks in a pair suggests bidirectional replication. However, replication forks are rarely seen in higher eukaryotes, but extensive regions of single-stranded DNA are often visible. Benbow and associates suggested that in higher eukaryotes the strands of duplex DNA may be separated throughout a whole looped domain of DNA. Replication could then occur with initiation at many points along each strand.

Replication reactions are similar in bacteria and eukaryotes, but some details differ. In eukaryotes at least two DNA polymerases, α and δ, are required. In budding yeast polymerase ε is also essential. Both polymerases δ and ε may replicate separate strands at the replication fork. Processing of Okazaki fragments also differs from that in bacteria, where either RNase H or the 5' to 3' exonuclease activity of DNA pol I removes the RNA primer (Fig. 9.14). This exo-nuclease activity is lacking in eukaryotic polymerases. Replication primers are removed in a two-step process by **RNase HI,** which makes an endonucleolytic attack that removes all but one nucleotide residue of the primer in a single piece, leaving a 5'-phospho group on the remaining ribonucleotides. That residue is removed by a 5' to 3' exonuclease designated RTH-1 nuclease (Eq. 9.9). This is a homolog of the yeast RAD27 protein. A polynucleotide kinase may then phosphorylate the 5' end of the DNA fragment.

Another difference between bacterial and eukaryotic replication is the presence of nucleosomes in eukaryotes. Sυme evidence suggests that nucleosomes may open and close to allow replication forks to pass through. Studies of SV40 minichromosomes indicate that passage of the replication machinery does destabilize nucleosomes, which must be partially reconstructed about 260 nucleotides past the elongation point. Another factor is the variable extent and location of modifications to histones, in particular to the H3 and H4 his tone tails. A code has been proposed according to which certain modifications would favor transcription or mitosis, while lack of modification would silence the genes.

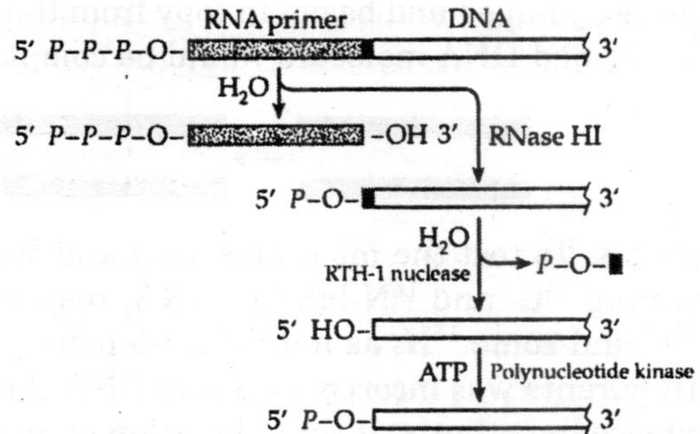

Much of the control of replication is at the initiation stage. Growth factors and other mitogenic stimuli acting at the plasma membrane can stimulate expression of such nuclear proteins as those encoded by the protooncogenes *c-myc, c-myb,* and *c-fos.* These may initiate a regulatory cascade (Fig. 9. 13) and trigger mitosis. As indicated in the 34-kDa protein kinase encoded by fission yeast gene *cdc2* (budding yeast CDC28) is essential for progression of the cell cycle through the Gl phase into mitosis (Eq. 9.3). A single oscillation in this kinase activity induced by a B-type cyclin can promote both replication and mitosis. However, in *S.cerevisiae* there are 14 different cyclinlike proteins, and their individual functions are not clear. The signal that is sent to the ORCs is likewise unclear. However, theoretical models involving Eq. 9.3 and many additional components have been proposed. Multiple phosphorylations may occvir, some on the RPA initiator protein. Many proteins required for replication, including DNA polymerase and primase, are associated with the nuclear matrix. The nuclear membrane may also be important in controlling replication.

INTEGRATION, EXCISION, AND RECOMBINATION OF DNA

The exchange of genetic information between chromosomes, plasmids, and viruses occurs in many ways, which are described collectively as recombination. Mutants of E. *coli* deficient in recombinational ability often have defects in genes designated *recA, B, C,* etc. (for recombination), or *ruvA, B,C* ... (for resistance to ultraviolet light). Some of these mutants are unusually sensitive to ultraviolet light because of their inability to repair damage to DNA. Several of the recombination enzymes are used for repair of ultraviolet damage and of double-strand breaks in DNA arising from other causes. In eukaryotes recombination occurs during meiosis. Many viruses, including phage, also carry genes for their own general recombination systems. In addition, the DNA of some viruses, such as the temperate phage *"k,* undergoes recombination with the host DNA. This can happen during the processes of integration of the viral DNA into the host genome or excision of the viral DNA during the lysogenic cycle of replication. Recombination occurs around specific sites in the chromosomes of both the virus and its host and is called **site-specific recombination.** Genetic recombination is essential to the development of genomic diversity, to the survival of a species, and to evolution.

Recombination Mechanisms

How can the homologous regions of two different DNA duplexes be brought together? As illustrated schematically in Eq. 9.10, the strand exchange must occur at exactly the same point in each duplex. An early attempt to explain this postulated a "copy choice" mechanism of replication. It was assumed that replication occurred along one DNA strand up to some random point at which the polymerase jumped and began to copy from the second of a pair of homologous chromosomes. The newly formed DNA molecule would be complementary to different parts of

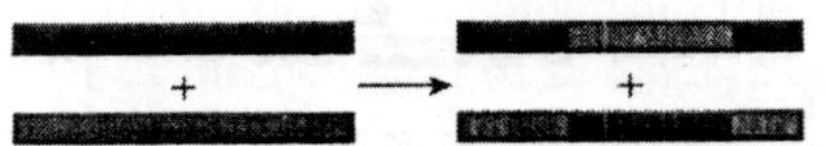

both parental DNA duplexes. To test the idea, Meselson and Weigle infected *E. coli* with two strains of phage *X* containing ^{13}C- and ^{15}N-labeled DNA, respectively. Recombinant DNA was found to contain some ^{13}C and some ^{15}N, as judged by density gradient centrifugation. It was clear that DNA from both parents was incorporated into DNA of recombinant progeny, a finding that ruled out the copy choice hypothesis. If recombination occurs instead by enzymatic cutting of two homologous duplex DNA molecules followed by rejoining, how is it possible to avoid inactivation of genes by addition or deletion of genetic material? Recombination cannot depend upon the random action of a nonspecific enzyme with random rejoining. Yet, general recombination can occur at any point and with a roughly constant frequency throughout the DNA chain. The explanation of these facts lies in the occurrence of base pairing between at least some short homologous regions of strands of the two different DNA duplexes.

The Holliday recombination intermediate. In 1964, Holliday suggested a recombination process that would give rise to chдracteristic H-shaped intermediates. Recom-bination could be initiated at special points on the duplexes, recognizable by a recombination enzyme. A short amount of unraveling would be followed by strand exchange with the two broken strands being rejoined by a ligase. The crossover points would then migrate up or down the chains as the two helices turned about their own axes. Long regions of heteroduplex DNA could be gen-erated in this way, and the process could be terminated at a random distance from the starting point, accounting for the observed unifor-mity of genetic recombination events. Chain cleavage and rejoining of two of the strands would terminate the process. If these were the same strands broken in the initiation event (cleavage at points *aa'*), genes lying outside the heteroduplex region would not be recombined, but cleavage of the other chains (at points *bb"*) would lead to their recombination. Intermediates of the type predicted by the Holliday model were soon observed by electron microscopy. Three-dimensional structures have been determined by X-ray crystallography and have been studied by atomic force microscopy.

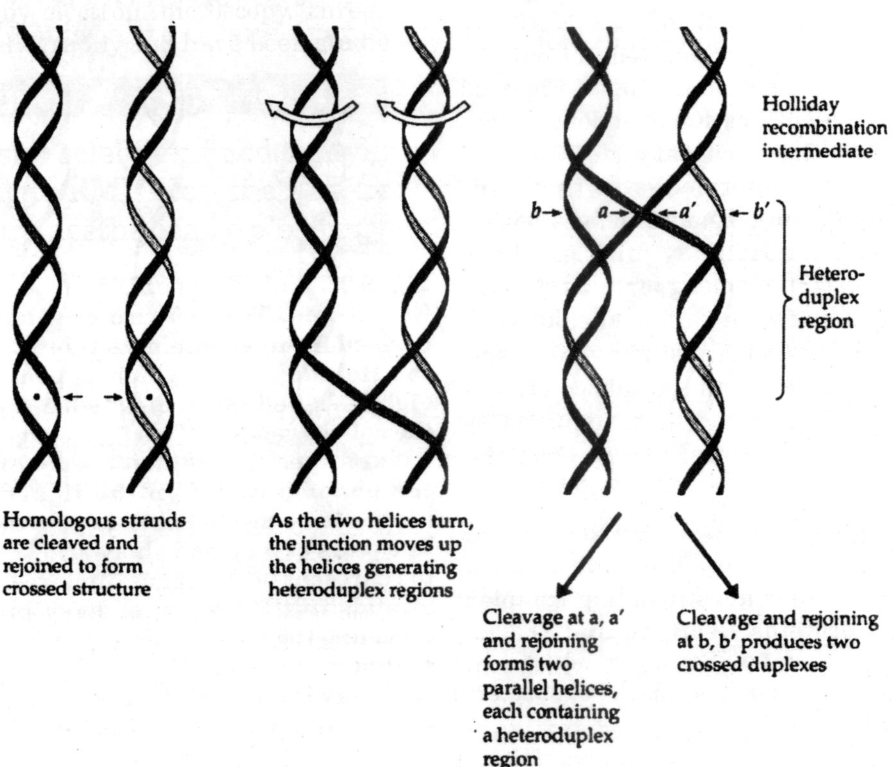

Fig. 9.22. A recombination mechanism involving single-stranded exchanges. After Holliday.

The cross-stranded structure can be formed with all base pairs in both duplexes intact. All that is required is formation of a nick in each of the two polynucleotide chains and a rejoining of the backbones across the close gap between the duplexes. This model also accounts for the cutting of the two crossed strands at exactly equivalent points to terminate the process. Various mechanisms of recombination exist, and most make use of the key

Holliday recombination intermediate. Such four-way junctions can arise in several ways. For example, a 3' or 5' single-stranded tail in a piece of dsDNA can "invade" another dsDNA that has a homologous sequence as indicated in Eq. 9.11. In this drawing a 5' tail has been displaced during repair of a gap in one strand. The resulting D loop may be trimmed out and a new connection made to give the Holliday intermediate. The cleavage points a, a′, b, b′ marked in Eq. 9.11 correspond to those in Fig. 9.22. Holliday junctions may also be formed in stalled replication forks and must be removed to allow replication and transcription to continue. The existence of the Holliday intermediate has been supported not only by electron micrographs such as that of Fig. 9.23 but also by the identification of endonucleases that carry out the necessary cleavages of synthetic Holliday intermedi-ates that have been made artificially. Endonucleases with a high specificity for Holliday junctions have been found in bacteria, among proteins encoded by viruses, and in a wide variety of eukaryotic cells. Additional proteins including helicases, DNA-binding proteins, and specialized **strand exchange proteins** are also required to catalyze the individual steps in the recombination process.

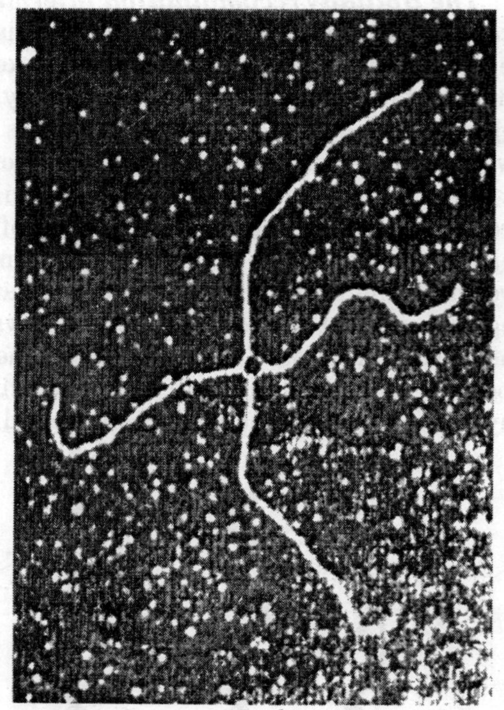

Fig. 9.23. A chi form of DNA from the colicin E1 plasmid. These forms are thought to be derived from recombination intermediates of the Holliday type, which appear as "figure eight"-shaped molecules twice the length of the colicin genome. This figure eight form was cut at a specific site that occurs only once in the genome (twice in the figure eight) by restriction enzyme EcoRl to give the chi form. The pairs of long and short arms are believed to represent homologous duplexes. The single strands in the crossover have pulled apart revealing the strand connections clearly. Such a structure would be expected from the Holli-day intermediate, e.g., if one of the two vertical duplexes were rotated end over end.

The main **RecBCD pathway** of recombination in *E. coli* depends upon a dsDNA nuclease and an unwinding complex consisting of proteins RecB, C, and D. The complex is a powerful *exonuclease,* which can digest the ends of a DNA duplex. It degrades the 3' ends most rapidly, leaving 5'-tails that can invade other homologous duplexes as in Eq. 9.11. The RecBCD complex is also an ATP-dependent helicase, which unwinds the DNA, preparing ssDNA for reaction with the strand-exchange protein **RecA.** The RecB-CD complex also functions to completely degrade foreign dsDNA such as that from invading bacteriophages. Why doesn't it also degrade the genomic DNA of the *E. coli* cell in which it functions? The answer lies in an eightbase DNA sequence, a recombination "hot spot" called chi (χ): 5-GCTGGTG G-3'. This χ sequence occurs 761 times in the leading strands for DNA replication in the *E. coli* genome. When the RecBCD complex reaches a χ sequence, when approaching it from the 3' end, the enzyme stops its exonuclease action by inactivating the nucleolytic activity of the D subunit and promotes recombination about five-to ten-fold as fast as at other sites.

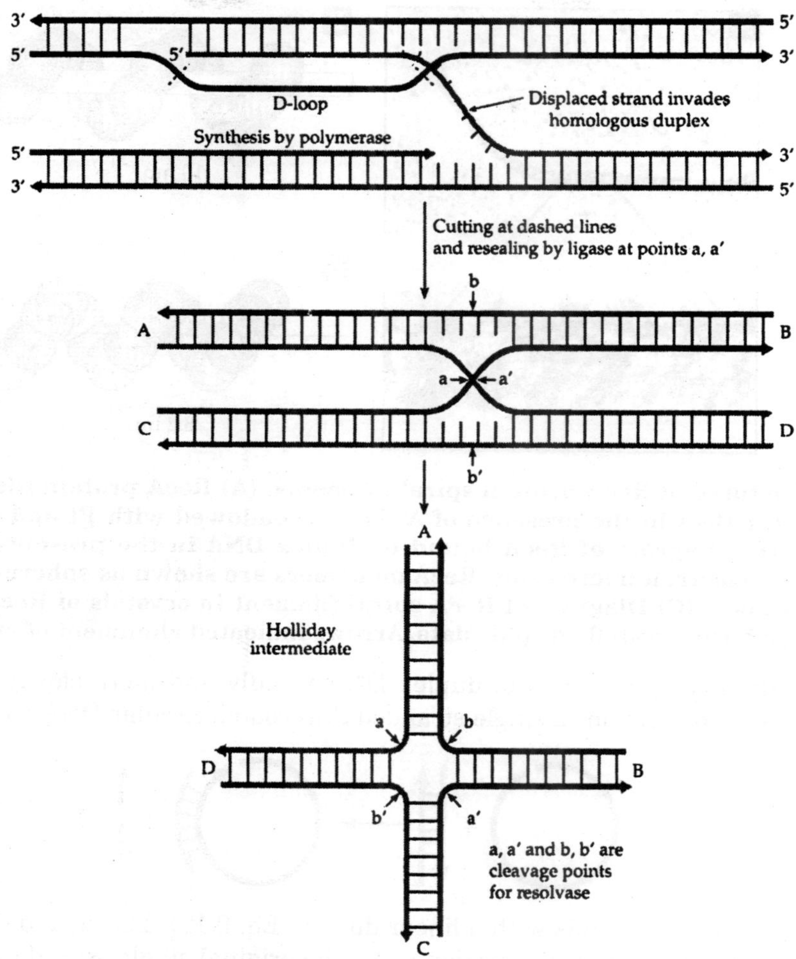

RecA and other strand-exchange proteins. The 352-residue product of the *E. coli RecA* gene is a multi-functional **recombinase,** which is required both for recombination and also for DNA repair. In its repair function the RecA protein acts as a DNA-dependent protease that cleaves a number of repressers in response to damage to DNA. It has a quite different role in recombination where it (1) brings a piece of single-stranded DNA (an end or a gap) together with a duplex; (2) locates homologous sequences; and (3) forms a synaptic complex in which strand exchange can occur. Electron microscopic observations show that the RecA protein binds to either single-stranded or duplex DNA in a cooperative manner to form long rodlike spiral filaments. Measurement of the lengths of RecA protein-covered duplexes shows that the DNA is underwound and stretched by about 50%. It contains ~18 nucleotides per turn. Similar filaments are formed with single-stranded DNA, 3–4 nucleotides being bound per RecA protein monomer. Formation of this ssDNA complex, which may be regarded as an initiation complex for recombination, requires MgATP and is faciliated by prior coating of the DNA with SSB protein. The RecA protein subunits are added in the 5' – 3' direction of the DNA, and SSB is displaced in the process.

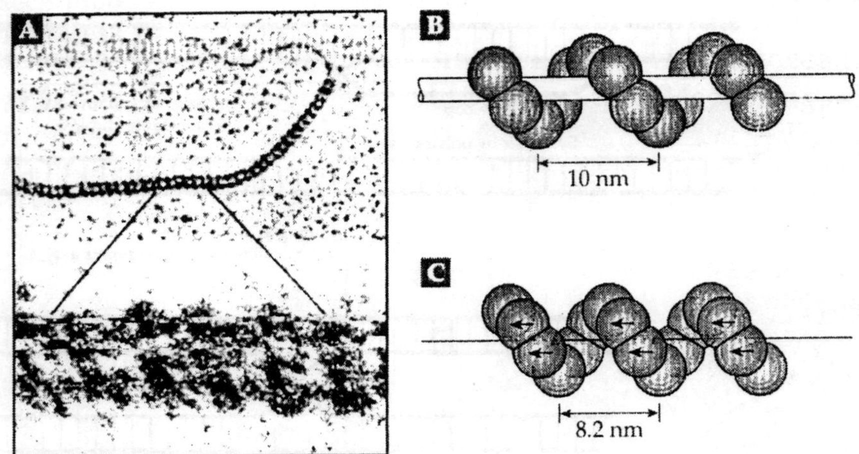

Fig. 9.24 Structures of RecA protein spiral filaments. (A) RecA protein filament formed on circular duplex DNA in the presence of ATP(γ-S), shadowed with Pt and seen by electron microscopy. (B) Diagram of RecA bound to duplex DNA in the presence of ATP(γ-S), as determined by electron microscopy. RecA monomers are shown as spheres, but their exact shape is unknown. (C) Diagram of RecA spiral filament in crystals of RecA protein free of DNA, based on X-ray crystallographic data. Arrows indicated alignment of monomers.

The initiation complex binds to duplex DNA rapidly and more slowly promotes strand exchange. In related reactions a single-stranded SSB-coated circular DNA will bind to RecA

protein, then exchange strands with a linear duplex (Eq. 9.12). The strand exchange requires ATP and advances in the 5' to 3' direction along the original single strand at the rate of a few bases / s. Strand exchange can also occur between two duplexes if there is a suitable gap in one strand, e.g., as is illustrated in Eq. 9.13.

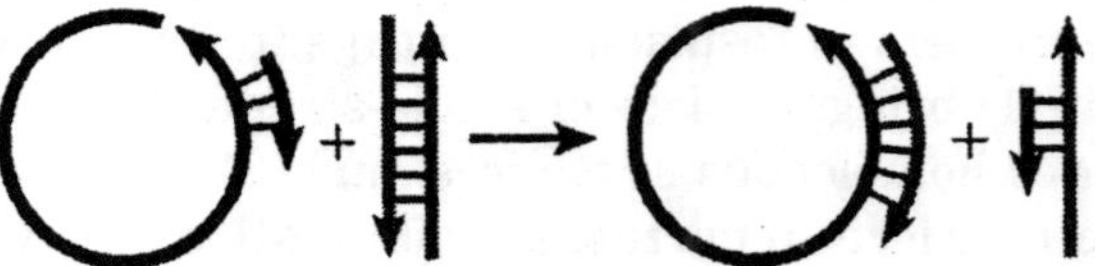

A possible mechanism of strand exchange is illustrated in Figs. 9.25 and 9.26. The RecA protein has binding sites that can accommodate nucleotides from two DNA molecules, one single-stranded and the other a duplex. It may also accommodate two DNA duplexes. As shown in Fig. 9.25, the RecA protein could test the hydrogen-bonding between many base pairs at once in a search of homologous regions. The two DNA chains would have to either slide past each other or repeatedly dissociate and reassociate until a homologous region was found. Then strand exchange could occur. The single strand may be wound into the major groove of the duplex to form an interwound triplex. The matching of hydrogen-bonding atoms may be an attractive way of searching for homology, but the actual search seems to substitute speed for precision. Base substitutions are quite permissive. The need for precise hydrogen-bonding has not been demonstrated, and the exact recognition mechanisms in homologous recombina-tion remain uncertain. Whole chromosomes must be aligned and checked rapidly in the homology search.

Many proteins similar to the RecA protein and with similar functions have been found. These include the products of gene *uvsX* of phage T4, the β protein of phage lambda, the yeast RAD51 and human RAD51 proteins, a meiosis-specific human homolog of the RecA protein, and corresponding proteins from plastids of higher plants. Both the UvsX protein of phage T4 and human RAD51 protein yield strands of coated DNA similar.

Processing the Holliday junction. Completion of the recombination process requires "resolution" of the Holliday intermediate by endonuclease action followed by ligation and perhaps by gap repair. The major recombination pathway in £. *coli* employs a binding protein, a nuclease, and a helicase encoded by genes *RuvA, B,* and C. **RuvA** is a DNA binding protein specific for symmetric Holliday junctions. **RuvB** is a closely associated ATP-dependent helicase. On the basis of genetic and X-ray crystallographic evidence it is now evident that some of the functions previously attributed to RecA are carried out by the RuvABC complex. As indicated, RuvA binds to the Holliday junctions, holding it in the symmetric square configuration in which branch migration is possible. Two molecules of the oligomeric RuvB helicase apparently rotate the DNA, causing branch migration and movement of the DNA through the RuvAB complex. Under some circumstances a different helicase, encoded by *E. coli* gene **RecG,** moves Holliday junctions in the opposite direction. In addition to RuvA a variety of other Holliday junction-binding proteins are known. These include p53 and the nuclear HMG proteins.

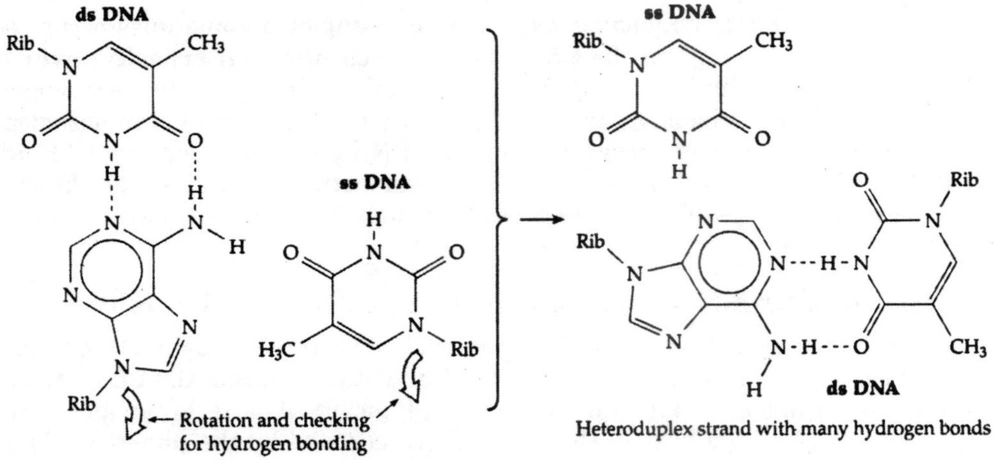

Fig. 9.25 A possible mechanism for homologous pairing of an ssDNA with a duplex DNA and strand exchange. The ssDNA (right) binds together with a hydrogen-bonded duplex (left). The RecA protein rotates the bases into the heteroduplex configuration, where hydrogen bonds may be formed in many of the base pairs.

RuvC is an endonuclease that is highly specific for Holliday junctions. It is a **resolvase** that cuts at either points a,a′ or b,b′ of Eq. 9.11 to form either "patched" or "spliced" recombiiiant DNA. Similar resolvases process bacteriophage DNA and have also been found in yeasts and in mammals. All are dimeric metal ion-dependent proteins.

Nonreciprocal Recombination and Unequal Crossing-Over

The phenomenon of **gene conversion** or **nonreciprocal recombination** was first recognized in genetic studies of fungi for which the four haploid miotic products can be examined individually (tetrad analysis). Instead of the normal Mendelian ratio of 2 : 2 for the gene distribution in the progeny at a heterozygous locus a ratio of 3 : 1 is sometimes ob-served . One of the recombinant chromosomes appears to have been altered to a parental type. A reasonable mechanism by which this can occur arises from the fact that heteroduplex regions, which are present in recombination intermediates, contain defects in base pairing. One strand of the heteroduplex will have a base that does not properly pair with the base in the other strand, or will have an extra base that loops out from the heteroduplex. Since cells contain repair mechanisms that search for defects and carry out a repair process, there is a likelihood that one strand in the heteroduplex region will be altered to restore perfect base pairing, thus causing the observed gene conversion. "Flanking" genetic markers outside of the heteroduplex region are unaffected by gene conver-sion, and during meiosis crossing-over between these markers occurs in about 50% of gene conversion events as would be predicted from the model of Fig. 9.21. Data from yeast show that nonreciprocal re-combination during meiosis may also result from double-strand breaks and gap formation followed by repair synthesis using both strands of the homologous chromosome as templates.

Recombination is not limited to meiosis but can occur between homologous chromosomes during mitosis, during the G_1 period preceding mitosis, or even during the G_2 period. Certain mutations in yeast abolish meiotic recombination but have much less effect on mitotic recombination. Thus, the two processes are not identical. It has been suggested that mitotic recombination is utilized to maintain sequence homogeneity between repeated eukaryotic genes.

Since DNA contains many repeated sequences, crossing-over sometimes occurs between locations that are not the same in the two duplexes. Such **unequal crossing-over** has the effect of lengthening one duplex and shortening the other. This may be very important in evolution. It may also, surprisingly, function to preserve homogeneity of chromosomes within a species. For example, tandem arrays of ribosomal RNA genes in yeast have 140 identical copies of their 9-kb repeat unit. Unequal crossing-over between either sister chromatids or homologous chromosomes, when repeated often enough, can lead statistically to a highly homogeneous population.

Site-Specific Recombination and the Integration and Excision

Recombination at specific sites in DNA is responsible for integration of DNA from viruses into the genome and for the cutting out of viral DNA and other pieces of DNA from the genome. The temperate bacteriophage λ, and the F factors and R factors of bacteria can all be integrated into the genomic DNA of the host in this way. Genes encoded by the phage or plasmid are required. In the case of phage λ the viral genes *int* and *xis* are required for integration and excision, respectively. These are not the same as the enzymes of the *rec* loci of the bacterium or the general recombination genes *exo* and *bet* of the phage. In addition, both integration and excision require an *E. coli* protein called integration host factor (IHF), a DNA bending protein resembling the DNA-binding HU.

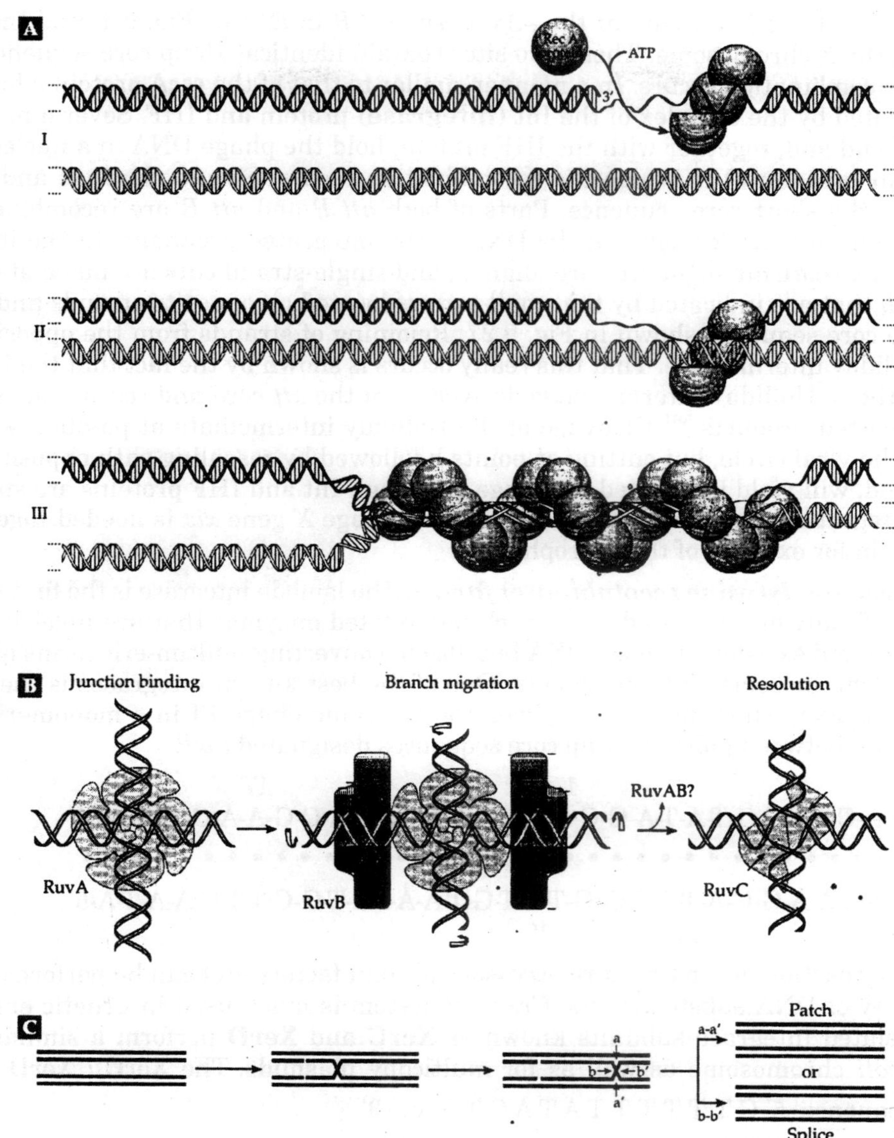

Fig. 9.26 (A) Model for genetic recombination proposed by Howard-Flanders *et al.* (I) RecA protein binds cooperatively to the single strand in a gapped duplex to form an initiation complex in preparation for pairing. (II) The initiation complex binds to the intact duplex, making transient contacts until a homologous site is reached. For clarity, the initiation complex is drawn with only a few protein monomers, but in reality it is likely to extend over hundreds or thousands of nucle-otides. (III) When homologous contacts are made and the strands become paired locally, the initiation complex acts as nucleus for further cooperative binding, which extends the RecA spiral filament around all three or perhaps all four interacting strands. (B) Arrangement of the proteins and DNA during three stages of recombination catalyzed by the RuvABC system. The two RuvB hexameric rings are shown in cross section with the DNA passing through their centers. (C) Scheme of DNA rearrangement during homologous recombination in *E. coli*

Integration of λ, DNA occurs at the ~25-bp site *att B* in *E. coli* (Fig. 9.4) and the ~240-bp site *att P* in the X chromosome. These two sites contain identical 15-bp core sequences within which the re-combination occurs. In a manner similar to that of the recA protein a homologous region is located by the complex of the Int **(integrase)** protein and IHF. Several molecules of Int protein bind and, together with the IHF protein, hold the phage DNA in a nucleosomelike structure (an **intasome**) in which the recombination occurs. Strand cleavage and rejoining occur within the short core sequence. Parts of both *att P* and *att B* are recombined to give sites *att L* (left) and *att R* (right) in the DNA of the integrated prophage. In the integration complex the two core *att* sequences are aligned, and single-strand cuts are made at one of the points *a* or *b* that are indicated by the small arrows located on opposite strands and seven bp apart in the core sequence shown in Fig. 9.27. Rejoining of strands from the opposite duplex yields a Holliday intermediate. That this really occurs is shown by the fact that the Int protein cleaves synthetic Holliday intermediates derived from the *att* core and reseals the strands to give the expected products.[581] Cleavage of the Holliday intermediate at points a will lead to excision of the viral circle, but cutting at points b followed by resealing with opposite strands, as is observed, will yield integrated prophage. Although Int and IHF proteins are sufficient to pro-mote integration, the **excisionase** encoded by phage *X* gene *xis* is needed together with the Int protein for excision of the *X* prophage.

The integrase (lyrosine recotnbinase) family. The lambda integrase is the first recognized member of a family of a hundred or more closely related enzymes that are involved not only in integration and excision of phage DNA but also in converting multim-eric forms of bacterial and plasmid chromosomes into monomers. One of the best known integrases is the 38.5-kDa **Cre recombinase,** which functions to keep the lysogenic phage PI in a monomeric form by re-combination between pairs of 34-bp core sequences designated *loxP.*

17' 1 1' 17'
5'-A-T-A-A-C-T-T-C-G-T-A-T-A-G-C-A-T-A-C-A-T-T-A-T-A-C-G-A-A-G-T-T-A-T-3'
• •
3'-T-A-T-T-G-A-A-G-C-A-T-A-T-C'-G-T-A-T-G-T-A-A-f-A-T-G-C-T-T-C-A-A-T-A-5'
1'

Since the reaction doesn't require accessory protein factors and can be performed *in vitro* with a variety of DNA substrates, the *Cre-loxP* system is much used in genetic engineering. A pair of related integrase subunits known as **XerC** and **XerD** perform a similar function for the *E. coli* chromosome as well as for multicopy plasmids. The XerC / XerD system is

Core sequence 5' C T T T T T T A T A C T A A G 3'
present in both 3' G A A A A A A T A T G A T T C 5'
att P and *att* B b

atypical because it utilizes a pair of integrase subunits rather than just one. However, the basic chemistry is the same for the entire family. As with the λ, integrase the XerC / XerD com-plex acts on a pair of identical core sequences that are aligned in an antiparallel fashion. Active sites in all of the integrases contain the conserved amino acid sequence Arg-His-Arg-Tyr. All of these residues are essential for catalysis. Staggered cuts are made sequentially in the core sequences, e.g., at points a and b in Fig. 9.27 and adjacent to the 5'-terminal thy-midylate residues in Fig. 9.28A. A transesterification reaction forms 3'-phosphotyrosyl linkages from the cut DNA to the integrase protein. The freed 5'-OH groups

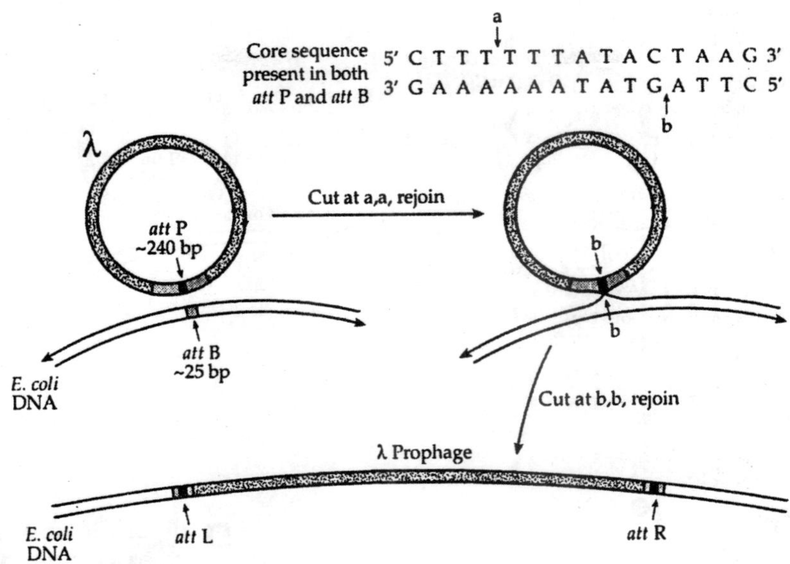

Fig. 9.27 Integration of the phage λ genome into the E. *coli* chromosome at site affB. The same recognition sequences are present at *attB* and *attP.* These are cut at points *a* with rejoining to give the structure at center right. This is cleaved at points *b* with rejoining to give the integrated prophage.

on the other cut end fold back and recombine with the bound 3' ends of the second duplex to generate a Holliday junction. In the mechanism proposed in this figure, three nucleotide units are involved in the folding back. The hydrogen bonds of their initial base pairs are broken, and new bonds are formed. This "base swapping" process is the equivalent of a short branch migration and verifies homology of the two recombination sites.

The Holliday junction can isomerize readily be-tween two forms. In one a base-stacked double helix runs between ends a and b in Fig. 9.28B and another runs between ends c and d. The core hexanucleotide sequences lie between the marked XerC and XerD cleavage sites. In the other isomer (lower drawing) one helix has ends a and d and the other b and c. Following the isomerization the XerD active sites act in two transesterification reactions (not shown but analogous to those in steps *a* and *b)* with base swapping of the trinucleotides at the cut ends. This generates the two separate recombinant duplexes. These might be two circular chromosomes or plasmids formed by recombination from a double length chromosome or plasmid. The previously men-tioned A, amd Cre recombinases appear to act by closely similar mechanisms.

Tyrosine recombinases of the lambda family also function in eukaryotes. Best known is the **FLP (Flip) recombinase,** which is encoded by the 2-µm plasmid of *Saccharomyces cerevisiac* and is thought to function in amplifying the number of plasmid copies. The 6.3-kbp plasmid contains a unique DNA sequence that lies between two 599-bp repeats in inverted orienta-tion. Embedded in each repeat is an FLP recombina-tion target (FRT) sequence, which is recognized by the plasmid recombinase. Each FRT segment includes inverted repeats 13 bp in length with an 8-bp spacer between them. As with other integrase systems the 8-bp spacer or **strand exchange region** is **asymmetric** and establishes the orientation of the recombination sites. The role of the recombinase is to invert one of the 599-bp repeats with respect to the other. This switches replication of the plasmid to a rolling circle pattern.

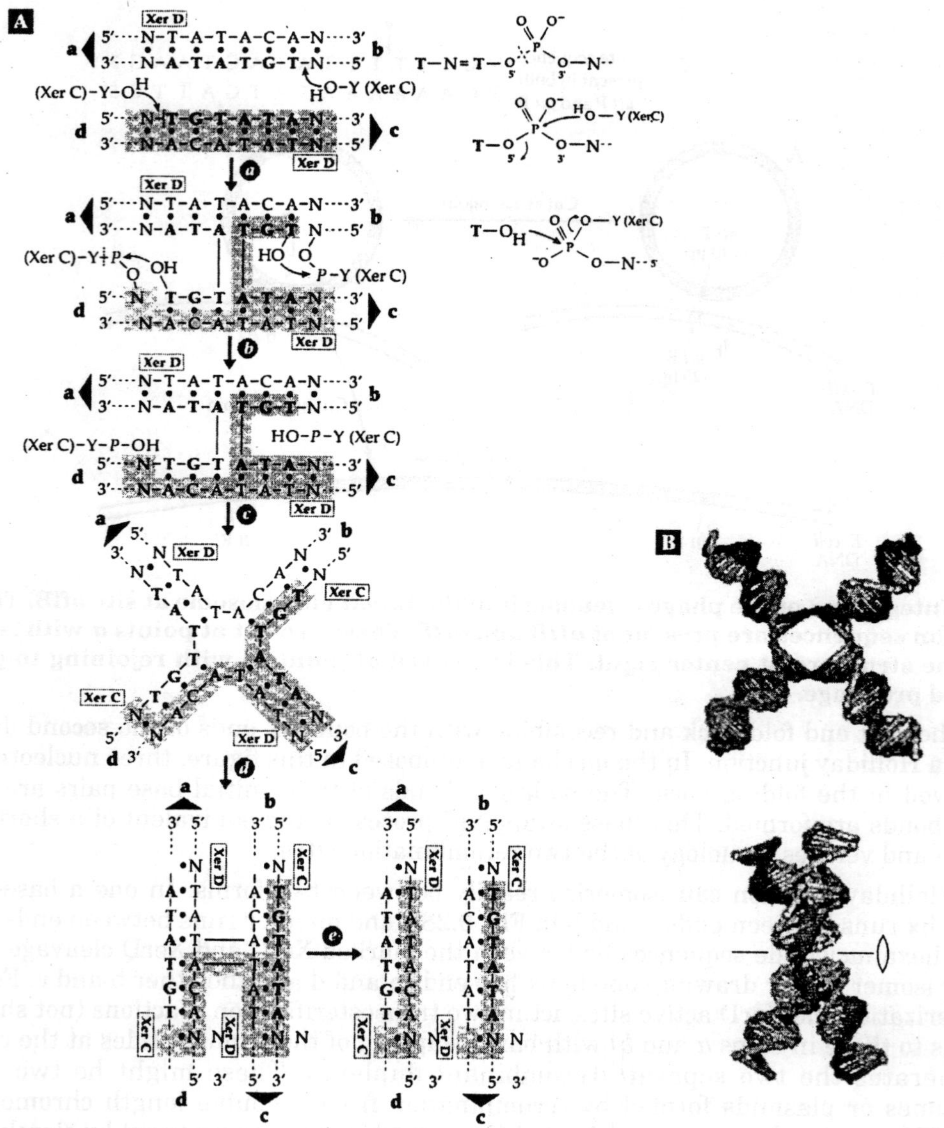

Fig. 9.28 (A) Action of the integrase XerC / XerD on a pair of *E. coli* sites, containing the central six bp sequences TATACA/ATATGT, which are shown in an antiparallel orientation. In step *a* the active site tyrosine hydroxyl groups (Y-OH) of a pair of XerC subunits carry out transesterification reactions on the 5'-terminal thymidylate residues of the central hexa-nucleotide sequences to yield 3'-phosphodiester linkages to the XerC tyrosines. In step *b* the cut 5'-ends, containing free thymidylate 5'-OH groups, fold back to form new base pairs, and the strands are resealed in a second transesterification. This generates a Holliday junction, which *isomerizes* (steps c + d) to an isome'ric species that is acted on by a second pair of transesterification steps *(e)* that are catalyzed by protein XerD, again with folding back of the central trinucleotides. After Arciszewska *et al.* (B) The two isoforms of an antiparallel stacked X Holliday junction are shown. These can be reached from the symmetric square form shown in Fig. 9.28 and schematically in (A) by folding into X-conformations in which all base pairs are stacked either in pairs a,b and c,d or a,c and b,d. From Eichman *et al.*

5′-G-A-A-G-T-T-C-C-T-A-T-T-C-T-C-T-A-G-A-A-A-G-t-A-T-A-G-G-A-A-C-T-T-C

• •

3′-C-T-T-C-A-A-G-G-A-T-A-A-G-A-G-A-T-C-T-T-T-C-a-T-A-T-C-C-T-T-G-A-A-G

FLP recombination target (FRT) sequence

The resolvase invertase family and invertible DNA sequences, A second large family of recombi-nases act by cleaving a target DNA sequence hydrolytically leaving a free 3'-OH end (Eq. 9.14, step *a).* This free end then attacks a phosphodiester linkage in a second strand of DNA, cleaving that strand with an in-line nucleophilic displacement (step *b).* Active sites usually contain a characteristic cluster of aspartate and glutamate (DDE) side chains, which probably act together with a metal ion, perhaps as in Fig. 9.13. Enzymes in this resolvase/ invertase family act either to resolve cointegrates in transposon action (next section) or to invert DNA sequences.

Hydrolysis *a*

Transesterification *b*

If recombination occurs within a piece of DNA at two homologous sites such as the *attL* and *attR* sites at the boundaries of the λ, prophage, the intervening DNA will be excised as a circular particle. In this instance the two homologous regions must be repeated in the same direction, as is indicated by the arrow structures in Eq. 9.15. If the homologous sequences are oriented .in opposite directions, i.e., they are inverted repeats, excision will not occur but the piece of DNA between the repeats will be inverted.

A number of such invertible DNA recombination systems are known and are some-times used to control specific genes. For example, many strains of *Salmonella* have two types of flagella, which are composed of flagellin subunits encoded by genes H1 and H2 respectively. On rare occasions an individual bacterium switches from one flagellar "phase" to the other. This occurs by recombination, as in Eq. 9.16, between two 26-recombination sites *hixL* and *hixR,* each of which contains a 14-bp inverted repeat. The 993-bp invertible segment encodes a recombinase gene called *hin* and a promoter, i.e., an mRNA initiation site, which has a specific

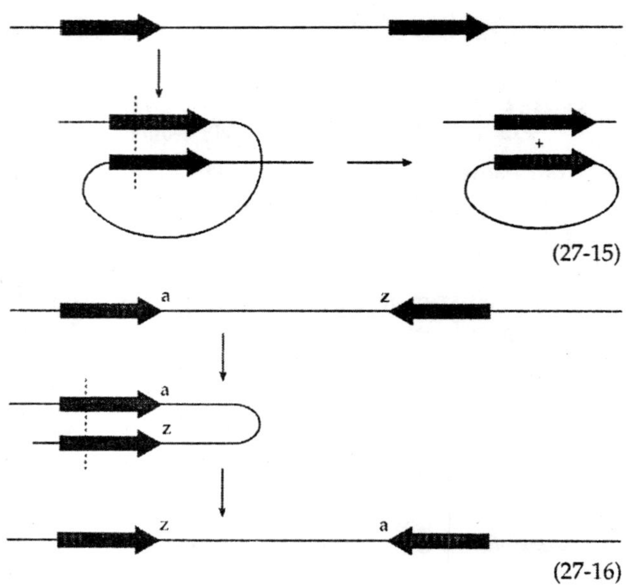

(27-15)

(27-16)

orientation. In one orientation mRNA is transcribed from a short operon that includes the right inverted repeat IRR, the H_2 flagellin gene and gene rh_1, which encodes a represser for flagellin gene H_1. Consequently, only gene H_2 is expressed. In the other orientation the RNA transcription is in the opposite direction so that neither H_2 nor rh_1 is expressed. However, H_1, which is located elsewhere, is expressed freely.

Two other proteins are required for efficient inver-sion by the Hin recombinase. A dimer of a 98-residue helix-turn-helix DNA binding protein called Fis (factor for inversion stimulation), a relative of protein HU, must bind to an enhancer, a 65-bp DNA segment. Binding of Fis to the enhancer helps to hold the supercoiled DNA and the recombinase in a correct orientation for reaction. Protein HU is also needed. The same Fis protein binds to an enhancer for a 3-kbp invertible DNA sequence, which controls alternative host preferences for bacteriophage Mu. The chemistry of the inversion reaction is related to that of the replicative transposons discussed.

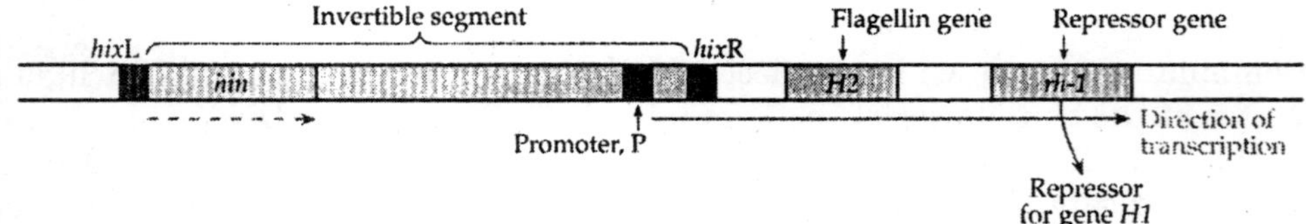

Microorganisms sometimes control the synthesis of surface proteins using segments of invertible DNA. The pathogenic bacterium *Campylobacter fetus* utilizes DNA rearrangements to allow one of a large family of surface layer (S-layer) proteins to be formed. The yeast FLP recombinase, mentioned in the preceding section, also inverts the sequence flanked by the 599-bp repeats.

Transposons and Insertion Sequences

The first evidence that some genes can move from one location to another within the genome came from studies of *Zea mays* by Barbara McClintock in the late 1940s. She concluded that the variegated kernels found in some colored maize were a result of **controlling elements,** which could move from place to place turning on or inhibiting expression of various genes including some of those determining anthocyanin pigment formation. Two of these systems have been studied especially intensively: the **"activator (Ac)-dissociation (Ds)"** system, discovered by McClintock and the "enhancer (En)-inhibitor (I)" system, discovered by Peterson and independently by McClintock.[603] Each contains two genetic elements (segments of DNA) of which Ac and En are autonomous, i.e., they can move by themselves. Ds and I, however, cannot move unless the other element of the pair is also present. Both Ds and En have now been cloned and sequenced.

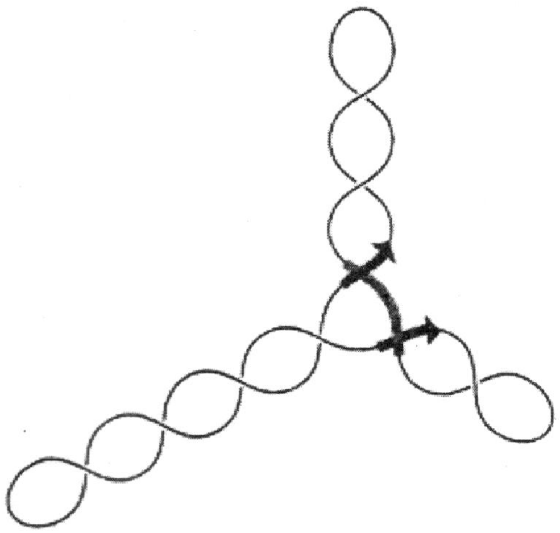

Fig. 9.29. Formation of a synaptic complex of a super-coiled circular DNA containing the sites *gix* (green), which pass over and under the enhancer (gray). The recombinase Gin and the enhancer-binding Fis form a synaptic complex with DNA in this form as seen directly by electron microscopy.

The general importance of transposable genetic elements was not appreciated by most molecular biologists until about 20 years after McClintock's discoveries. Then several moveable **insertion sequences** (Is elements) were found in enteric bacteria. Like the controlling elements of maize these small (0.8–1.4 kb) sequences can move and insert themselves at many points in the genome, often inactivating genes which they enter. The *E. coli* genome contains eight copies of IS1, and five of IS2, as well as several others. Most species of *Shigella* contain more than 40 copies of IS1, Mobile elements similar to those present in maize also exist in archaea.

Following the discovery of the IS elements it was found that transposable elements named **transposons** could transfer resistance to antibi-otics between bacteria. All of these transposable elements have inverted repeat sequences at the ends. For example, IS1 contains the following sequence at both ends but with opposite orientation as if in a palindrome. Some complex

C C
CAATT T TGAT

transposons have an IS sequence at each end. For example, Tn10 contains IS10, a relative of IS1, at both ends. These provide the characteristic inverted repeat termini. Fig. 9.30 shows a schematic drawing of a bacterial drug resis-tance plasmid containing IS1, IS2, and IS10 as well as transposons Tn3, Tn10, and Tn55. The two IS1 elements surround the large resistance determinant, which can be transferred as a block. To be autono-mously mobile a transposon must contain a **transposase** that enables it to be transferred. It usually carries other genes as well and may also contain one or more signals for transcriptional regulation such as promoters.

The chemistry of transposition is more complex than that of simple site-specific recombination. Trans-position can occur at many sites in a genome, and no homology with

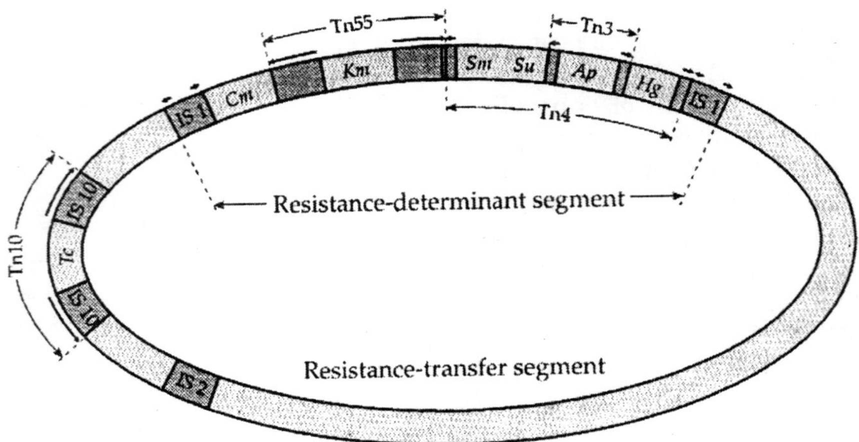

Fig. 9.30 Transposons in an antibiotic-resistance plasmid. The plasmid appears to have been formed by the joining of a resistance-determinant segment and a resistarke-transfer segment; there are insertion elements *(IS1)* at the junctions, where the two segments sometimes dissociate reversibly. Genes encoding resistance to the antibiotics chloramphenicol *(Cm)*, kanamy-cin *(Km)*, streptomycin *(Sin)*, sulfonamide *(Su)*, and ampicillin *(Ap)* and to mercury *(Hg)* are clustered on the resistance-determinant segment, which consists of multiple transposable elements; inverted-repeat termini are designated by arrows pointing out-ward from the element. A transposon encoding resistance to tetracycline (Tc) is on the resistance-transfer segment. Transpo-son Tn3 lies within Tn4. Each transposon can be transferred independently.

the transposon termini is required. Transposition is accompanied by *duplication of a short sequence of the recipient DNA* exactly at the ends of the transposon. Usually 5, 9, or 11 base pairs are duplicated as in the hypothetical example of Eq. 9.17. This happens because staggered cuts 5, 9, or 11 bp apart in the recipient DNA are made during recombination. These are indicated by the small arrows in Eq. 9.17. Transposons causing 5-bp duplications are Tn3, Tn7, γδ, phage Mu of *E. coli,* and Tyl of yeast; IS1, Tn10, and Tn5 of E. *coli* cause 9-bp duplications. When a transposon of one major group moves to a new location, the original copy remains. In this case transposition involves a combination of replication and site-specific (for the transposon) recombination. As a consequence, a circular DNA molecule containing the transposon will often react with a second circular DNA to form a large circle, a **cointegrate,** which con-tains two copies of the transposon. However, another group of transposons utilize a “cut-and-paste” mechanism that doesn’t require extensive DNA duplication.

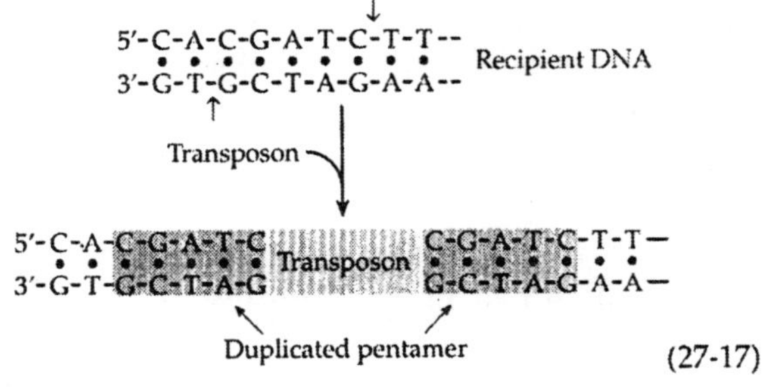

(27-17)

Cut-and-paste (nonreplicative) transposons,The transposases of *E. coli* Tn5, Tn7, and Tn10 act by hydrolytically cutting both strands of duplex DNA at the transposon ends leaving the phospho groups attached to the 5' cut ends, as is depicted in detail in Eq. 9.14, step *a*. The two 3' ends then carry out transesterification reactions, as in Eq. 9.14, step *b*. These two steps are used to nick both strands of the DNA carrying the transposon and to join them to a target DNA sequence to give a branched intermediate (Eq. 9.18, step *a*). Nonreplicative transposons apparently cut off two arms, e.g., A and B, and heal up the small gaps by repair synthesis, leaving the transposon in a new location between C and D. The gap repair accounts for the duplication of the end sequences of the cut target DNA.

Replicative transposons. In 1979 Shapiro proposed the mechanism illustrated in Fig. 9.31 for replicative transposons. The two inversely repeated segments (green) at the ends of the transposon are aligned with the recipient DNA whose ends are la-beled C and D. In fact, the recombining DNA mole-cules must be supercoiled. Staggered cuts are made in the recipient DNA at points *a* and *b*, which are 5, 9, or 11 bp apart, depending upon the specific recombinase. Nicks are also made in the transposon ends. The 3' ends from the transposon are resealed with the 5' ends from the recipient DNA (step *a*) to give a structure that in effect has two replication forks. Replication (step *b*) yields the cointegrate, which contains two copies of the transposon as indicated. In a third step (step c) recombination between the two integrated transposons yields a copy of the original transposon-containing donor and the recipient DNA, which now also contains a copy of the transposon.

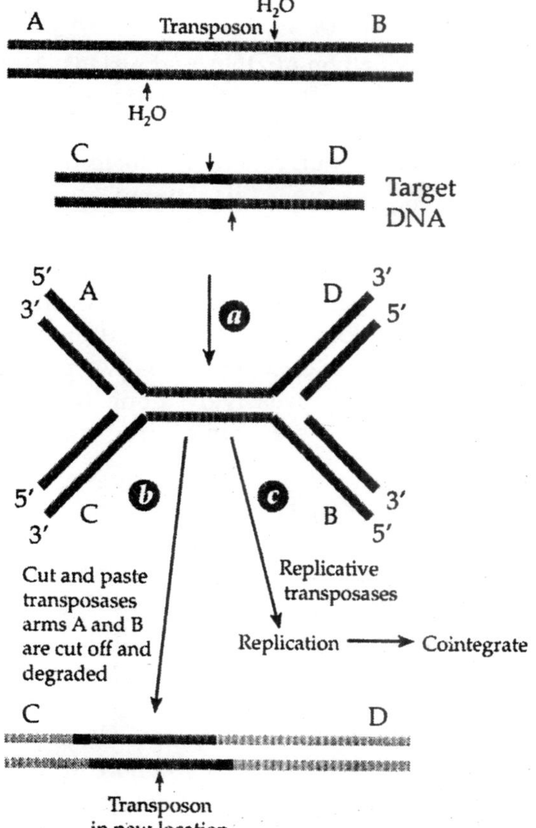

When a transposon reacts with another part of the same DNA circle there are two possibilities. The piece of DNA lying between the transposon and the recipi-ent site may be excised as a circle containing a copy of the transposon. Alternatively, there will be inversion of that sequence as well as replication of the transposon.

The closely related transposons Tn3 and γδ are understood best. They contain not only a transposase gene but also a **resolvase** gene. The transposase carries out the recombination reactions that yield the cointegrates, while the resolvase catalyzes site-specific recombination between the two transposons in the cointegrate to complete the transposition. Several subunits of the resolvases bind the two **resolution (res) sites** of the supercoiled DNA in a parallel orien-tation with the DNA supercoiled as in Fig. 9.29. However, no enhancer-binding protein is needed, and the two *res* sites must be supercoiled. Purified y§ resolvase uses hydroxyl groups of the serine-10 side chains as the nucleophiles to cleave the DNA by displacement at specific *res* sites to give transient enzyme-bound phosphodiester linkages.

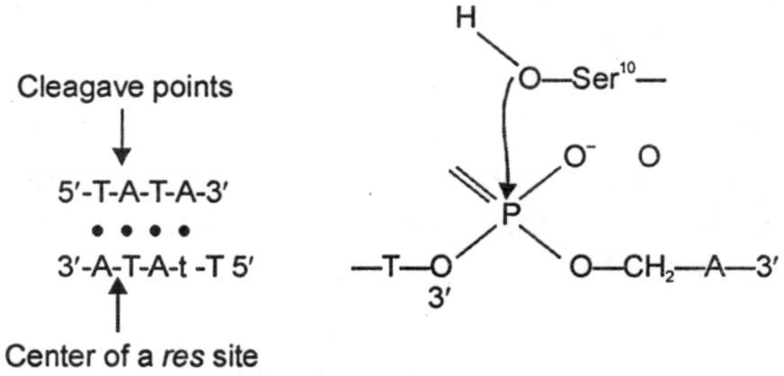

Cleagave points

5′-T-A-T-A-3′

• • • •

3′-A-T-A-t -T 5′

Center of a *res* site

H—O—Ser^{10}—

O^- O

P

—T—O 3′ — O—CH_2—A—3′

The synaptic complex contains 240 bp of DNA and at least two resolvase dimers. All four DNA chains are cut to give eight ends. Four of these are bound to the serine side chains in phosphodiester linkage. In the second step the freed 3'–OH groups react with the bound ends of the other duplex via a transesterification reaction to form the recombinant chains.

The resolvases act on supercoiled cointegrated DNA molecules that contain two directly repeated res sites to produce two singly linked circles (which are still supercoiled) each containing one res site. The two *res* (resolution) sites within the transposons are aligned, the open circle of DNA shown at the upper left being folded as shown in the lower part of the drawing. .The DNA substrate is not knotted. However, after recombination it is catentated and will require action of a topoisomerase to separate the two products. Occasionally additional recombina-tion events occur, perhaps processively along inter-wound double helices. This produces various knotted products. An electron micrograph of one of these is shown in Fig. 9.17.

The temperate bacteriophage Mu, The most efficient transposon known is the 37-kb genome of the **mutator bacteriophage** Mu. Once it becomes integrated into a bacterial chromosome, it replicates by repeated rounds of transposition within the host bacterium. During the lytic cycle some of the replicating DNA is excised as extrachromosomal circles of various sizes, which are packaged into virus particles by a headfull mechanism. The circles contain copies of some host DNA, but this is left behind when a virus particle infects a new cell. The phage Mu DNA is integrated into host DNA by a cut-and-paste mechanism in a transpososome that contains a tetramer of the virally encoded transposase (MuA protein) bound to a supercoiled DNA. The transpososome resembles that in Fig. 9.29 and contains cleavage sites at the ends of the transposon and also an enhancer sequence. Several steps involving conformational alterations occur in the transpososome. One is an ATP-dependent action of a second Mu-encoded protein MuB. During the lytic cycle replicative transposition predominates.

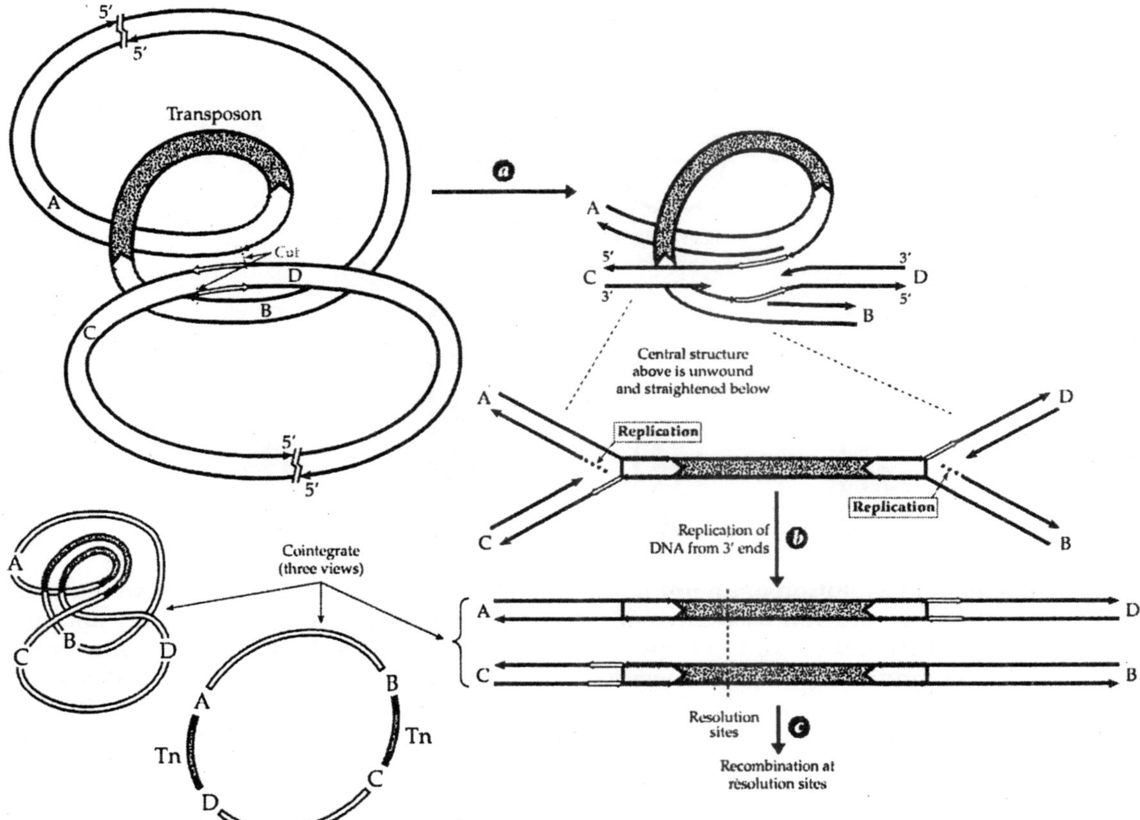

Fig. 9.31 Scheme for integration of a transposon (stippled duplex) present in a piece of DNA with ends A and B into another piece of DNA with ends C and D and containing a suitable recognition sequence (open bars). Inversely repeated sequences in the transposon are shown as solid bars with a direction arrowhead. Arrowheads point toward 5' strand ends. Cleavage and rejoining at points a, a and b, b yield an intermediate with two replication forks. Replication through the transposon yields one unchanged DNA segment with ends A and B and a transposon inserted into the other DNA segment. If A is continuous with B, a cointegrate structure is formed.

Some other transposons, Transposons have a variety of biological functions. For example, haploid cells of the yeast *S. cerevisiae* exist as one of two mating types *a* or α. The mating type is established by trans-position of one of two "cassettes" of genes from two different "silent" locations to a location from which they can be expressed.

One of the best known eukaryotic transposons is the P element of *Drosophila,* which transposes only within the germ line cells of developing embryos, somatic cells being unaffected. It belongs to the same family as Barbara McClintock's Ac element of maize and the *Tel* family of nematodes.[627] P ele-ments use a nonreplicative cut-and-paste method of transposition. The 87-kDa transposase protein requires GTP and Mg^{2+} for activity. It was only in the past few decades that P elements have been found in *D. melanogaster.* They may have entered this fruit fly from another species, possibly transferred by a mite.

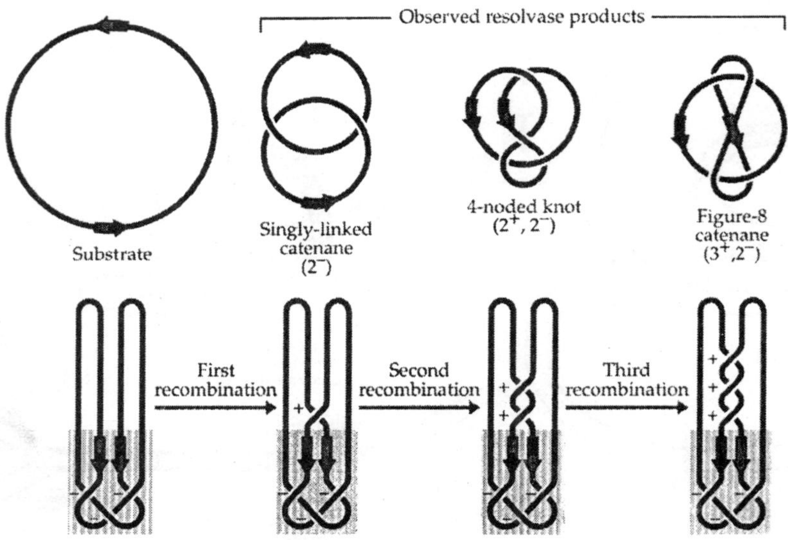

Fig. 9.32 Scheme for resolution of an unknotted cointegrate molecule by a resolvase that cuts the transposons at resolution (*res*) sites and recombines them. The resolvase may act only once or repeatedly as shown. In the upper row, the duplex DNA substrate and products are represented in standard topological form as they might appear after nicking. In the lower row the DNAs are depicted as folded forms bound to the resolvase with the two directly repeated *res* sites (thick arrows) dividing the substrate into two domains (thick and thin regions). The substrate at synapsis has three (–) supercoils that entail crossing of the two domains. Successive rounds of recombination, each introducing a single (+) interdomainal node, are drawn in the lower row. Bound resolvase maintains the three synaptic supercoils. After dissociation from the resolvase at any stage, the product supercoil nodes either cancel with ones of opposite sign or are removed by subsequent nicking. The node composition is indicated in parentheses.

A second *Drosophila* transposon called **mariner** typifies the *mariner* / *Te*1 transposon superfamily, which also contains members from nematodes, other invertebrates, fishes, amphibia, and possibly human beings. These transposons encode a transposase containing a D, D, D or D, D, E motif but no other proteins. They contain short ~30-bp terminal inverted repeats and become inserted into host TA sequences. Movement of some repetitive sequences of the LINE and SINE families within the human genome may be assisted by *mariner* transposons.

The maize transposon *ac* is widely used as a means of inactivating genes and placing a "tag" that can be used to map the gene and to permit it to be cloned and sequenced. Although initially of use only in maize the method has been extended to other plants, and genetically engineered transposons have allowed it to be utilized in animals.

A different kind of transposition controls self-sterility in maize. The cause of self-sterility in one strain has been traced to the presence of two linear episomes called S–1 (614 kb) and

S–2 (514 kb) within mitochondria. These have inverted terminal 208 bp repeats. On rare occasions they recombine with the circular mtDNA converting it to a linear form with the episomes covalently linked to one end. The change is accompanied by reversion to fertility.

As described already, mammalian DNA contains many **retrotransposons** (retroposons) that lie within short direct repeats characteristic of transposons. However, they contain a poly(A) tail at the 3' end, an indication of their relationship to RNA transcripts, and are discussed in.

Other Causes of Genetic Recombination

Because their integration sites in host DNA do not depend upon homology with the transposon ends, transposition is sometimes called **"illegitimate recombination."** Although no homology is required, there are prefered sites for integration. For example, Tn10 is transposed most readily into certain "hot spots" in the *Salmonella* DNA among which is the sequence 5'-GC^{m5}CAGGC. Illegitimate recombina-tion can also be induced by other processes that in-volve DNA chain cleavage, e.g., by topoisomerases. Whenever a DNA chain breaks, it must be repaired, a process that often also involves recombination. Recombination is often observed to occur between direct repeat sequences, which are a major cause of instability in the genome.

Under some circumstances selected segments of the genome are **amplified** by repeated replication of a gene or genes. Amplication of specific genes occurs in viruses, in bacteria where it may provide for adaptation to conditions of stress, and in eukaryotes. In oocytes of amphibia, such as *Xenopus,* excess DNA accumulates around the nucleoli and later breaks up to form 1000 or more separate nucleoli. As many as 3000 copies of the rDNA (which forms a dis-tinct satellite band upon centrifugation) may be present. Much of this DNA exists as extrachromosomal rings containing 1–20 rDNA units. Using these genes as many as 10^{12} ribosomes per oocyte are synthesized.

Tetrahymena contains only one set of rRNA genes per haploid genome in its diploid **micronucleus.** Following sexual conjugation the chromosomes of the micronucleus undergo multiple replications to form polytene chromosomes containing thousands of copies. However, about 5% of the resulting DNA is ex-cised as linear pieces with characteristic inverted repeat sequences and 3' single-stranded tails:

$$5'\text{--}\overrightarrow{C_4A_4C_4A_4C_4}\text{---//---}G_4T_4G_4T_4G_4T_4G_4T_4G_4\text{--}3'$$

$$3'\text{--}G_4T_4G_4T_4G_4T_4G_4T_4G_4\text{---//---}\underleftarrow{C_4A_4C_4A_4C_4}\text{--}5'$$

These tails are joined by a protein to form the circles, which are segregated in the **macronucleus.** The other 95% of the amplified DNA is degraded. At the next meiosis the macronucleus is discarded entirely, and a new one is formed at the next diploid stage.

Toxic drugs often cause cells to amplify genes that help resist the drug. This can be a major problem in the chemotherapy of cancer. For example, a culture of human leukemia cells grown in the presence of in-creasing concentrations of methotrexate increased its level of dihydrofolate reductase 240–fold. The cause is an increase in the number of copies of a chromosomal region containing the gene. Cancer cells tend to amplify oncogenes such as *c-myc*.

There are several mechanisms for gene amplification. The formation and breakup of polytene chromosomes in *Tetrahymena is* one. The circular copies of rDNA in *Xenopus* and many other species are generated by a rolling circle mechanism similar to that in Eq. 9.7. If each circle excised from the original chro-mosomes contains an origin of replication, many copies can be formed. Another possible mechanism is replication of a local region of DNA several times followed by excision of pieces of the DNA. A mechanism that may give rise to homogeneously staining regions is unequal crossing-over repeated several times within a gene cluster. Transposition can cause excision of DNA in a circular form, which can be amplified by a rolling-circle mechanism.

DAMAGE AND REPAIR OF DNA

A characteristic of living things is their high degree of mutability. Harmful mutations take a toll of human life at an early age, and the very high incidence of cancer in older persons is largely a result of the accumulation of somatic mutations. Mutations are a major factor in aging and are continuously introducing new genetic defects into the population. Mutations can be described as **base substitutions, deletions, or additions.** Base substitution mutations are classified as **transitions,** in which a pyrimidine in one strand is replaced by a different pyrimidine. In the complementary strand a purine is replaced by the other purine, e.g.,

$$C \bullet G \rightarrow T * G \xrightarrow{\text{Replication}} T \bullet A \quad C \bullet + \quad G \qquad (27\text{-}19)$$

$$A \bullet T \rightarrow \underset{\text{Mispair}}{G * T} \longrightarrow \underset{\text{Mutation}}{G \bullet C} \quad A \bullet + \quad T \qquad (27\text{-}20)$$

In a *transversion* a purine in one chain is replaced by a pyrimidine, while the pyrimidine in the complementary chain is replaced by a purine:

$$C \bullet G \rightarrow A * G \longrightarrow A \bullet T \quad C \bullet + \quad G \qquad (27\text{-}21)$$

Here the central asterisks designate mispairs and the green shade marks mismatched bases and also the resulting mutant base pair formed in the next replica-tion cycle.

Causes of Mutations

DNA can be damaged in many ways. Sponta-neous hydrolysis of the glycosidic bonds between nucleic acid bases and the deoxyribose to which it is connected cause the loss of ~ 10^5 purines and pyrimi-dine rings per day from the DNA in a mammalian cell. About 100 residues per day of cytosine are deaminated by such agents as nitrite or bisulfite to form uracil. Like thymine, uracil will pair with adenine causing $C \bullet G \rightarrow T \bullet G$ a transition mutation as in Eq. 9.19. A few adenines per day are also deaminated to form hypoxanthine. Oxidized bases are formed by attack of HO• radicals and other species of reduced oxygen. Alkylating agents from the external environment as well as *S*-adenosylmethionine carry out slow, nonspecific alkylation of purine and pyrimidine bases. Polycyclic aromatic hydrocarbons and other carcinogens are converted to metabolites that alkylate DNA, and alkylated bases often mispair during replication. Ultraviolet light induces formation of photohydrates, pyrimidine dimers, and other photochemical products. X-rays and gamma rays cleave nucleic acid bases and break chromosomes. Natural radioactivity has a measurable effect. Mistakes caused by mispairing or misalignment are made during replication and recombination and even during repair of DNA. Errors in replication are especially numerous in highly repetitive DNA sequences. Some errors probably arise as a result of tautomerization and others from incorporation of uracil in place of thymine. Thus, to keep its DNA in repair a cell must continuously deal with missing bases, wrong bases, altered bases or sugars, pyrimidine dimers and other crosslinkages, deletions, and insertions.

Fidelity of Replication

During DNA replication in *E. coli* only one mistake is made on the average during polymerization of 10^9–10^{10} nucleotides. The rate varies among different sites in the genome. In eukaryotes the error rate may be only 1 in 10^{-10} base pair or less per generation. Early workers often attributed the specificity in base pairing and the resultant high precision of replication entirely to the strength of the two or three hydrogen bonds formed together with the stabilization provided by the adjacent helix. However, the Gibbs energy of formation of the base pairs is small, and the additional energy of binding to the end of an existing helix is insufficient to account for the speci-ficity of pairing. Thus, according to Eq. 9.30 a difference in ΔG of binding between the correct nucle-otide and an incorrect one of 11 kJ/mol would give an error rate of 1 in 100.

The role of polymerases. The polymerase enzymes play a major role in ensuring correct pairing during replication, transcription, and protein synthesis. RNA and DNA polymerases are large molecules. The binding site on the enzyme can completely surround the double helix. Water will be excluded as the enzyme folds around the base pairs. This may have the effect of greatly increasing differences in the Gibbs energies of binding and thereby enhancing selectivity. It has traditionally been assumed that formation of proper hydrogen bonds is essential for the base selection. A hypothetical active site is portrayed in Fig. 9.33. A guanine ring of the template strand of DNA is portrayed at the point where the complementary DNA strand is growing from the 3' end. The proper nucleoside triphosphate must be fitted in to form the correct GC base pair before the displacement reaction takes place to link the new nucleotide unit to the growing chain. Let us suppose that the enzyme possesses binding sites for the deoxyribose unit of the template nucleotide and for the sugar unit of the incoming nucleoside triphosphate and that the two binding sites are held at a fixed distance one from the other. As indicated in Fig. 9.33, some group H-Y might also be present at each binding site to hydrogen bond to the nitrogen or oxygen indicated by the heavy

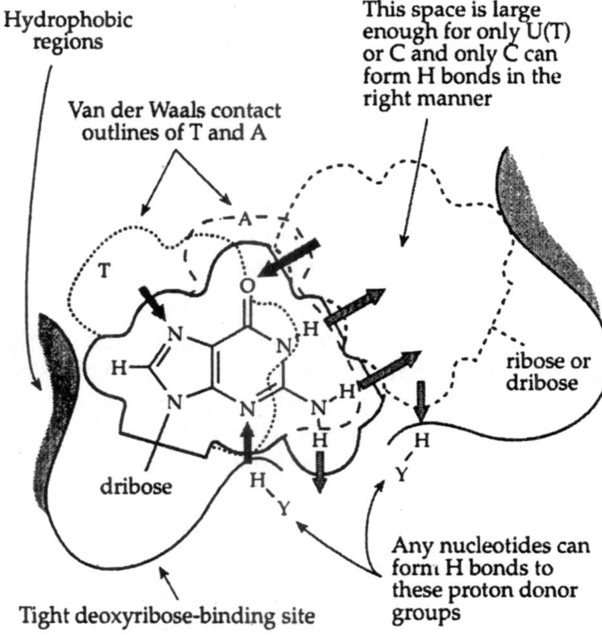

Fig. 9.33 Selecting the right nucleotide for the next unit in a growing RNA or DNA chain. A deoxyguanosine unit of the template chain is shown bound to a hypothetical site of a DNA polymerase.

green arrows. All four of the bases could form hydrogen bonds of this type in the same position. Hydrophobic interactions could provide additional stabilization. With such an arrangement the correct nucleoside triphosphate could be selected no matter which one of the four bases occupied the binding site on the left side of the figure. (The outlines of the thymine and adenine rings have been drawn in with dotted and dashed lines, respectively.) If a purine is present on the left side, as is shown in the drawing, there is room on the right side only for a pyrimidine ring. Thus, A and G are excluded, and the choice is only between C and U (or T). However, U will be excluded because the'dipoles needed to form the hydrogen bonds point in the wrong direction. These dipolar groups are hydrated in solution. They are unlikely to give up their associated water molecules unless hydrogen bonds can be formed within the base pair. Not only would a molecule of U (or T) be unable to form the stabilizing hydrogen bonds within the vacant site but also the electrostatic repulsion of the like ends of the dipoles would tend to prevent associa-tion. This would lower the affinity of the polymerase for mispaired bases. Verification that the proper base pair has been formed could be accomplished by using its tautomeric properties to sense an electronic displacement through the hydrogen-bonded network. Considerable experimental evidence suggests, however, that hydrogen bonding may not be important to the selectivity of DNA polymerases. For example, the triphosphate of the following analog of thymidine is efficiently incorporated by *E. coli* DNA polymerase I into DNA opposite the thymine base;pair partner adenine.

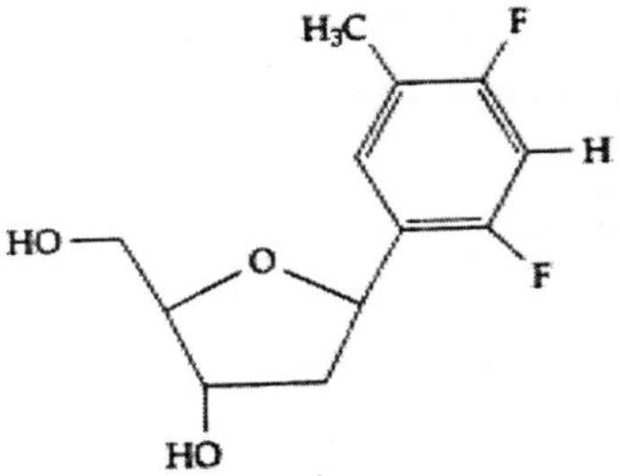

H_3C F H HO F HO

Structural analog of thymidine

Similar results have been obtained for a variety of other analogs with poor hydrogen bonding character-istics. Apparently *shape* is most important. Binding strengths in transition state structures must also be considered.

Editing (proofreading). As is discussed, base-pairing is checked after a new monomer is added at the 3' end of a growing polynucleotide chain, as well as before polymerization occurs. If the wrong base pair has been formed, the newly created linkage is hydrolyzed, and the incorrect nucleotide is released. It has been estimated that in E. *coli* the error rate for DNA polymerase III holoenzyme is ~1×10^{-7} per base pair of which proofreading by the ε subunit may provide ~10^{-2}. Additional mismatch repair reduces the error rate 200- to 300-fold to give the overall error rate of <10^{-9}. Fidelity of replication, which seems to be higher on the lagging strand than on the leading strand, may also be improved by operation of other proteins, such as the sliding clamp, that are part of the replication apparatus. Replicative proteins also appear to be designed to minimize **frameshift mutations** that could arise by slippage of the template versus the replicated strand in long runs of dA•dT pairs.

Repair of Damaged DNA

A final check of the fidelity of replication is made after a new strand has been formed. Mismatched base pairs are identified, and the incorrect nucleotides are cut out and replaced by correct ones. Some of the thymine dimers created by the action of light are also repaired photochemically by photolyases. **Photoreactivation** was the first DNA repair process recognized. However, most thymine dimers in human DNA must be excised and replaced. Loops, gaps, and double-stranded breaks are also sensed, and appropriate corrections are made. If necessary the sequence of a badly damaged segment of DNA may be copied from the DNA in a sister chro-matid before mitosis is completed or from the homologous chromosome.

Much of our understanding of DNA repair comes from investigations of mutant strains of *E. coli,* which display an elevated incidence of mutations. On this basis, the "mutator" genes *dam, mutD, mutH, mutL, mutS, mutU* (also called *uvrD),* and *mutY* were implicated in DNA repair. An additional series of genes, *uvrA, B, C,* and *D,* were identified by their participation in resistance to ultraviolet radiation damage and were also recognized as being involved in DNA repair. Many corresponding genes were found later in the yeast, *Drosophila,* and human genomes. The study of human DNA repair has also been facilitated by the recognition of a group of human inherited defects (Box 9-A). There are 130 known human DNA repair genes.

Methyl-directed and other mismatch repair. In E. *coli* the methylase encoded by gene *dam*, within ~7 min of replication, methylates adenine at N-6 in the sequence GATC, which occurs at many locations. This methylation provides a label for the template strand in a newly replicated duplex. The GATC sites in the newly synthesized strand will not yet be methylated, and repair enzymes recognize it as the strand on which to carry out excision repair. An other system, which may function during recombination, acts on fully *dam* methylated DNA. Since eukaryotes do not use methylation of GATC sites to distinguish the old template and newly replicated DNA strands, other mechanisms must operate. Ten proteins are required for methyl-directed mismatch correction. Of the required *mut* genes, *mutD* (also called *dnaQ)* was found to be the structural gene for the proofreading subunit e of DNA polymerase III. *MutH, L,* and *S* as well as ATP are also needed, as is a helicase, an exonuclease, DNA pol III, and a ligase. Homodimers of proteins MutS and MutL preferentially bind to DNA with mismatched base pairs such as Pur-Pur (e.g., A•G) or Pyr•Pyr (less well recognized). Some of these mismatched pairs may contain modified bases. After **recognition** by the MutS•MutL complex the endonuclease MutH is activated and cuts the DNA chain at the nearest unmethylated GATC sequence. An exonuclease then degrades the chain past the mismatched base leaving a gap that must be filled by the action of DNA polymerase and DNA ligase. This **long-patch mismatch repair** (MMR) is utilized also by eukaryotes, using proteins homologous to E. *coli* proteins MutS and MutL. However, methylation is not involved. In both yeast and human beings there are at least six MutS homologs, some of which have functions other than repair. For example, in yeast MutS- and MutL-related proteins are essential for normal levels of meiotic crossing-over. Defects in human mismatch proteins may cause hereditary non-polyposis colorectal cancer.

Excision repair. The *E. coli* mismatch repair is a type of excision repair. However, a different **nucleotide excision repair** system (NER) is utilized by all organisms from bacteria to human to remove a variety of defects. These include thymine dimers, photohydrates, oxidized bases, adducts of cisplatin (Box 5-B), mutagens derived from polycyclic aromatic compounds, and poorly recognized OC mismatched pairs. In *E. coli* this excision repair process depends upon proteins encoded by genes ***UvrA,* B, C,** and **D** and also DNA polymerase I and DNA ligase. A dimer of protein UvrA forms a complex with helicase UvrB.

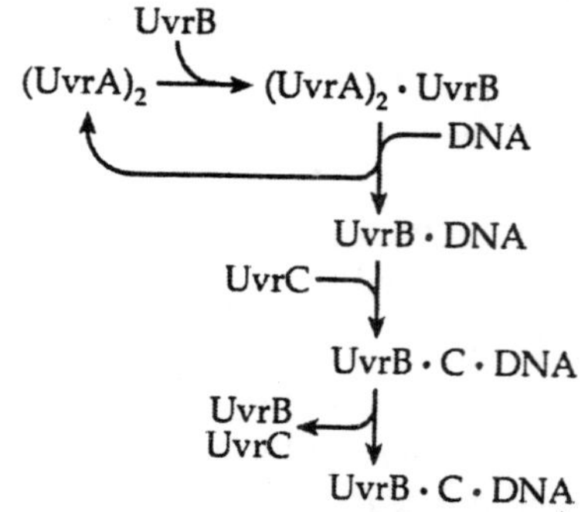

UvrB

$(UvrA)_2 \rightarrow (UvrA)_2 \cdot UvrB$

DNA

UvrB · DNA

UvrC

UvrB · C · DNA

UvrB
UvrC

UvrB · C · DNA

The helicase, driven by ATP hydrolysis, may move along the DNA chain with its associated UvrA protein in search of defects. When one is located UvrB binds tightly. UvrA then dissociates, and the nuclease UvrC binds and cleaves the DNA chain in two places as in the following scheme. One is at the fourth or fifth phosphodiester linkage in the 3' direction from the defect. The other is at the eighth phosphodiester on the 5' side. The resulting gap is filled by a DNA polymerase and DNA ligase. Another helicase, encod-ed by *UvrD,* is also required.

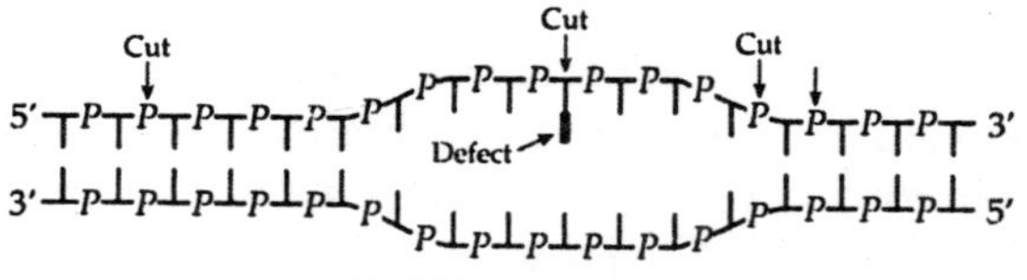

Cut Cut Cut

Defect

5' 3'

3' 5'

UvrBC preincision complex

Both yeast and human cells have similar but more complex systems of nucleotide excision repair. The dual incision steps require at least six components. Several of these (XPA, C, F, G) have been found defective in various forms of the inherited disease **xeroderma pigmentosum** (Box 9.A), in which there is a high incidence of UV-induced skin cancer. Replication protein A (Section C,10), a three-subunit ssDNA binding protein, is also essential. In eukaryotic excision repair the DNA is cleaved at the same position as in the bacterial preincision complex (preceding scheme), on the 3' side of the defect but further out on the 5' side, the excised polynucleotide being –24–32 nucleotides in length. Repair of the resulting gap in the DNA duplex is accomplished by synthesis mediated by DNA polymerase 5 or e and action of a DNA ligase.

One of the lesions removed by nucleotide excision is the thymine photodimer. In the fission yeast S. *pombe* an **alternative excision repair** system, special-ized for removal of thymine dimers and 6-4 photo-products (Chapter 23), produces two-nucleotide gaps with 3'-OH and 5'-phospho-group ends. Alternative NER pathways are also employed by bacteria.

Base excision repair (BER), The N-glycosyl linkages of the purine and pyrimidine bases to the deoxribose residues of the sugar-phosphate backbone of DNA are subject to spontaneous hydrolysis, one important source of damage to DNA. Similar hydrolytic reactions are catalyzed by **DNA glycosylases,** which remove many mismatched or damaged bases. At least seven enzymes of this type are present in cells of E. *coli.* One of them, a **uracil-DNA glycosylase,** hydrolyzes the glycosyl linkage wherever uracil has been incorporated accidentally in place of thymine or has been produced by deamination of cytosine. Acting via a nucleophilic displacement mechanism, it removes the uracil, leaving the DNA backbone intact but with an **apyrimidinic site in** one chain. The resulting apyrimidinic (apurinic sites) are recognized and cleaved by

apurinic /apyrimidinic DNA endonucleases (AP nucleases). These enzymes cut the DNA backbone, some leaving a 3'-phosphate group and a 5'-OH terminus and others a 3'-OH and 5'-phosphate. The resulting gap can be filled, for example by the action of DNA polymerase I and a ligase. To prevent the eliminated uracil from being converted to dUTP and reincorporated into DNA a **deoxyUTPase,** essential to both *E. coli* and yeast, hydrolyzes dUTP to dUMP and inorganic phosphate, decreasing the concentration of dUTP. Other enzymes, known as **DNA glycosylase/AP** lyases, use an amino group of an enzyme side chain as a nucleophile in a ringopening reaction that is followed by P elimination of the nucleotide base with formation of a Schiff base intermediate (green arrow, Eq. 9.23). An example is the bacteriophage T4 enzyme **T4 endonuclease V.**

Others include the *E. coli* **endonu-clease III,** encoded by the **Nth** gene, its yeast homolog **Nth-Spo**, and human DNA pol ι (iota).[718a] After the Schiff base is formed, the DNA back-bone is cut on the 3' side of the abasic site by a second β elimination reaction. A δ elimination may also occur, cleaving the DNA chain on the 5' side of the lesion. Alternatively, a hydrolytic cleavage is possible as is also indicated in Eq. 9.23. Either type of DNA glycosylase forms a single-nucleotide gap or a gap missing just a few nucleotides. This gap can also be filled by polymerase and ligase action.

Both NER and BER forms of exci-sion repair remove a great variety of defects, many of which are a result of oxidative damage. Most promi-nent among these is **7,8-dihydro-8-oxoguanine** (8-OG), which is able to base pair with either cytosine (with normal Watson-Crick hydrogen bonding) or with adenine, which will yield a purine-purine mismatch and a C•G → A•T transversion mutation, a frequent mutation in human cancers.

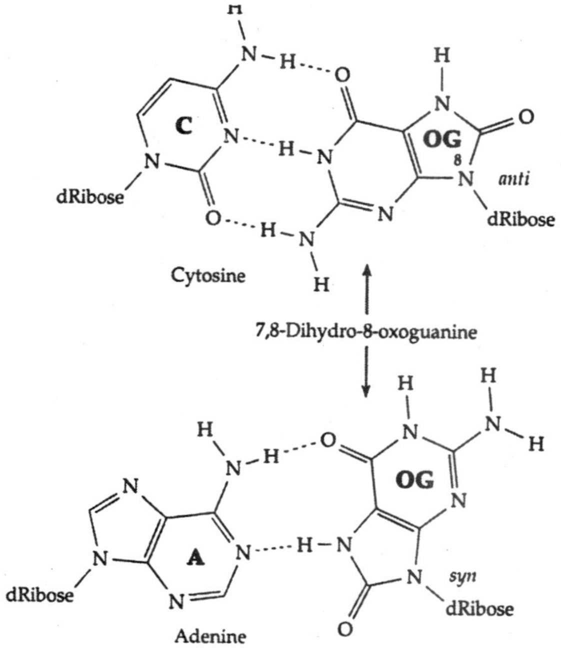

H
N–H····O
H
N
C
OG
O
N····H–N
8
N
N
anti
dRibose
O····H–N
N
dRibose
Cytosine
H
7,8-Dihydro-8-oxoguanine
H
H
H
N
N
N–H····O
H
N
OG
N
N
A
N·····H–N
N
syn
dRibose
N
dRibose
Adenine
O

In *E. coli* three enzym,es protect against 8-OG DNA mismatches. A glycosyltransferase encoded by the *mutM (orfpg)* gene removes 8-OG from DNA. Because some A*OG base pairs may remain, a second glycosyltransferase (**MutY**) removes adenine from them. Interestingly,

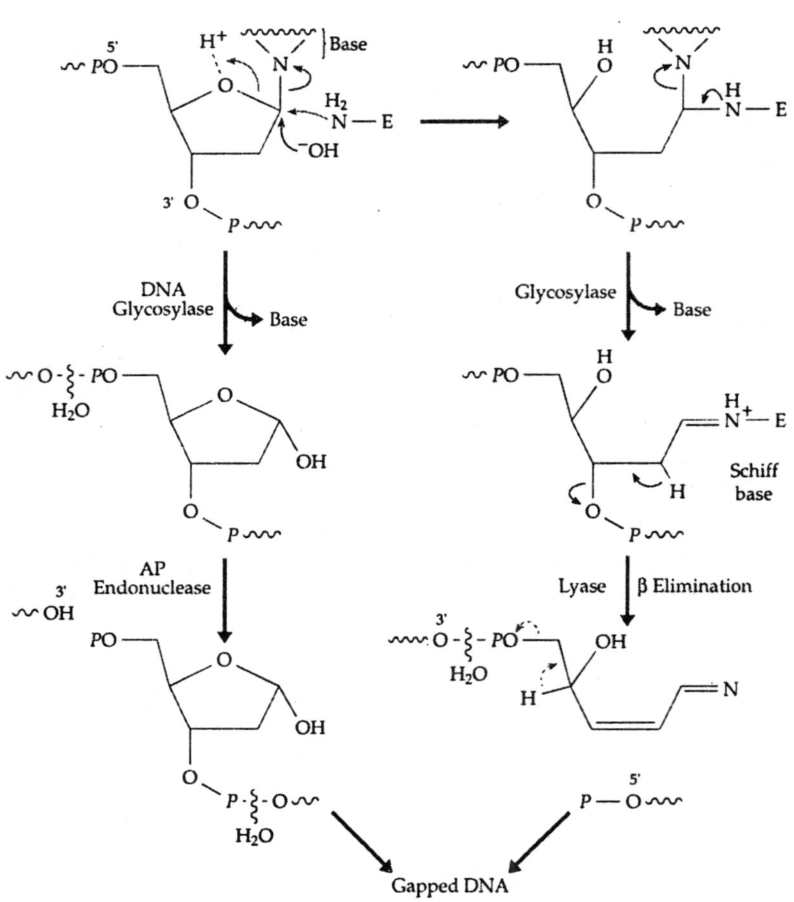

MutY contains an Fe_4S_4 cluster, essential to its function. Both yeast and human cells have a corresponding enzyme. The third *E. coli* enzyme **(MutT)** is a nucleoside triphosphate pyrophosphohydrolase, similar to the previously mentioned dUTPase. It preferentially hydrolyzes free 8-oxo-dGTP, preventing its incorporation into DNA.

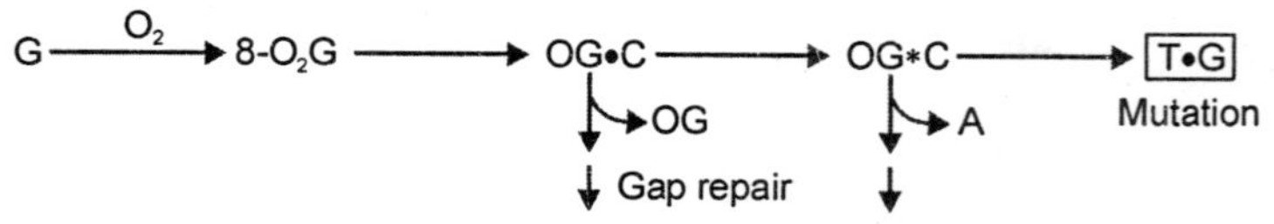

$$G \xrightarrow{O_2} 8\text{-}O_2G \longrightarrow OG{\bullet}C \longrightarrow OG{*}C \longrightarrow \boxed{T{\bullet}G}$$

OG•C → OG released → Gap repair; OG*C → A released; T•G: Mutation

Other DNA glycosylases remove thymine rings that have been converted to saturated or fragmented forms by oxidizing agents or ionizing radiation. Among these are 5,6-hydrated thymine and urea, which are still attached in ribosyl linkage. Purines such as hypoxanthine and 3-methyladenine can all be removed by glycosylases.

Reactions of alkylated bases. The O^6 alkylation of guanine rings in DNA is highly mutagenic, presum-ably because mispairing with thymine will cause C•G → T•A mutations.

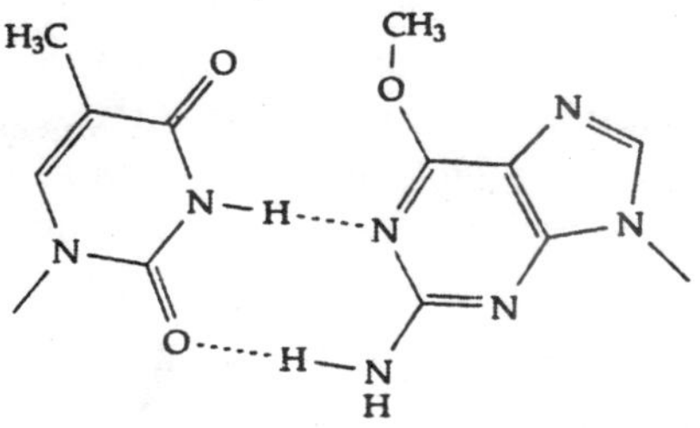

Thymine–O⁶-methylguanine pair

Both bacteria and higher eukaryotes have an enzyme-like **O^6-methylguanine-DNA methyltransferase,** whose synthesis may be induced by culture of cells in the presence of alkylating agents. This 354-residue Zn^{2+}-containing **Ada protein** acts as both acceptor and catalyst for transfer of the methyl group off from the O^6-methylguanine onto the sulfur atom of a cysteine side chain in the protein. The presence of alkylating agents also induces synthesis of DNA glycosylases that remove 3-methyladenine and other alkylated bases. 5-Methylcytosine, present in CpG islands in human DNA, will occasionally be hydrolyzed to thymine, giving T*G mismatches. The mismatched thymines are removed in cells of *E. coli* by a specialized endonuclease (**vsr** gene), which hydrolyzes the phosphodiester linkage preceding the mismatch. In eukaryotic cells a **thymine-DNA glycosylase** accomplishes the removal.

Repair of double-strand breaks, Following replication both in bacteria and in eukaryotes, recombinational exchanges may occur between sister chro-matid duplexes or between homologous pairs of chromosomes. A newly replicated chain may have gaps because of defects in the template strand from which it was copied. Copying from a sister duplex may permit assembly of a correct chromosome se-quence and survival for the cell. Recombinational repair is also a major mechanism for preventing loss of genetic information from **double-strand breaks.** Such breaks are induced by ionizing radiation or chemical damage. They are often created at replica-tion forks, stalling DNA synthesis with potentially disastrous consequences to the cell. Repair of these breaks, which is required for completion of replication, depends upon a large complex of replication and recombination proteins. In *E. coli* this includes protein RecF. Repair of double-strand breaks by homologous recombination may be the most frequently employed type of DNA repair in bacteria, yeasts such as *S. cerevisiae* and also in mammals and other organisms.

A second repair pathway, called **nonhomologous endjoining (NHEJ),** is also utilized and may be relatively more important in higher organisms. The NHEJ mechanism repairs double-strand breaks caused by ionizing radiation and is also employed in specialized genomic rearrangements of developing B and T cells of the immune system. This V(D)J recombination is used to create the huge array of antibodies and antigen receptors required for immunological recognition and protection. NHEJ doesn't depend upon templates but simply rejoins duplex ends. Errors can be made if the ends have been damaged or if the wrong ends are joined. Several special proteins are required. One is a DNA-activated protein kinase (DNA-PK) and a hetero-dimeric **endbinding protein** with subunits **KU70** and KU80. The serine / threonine protein kinase ATM is also activated and phosphorylates protein p53 in response to γ-irradiation of cells. This is thought to be part of a signaling pathway for control of the cell cycle following DNA damage. Mitosis may be delayed while repair is completed. Alternatively, the damaged cell may be killed by apoptosis.

Subjects of current interest are the mechanisms by which completion of DNA synthesis is signaled at the various checkpoints in the cell cycle. These include points in addition to those marked in Fig. 9.15. At every point signals must accumulate to indicate that replication must be delayed to allow repair or that the cell must be allowed to die by apoptosis. Failure to accomplish repair may lead to cancer. Even one double-strand break will prevent the completion of mitosis. Among other problems faced by a dividing cell are slow replication and stalling at replication forks. A range of chemical alterations in chromatin and other proteins are associated with DNA damage and repair. Among these are conjugation of PCNA with ubiquitin and SUMO.

The SOS response and translesion repair. If cells of bacteria or eukaryotes are heavily damaged by UV, X-ray irradiation, or mutagenic chemicals, an emergency or **SOS response** is initiated. In *E. coli* the two proteins specified by genes *lexA* and *recA* initiate the response. The *lexA* protein is a represser that prevents transcription of a group of SOS genes. It is thought that some product from damaged DNA activates the RecA protease activity. The activated RecA protein then cleaves the lexA protein allowing transcription of the SOS genes. The SOS response is transient but complex. It includes increased recombinational activity, alterations in repli-cation initiation, inhibition of nucleases, and induction of an **errorprone DNA synthesis.** The cell now replicates DNA more rapidly than normal but with an increased frequency of errors. For example, it will bypass thymine dimers and other defects, and it will continue the DNA synthesis even though incorrect bases have been put into the chain opposite the thymine dimer. Later the errors may be corrected by recombinational repair or by photoreactivation. In *E. coli* this translesional repair depends upon DNA polymerase III, the RecA protein, and two proteins encoded by genes *umuC* and *umuD*. These genes were recognized as providing resistance to mutations induced by UV radiation. Study of similar genes in yeast led to the discovery of two new and unusual DNA polymerases essential for translesional repair. DNA polymerase ζ (zeta), encoded by gene *REV3* and *REV7* subunits, bypasses abasic sites by inserting a dCMP residue into the growing DNA chain. Because C may be the wrong base the process is "errorprone." Polymerase η (eta) bypasses thymine photodimers by placing two consecutive dAMP residues in the growing strand. This is an errorfree process. Genes for human polymerases ζ and η have also been identified, and additional polymerases have been found in yeast and other organisms. The human XP-V gene for polymerase η is defective in xeroderma pigmentosa variant type. Polymerase ι may act sequentially with pol η to bypass highly distorting lesions. The very imprecise pol θ may also be used to bypass hard-to-rernove lesions. Polymerase λ is thought to play a role in DNA repair during meiosis, while pol κ is needed in some way to provide cohesion between sister chromatids. Some of these DNA polymerases are very inaccurrate. It has been suggested that human pol i may participate in hypermutation of immunoglobin variable genes. Polymerase ϕ of yeast may be required for synthesis of ribosomal RNA.

Poly (ADP-ribose). A eukaryotic peculiarity, which is not well understood, is the synthesis of poly(ADP-ribose) chains attached to many sites in nuclear proteins. Increased synthesis is observed following damage to DNA. The poly (ADP-ribose) polymerase binds to DNA near strand breaks or nicks and, using NAD^+ as a substrate, synthesizes the highly branched polymer attached to a small number of nuclear target proteins.

MUTAGENS IN THE ENVIRONMENT

More than 500 new chemicals are introduced into the environment industrially each year. Some widely used drugs, e.g., **hycanthone**, are mutagen-ic. Powerful mutagens are present naturally in some foods. Others have been added through ignorance. Although many of these have now been removed, the problem cannot be ignored.

One way in which chemical compounds can induce base substitution mutation is through their incorporation into the structure of DNA itself. Thus, 5-bromodeoxyuridine (or bromouracil) can replace thymidine in DNA, where it serves as an efficient mutagenic agent. 2-Aminopurine, an analog of adenine, pairs with thymine, just as does adenine when incorporated into DNA.

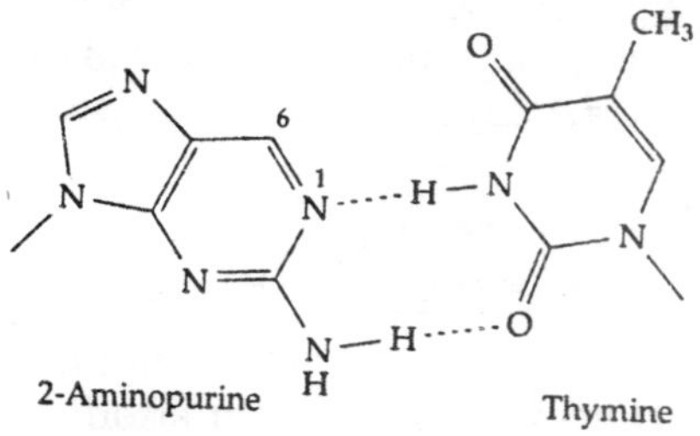

2-Aminopurine Thymine

However, if it protonates on N-1, it can pair with cytosine, causing mutation.

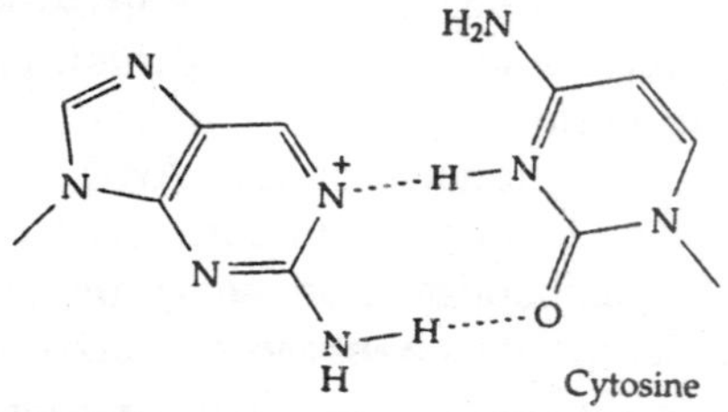

Cytosine

Another mutagenic base is N^6-hydroxylaminopurine.

Many alkylating agents are powerful mutagens. Alkylation can occur at many places in DNA, but the N-7 position of guanine is especially susceptible (Eq. 9.18). The resultant positive charge on the imidazole ring portion of the alkylated guanine causes hydroly-sis of the N-glycosyl linkage and depurination. However, this may be a lethal rather than a mutagenic event. The previously mentioned methylation on O-6 of guanine is probably more important in inducing mutations.

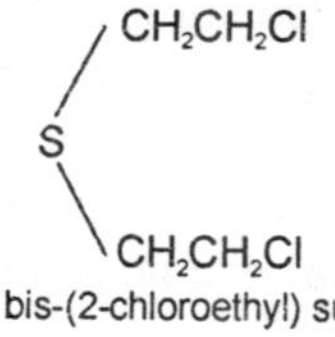

$S(CH_2CH_2Cl)_2$

bis-(2-chloroethyl) sulfide

Among the most biologically reactive alkylating agents are the nitrogen and sulfur "mustards" such as bis-(2-chloroethyl)sulfide. These toxic bifunctional compounds cause lethal crosslinking of DNA chains. The monofunctional half-mustards are mutagenicfaut are less acutely toxic. Another group of alkylating agents, the **nitrosamines**, are highly mutagenic. Much used in the laboratory as a mutagen is **N-methyl-N'-nitro-N-nitroso-guanidine:**

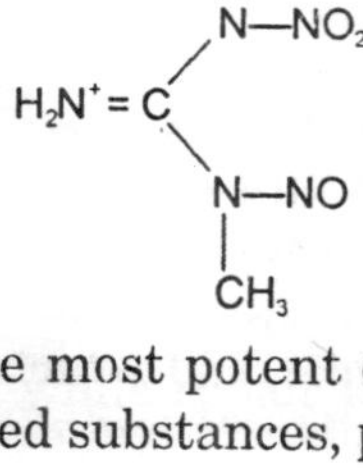

$H_2N^+=C(NH{-}NO_2)(N(CH_3){-}NO)$

N-Ethyl-N-nitrosourea is one of the most potent car-cinogens known. These compounds, as well as diazomethane and other related substances, probably act via a common intermediate: $CH_3N_2^+$ or $C_2H_5N_2^+$

Any secondary amine will react with nitrous acid to form a nitrosamine (Eq. 9. 25). Tertiary amines can also react with loss of one alkyl group. This can occur in the stomach, and the nitrosamines may be absorbed into the system. All plants contain some nitrate and some, such as spinach and beets, have large amounts. Bacon and other cured R' R meats contain both nitrites and nitrates, and many drugs and natural food constituents are secondary amines. Cigarette smoke also contains a nitrosamine. There is a possibility that these substances may induce human cancer.

$$R(R')NH \xrightarrow{HNO_2 \quad H_2O} R(R)N-N=O$$

Nitrosamine

About 3% of residents of North America excrete feces that contain mutagenic unsaturated glyceryl ethers, which have been named fecapentaenes. They could be among the causative agents of colorectal cancer. Protonation of a double bond will produce a reactive carbocation, which may be an active alkylating agent.

HO H
HO O

A fecapentaene

Many halogenated compounds are carcinogenic. Among these is 1,2-dibromoethane, which has been produced in the United States in quantities as great as 10^8 kg / year. It has been widely used as a fumigant for foods, as an industrial solvent, and as an additive to gasoline. It is a **procarcinogen,** whose carcinogenic properties are expressed only after **metabolic activa-tion.** One pathway of activation is reaction with glutathione to form a thioether, which can cyclize to a sulfonium compound. The latter alkylates nucleophilic groups in DNA. Other bifunctional electrophiles such as acrolein, malondialdehyde, vinyl chloride, and urethane are procarcinogens. Styrene, another procarcinogen, is converted to the carcinogenic styrene oxide by action of a cytochrome P450. Metabolites of glucose such as **pyruvaldehyde, methylglyoxal,** and other reactive aldehydes can also attack DNA.

Br — — Br

G−SH → Br^- Glutathione S-transferase

Br — — S—G

→ Br^-

S^+—G → Br — — S—G

Nui

Less common than base pair exchanges are frame-shift mutations. A characteristic of such mutations is that they do not revert (back mutate) as readily as do base substitution mutations, and reversion is not induced by chemicals known to cause base substitutions. However, reversion of frame-shift mutations is induced by acridines and other flat molecules that are known to act as intercalating agents in DNA helices and which promote frame-shift mutations. They are especially effective in causing mutations of regions in which long repeated sequences of a single base such as $(A)_n$ or $(G)_n$ occur. For example, deletion of two base pairs from a "hot spot" (site of frequent mutation) in the *Salmonella* histidine operon with the following sequence is induced by 2-nitrosofluorene and causes reversion of a (-1) histidine-requiring mutant: 5'-CGCGCGCG. Whereas simple intercalating agents are often not very mutagenic, compounds that are both an intercalating agent and an alkylating agent are especially potent. An example is the following compound, which contains a half-mustard side chain:

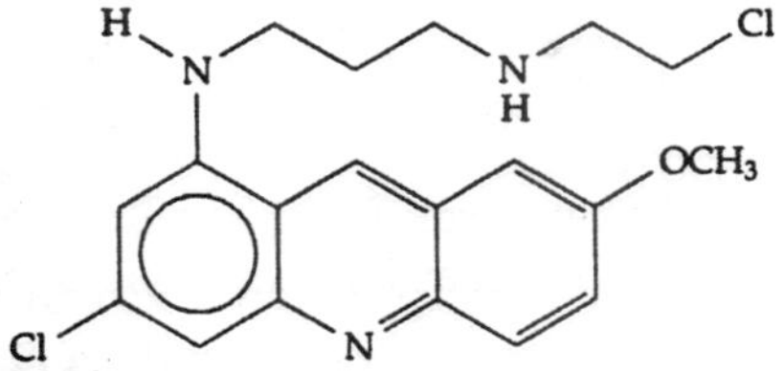

When the CH_2Cl group of the side chain is replaced by CH_2OH, the compound is 100 times less mutagenic.

The carcinogenic **aflatoxins,** which are produced by *Aspergillus flavus,* may be present in infected pea-nuts and other foodstuffs. Like many other compounds that are carcinogenic or

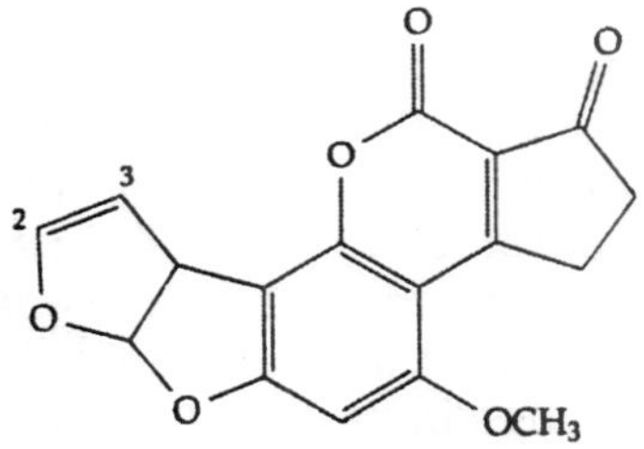

Aflatoxin B_1

mutagenic, the aflatoxins are not unusually reactive chemically. However, they are activated in the animal body by hydroxylation and formation of 2,3-epoxides. The latter may react with N-7 of guanyl residues in DNA.

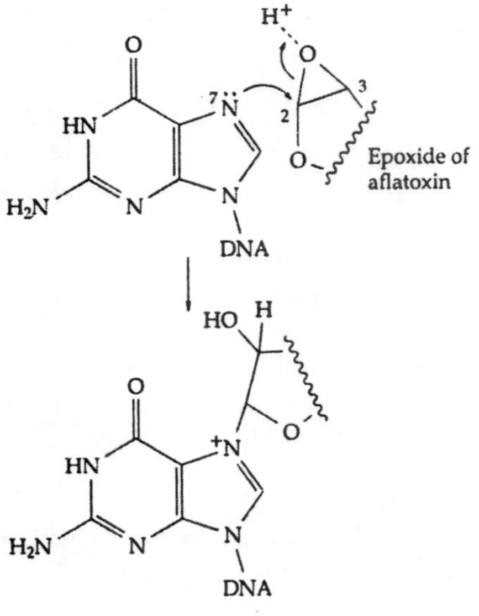

Epoxide of aflatoxin

Benz(a)anthracene is also converted in the body to a carcinogenic epoxide. Benzo(a)pyrene was isolated from coal tar in 1929 and in 1930 provid-ed the first demonstration

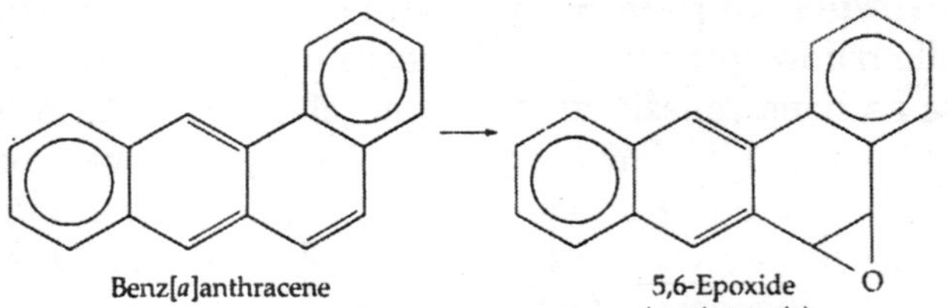

Benz[a]anthracene

5,6-Epoxide (carcinogenic)

of the carcinogenicity of a pure chemical compound. It can also be activated by conversion in the ER to the 7,8-dihydrodiol 9,10-epoxide. Hydroxylation activates 2-acetylaminofluorene to a carcinogenic N-sulfate, which reacts with guanine rings.

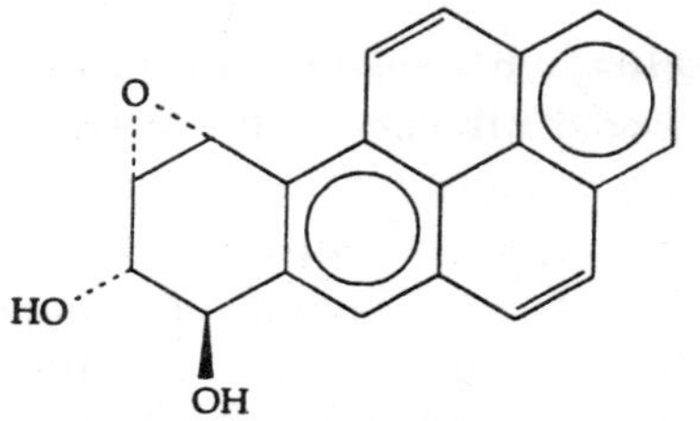

Benzo(a)pyrene diol epoxide

Although it seems incongruous, many of the most potent antitumor drugs are powerful mutagens. Among these are intercalating agents such as daunomycin, neocarzinostatin, and bleomycin. These are alkylating reagents or attack DNA in other ways. The fact that such compounds are in use for chemotherapy emphasizes the need for new approaches to cancer treatment.

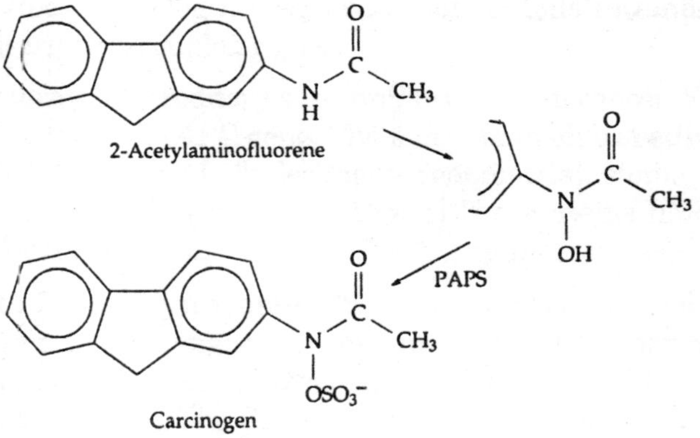

2-Acetylaminofluorene

PAPS

Carcinogen

How can compounds be recognized as mutagenic? This is an important question that is difficult to answer and which has led to many controversies. For exam-ple, how dangerous is formaldehyde in the environment? Butadiene? Dioxin? Is fluoride a carcinogen?

A quick test for mutagenic activity makes use of **tester strains** of bacteria developed by Ames and associates. These are *Salmonella* mutants that are unable to synthesize their own histidine, but which can grow when a mutagenic agent produces a back mutation. One of the strains can be mutated by agents causing base exchanges, while the other three, which contain different types of frame-shift mutations, are affected differently by various mutagens. About 10^9 bacteria are spread on a Petri plate, and a small amount of the mutagenic chemical is introduced in the center of the plate. Where back mutation has occurred, a colony of the bacteria appears. The strains all carry a mutation in the main DNA excision-repair system so that most mutations are not repaired, and the test is very sensitive. Addition of a liver homoge-nate plus an NADPH-generating system to the sample tested allows activation of many aromatic chemicals by hydroxylation. Feeding of mutagens to *Drosophila* eye color mutants permits testing for back mutation in a eukaryote.

The bacterial tests have been widely used and have been of great value. For example, they revealed that certain chemicals that were being used as flame retardants in children's clothing are mutagens, and that mutagens can be generated during cooking of meat and other foods. However, there are good reasons for using other methods of identifying mu-tagens together with the bacterial tests. Government regulatory agencies in the United States have relied largely on tests with rodents. Compounds are tested at very high doses with these short-lived ani-mals. Extrapolation to human exposures at very low levels has been criticized. However, rodent tests have identified many true carcinogens. A broader range of tests are now in use. Direct moni-toring of the accumulation of defects in animal and human cells is now possible. For example, in the **^{32}P-postlabeling technique** DNA is enzymatically di-gested to the nucleotide level, and the adducts with mutagens are labeled with ^{32}P, separated, and their quantities measured. Laser-excited fluorescence from such adducts, GC / mass spectrometry, and immunological methods can also be used to identify DNA adducts. Careful vigilance is needed to keep our mutation rate at as low a level as possible.

A DEFICIENCIES IN HUMAN DNA REPAIR

Problems with human repair systems account for several serious diseases. One is the rare skin condition **xeroderma pigmentosum** (XP), an autosomal recessive hereditary defect. Homozygous individuals are extremely sensitive to ultravio-let radiation and have a high incidence of multiple carcinomas. Defects in at least seven human genes, listed in the accompanying table, can cause XP. Most of the genes have been cloned and found to be homolougs of genes for nucleotide excision repair (NER). In rodents the *ERCC* genes have the same functions as do genes for resistance to ultravi-olet light in yeast *(RAD* genes). This information has elucidated the functions of the XP genes and the basis for this group of human diseases.

The XP-A protein (see table) is a zinc-finger-containing DNA-binding protein, which interacts with human replication protein A. It may also interact with proteins corresponding to the rodent ERCC1 and ERCC4 proteins to form a complex that recognizes DNA defects. Proteins XPC and XPE may be specialized DNA binding proteins for cyclobutane dimers and other photoproducts· The XPG protein is an endonuclease that probably cuts on the 3' side of the DNA defect. In human cell lines with a variety of XP mutations the DNA problems have been corrected by transfer of the corresponding normal genes into the cells.

While defects in protein XPD often cause typical XP symptoms, some defects in the same protein lead to **trichothiodystrophy** (TTD, brittle hair disease). The hair is sulfur deficient, and scaly skin mental retardation, and other symptoms are observed. Like their yeast counterparts (proteins RAD3 and RAD25), XPB and XPD are both DNA helicases. They also constitute distinct subunits of the human transcription factor TFIIHP, which is discussed in. It seems likely that XPD is involved in **transcription-coupled repair** (TCR) of DNA. This is a subpathway of the nucleotide excision repair (NER) pathway, which allows for rapid repair of the transcribed strand of DNA. This is important in tissues such as skin, where the global NER process may be too slow to keep up with the need for rapid protein synthesis. Transcription coupled repair also appears to depend upon proteins CSA and CSB, defects which may result in the rare **cockayne syndrome.** Patients are not only photosensitive but have severe mental and physical retardation including skeletal defects and a wizened appearance.

In a variant form of XP, designated XP–V, nucleotide excision repair is normal, but DNA replication is very slow. Postreplicational translesional repair (bypass repair) is also slow, and patients are cancerprone. The recently discovered DNA polymerase η may be defective.

Children with **ataxia telangiectasia (AT)** have progressive neurological problems, a weak immune system, premature aging, and a high incidence of cancer. Their skin fibroblasts are deficient in the ability to repair X-ray damage, which causes many double-strand breaks: Apparently in this disease cells do not wait until DNA repair has been carried out after exposure to ionizing radiation but attempt to replicate the damaged DNA. The defective protein ATM (ataxia telangiectasia, mutated) has been identified as a large 370-kDa Ser/Thr protein kinase with a carboxyl terminal domain similar to phosphatidylinositol 3-kinase. It appears to play a crucial role in the cell cycle DNA damage check-points by participating in the detection of double-strand breaks and in delay of replication, while they are repaired by homologous recombination. Although AT is an autosomal recessive disease, women heterozygous for the ATM gene have an increased susceptibility to breast cancer. This observation led to the discovery that the proteins encoded by the well known breast cancer genes *Brcal* and *Brcn2* form a complex with ATM. Phosphorylation of BRCA1 by ATM may initiate a signaling pathway through the p53, c-Abl, and Chk2 proteins that cause cell cycle arrest. BRCA1 and BRCA2 also form a complex with protein RAD51, a RecA homolog necessary for homologous recombination. BRCA1 also may be essential to transcription-coupled repair.

There are two major forms of hereditary susceptibility to colon cancer. Familial adenomatous polyposis is caused by defects in the *APC* gene. The more common **hereditary non-polyposis colorectal cancer** (HNPCC), which includes many endometrial, stomach, and urinary tract tumors, results from defects in DNA mismatch repair. The proteins hMSH2 and hMSLl are homologs of the *E.coli* MutS and MutL.

Cells of patients with **Bloom syndrome** (BS) have many chromosome breaks and a high frequency of sister chromatid exchanges, perhaps in an effort to correct these breaks. The body is small but well-proportioned. A somewhat similar disease, the **Werner syndrome** (WS), is associated with premature aging. The Bloom's protein BLM and the WS gene product WRN are both helicases related to *E.coli* RecQ. Protein BLM colocalizes with replication protein A as discrete foci in the meiotic synaptonemal complex. Protein WRN also seems to be associated with DNA replication. Defects appear to increase homologous and illegitimate recombination. Both proteins may also function in transcription.

Many other diseases leading to a high incidence of cancer are known. Among them is the Nijmegan breakage syndrome, in which chromosomes are hypersensitive ro oreakage by ionizing radiation. The gene has been identified[1] by positional cloning, and its protein is apparently involved in repair of double strand breaks.PP **Fanconi anemia, Gardner syndrome,** and hereditary **retinoblastoma** may also involve defects in DNA repair.

Some Human Hereditary Defects of DNA Repair

Human diseease	*Human gene invloved*		*Yeast gene*	*Function*
Xeroderma pigmentosum (XP)				
XP – A	XPA		RAD14	DNA-binding, damage recognition
XP–B	XPB	ERCC3	RAD25(SSL2)	DNA helicase
XP–C	XPC		RAD4	DNA-binding, thymine dimers
XP–D	XPD	ERCC2	RAD3	DNA helicase
XP–E	XPE		RAD16	
XP–F	XPF		RAD10	DNA nuclease complex
XP–G	XPG		RAD2	DNA nuclease
XP–V	XPV		RAD30	DNA polymerase η
Cyckayne sundrome (CS)				
CSA	CSA			Interaction with helicase CSB and with TFIIH
CSB	CSB			DNA helicase
Tricthiodystrophy (TID, brittle hair disease)				
	XPDE			
Ataxia telangiectasia (AT)				
	ATM			Cell cycle delay for repair of ds breaks
Human nonpolyposis colorectal cancer (HNPCC)				
	hMSH2	MSH2	Mismatch repair	
	hMSL1		MSL1	
	GTBP			

10 ARTIFICIAL REARING

Fetal alcohol spectrum disordgers (FASDs) include a range of physical, neurological, and behavioural alterations associated with prenatal alcohol exposure. Studies using animal model systems are conducted to better identify risk factors associated with increased vulnerability to prenatal alcohol, mechanisms by which alcohol disrupts development, and potential interventions/treatments. Numerous animal models have been used, including ovine and nonhuman primate models. Rodents are among the most frequently studied model for fetal alcohol effects. Artificial rearing has been used to examine the effects of alcohol exposure during the third trimester equivalent "brain growth spurt." The brain growth spurt is characterized by rapid brain development and is a period during which the brain is particularly vulnerable to many insults, including alcohol exposure.

In humans, the brain growth spurt begins at the end of the second trimester and continues through the first 2 postnatal years. Rats, however, are altricial relative to humans, and so the brain growth spurt begins after birth. Therefore, to model the brain development that occurs during the third trimester in humans, the rat is exposed to alcohol during the early postnatal period, usually from postnatal days (PD) 4 through 9. Many experimental paradigms have been used to expose rat pups to alcohol during this period, including intragastric intubation and vapour inhalation. Artificial rearing, commonly referred to as the "pup-in-a-cup" technique, was first used for examining the effects of early alcohol exposure by Diaz and Samson. It has been used by several groups to examine the effects of devel-opmental alcohol exposure on brain structure and function, as well as behaviour. The decision to use artificial rearing as an administrative technique depends on several factors, including the level of alcohol exposure, the exposure paradigm (chronic vs binge), and outcome measures.

Exposure to high levels of alcohol in rat pups can inhibit suckling, leading to potential nutritional confounds and changes in maternal-pup interactions. Artificial rearing allows the investigator to control for and manipulate nutritional variables to ensure adequate growth of the pups. In addition, dams may interact with the ethanol-treated pups differently from control pups. The artificial rearing procedure eliminates the impact of such differential interactions on pups' development. However, the major negative aspect of the artificial rearing procedure is the separation of the pup from the mother, which could potentially affect the pup's brain and behavioural development. Therefore, it is important to always include two control groups: an artificially reared control and a normally reared control group.

These control groups will allow the investigator to determine the effects of alcohol as well as any potential effects of artificial rearing. Surprisingly, there are relatively few reported differences between artificially and normally reared subjects. Before using this procedure, various materials, such as the artificial milk diet and surgical tools, must be prepared. For artificial rearing, rat pups are surgically implanted with an intragastric cannula to allow the artificial milk diet to be delivered via a time-controlled infusion pump. Ethanol is mixed with the milk diet and administered to the rat pups directly via the intragastric cannula. The pups remain in a temperature and humidity controlled artificial rearing environment throughout the ethanol treatment period and are then fostered to a lactating dam.

MATERIALS

Milk Diet

Stock Milk Diet

The milk diet is intended to mimic maternal milk, providing the artificially reared pups with a nutritionally balanced diet, similar to that obtained by their suckle control siblings. To prepare approximately 2 liters of diet, the following ingredients are needed:

1. 1500 mL of Carnation evaporated milk.
2. 450 mL of sterilized double-distilled H_2O.
3. 70 g of Purina™Protein 710.
4. 120g(~133mL)of corn oil.
5. 2.0 g of methionine.
6. 1.0g of tryptophan.
7. 10.0g of ICN Nutritional Vitamin Diet Fortification Mix (refrigerated).
8. 11.0 g of calcium phosphate.
9. 0.2 g of deoxycholate acid.
10. To make the mineral mix, add the following: 0.24g of zinc sulfate, 0.24g of cupric sulfate, 0.24 g of ferrous sulfate, 4.0 g of magnesium chloride, 4.0 g of potassium chloride.

Treatment Diet

1. 95% ethanol.
2. Maltose-Dextrin.

Gastrostomy Surgery

1. Stainless steel rods (0.25mm x 76mm, A-M Systems, Inc.)
2. Silastic tubing (small; 0.30mm ID; 0.64mm OD, Dow Corning, Midland, MI).
3. Polyethylene tubing (small; PE 10, 0.28-mm ID; 0.61-mm OD, Clay Adams).
4. Polyethylene tubing (large; PE 50, 0.58-mm ID; 0.965-mm OD, Clay Adams).
5. Thin plastic sheet (e.g., polyethylene sandwich bag).
6. Thick plastic.

Artifical Rearing

1. Flexible plastic tubing for feeding lines (Tygon tubing; 0.60-mm ID, Tygon, Saint-Gobain PPL. Corp., Bridgewater, NJ).
2. Silastic tubing (large; 0.76-mm ID; 1.65-mm OD, Dow Corning, Midland, MI).
3. Syringes.
4. Soldering iron.

Artificial Rearing Components

1. Tank : Throughout the procedure, rat pups are housed in cups that float within a tank [made of stainless steel, Plexiglas, or glass (e.g., a fish tank or other spe-cialty created tank will do)] filled with temperature-controlled water. The temperature of the water is calibrated so that the temperature inside the cups housing the pups is maintained close to 38°C. A cover above the tank may be used to help stabilize the water temperature.

2. Water circulator/heater : The water in the tank is maintained at a constant temperature and circulated, using an immersion water circulator/heater. This provides the cup environment with the desired temperature and offers additional vestibular stimulation to the isolated pups.

3. Pump : An infusion pump able to hold several (8-30) syringes should be located in a slightly elevated position relative to the tank. The elevation helps reduce the possibility of air bubbles entering the feeding lines. The infusion pump should be calibrated with the same syringe size that will be used for the diet feedings and an infusion rate chart should be created. The infusion pump should be periodically calibrated to ensure accuracy.

4. Timer : The activation of the infusion pump is controlled by a programmable timer. Although some infusion pumps come equipped with an internal timer control, this component may need to be added separately

5. Graduated cylinder: A graduated cylinder (accurate to 0.01 mL) is placed next to the tank to measure the actual volume of milk diet delivered.

6. Cups : Pups are individually housed in cups (i.e., clean plastic or Styrofoam 16-oz containers), each covered with a lid containing punched air holes. To keep the cups afloat, each cup is placed within a ballast cup partially filled with sand or other substance to provide enough weight to maintain the cup in an upright position within the circulating water.. Adding a grid of strings to reduce movement of the cups throughout the tank is suggested. Alternatively, the tank can be covered with a Plexiglas (or other material) top with holes for each cup to limit cup movement. To reduce the effects of isolation, the cups are provided with elements that resemble the maternal niche, including cage bedding and a piece of artificial fur attached to the inside of the cup to provide tactile stimulation, simulating maternal contact. Olfactory stimulation with familiar scents can be provided by adding a small amount of bedding from the maternal cage.

METHODS

Milk Diet

Stock Milk Diet

The milk formula used to feed the pups should be prepared before the commencement of the artificial rearing. Typically, the diet is prepared days or even months before being used and kept frozen until needed.

1. To make the mineral mix, add ingredients 10 – 14 in 100mL of sterilized double distilled water. Place on a stir plate at low heat until the minerals dissolve.
2. Add the mineral mix to ingredients 1 – 9 in a large con-tainer for mixing.
3. Blend for 1.5 h.
4. Allow the diet to settle for 0.5 h.
5. Siphon a known quantity of diet using sterile pipettes into sterilized bottles and seal the bottles.
6. Place the bottles in a water bath at 60-65°C for 30min.
7. Place the bottles in the freezer until needed.

Treatment Diet

Ethanol is added to the stock milk diet at a dose determined by the investigator. To control for increased calories associated with ethanol, an isocaloric maltose-dextrin solution is added to the stock milk diet for artificially reared control subjects. That is, 1 mL of the maltose solution contains the same number of calories provided by 1 mL of 95% ethanol. To make isocaloric maltose-dextrin diet:

1. Add 27.415 g of maltose dextrin to 20 mL of double-distilled water.
2. Place the solution on a stir plate with a low level of heat until the maltose is completely dissolved.
3. The maltose solution should be stored in the freezer or refrigerator.

Gastrostomy Surgery

Gastrostomy Surgery Tools

The gastrostomy tools are used for implantation of the intragastric cannula. These tools are not commercially available, but can easily be constructed in the lab. Tools should be stored in an antiseptic substance (70% ethanol) in an airtight container and flushed with sterilized water before use.

Guide wire

A stainless steel rod is covered with the small silastic laboratory tubing. The silastic tubing should be approximately 5-8 mm longer than the wire to cover the sharp end and avoid perforation of the internal organs as the wire is introduced.

Intragastric Cannual

The intragastric cannula is ~15 cm of PE 10 tubing with a small plastic flange at one end. To make this tool, the end of the PE 10 tubing is placed close to a heat source (e.g., soldering iron) until a small portion at the tip of the tubing softens. The melted plastic is immediately softly pressed against a clean cold metal surface, flattening the end without closing off the tube. A circular piece of soft and flexible plastic (~5 mm diameter; can be created with a hole punch) is needled onto the tubing and pressed onto the flat end of the intragastric cannula. The flange and washer prevent the intragastric cannula from coming out of the stomach.

Washers

To anchor the intragastric cannula, a washer is placed outside the abdominal wall, where the cannula exits the pup. The washer consists of a circular piece of soft and flexible plastic (5 mm diameter) similar to the one anchoring the cannula from the inside of the stomach. This washer sits next to the skin and is held in place by a short piece (3 – 4 mm) of PE 50 tubing that fits tight on the PE 10 intragastric cannula. One end of the PE 50 tubing is heated to create a flange similar to that on the intragastric cannula. In preparation for the surgery, the washer is needled onto a stainless steel rod. One of the neck washers should be placed on the same rod for greater efficiency during the gastrostomy surgery. This tool is referred to as the "*washers*" tool.

Neck Washers

Two pieces of thick plastic (-5 mm) without sharp edges are needed for each pup. To prepare for surgeries, several washers can be pierced onto a 21-gauge needle to facilitate their use during the surgical procedure.

Gastrostomy Surgery

All artificially reared subjects will undergo gastrostomy surgery, whereas suckle controls will be separated from the dam during the surgery time and then replaced with a foster dam. Litter size of the foster dam should be kept constant by including nonexperimental rat pups.

1. Subjects are anesthetized via halothane or isoflourane. Young pups are some-what resistant to anesthesia and must be carefully monitored.
2. Once the pup is fully anesthetized, hold the pup loosely by the sides of the head, with the nose facing up, to create a clear and straight path through the esophagus to the stomach.
3. Dip the silastic-covered guide wire into corn oil'for lubrication. Slowly place the silastic-covered guide wire down the esophagus, ensuring that the guidewire follows the dorsal side of the airway so that the wire enters the esophagus and not the trachea toward the lungs. If strong resistance is experienced, it is likely that the wire has been misplaced.
4. Once the guide wire is in the stomach (the stomach can be visible via the milk band), pressure is placed on the wire to puncture through the abdominal wall, so one end of the guide wire protrudes from the mouth and the other from the abdominal wall.
5. Attach the intragastric cannula to the end of the guide wire protruding from the mouth (this can also be done prior to the placement of the guidewire into the stomach). Ensure that the intragastric cannula overlaps with the silastic on the guide wire; otherwise, it may fall off of the guide wire during the procedure.
6. Slowly pull the guide wire and intragastric cannula from the abdominal wall. The washer at the flange end of the intragastric cannula will collapse; however, some slight resistance may be felt when the flange enters the esophagus and when the flange enters the stomach.

7. Once in the stomach, the outline of the washer on the intragastric cannula should be visible in the stomach.
8. Place antibiotic cream on the abdominal wound.
9. Remove the guide wire from the intragastric cannula and replace with the "washers" wire.
10. Slide the washers onto the intragastric cannula and slide to the outside of the abdominal wall Ensure that the washers keep the intragastric cannula in place, but are not putting undue pressure on the abdomen.
12. Slide the neck washer onto the gastrostomy tube.
13. Wipe the back of the neck with antiseptic.
14. The intragastric cannula will be placed through the skin on the neck to minimize the likelihood that the tube is pulled out of the stomach as the pup moves around. Pull the skin up from the back of the neck and use the "*washers*" wire to puncture through the skin.
15. Place the final neck washer onto the wire and onto the intragastric cannula. Ensure that the neck washers do not pinch the skin.
16. The distance between the washers at the abdominal wall and the neck needs to be long enough to allow full mobility of the rat pup, but not too long so that the tube can get easily caught and pulled. This distance will need to be' adjusted as the rat pup grows.
17. After gastrostomy surgery, subjects are allowed to recover in an appropriately heated environment (e.g., a heating pad) and are then placed in the artificial rearing environment. A piece of large Silastic tubing (~ 3 cm long) is placed on the end of the intragastric cannula to prevent twisting and kinking of the intra-gastric cannula as the subject moves around.

Artificial Rearing

The pups are maintained in the artificial rearing environment during the ethanol treatment and typically for 2 additional days to allow the pups to go through any withdrawal before fostered back to a lactating dam. The pups will be housed in the cups within the artificial rearing apparatus and fed automatically.

Preparation

1. The artificial rearing environment should be prepared the night before use. The tank should be filled with water and warmed to the appropriate temperature so that the inside of the cup is 38°C.
2. Place the cups with fur in the tank so temperature is stabilized.
3. Feeding lines that will deliver the milk diet from the syringe on the pump to the intragastric cannulae are prepared. Each feeding line comprises a length (1.1.5m) of flexible plastic tubing (Tygon tubing) connected to a needle (21-gauge that will be placed on the milk syringes. Feeding lines should be stored in an antiseptic at least overnight and flushed with sterilized water before use.

Determination of Diet Infusion Rate

1. Each morning, pups are weighed and the mean body weight of the pups in the artificial rearing environment is determined and recorded. Suckle control subjects should also be weighed each morning, but not included in the mean.
2. The amount of milk diet to be delivered within a 24-h period (12 feedings; a feeding every 2h) is determined as follows: Mean body weight (g) × 0.33 = mL of diet/day. To determine the amount of milk diet to be delivered during each feeding, divide this number by 12. Refer to the pump calibration chart to determine the infusion rate.

Delivery of Milk Diet

The milk diet is delivered into each intragastric cannula every 2h via the timer-controlled infusion pump. Syringes filled with milk are typically changed twice a day, but this will depend on the alcohol treatment protocol.

1. Remove the milk diet from the refrigerator and warm to room temperature.
2. Record the amount of milk infused during the previous delivery period, as measured by the water in the graduated cylinder.
3. Disconnect pups from the feeding lines and remove the syringes from the infu-sion pump.
4. Rinse each feeding line with approximately 2 mL of sterilized water.
5. Fill syringes with the appropriate diet, color coding the syringes if more than one treatment diet is being delivered. Fill the syringe with at least 2 extra mL of milk to allow for priming.
6. Place the syringes in the pump and attach the feeding lines (attach the needles securely, but do not overtighten). Center the syringes evenly on the pump to ensure that pressure on the syringes is evenly distributed. Fill an additional syringe with water to measure the volume actually delivered.
7. Prime the feeding lines so they are filled with milk diet and ensure that there are no air bubbles present in the syringes or feeding lines.
8. Set the pump according to the appropriate infusion rate for the day.
9. Place the feeding line from the water syringe into the dry graduated cylinder to measure the actual volume of infusion.
10. Measure and record water temperature and the temperature inside the cups.
11. Re-connect the feeding lines to the intragastric. cannulae. The feeding line should press-fit and stay attached to the intragastric cannula.

Care of Artificially Reared Pups

The investigator is responsible for the care of the pups; thus, it is critical that the pups be checked regularly to ensure their health and comfort. For example, the pups' ano-genital area should be gently stroked to facilitate urination and defecation. In addition, every morning and afternoon, pups will be bathed and the intragastric cannulae flushed with sterilized water to keep them patent. Typically, bathing will co-occur with the changing of milk syringes.

1. Fill a cup with clean warm water (~38°C) for bathing the pups. Dip the lower half of each pup in the cup of water and wipe dry with a soft tissue. Gently stimu-late the ano-genital region with a cotton ball to facilitate urination and defecation.
2. Flush each intragastric cannula with 0.1 mL of sterilized water. The water should flow easily. If the cannula gets clogged, insert a small wire into the cannula and then try flushing with water again.
3. Return the pup to the artificial rearing cup. Attach the feeding line to the intragastric cannula of each pup. The feeding line should pressfit and stay attached to the intragastric cannula. Check to ensure that each pup is getting the appropriate diet.
4. Ensure that the lid on the cup is securely attached.

Fostering of Pups to Lactating Dam

Once the artificial rearing period is complete, pups are removed from the artificial rearing environment and fostered to a lactating dam.

1. Intragastric cannulae are trimmed close to the abdominal wall and sealed with a soldering iron.
2. A small amount of milk diet is applied in the mouth via silastic tubing to stimu-late suckling behaviour.
3. To minimize possible maternal rejection, pups are covered in a slurry of the mother's feces before being placed in the dam's cage.

NOTES

1. Various research groups have used slightly different diets for artificial rearing. A commonly used milk diet is that proposed originally by West et al as a modification from the one used in earlier studies. This milk formula includes more water and protein than the original one, aiming to increase the body weight gain in artificially reared rat pups. Wainwright et al. who have examined the effects of varying dietary components in artificial milk, use a slightly different milk stock solution and report no significant differences in body weight gain between artificially reared and normally reared pups. Thus, although not used so extensively in the literature, this milk diet should also be considered.
2. If the tools are left in the 70% ethanol for a period of months, the plastic may shrink or harden; thus, it is suggested that they be stored in the ethanol for no more than one month.
3. Alternatively, washers can be created with PE 50 tubing.
4. Some investigators may decide to have the suckle controls undergo a sham surgery.
5. Occasionally, during the gastrostomy surgery, the guide wire pokes through the silastic covering when pressed out of the stomach. If this occurs, massaging the area near the exit wound while gently pushing downward on the silastic at the top of the guide wire usually will cause the silastic to emerge. The silastic can not be simply taken off of the guide wire, as the press-fit of the silastic holds the intragastric cannula onto the guide wire. If this occurs, the silastic covering on the guide wire should be replaced before the next surgery, as it will likely re-occur.
6. Many investigators set the pumps to deliver the milk formula every 2 h for a 20-min delivery period, but this may vary according to the experimental design.
7. Some investigators will begin artificial rearing with a lower infusion rate

8. Occasionally, an individual pup may fail to gain weight or even lose weight. This typically occurs if the intragastric cannula becomes unhooked from the feeding line during the evening hours. The exclusion of these subjects from the study is a decision that needs to be made based on the particular conditions and history of the pups under consideration. If the subject remains in the study, the body weight should not be added to the mean as it will slow growth of all pups on the pump.
9. Some investigators have kept the milk syringes in a refrigerator during the milk infusion, allowing the milk to warm as it flows through the feeding lines.
10. If, as the pup matures and becomes more active, the feeding line becomes unhooked, the connection may be taped.
11. While in the home cage, the dam licks the pups, stimulating' their ano-genital area to induce urination and defecation. The pups in isolation will have difficulty excreting by themselves. Softly stimulating the anogenital area with a cotton ball or a piece of soft tissue wet with warm water will stimulate excretion. This task needs to be performed with regularity (at least twice a day). Some investigators stimulate the pups three to four times a day, whereas others only once a day when the pups gets older.
12. Artificial rearing should be monitored periodically throughout the day. One should check that feeding lines do not become disconnected from the intragas-tric cannulae, that the pump is accurately set and 'operational, and that the tem-perature is stable. Pups should be checked to monitor health and to ensure that the cannulae and/or feeding lines do not become twisted.
13. Occasionally, pups will become bloated. Bloating may be related to etha-nol-induced reductions in gastric motility or for other reasons. Ano-genital stimulation is critical for allowing the pups to urinate and defecate. In addition, bloating may occur if the temperature within the environment is reduced. If a subject becomes bloated, a warm bath may help stimulate defecation.
14. Some investigators apply cornstarch baby powder to the anogenital region of the pup after bathing to reduce irritation during urination.
15. As the rat pups get older, they become more active and it is important that the lid is on securely so that the pup can not exit the artificial rearing cup.

11 ETHANOL IN CULTURE

Studies relying on animal models demonstrate that the development and organization of the brain are profoundly effected by ethanol. In vivo models identify targets of ethanol action and provide context to the effects, however, they are limited in their utility as a substrate for exploring molecular, intracellular, and intercellular mechanisms. In vitro systems lend themselves to such analyses because the composition of the culture, the environment of the cells, and the manipulation of the environment can be carefully controlled. This is particularly important in examinations of growth factor actions, intracellular signal transduction, and genetic manipulations such as electroporation and blocking RNAs.

TWO CULTURE MODELS

Two types of cultures have been particularly useful in alcohol studies. These are 1) primary cultures in which samples are derived directly from animals and 2) cultures of cell lines (e.g., cancer cells and genetically engineered/selected -lines). The present chapter discusses the primary cultures. Various types of primary cultures have been used in alcohol studies: dissociated cell cultures and organotypic explants. Each has its uses and limitations.

Dissociated Cell Cultures

Primary cell cultures are composed of neurons, glia, or stem/progenitor cells. The constituents are relatively homogeneous, nontransformed cells. Common cell types used in these cultures are cortical neurons, cerebellar granule neurons, and astrocytes. These cultures are enriched by various procedures so that the cell of choice comprises 90-95% of the cultured cells.

The enrichment method varies with the type of cell to be examined, and the present chapter describes the method tor obtaining primary cultures of cortical neurons. For example, procedures used for generating cultures of astrocytes and cerebellar granule neurons can be found elsewhere. Dissociated cultures can be criticized for their nonphysiological arrangement, i.e., removing cells from their normal microenvironments; Neurons are intimately dependent on interactions with other cells, thus, it becomes particularly important to preserve in vivo cellular organization before extrapolating the effects of exogenous substances on neuronal development. This need drove investigators to develop systems in which quasi-normal relationships are maintained organotypic (slice) cultures.

Organotypic Slice Cultures

Many studies have focused on the developing cerebral cortex. Neurons in the cer-ebral cortex are derived from two proliferative zones, the ventricular zone (VZ) 10 Generation and Use of Primary Rat Cultures for Studies of the Effects of Ethanol and the *subventricular zone* (SZ).

Intrinsic activities of cells in the proliferative zones proceed normally in slice preparations. This includes interkinetic nuclear migration of VZ cells and the migration of postmitotic cells. Indeed, these metrics describing these activities are similar in slices and in vivo preparations. Neurogenesis in the neocortical VZ is influenced by both intrinsic intracellular factors and extrinsic intercellular factors. Using such explants has provided unique insight into the mechanism of ethahol toxicity. Chief criticisms of the cortical explant model are that not all relationships in the developing cortex are maintained.

The formation of the slice is likely to disrupt paths of migration or axonal trajectories. Long-distance projections such as the ascending monoaminergic and thalamic afferents are often severed in the production of the slice. Likewise, under normal circumstances, many cortical local circuit neurons are generated in another site, the ganglionic eminence (GE). In generating the slices, either the GE is eliminated from the preparation or the path-ways connecting the GE to the developing cortex may be eliminated. Thus, the contribution from the young local circuit neurons could be eliminated from the developing cortex of the slice.

A consequence of producing the slice is that cells on the surface of the slice are damaged. The response is the death of many of these cells and in the formation of a glial coat. These can lead to an altered microenvironment of cells near the surface of the slice and possibly a change in the microenvironment of other strata of the slice. Conceivably, the increased amount of glia are elaborating growth factors that are expressed in higher amount than would be expressed in vivo.

MATERIALS

Animals

Timed pregnant Sprague-Dawley rats: Depending upon the issue to be addressed, rats may be anywhere from 13 to 19 days pregnant. Commonly, fetal rats are harvested on gestational (G) day 16. The first day a sperm-positive plug is detected is designated as G1.

Primary Cultures of Cortical Neurons

Surgical Equipment

1. Operating scissors.
2. Microdissecting scissors.
3. Microdissecting forceps.
4. Microdissecting tweezers #5 (× 2).
5. Spatula.
6. 60-mm or 10-mm round tissue culture dishes.

7. Dissecting microscope.
8. 1.5-mL microcentrifuge tubes.
9. Glass aspirating pipets.
10. Alcohol burner.

Reagents

1. Hank's balanced salt solution (HBSS, Gibco).
2. 70% ethanol in water.
3. Poly-D-lysine molecular (PDL) weight 70 – 100kDa (Sigma, St. Louis, MO) is dissolved in sterile water at 10 mg/mL to make a 100 x stock solution that is aliquoted and frozen at –20°C. Working solutions of PDL are prepared by dilution in phosphate-buffered saline (PBS, Gibco), to a final concentration of 100 μg/mL.
4. Eagle's Minimal Essential Medium (MEM) containing glutamine (Gibco) and supplemented with 10% fetal bovine serum (FBS, Gibco), 33 mM dextrose, and penicillin/streptomycin (Gibco).

Culturing Methods

Preparation of the Culture Dishes

1. Coat the experimental tissue culture plates with PDL sufficient to cover the plating surface for a minimum of one hour at room temperature.
2. Chill two 100-mm tissue culture plates containing 10mL each of HBSS on ice.
3. Chill a 1.5-mL centrifuge tube containing 10mL of HBSS on ice.
4. Sterilize surgical instruments in 70% ethanol.

Dissection of the Rat Cortices

1. Anesthetize the timed pregnant G16 dams with a cocktail of ketamine (1.0 mg/kg) and xylazine (1.0 mg/kg). Open the abdominal cavity to expose the uterine horns and extract the pups, one at a time.
2. Hold the pup between the thumb and index finger and cut a shallow incision from the upper spine to the base of the brain, and then laterally around the head from the base of the brain to the eye. Use the forceps to peel the skin and skull from left to right up over the head and expose the brain. Scoop out the entire brain with a spatula and place in one of the culture dishes with chilled HBSS to rinse blood away. Repeat this pro-cedure for all pups. Transfer the brains to a fresh culture dish with chilled HBSS.
3. Under the dissecting scope, use the sharp #5 tweezers to remove the meninges and vasculature from the cortex as fully as possible. After removal of the meninges, use the tweezers to pinch off the cortical wall by removal of the olfactory bulb, the striatum, and the hippocampus.
4. Place all cortices in a 1.5-mL centrifuge tube with BBSS. Use sterile glass pipets to gently triturate the cortices and dissociate the cells.

5. After the initial cell dissociation, allow heavier fragments (meninges, tissue clumps) to settle for a few minutes to the bottom of the conical tube. Carefully transfer the cell suspension to a 15 mL conical tube leaving the fragments in the bottom of the centrifuge tube. For optimal cell yield, add 1 mL of BBSS to the fragments and repeat the trituration, breaking up as many tissue chucks as possible, while leaving any meninges present following dissection to settle out of the cell suspension. Transfer the cell suspension to the 15 mL conical tube.
6. Centrifuge the cells at 600 x g for 3 min to pellet cells. Add 2.0 mL of 10% serum containing MEM and resuspend cells. Dilute the cell suspension to a concentra-tion suitable for counting.

Yield and Plating of the Cells

The total number of primary neuronal cells recovered after the dissociation of G16 cortices is highly dependent on a variety of factors. These factors include the number of pups in utero, the extent to which areas other than the cortical wall are included for culture, and the quality of dissociation. Two pregnant dams with 12 pups each generate approximately 100 million cells with considerable flux depend-ent on the variables stated. This cell yield would result in approximately four to six 100-mm tissue culture plates, if treatment with growth factors and ethanol with intent to generate a protein lysate, is planned. Plating 2 x 10^7 cells per 100mm tis-sue culture dish yields more than 1.0 mg of total protein and is approximately 70% confluent.

An Alternative Method for Obtaining Primary Cultures of Cortical Neurons *Reagents*

1. HEBSS: 25mM HEPES, 13.8rMMaCl, 5.0nM KCl, 4.2mM $NaHCO_3$, 1.0mAf NaH_2PO_4, 0.01% phenol red.
2. $MgSO_4$ stock solution (3.82%): 1.14g $MgSO_4$ in 30mL of sterile water.
3. Solution A: 50 mL of HEBSS containing 2.5mg/ml dextrose, 0.30% bovine serum albumin fraction V (BSA), and 0.50 mL of $MgSO_4$ stock solution.
4. Solution B: 12.5mL of HEBSS containing 4.0% BSA and 0.10 mL of $MgSO_4$ stock solution.
5. Solution C: 10mL of Solution A containing 2.5mg trypsin (1000-1500U/mg; Sigma).
6. Solution D: 10mL of Solution A containing 0.10ml $MgSO_4$ stock solution, 7.5mg trypsin inhibitor (Sigma), 130kU/ml DNase (Sigma).
7. Solution E: Add 8.4 mL of Solution A to 1.6ml of Solution D.

Methods

1. Prepare all solutions described in Subheading 2.3 (A through E) before perform-ing dissection. Filter Solutions A to E through a syringe with a 0.22-io,m filter attached.
2. Dissect out the cerebral cortices of all pups with a spatula and place the corti-cal tissue in a 1.5-mL microcentrifuge tube containing 1.0 mL of solution A.3. In a sterile hood, carefully pour the contents of the tube onto the frosted side of a sterile frosted glass plate. Use a sterile razor blade to homogenize tissue.

3. Pipet the homogenized tissue into a 15-mL conical tube. Add Solution C to the conical tube and place in a hot water bath at 37°C for 15 min. Swirl occasionally to mix.
4. Add Solution E to the conical tube and swirl gently. Centrifuge at 2000rpm for 5 s to pellet cells.
6. Decant the supernatant and add Solution D. Triturate 10-15 times with a pipet for maximum cell yield. Let the solution sit for 5 min to allow debris and clumps to settle. Remove the supernatant and place in a fresh 15-mL conical tube.
7. Add Solution B to the bottom of the conical tube. There will be a visible separa-tion between Solution B and the cell-rich supernatant. Centrifuge at 2,000 rpm for 5 min to pellet the cells.
8. Decant the supernatant and add MEM medium containing 10% serum. Mix gen-tly and triturate to resuspend the cells for counting.
9. Plate cells

Organotypic Slice Cultures

Equipment

1. Macllwain Tissue Chopper.
2. Razor blade.
3. Plastic disc for tissue chopper.
4. 70% ethanol.
5. Tube of commercial Krazy Glue.
6. Glass Pasteur pipet.
7. 4-inch microdissecting forceps with 0.8-mm tip width.
8. Fine-tip paintbrush with trimmed bristles.
9. Spatula.
10. Plastic transfer pipet.
11. Ice.
12. 60-mm tissue culture dishes (Falcon, Lincoln Park, NY).
13. Millicell filter inserts with 0.40 uAf pores.
14. Plastic container with lid.

Reagents

1. Neurobasal medium (Gibco, Carlsbad, CA) supplemented with B27 (Gibco), 200 μ*M* L-glutamine (Gibco), 100 μ*M* penicillin/streptomycin (Gibco), and 100 *mM* dextrose.
2. Krebs buffer (10x) containing'Nad (73.6g/L), KC1 (1.87g/L), NaH_2PO_4–H_2O (1.66g/L), MgCL, (2.44g/L), CaCl, (3.68 g/L). This can be stored at 4°C for 2-3 months.

3. Krebs buffer with additives containing 10 × Krebs buffer (final concentration): 126mM NaCl, 2.5 mM KCl, 1.2mM NaH_2PO_4, 1.2mM MgCL,, 2.5mA/ $CaCl_2$), 11 mM dextrose, 25 mM $NaHCO_3$, lOmM HEPES, double-distilled H_2O.
4. 4.0% (w/v) paraformaldehyde in PBS (pH 7.4) at room temperature, freshly prepared.

Methods

Dissection

Pregnant C57/B6 mice are anesthetized on G15 or G17 with a cocktail of keta-mine (1.0mg/kg) and xylazine (1.0mg/kg). The fetuses are collected and their brains are removed. The forebrains are dissected out with forceps and a spatula and placed into 60-mm tissue culture dishes containing Krebs buffer with additives on ice.

Preparation and Culture of Mouse Cortical Slices

A brain is removed from a tissue culture dish and prepared for cutting. Krazy glue is used to affix the rear of the brain to the plastic disc to prevent movement while slicing. A reservoir of Krebs buffer with additives is placed around the perimeter of the brain to allow slices to float away from the uncut brain during the slicing process. A trimmed paintbrush, which allows more control for gentle manipulation, is used to gently push down on and flatten out each hemisphere to ensure a uniform thickness of slices. The brain is cut into 300- to 400 μm coronal slices with the Macllwain Tissue Chopper. The slices are gently retrieved with a plastic transfer pipet containing Krebs with additives and transferred to the original Krebs buffer containing 60mm tissue culture dish. Approximately six coronal slices are obtained per fetal brain.

Slices from each brain are transferred to 3.0mL of neurobasal medium supple-mented with B27, L-glutamine, penicillin/streptomycin, and dextrose, as indicated previously, in a 60-mm tissue culture dish containing a 0.40 μ*M* Millicell filter insert. Slices are carefully arranged flat with forceps in the center of the insert. The residual Krebs buffer remaining in the insert, following the transfer of the slices, is slowly removed and replaced with neurobasal medium from the surrounding dish. The level of medium is then lowered to form a meniscus just above the slices. Slices are cultured in an incubator for 2h at 37°C with 95% O_2 / 5.0% CO_2 to allow for recovery before experimental treatments.

Preparation and Culturing of Rat Cortical Slices

There are slight variations in the preparation and culturing of rat cortical slices that are worthy of mention. Rat cortical slices do not tolerate Krebs buffer well. Therefore, to preserve the integrity of the cortical slices, artificial cerebrospinal fluid (ACSF) is substituted for Krebs. Brains are dissected out of fetuses and placed in ACSF containing 124mM NaCl, 5.0mM KCl, 1.23mM NaH_2PO_4~H_2O, 18.5mMMgSO_4, 2.0mM $CaCl_2$, 10mM dextrose, and 26 mM $NaCO_3$ in sterile water. Slices are cultured in neurobasal medium containing L-glutamine, penicillin/streptomycin and 20% FBS. Neurobasal medium supplemented with B27 or N2, 5.0mM L-glutamine, and peni-cillin/streptomycin without serum. After 2h, the serum-free medium is replaced with neurobasal medium supplemented with B27 or N2, 5.OmM L-glutamine and penicil-lin/streptomycin without serum. For shorter ethanol treatments (less than 90min), slices are cultured in a serum-free medium for the duration of the experiment.

Controlled Ethanol Exposure

A potential confounding factor for studies of ethanol exposure is the volatility of ethanol. This problem is overcome by incubating primary cultures within a sealed chamber containing ethanol. These chambers maintain a consistent vapor pressure established by ethanol evaporation, prevent loss of ethanol from the culture medium, and can stabilize the concentration of ethanol for more than 72 h.

Equipment

1. 1.6-liter Rubbermaid "Flex and Seal" food storage containers.
2. "Picnic table" plastic 24-slot microcentrifuge tube rack with the handles snapped off to make a flat surface for incubation of tissue culture dishes and plates.
3. Syringe (60 mL) with 6-inch tubing attached for CO_2 injection.
4. GM7 Micro-Stat Analyzer

Reagents

1. Medium with MEM supplemented with 33mM dextrose, penicillin/streptomycin, and serum as needed.
2. 95% ethanol.
3. Compressed carbon dioxide gas (CO_2).

Methods

1. Set up the chambers prior to initiating ethanol treatment. Each experimental concen-tration of ethanol requires a separate chamber. For example, to treat neuronal cultures with ethanol at 0,200, and 400mg/dl, a minimum of three chambers is required. Add 200 mL of water to each chamber and set a rack inside the chamber. The 1.60-liter chamber system described here holds up to two 100-mm culture dishes per rack.
2. The amount of ethanol added to the treatment medium (in ij,L/mL) is determined by the following equation: (molar concentration of ethanol treatment) x (46.07 g/ M)/ (ethanol concentration = 0.95) x (specific gravity of ethanol = 0.7750 g/mL). For exposure to 400mg/dL (86.6mA/) of ethanol, 5.43 uL/mL of 95% ethanol is added to both the culture medium and the water in the chamber (AUO NuUr4). Cell cultures not subject to ethanol exposure are incubated in identical chambers, but without ethanol addition.
3. Place the culture dishes securely on the rack in the ethanol concentration appro-priate chamber. Make sure the dishes are level and that medium adequately cov-ers the cells. Position the lid on the chamber and secure all but one corner. Fill a syringe with 60 mL of CO_2 by using a short tubing attachment to connect to the valve on the CO_2 tank. Pinch off the tubing with a hemostat and, removing the tubing from the valve, insert the tubing into the unsecured corner of the chamber. Release the constriction

on the tubing and plunge the CO_2 from the syringe into the closed chamber. Quickly close the open corner to tightly seal off the chamber and place the chamber in the incubator. These instructions assume the use of a 37°C incubator with 6.0% CO_2 and saturating humidity.

APPLICATION OF EXOGENOUS SUBSTANCES AND HARVESTING -THE CELLS/TISSUE

Fresh Samples

Primary neuronal cell cultures or organotypic slice cultures are easily manipulated for a variety of biochemical assays. Growth factors, biological inhibitors, or other modulatory substances may be added to the medium alone, or in combination with ethanol, using the chambered treatment system. Conditioned medium from the cultures may be saved for detection of cell released cytokines by enzyme-linked immunosorbant assay (ELISA).

Brief exposure of cell cultures to 0.25% trypsin/EDTA (Gibco) generates a dissociated cell suspension suitable to process for cell sorting by flow cytometry. Protein expression profiles from slice cultures or dissociated primary cultures may be deter-mined by Western immunoblot or ELISA after lysis in an appropriate lysis buffer.

Fixed Samples

Cortical slices are best prepared for immunohistochemistry by fixation. Slices are tixed in freshly made 4% paraformaldehyde for 35 min at room temperature after experimental treatments. Fixative is gently removed from the filter and culture dish and slices are washed once with 0.10MPBS. Slices are transferred to a 12-well plate (each well labeled with the original treatment condition) containing 10% sucrose and incubated for 24 h. We have found it useful to use a spatula with the tip bent to a 90° angle to remove slicǝs from the culture dish and transfer them to the 12-well plate. After 24 h in 10% sucrose, slices are incubated in 15% sucrose for at least 3 h before freezing in prerjaration for cutting. Qn preparation for cryosectioning, slicesare removed from the sucrose and trans-ferred to a precleaned Superfrosti'/Plus microscope slide.

Typically three slices or groups of slices are placed on each slide. Sucrose is aspirated off as much as possible. Slices are covered with just enough Optimal Cutting Temperature (OCT; Sakura Finetechnical, Tokyo, Japan) embedding medium to allow for complete coverage of the slices. The slide is pushed down firmly into a flat block of dry ice until completely frozen. Slides are wrapped in aluminum foil and placed in –20°C freezer until ready for cryosectioning in 12μm increments.

We have found that slice integrity and antigenicity diminish if left in the freezer for longer than a week. Before immunonistochemistry, sections are incubated in U.01UM citric acid in sterile ddH_2O (pH 6.0) for 20min followed by two rinses in PBS. Standard protocols are then used for the remaining immunolabeling process.

NOTES

1. Samples of neurons from pups and older animals are less likely to be viable, though allowances (mostly in the composition of the medium) can be made.
2. This takes practice since the cortex is thin and will tend to peel away with the meninges. One method that seems to work well is to start at the olfactory bulbs or the thicker striatal areas.
3. Overly vigorous trituration will result in low cell recovery. Heat the glass pipette tip over an alcohol burner to serially narrow the pipet tip for complete dissocia-tion of cells (cool the tip before triturating).
4. It is recommended that the concentration of ethanol in the medium be measured with an Analox GM7 Micro-Stat Analyzer (or another similar instrument) before and after treatment. Ethanol concentrations will vary with chamber selec-tion and culture conditions and may need to be optimized experimentally. The calculation given for ethanol addition is intended as a starting point to achieve the desired ethanol concentration. In practice, the mathematically determined ethanol concentration generally results in a lower measured ethanol concentra-tion for the chambers described. Increasing the ethanol added to the chamber water and the treatment medium by 10-20% attains the desired ethanol concen-tration. Samples collected for ethanol measurement may be frozen at -20°C for several weeks without detectable loss of ethanol.

12 APOPTOSIS ANALYSIS

CYP2E1, an ethanol-inducible form of P450, metabolizes and activates many toxicologically important substrates, including ethanol, carbon tetrachloride, acetaminophen, and N-nitrosodimethylamine to more toxic products. Induction of cytochrome P450 2E1 by ethanol appears to be one of the central pathways by which ethanol generates a state of oxidative stress.

In the intragastric model of ethanol feeding, significant alcohol injury occurs. In these models, large increases in lipid peroxidation have been observed, and the ethanol-induced liver pathology has been shown to correlate with CYP2E1 levels and elevated lipid per-oxidation, and to be blocked by inhibitors of CYP2E1. Although there is disagreement of the role of CYP2E1 in alcohol liver injury, convincing evidence has been presented that CYP2E1 in the hepatocyte is a key source of reactive oxygen production and in DNA damage. Anapproach that our laboratory has used to try to understand basic effects and actions of CYP2E1 is to establish cell lines that constitutively express human CYP2E1 . HepG2 cell lines, which overexpress CYP2E1 , were established either by retroviral infection methods (MV2E1-9 cells, or E9 cells) or by plasmid transfec-tion methods (E47 cells).

Hepatotoxins such as CC14 or acetaminophen were more toxic in E9 cells than control HepG2 cells validating the utility of the model to study CYP2E1 -dependent toxicity. We have characterized the toxicity of ethanol, polyunsaturated fatty acids (PUFA) such as arachidonic acid (AA) and iron in the E9 and E47 cell lines. Concentrations of ethanol or AA or iron that were toxic to the CYP2E1 -expressing cells had no effect on control HepG2 cells not expressing CYP2Elor to HepG2 cells expressing a different P450, CYP3A4 (3A4 cells). Toxicity to CYP2E1 -expressing cells was found when glu-tathione (GSH) was depleted by treatment with 1-buthionine sulfoximine. Inhibitors of CYP2E1 prevented the toxicity by the aforementioned treatments. Antioxidants such as vitamin E, trolox, and catalase also prevented toxicity. The aforementioned treatments of CYP2E1 -expressing cells with ethanol, AA, iron, or BSO resulted in an increase in oxidative stress to the cells

In other experiments, damage to mitochondria, e.g., decline in mitochondrial membrane potential, was found in CYP2E1-expressing cells treated with AA or iron or BSO. Please refer to Caro and others for recent reviews of this work and for CYP2E1 biochemistry and toxicology. This chapter describes protocols used with the HepG2 cells developed to express CYP2E1 to study CYP2E1-dependent toxicity, mitochondrial damage, and oxidant stress. Assays for CYP2E1 expression, for loss of cell viability, for apoptosis, for decline in mitochondrial membrane potential, for lipid peroxidation and generation of reactive oxygen species (ROS), and for activation of mitogen activated protein kinases (MAP kinases) when the CYP2E1-expressing HepG2 cells are challenged with pro-oxidants will be described herein.

MATERIALS

Unless otherwise mentioned, solutions are prepared in distilled, deionized water and, except for buffers, prepared fresh.

Cell Line Maintenance ana Storage

1. Dulbecco's Modified Eagle's Medium (DMEM, Sigma, St Louis, MO), supplemented with 10% fetal bovine serum (FBS, Sigma, St Louis, MO) and 1% penicillin-streptomycin-glutamine (GIBCO, Grand Island, NY).
2. Trypsin (0.25%) and ethylenediamine tetraacetic acid (EDTA; 1 *mM)* from GIBCO.
3. G418 antibiotic (GIBCO, Grand Island, NY).
4. Cell freezing medium, DMEM containing 20% FBS, 10% dimethyl sulfoxide (DMSO, Sigma), 1% penicillin-streptomycin-glutamine mixture.

CYP2E1 Oxidation Activity

1. 1*M* KH_2PO_4/K_2HPO_4 buffer, pH 7.4.
2. 10 mM/p-nitrophenol (FLUKA) in water.
3. 10 mM NADPH (Sigma) in water.
4. 20 % (W/V) Trichloroactic acid (Sigma).
5. 10N NaOH.

Cell Toxicity

1. MTT Test Solution, Thiazolyl' Blue Tetrazolium (Sigma), 3.3mg/mL in PBS, filter before use. N-propanol (Sigma).
2. LDH test solution: A. 50m*M* KH_2PO_4/K_2HPO_4, pH. 7.3; B. NADH, 4.1 mg/mL in solution A; C. pyruvic acid, 2.1 mg/mL in solution A. In 100mL LDH Test Solution: Solution A 94mL, Solution B 3mL, Solution C 3mL.
3. Trypan Blue Exclusion Test, 0.4 % Trypan Blue Solution (Sigma).

Measurement of Malondialdehyde and ROS

1. Stock TCA-TBA-HCl reagent: 15% w/v trichloroacetic acid, 0.375 % w/v thio-barbituric acid, 0.25 N hydrochloric acid. 0.1m*M* butylated hydroxy toluene (BHT). This solution may be mildly heated to assist in the dissolution of the thiobarbituric acid.
2. Dichlorofluorescin diacetate: 5mA/ stock solution in DMSO (DCF-DA).

Determination of GSH

1. Buffer A, 0.1 M sodium phosphate, pH 7.4, *5µM* EDTA, 0.6m*M* 5,5'-dithiobis (2-Nitrobenzoic acid) (DTNB, Sigma), 0.2m*M* NADPH (Sigma), pH 7.5.
2. Buffer B, 10U/mL glutathione reductase (Roche) in 0.1 *M* sodium phosphate buffer, pH 7.5.
3. 10% trichloroacetic acid (TCA).
4. Reduced-form GSH (Sigma).

Apoptosis Analysis

1. Cell lysis buffer: 10mM Tris-HCl, pH 7.5, 10m*M* NaCl, 10m*M* EDTA, 100 µg/ mL proteinase K, and 0.5 % sodium dodecyl sulfate.
2. Chloroform/isoamyl alcohol (24:1) (Sigma), 0.1 µ.g/mL RNase (Sigma), µg/mL ethidium bromide (Sigma),
3. Rhodamine 123 (5 µg/mL final), propidium iodide (25 µg/mL final) (Sigma).

Caspase Activity

1. Lysis buffer: 50m*M* Tris HCl, pH 7.4, 50m*M* β-glycerophosphate, 15mM $MgCl_2$, 15m*M* EDTA, 10 µ*M* phenylmethy sulfonyl fluoride; ImA/dithiothretol; 150 µg/mL digitonin.
2. Caspase 3—substrate II, AC-DEVD-AMC.
3. Caspase 8 substrate II, fluorogenic Z-IETD-AFC.
4. Caspase 9 substrate 1, fluorogenic AC-LEHD-AFC (CalBiochem).
5. Prepare 50m*M* stock solution in DMSO.
6. Reaction buffer: 100m*M* HEPES (pH. 7.5), 10% sucrose, 10m*M* dithiothreitol 0.5m*M* EDTA (Sigma).

Mitogen- Activated Protein Kinase (MAPK) Phosphorylation

1. pp38/p38, p-JNK/JNK, p-ERK/ERK antibodies (Santa Cruz Biotechnology, Inc., Santa Cruz, CA).
2. p38 inhibitor SB203580, JNK inhibitor SP600125, ERK inhibitor U0126 (CalBiochem). Stock solutions of 5 mM dissolved in DMSO. Horseradish conju-gated anti-IgG antibody (Sigma)
3. Enhanced chemiluminescence immune blot detection reagent (ECL, Amersham Biosciences Inc.).

Methods

The HepG2 E47 cell line which constitutively expresses human CYP2E1 was established as follows. Human CYP2E1 cDNA, excised from a plasmid p91023(B)-2El (kindly provided by Dr. F. J. Gonzalez, National Cancer Institute, Bethesda, MD) was inserted into the EcoRI restriction site of pCI-neo expression vector (Promega, Madison, WI) in the sense and antisense orientation to form the plasmids pCI-2E1 and pCI-as-2El. The empty pCI plasmid was used to prepare the control, non-CYP2El-expressing HepG2 C34 cells. Transfections of HepG2 cells were carried out with the use of LipofecAMINE reagent (Life Technologies, Grand Island, NY). Cells were then selected with 0.8mg/mL geneticin in the medium, colonies formed from the survivor cells were grown to large scale, and subjected to Western blot analysis and PNP oxidation activity assay to determine CYP2E1 activity. Positive clones were subjected to another two rounds of limited dilution screen-ing to create stable cell lines.

Cell Culture and Storage

1. HepG2 E47 or HepG2 C34 cells are maintained in DMEM containing 10 % FBS and 1 % penicillin-streptomycin-glutamine mixture and 0.1 mg/mL G418. Cells are passaged when approaching confluence by splitting with trypsin/EDTA in 1:5 ratio to provide new maintenance cultures.
2. An aliquot of cells is centrifuged at 600g for 5min, the pellet is resuspended in the cell freezing medium at a concentration of 1×10^6 /mL and aliquoted into cryogenic vials in 1 mL each, and the vials are stored at –20°C for 2 hi and then at –80°C for 1 h and finally the vials are stored in a rack in liquid nitrogen.
3. When experiments are to be initiated, one or two frozen vials of cells are taken out from the liquid nitrogen and immediately put in a 37°C water bath for 5 min. Cells are centrifuged at 1000rpm for 5min, the supernatant is discarded and the cells are resuspended in DMEM in two 75-mL culture flasks. Cells are cultured for about 1 wk, and are now ready to set up new experiments.

CYP2E1 Activity

1. When the culture of HepG2 E47 or C34 cells grown in the 75-mL flask approaches confluence, cells are harvested with a cell policeman and centrifuged 600g for 5min. The cell pellet is resuspended in 5mL of 0.15M KC1 and soni-cated for 10s in an ice bath with a W-375 sonicator (50% duty cycle, output at 4 W) to obtain the cellular lysate.
2. Cell lysate is centrifuged at 9000 g for 15 min, and the supernatant is further centrifuged at 105,000 g for 1 h. Afterward, the microsomal pellet is collected.
3. The microsomal pellet is resuspended in 200 l iL of 0.15M KC1 with a miniho-mogenizer, and the protein concentration is determined by a protein assay kit from BioRad (Hercules, CA).
4. 100 µg of microsomal protein is incubated in 100- µL reaction buffer containing 0.1*M* KH_2PO_4/K_2HPO_4, pH 7.4, 0.4 *mM* p-nitrophenol, 1*mM* NADPH in a 37°C shaking water bath for 1 h.
5. At the end of this period, 30 µL of 20% TCA is added to stop the reaction and the mixture is centrifuged at 10,000 g for 10 min. The supernatant is collected and mixed with 10 µL of 10 *N* NaOH. The absorbance of the pink-yellow colour supernatant is determined in a spectrophotometer at 510nm, and the CYP2E1 activity is calculated using an extinction coefficient of 9.53×10^5 M^{-1} CM^{-1}. The activity is expressed as pmoles/mg/min.

6. HepG2 E47 cell CYP2E1 protein content can be determined using a regular Western blot method with a CYP2E1 antibody.

MTT Assay to Determine Cytotoxicity

1. HepG2 E47 and control C34 cells are seeded onto 24-well plates at concentrations of 1×10^4/1 mL DMEM / well. Cells are cultured for 24 h and the medium is replaced with fresh DMEM.
2. Arachidonic acid (30-60 µM, diluted with serum to prepare 5 *mM* stock solution, aliquot and keep in -20°C), BSO (0.1 mM), (prepare 0.1M stock solution with IX PBS), Ethanol (50-100mM) or iron (25 µM, prepared as Fe/nitrilotriacetate = 1:3 couple at concentration of 25 mM stock solution) are added to the cells for 24 h, or varying time points
3. Two hours before ending the treatment, 75 µL of MTT solution (tetrazolium salts) is added onto each well and continue to incubate at 37°C for 2h. The medium is then discarded followed by adding 1 mL of n-propanol to stop the metabolism, place in room temperature for 1 h in the dark.
4. The absorbance at 570 and 630 nm are recorded. The net $A_{570}–A_{630}$ is taken as the index of cell viability. The net absorbance from the wells of cells cultured with control medium is taken as the 100% viability value. The percent viability of treated cells is calculated by the formula $(A_{570} - A_{630})_{sample} / (A_{570} - A_{630})_{control} \times 100\%$.

LDH Assay to Determine Cytotoxicity

1. HepG2 E47 or HepG2 C34 cells are seeded onto six-well plates at a concentration of 5 × 10V2mL DMEM/well for 24 h. The medium is replaced with fresh medium and treated.
2. At the end of treatment, the culture medium and the cells are collected separately.
3. The cells are resuspended in 0.5 mL of PBS and sonicated for 5 s in an ice bath (d cycle 50 %, output control 4W).
4. Mix 50 µL of medium or cell lysate with 950 µL of LDH reaction solution and measure LDH activity (A Absorbance per min) due to NADH formation by scan-ning in a spectrophotometer at 340 nm, 35°C for 1 min. The Cytotoxicity index is expressed as the ratio of $LDH_{medium} / LDH_{lysate + medium} \times 100\%$.

Trypan Blue Exclusion Assay to Determine Cytotoxicity

1. HepG2 E47 or C34 cells are seeded onto six-well plates at a concentration of 5×10^5 2mL DMEM per well for 24 h. The medium is replaced with fresh medium.
2. Treat the cells.
3. At the end of treatment, collect the medium and cells (trypsin/EDTA harvesting).
4. Collect cells by trypsinization and resuspend the cells in 1 mL of PBS and treat the cells with 0.1 mL of 0.05 % trypan blue for 5min.

5. Count the excluding non-staining or staining cells in a hemocytometer under a light microscope. The cell toxicity is expressed as: Staining $_{total}$/Staining$_{total}$ + Excluding$_{total}$ × 100%.

Lipid Peroxidation Analysis

1. HepG2 E47 or C34 cells are seeded onto six-well plates at a concentration of 5 × 10^5 / well and cultured for 24 hours.Fresh medium is added and the cells are then treated with different toxins
2. Cells are harvested by trypsinization and resuspended in 0.5 mL of PBS containing 0.1 *mM* butylated hydroxytoluene (BHT) and sonicated for 10 s in an ice bath using an Ultrasonic Cell Disrupter (50% duty cycle, output at 6). Protein concentration of the lysate is determined as mentioned previously.
3. Mix 0.5 mg of lysate protein with 0.5 mL of trichloroacetic acid-thiobarbituric acid-HC1 solution (containing 0.1 *mM* BHT) and heat in a 100°C water bath for 1 h.
4. Centrifuge the reaction solution at 14,000rpm in a microcentrifuge for 10 min-utes and collect the supernatant.
5. The formation of TBA-reactive substances (TEARS) is determined by measuring absorbance at 535 nm and using an extinction coefficient of 1.56 × 10^5M/cm to calculate malondialdehyde equivalents (nmoles/mg protein).

Measurement of ROS Production

1. HepG2 E47 or C34 cells are seeded onto six-well plates at a concentration of 5 x 10^5 / 2mL DMEM per well and cultured for 24 h. Replace with fresh medium and treat with different toxins as mentioned in Subheading 3.3., step 2.
2. One-half hour before ending treatment, 2',7'-dichlorofluorescin diacetate is added at a final concentration of 2 µg/mL for 30min.
3. Cells are harvested by trypsinization and resuspended in 3mL Ix PBS (avoid direct exposure to strong light).
4. Fluorescence is immediately determined in a Perkin-Elmer 650-105 fluores-cence spectrometer using wavelengths of 490 nm for excitation and 525 nm for emission. Results are expressed as arbitrary units of fluorescence per milligram of protein.

Measurement of GSH Levels

1. HepG2 E47 or C34 cells are seeded onto six-well plates at a concentration of 5 × 10^5 / 2mL DMEM per well for 24 h and treated with different toxins.
2. Cells are scraped and harvested in 0.5mL Ix PBS and mixed with 0.5mL 10% TCA solution and vortexed briefly.
3. Centrifuge the cell mixture solution at 10,000 rpm for 5min and collect the supernatant.
4. In a 1-mL cuvette, add 900 µL of Buffer A, 100 µL of Buffer B and 50 µL of sample and determine absorbance at 412nm for 2min.

5. Prepare a standard curve with four different concentrations of reduced form of glutathione from 0 to 100 μM in 5% TCA and determine the absorbance at 412 nm for 2min as step 4 and evaluate the GSH levels (nmoles/mg protein) by comparing with the standard curve.

Mitochondrial Membrane Potential Analysis

1. Mitochondria membrane potential is analyzed from the accumulation of rhodamine 123. Cells are seeded onto six-well plates at a concentration of 5×10^5 well for 24 h and treated as detailed previously.
2. One hour before ending treatment, rhodamine 123 is added at a final concentration of 5 μg/mL for 1 h.
3. Cells are harvested by trypsinization and resuspended in 1 mL of PBS containing 25 μg/mL PI and the intensity of the fluorescence from rhodamine 123 and PI is determined by Fas-Scan Flow Cytometry. Cells with high mitochondrial potential and viable are mainly in the lower right quadrants of the PI vs rhodamine 123 curves.

Apoptosis Determination by Flow Cytometry

1. E47 or C34 cells are grown in 6 well plates and treated with different toxins as mentioned previously.
2. Harvest the cells by trypsinization and fix with 70% ethanol for 15 min at room temperature.
3. Cells are washed with PBS and incubated with 50 μg/mL RNase A for 2h at room temperature.
4. Stain the cells with 50 μg/mL PI for 30 min and the fluorescence (excitation, 488 nm; emission, 575 nm) is measured with a flow cytometer. The cells under-going apoptosis are calculated from the sub-G hypodiploid area programmed by the analyzer.

Apoptosis Analysis by Determination of DNA Fragmentation (DNA Ladder)

1. E47 or C34 cells at concentration of 5×10^5/2mL DMEM per well are grown on six-well plates and treated as detailed previously.
2. The cells are harvested and resuspended in 1 mL of'lysis buffer and incubated for 1 h at 50°C.
3. The lysis samples are then extracted with 2mL of phenol (neutralized with TE buffer, pH 7.5), followed by extraction with 1 mL of chloroform/isoamyl alcohol (24:1).
4. The aqueous supernatants are precipitated with 2.5 volumes of ice-cold ethanol plus 10 % volume of 3M sodium acetate, pH 5.2, at –20°C overnight.
5. Centrifuge the samples at 13,000g for 10 min and the pellets are airdried and resus-pended with 50μL of TE buffer, pH 7.5, supplemented with 0.1 μg/mL RNase A.

6. Samples are electrophoretically separated on a 1.5% agarose gel in 0.5 × TBE buffer containing 1 µg/mL ethidium bromide,at 100 V for 2h.
7. Pictures of the separated DNA on the gel are taken under UV transillumination

Immunohistochemistry Analysis to Determine Nuclear DNA Fragmentation

1. Nuclear DNA fragmentation is determined by a DNA Fragmentation Detection Kit from CalBiochem. HepG2 E47 cells or C34 cells are grown on a cover slide in six-well plates and treated with different toxins as mentioned previously.
2. After treatment, the slides are fixed with 4% para-formaldehyde at room temperature for 15 min and are dehydrated by using ethanol (concentrations are from 70% to 95%).
3. Permeabilize the cells with proteinase K. Inactivate endogenous peroxidase by incubating the cells on the coverslide with 3 % H_2O_2.
4. The fragmented DNA is labeled with labeling solution of the kit and the labeled DNA is mixed with a conjugate solution provided by the kit.
5. Incubate the cells on the slide with 3, 3'-diaminobenzidine (DAB)/H_2O_2/urea solution (provide by the kit).
6. Nuclear DNA fragmentation is detected by stained dark-brown cell nuclei and the number of stained cells is counted under a light microscope.

Caspase Activity

1. Cells are grown on six-well plates at the concentration of 5×10^5 / 2mL DMEM per well and treated with different toxins. Cells are scrapped and harvested.
2. The cell pellet is suspended in 300 µL of lysis medium and incubated for 30 min at room temperature.
3. The cell suspension is centrifuged at 200g for 5mins and the supernatant is collected.
4. 100µg of supernatant protein is added into 0.5 mL of the assaying solution con-taining 50 μM caspase substrates (caspase 3- substrate II, AC-DEVD-AMC; casepase 8 substrate II or Granzyme B substrate II, Z-IETD-AFC; caspase 9 substrate I, fluorogenic, AC-LEHD-AFC respectively, CalBiochem) and incubated for 12h in a 30°C shaking water bath overnight.
5. Caspase activities are determined by measuring the. fluorescence associated with released AMC (7-amino-4-methyl-coumarin, excitation at 380nm, 'emission at 460 nm) or AFC (excitation at 400 nm, emission at 505 nm) in a spectrofluorim-eter. Results are expressed as arbitrary units per 100(J.g cell lysate protein.

Activation ofp38 MAPK

1. E47 or C34 cells are grown on six-well plates for 24 h, and fresh medium is added.
2. The cells are treated with either 60 μM AA or 0.1 mAf BSO in the absence or presence of 5 μM of the p38 MAPKJnhibitor SB203580 for 0.5, 1.0, 2.0, 4.0, 8.0, or 12h, respectively

3. Collect the cell pellets, resuspend in 200 µL of PBS and sonicate for 10s in an ice bath (50% duty cycle, output at 5 – 7 W).
4. Prepare a 8% SDS-polyacrylamide gel, loading 20 µg of each lysate protein onto the gel and carry out the gel electrophoresis for 2 h at 50 V.
5. The blotted membranes are incubated with pp38 MAPK antibody (Santa Cruz, 1:200 diluted) for 4h at room temperature (or overnight in the cold room).
6. Wash the blotted membranes with PBS containing 0.05 % Tween 20 and incu-bate with anti-IgG antibody conjugated with horseradish peroxidase (Sigma, 1:10,000 diluted) for Ih.
7. Wash the membrane and develop the fluorescence using the enhanced chemiluminescence immune blot detection agent. The arbitrary units of density of each lane are determined after scanning with computer software Image! (National Institutes of Health).
8. The blotted membranes are washed in a Re-Blot Plus Strong Solution for 15 min-utes to wash out the previous bound pp38 MAPK antibody and re-incubate the membrane with p38 MAPK antibody, and repeat step 6 and 7 to determine the total p38 MAPK. The final pp38 MAPK activation levels are expressed as the ratio of pp38 MAPK/p38 MAPK.
9. Similar procedures can be carried out to determine the activation of JNK or ERK or Akt by use of specific antibodies recognizing these kinases.

NOTES

1. HepG2 E47 cells are a human CYP2E1 cDNA transfected cell line; because of different growth conditions at different times and altered cell density, medium nutrition, and other conditions, the CYP2E1 expression levels may vary considerably. To insure the cells are expressing high CYP2E1 before initiating the experiment, the CYP2E1 PNP oxidation activity should be determined regularly (at least weekly)

   Because the preparation of microsomes from the cells usually is time consuming, in such case, cell lysate can be used for quick determination of CYP2E1 activity but the lysate protein used for this measurement should be three to five times more than the microsomal protein (100 µg per 100 µL) level. To validate that the oxidation of pnitrophenol is via CYP2E1, one can evaluate whether an inhibitor of CYP2E1 such as 1 *mM* diallyldisulfide or 0.1 *mM* diethyldithiocarbamate or anti-CYP2El antibody blocks this oxidation.
2. The medium should contain high concentrations of G418 (0.1-0.4 mg /mL) to exclude those cells which do not express CYP2E1. Because G418 is a very expensive antibiotic, for long-term use of these cell lines according, our experi-ence, G418 can be used for 1 wk per month in the culture medium to maintain this cell line.
3. We generally find that 2wk after thawing out fresh cells is an appropriate time when CYP2E1 levels are high to carry out experiments. We typically discard cultures 4 to 6 wk after thawing. Thus, it may be helpful to thaw out new aliquots and start new cultures about a week before discarding the old culture to keep experiments moving ahead.
4. To use E47 cells for toxicology studies, the initial seeding amount of the cells should be controlled carefully. Too many cells increase the density which will enhance resistance to the added toxin

which will lower the toxicity index. In many cases, if the treatment period is not longer than 24 h, low serum medium can be used (2 %). This will enhance sensitivity of the cells to added toxins. In this case, control groups should be carefully set up.

5. To validate a role for oxidant stress or caspase or p38 MAPK activation in the cytotoxicity produced by the addition of ethanol or AA or iron or BSO to the E47 cells, the decline in viability as reflected by the decrease in MTT reduc-tion or the increase in the LDH out/LDH total ratio or the increase in trypan blue uptake should be blocked by, e.g., trolox and oc-tocopherol, or N-acetyl-cysteine and GSH ethyl ester or by SB203580 or Ac-DEVD- FMK and Z-VAD-FMK. Inhibitors of other MAPK, e.g., PD98059 for ERK 1/2 or SP600125 for JNK can be studied to assess any specificity for p38 MAPK in the overall toxicity.
6. For the determination of lipid peroxidation, 0.1 *mM* BHT should be added in the PBS used to sonicate cells to prepare cell lysate, and in the incubation buffer when heating the reaction in boiling water in order to prevent artifactual lipid peroxidation. Besides the cells, lipid peroxidation in the culture medium can be determined because peroxidized lipid products, e.g., lipid hydroperoxides and malondialdehyde, can be secreted from the cells into the medium. We have found approx 15% of total lipid peroxidation products are present in the medium when E47 cells were treated with 60*iM* AA for 24h. To validate that the TEARS assay is reflecting lipid peroxidation, antioxidants such as 0.1 *mM* trolox, or a-tocopherol can be added to inhibit the absorbance at 535 nm. To validate the 2',7'-dichlorofluorescin fluorescence reflects ROS production; anti-oxidants such as 1 to 5 mM N-acetylcysteine or GSH ethylester can be added to inhibit the arbitrary units of fluorescence.
7. Mitochondrial membrane potential is analyzed by double staining with rhodam-ine 123 and PI and fluorescence of the two dyes is detected by flow cytometry. When preparing the sample for determination by flow cytometry, make sure to prepare four additional samples that are nonstained, stained by rhodamine alone, stained by PI alone, and stained by both rhodamine and PI as reference control to adjust the flow cytometer for the following measurements.
8. To validate that the DNA fragmentation or the PI fluorescence reflects apoptosis, inhibitors of specific caspases, e. g., 0.05 mM Ac-DEVD-FMK (caspase inhibi-tor) or a pan caspase inhibitor such 0.05 mM Z-VAD-FMK can be added. These inhibitors should also prevent any increases in caspase 3 (or other caspase) activi-ties. Moreover, these inhibitors would provide evidence that any apoptosis is caspase-dependent, since caspase-independent apoptosis may also occur.
9. MAPK in E47 cells can be activated at very early time periods (as early as 15min) depending on different toxins. P38 MAPK can be activated by treatment with 60µM AA in 0.5 h, but treating the cells with 0.1 mM BSO requires 2h or longer to activate p38 MAPK. To study the activation of MAPK in E47 cells by different toxins, a time and dose course study should be conducted to select the correct conditions that may active MAPK in E47 or C34 cells.

Index

A

Absorption isotherm, 2, 3
Acetylene reduction test, 117
Active chromatin, 225
Active glutamine synthetase, 130
Adenoviruses, 23
Adenylyltransferase, 131
Affymetrix arrays, 81
Affymetrix GeneChip, 81, 85
Affymetrix microarray, 81
Agmatine cycle, 145
Allosteric effectors, 38
Allosteric interface, 36
Allosteric regulators, 38
Amiloride, 100
Amino, 39
Animal models, 80
Ankyrin repeat structure, 112
Anticooperativity, 7
Antizyme, 147
Apoptosis analysis, 322-329
Apple, 46
Arabidopsis, 157
Arginine, 137, 138, 142
Argininosuccinate, 140
Arigininosuccinic aciduria, 141
Arrestin, 105
Artificial chromosomes, 265
Artificial rearing, 304-311
    artificial rearing, 306
    components, 306
    fatal alcohol spectrum disodgers, 304
    gastrostomy surgery, 307-311
    milk diet, 305
Asparagine synthetase, 128
Asparagines, 148, 149
Aspartame, 99
Aspartate carbamoyltransferase, 26
Aspergillus, 145
A. flavus, 299
Assimilatory nitrate reductases, 126
Atomic motion of retinal, 108
Autoimmune diseases, 218
Azospirillum brasilense, 130
Azotobzcter vinlzndii, 125

B

Bacillus subtilis, 219, 247
Bacillus, 151
Bacterial pili, 15-16
Bacteroids, 117
Baseplate, 56
B-cell receptor, 212
Benzimidazole, 57
Binding constant, 7
Biosynthesis of glutamate, 135
Bloom syndrome, 248
Bohr effect, 38, 39
Bohr proton, 41
Boltzmann's constant, 1
Brain, 80

C

Caenorhabditis elegans, 230
Calcineurin, 209

Camptothecin, 251
Cancer, 218
Capsaicin receptor, 113
Carbamoyl phosphate synthetases, 137
Carbihydrates, 99
Carbon dioxide, 39
Carbon monoxide, 40
Carcinoembryonic antigen, 218
Carnitine, 151
Catabolism of arginine, 142
Cell culture, 80
Cell mediated immunity, 211
Central nervous system, 80
Centromeres, 231
CG doublets, 236
Chaperonins, 52
Chlorocruorins, 42
Citrulline, 137, 138, 140
Clostridium ammobutylicum, 134
C. pasteurianum, 117
C. propionicum, 166
Cochlea, 108
Coiled peptide chain, 11
Colour blindness, 111
Concanavalin, 19
Cone-pigment absorption spectra, 110
Cooperative binding, 10
Cooperative processes, 9-11
Creatine, 143
Cyanide, 40
Cyclocreatine, 143
Cyclosporin, 209
Cysteine desulfurase, 181

D

Decarboxylation, 132
Defensive system, 197-220
  antibodies, 203, 210
  antibody diversity, 207
  – like proteins, 212
  antigen-binding units, 198
  antigens, 203, 205
  B cells, 197
  Class II MHC proteins, 214
  combinatorial association, 208
  cytotoxic T cells, 214
  diverse MHC proteins, 216
  foreign peptides, 215
  gene rearrangements, 204
  HIV, 216
  hypervariable loops, 200
  ITAM, 208
  light chain expression, 211
  major histocompatibility complex proteins, 210
  oligomerization of antibodies, 208
  peptides, 211
  principal of evolution, 197
  self antigens, 217
  somatic mutation, 208
  specific molecules, 201
  sulfur compounds, 219
  X-ray analysis, 202
Dexyhemo-globin, 39
Difluoromethylornithine, 148
Dihedral symmetry, 18-19
Dissimilarity nitrate reductases, 126
Dissimilarity nitrite reductases, 126
Dissociation constant, 6, 7
DNA, 11, 15, 20, 43, 45, 80, 221, 222, 223, 313-321
  base excision repair, 292
  causes of mutations, 288
  damage, 288
  fidelity of replication, 289
  gyrase, 251
  human hereditary defects of DNA repair, 303
  ligases, 246
  of organelles, 235
  poly(ADP-ribose), 296

reaction of alkylated bases, 294
repair, 288
– of damaged DNA, 290
– – double strand breaks, 295
SOS response, 296
Dogfish, 139
Dreyer, William, 206

E

E. coli, 16, 26, 126, 130, 131, 139, 148, 149, 216, 221, 246, 247, 248, 249, 250, 252, 254, 255, 257, 270, 274, 275, 289, 291, 293, 296, 302
Effector functions, 198
Erythrocrurins, 42
Escorts, 43
Ethanol in culture, 313-321
application of exogenous substances, 320
culture models, 313
culturing methods, 315
dissociated cell cultures, 313
organotypic slice cultures, 314, 317
primary cultures of cortical neurons, 314
yield and plating of the cells, 316
Ethanol, 80
Eukaryotic viruses, 265
Excretion of ammonia, 141
Excretory system, 116-180
alanine, 160
amidino transfer, 143
amino acids, 164
arginine, 133
assembly of iron-sulfur centers, 180
bile pigments, 175-177
biosynthesis, 145
branched chain amino acids, 160
breakdown, 147
carnitine, 149
catabolism, 162-164
– amino acids, 132
– glutamate, 132
– glutamine, 132
– of lysine, 152, 154
chlorophyll, 174
compounds derived from aspartate, 148
control of biosynthetic reactions of aspartate, 149
corrins, 173-174
creatine synthesis, 143
cysteine, 177, 179-180
dipicolinic acid, 149
fixation of N_2, 116
glucogenic, 164
glutamate dehydrogenase, 129-130
– synthetase, 129-130
glutamine synthetase, 130
glycine, 164, 167, 168-170
homocysteine, 155
incorporation into amino acids, 127
– of ammonium ions, 126-127
– – NH, 127
– – proteins, 127
interconversion of nitrate, 126-127
– – nitrite, 126-127
isoleucine, 165
ketogenic, 164
leucine, 165
lysine diaminopimelate, 149
mercaptopyruvate, 180
methionine, 155
nitrogen cycle, 116
genetic engineering, 125
legume nodules and cyanobacterial heterocysts, 123-125
mechanism of nitrogenase action, 120-123
nitrogen fixation genes, 123
nitrogenases, 118
reduction of elemental nitrogen, 116-118
ornithine, 133
plant hormone athylene, 157
polyamines, 133, 144, 145

porphobilinogen, 170
porphyries, 174
porphyrins, 170, 171-173
praline, 133
related substances, 170
salvage pathway, 157
serine, 164, 166, 167
sulfur metabolism, 177
synthesis and catabolism of praline, 136
– of arginine, 137
– – ornithine, 137
– – urea cycle, 137
systeine sulfinate, 180
taurine, 180
thiosulgate, 180
threonine, 159
uptake of amino acids by cell, 129
valine, 165
Exons, 235

F

FeMo-coenzyme molecules, 119
Fermentation of glutamate, 133
Fetal alcohol spectrum disorder, 58
Fibrous tissue kit, 85
Filamentous bacteriophages, 14-15
Filaments of muscle, 16
Fingerprinting, 41
Flagella, 25
Flavonoid compounds, 124
Functional magnetic resonance imaging, 95
Functionalis glands, 78
Functionalis stroma, 49

G

Gene expression in a complex tissue, 65-79
animal model, 66
BAT2, 78
cDNa amplification strategy, 71
cloning, 72
differential display, 72
gene microarray analysis, 76
inadequate secretory phases, 77
laser capture microdissection (LCM), 67-69, 78
analysis of endometrial tissue, 69-70
and differential display, 77
microarry analysis of endometrial cells, 78
primate endometrium, 65
spstiotemporal regulation, 65-66
sequencing, 72
serum P level, 76
SLPI, 78
syncytin, 78
synthesis and amplification of cDNA populations, 70
WFDC2, 78
window of receptivity, 73-76
Gene expression, 80-90
affymetrix specific quality control metrics, 85
scaling factor, 85
statistical difference thresholds, 86
per cent present, 86
GAPDH control expression ratios, 86
amount of biological material, 83-84
experimental technique, 84
bioinformatics and network analysis, 89
DNA microarrys, 81
experimental design, 82
generalized control measures, 86
box-and-whisker plot, 86
microarry quality control, 85
multivariant analysis, 88-89
number of replicates, 82-83
primary analysis, 87
protocol, 85
randomization, 84
RNA quality control, 84
scatter plot, 86
reagents, 82
statistical filtering, 87-88

validation of candidate genes, 89
Gestation, 98
Giardia, 142
Gibbs energy, 1
G-segment, 251, 252
Guanisinoacetic acid, 143
Gustducin, 99

H

Hair cells, 108, 111
Hair cell transtuction, 112
Helical structures, 13-14
Helicases, 247
Hemerythrins, 42
Hemocyanins, 42
Hemoglobin Creteil, 41
Hemoglobin Suresnes, 41
Hemoglobins M, 41
Hemoglobins S, 41
Hill plot, 10, 11
Homozygote, 41
Human leukocyte antigen, 211
Hyperbolic, 3
Hypusine, 148

I

Icosadeltahedra, 23
Icosahedron, 19
Imidazole, 39
Immune system, 218
Immunoglobulin domains, 199
Immunoglobulin, 199, 200, 203, 234
Immunological responses, 183-196
adoptive transfer, 185
antibody incubation, 187
B cell depletion of splenocytes, 188
B cells, 187
column fractionation of spleen cell suspension, 187
– purification, 187
culture plate, 189
cytokine response of cell, 184, 189
flow cytometry staining, 184, 188
isolation, 186
lymph node, 184
magnetic column elution, 187
microbead incubation, 187
spleen, 184
– cell suspension, 184
T cells, 184, 190
thymus, 184
Initiator proteins, 252, 253
Insecticidai analogs of arginine, 142
Insovaleric academia, 219
Insulin, 21
Intercalation, 56
Intragastric incubation, 58-64
alcohol administration, 59-63
evaluation of alcohol administration, 60
measurement of BACs, 59-63
Intrinsic binding constant, 5
Introns, 235
Inversin, 79
Iron protein, 118
Isologous bond, 17-18

J

Jamaica, 219

K

Killer plasmid, 145
Kinetic energy, 1
Klebsiella pneumoniae, 124

L

Lachrimatory factor, 220
Lactate dehydrogenase, 18
Lactic academia, 219
Lamina, 230
Lamprey hemoglobin, 31
Leghemoglobin, 117
Leghemoglobins, 42
Leishmania, 236

Lipoic acid, 162
Listeria monocytogenes, 183, 185, 190
Lock washer, 25
Long interspersed repeat sequences, 233
Lucina pectinata, 41
Lysine, 148

M

Macromolecules, 1-57
 abnormal human hemoglobins, 41
 activated cell surface receptors, 46
 allosteric regulators, 38
 anticancer drugs, 57
 bacteriophages, 43
 binding equilibria, 1-4, 29
 Bohr effect, 38
 comparative biochemistry of hemoglobin, 42
 complementarity, 11
 confirmation state, 31
 conformational change, 28
 cooperative changes in conformation, 26
 cytoskeleton, 47
  intermediate filaments, 47
  microfilaments, 48
  microtubules, 49
 electrostatic repulsion, 7
 higher oligomers, 32
 induced fit model, 31
 induced fit, 27
 isologous dimmers, 28
 kringles, 46
 life and death for proteins, 52
 malaria, 53-55
 microscoping binding, 5, 8
 mitosis, 57
 multiple binding sites, 4-9
 MWC model, 29
 oligomers with two-fold (Dyad) axes, 17
 oxygen-carrying proteins, 33
  binding of oxygen, 34
  hemoglobin, 33-36
  myoglobin, 33-36
  oxygenation of hemoglobin, 36
  structutal changes, 36
 packing, 11-26
 rings helices, 12
 scaffolding, 44
 self-assembly of macromolecular structures, 42
 sickle cell disease, 53-55
 statistical effects, 5
 substitutes, 53-55
 tetraploid, 57
 T-even bacteriophages, 56
 unequal binding of substrate, 27
 unsymmetric dimmer, 28
Magnetosomes, 113
Maple syrup urine disease, 219
Metal sulfide clusters FeMo-protein, 121
Methemoglobin, 40, 41
Methionine, 148
Microarry, 80
Microdissection, 80
Microtubule organizing, 50
Microtubule-associated proteins, 50
Molecule, 26
Molecules with cyclic symmetry, 12
Multienzyme complexes, 26
Mycoplasma genitalium, 125

N

N-acetyl-glutamate, 139
Nasal epithelium, 96
Nerve impulses, 96
Neural cell, 80
Neurospora, 145, 148
Nif genes, 124
Nitric oxide, 40
Noxious stimuli, 112
Nuclear estrogen receptor, 66
Nuclear membrane, 230

Nucleic acid, 221-303
bacterial chromosomes, 222
cell nucleus, 228
chloroplast DNA, 263
chromosome ends, 261
deficiencies in human DNA repair, 301-303
DNA, 268
recombination, 268-274
DNA in viruses, 221
Drosophila, 285, 286
Drosophila, 301
elongation of DNA chains, 256
environment of DNA, 221
eukaryotic nuclear DNA, 264
families of DNA polymerases, 242
gene clusters, 234
genes for ribosomal RNA, 233
Genetic recombination, 287
histones, 223
insertion sequences, 280
linker histones, 224
mariner, 286
MCM, 266
mitochondrial DNA, 263
mutagens in the environment, 296-301
neurospora, 232
non-histone nuclear proteins, 229
non-reciprocal recombination, 274
nuclear matrix, 228
nucleosomes, 223
ORC, 266
organization of DNA, 230
packaging , 260
plasmids, 222, 260
protamines, 223
pseudogeges, 234
repetitive DNA, 231
replication, 238
autoradiography, 239
chemistry of DNA polymerization, 240
DNA polymerases, 241
editing, 241
exonuclease activities, 241
proofreading, 241
RNA primers, 240
– of bacterial DNA, 253-259
– – viral DNA, 259-260
termination of replication, 256
topology of DNA, 221
transposons, 280, 285
viral DNA, 264
Nucleolus, 230
Nucleosomes, 224

O

Odorant receptors, 96, 97
Odorant, 94
Okazaki fragments, 240
Olfaction, 91
Olfactory neuron expression, 94
Olfactory receptor activation, 97-98
Oligomers
asymmetry and quasi equivalence, 20-23
with cubic symmetry, 19
Ornithine decarboxylase, 148
Overlapping genes, 235
Oxoacid dehydrogenases, 26
Oxygen binding, 36

P

Paired interactions, 17
Pinospin, 106
Polarity, 50
Polyclonal stimulation, 189
Polyhedra, 19, 23
Polymerases tappisomerases, 247
Polymerases, 244-245
Pores, 230
Primuses, 252
Proline oxidase, 136
Protesomes, 52

Protomer, 12
Protons, 8, 9, 11, 39
Pseudomonas putida, 142, 153
Putrescine, 56, 146
Pyrimidines, 148

Q

Quasi-equivalence in virus coats, 23

R

Regulatory subunits, 26
Replication of bacterial DNA, 253-259
Replication of nuclear DNA, 265
Retinal-lysine linkage, 108
Reverse gyrase, 251
Rhizobium, 120
Rhodopsin absorption spectrum, 107
Rhodopsin kinase, 105
Ribonucleases, 252
RNA, 11, 20, 25, 81, 233
RNA fatty and fibrous tissue kit, 85
RNA molecules, 233
Rod-shaped plant virus, 15

S

Sacchammyces cerecisiae, 53, 232
Saccharin, 99
Saccharomyces cervisiae, 141
Salmonella typhimurium, 44
Salmonella, 300
Satellite tobacco, 20
Saturation fraction, 2, 3
Saturation, 2
Scapharca inaequivalvis, 42
Scatchard plot, 3
Segmental flexibility, 199
Selective stickness, 11
Selenophosphate, 182
Self-assembly, 11
Sensory systems, 91-115
  anosmias, 93
  bitter receptors, 99-100
  color perception, 91
  combination mechanism, 94-95
  – of senses, 98
  effects of hydrogen, 101
  glutamate receptor, 101
  magnetic resonance imaging, 95
  mechanical stimuli, 115
  olfaction, 114
  organic compounds detected by olfaction, 91
  photoreceptor molecules, 102
    color vision, 106
    detection of mechanical stimuli, 107
    green pigments, 106
    light absorption, 104-105
    light-induced the calcium level, 105
    mechanosensory channels, 109
    red pigments, 106
    rhodopsin, 102-104
    stereocilia, 108-109
  photoreceptor, 114
  salty tastes, 100
  sensory connection, 92
  seven-trans-membrane-helix receptors, 93-94
  signal transduction, 114
  sweet compounds, 100
  Taste, 114
  touching sensing, 111-113
  touching sensing: painful stimuli, 112-113
  touching sensing: subtle sensory systems, 113
Short interspersed sequences, 232
Sialomucin, 79
Sigmoidal, 3, 9
Significance analysis of microarrays, 88, 89
Single strand binding proteins, 246
Siroheme, 126
Southern bean mosaic virus, 23
Spermidine, 56
Squalus acanthias, 139
Stepwise formation constant, 6
Stereocilia, 108, 111

Streptococcus infections, 218
Streptomyces griseus, 142
Streptomyces thermoautotrophicus, 125
Sucrolose, 99
Sulfide ions, 41
Sulfonylureas, 162

T

Target average intensity, 85
Taxol, 57
T-cell receptors, 197, 200, 213
Telomeres, 232
Temperate bacteriophage, 284
Thermoplasma acidophilum, 53
Thermus aquaticus, 243
Thiamin, 9
Thiosulfate sulfurtransferase, 181
Thymus, 217
TIGR multi-experiment viewer, 88, 89
Tomato bushy stunt virus, 23
Treadmilling, 50
Trypanosomes, 148
Trypanosorna, 236
T-segment, 251, 252primosome, 253
Tublins, 50

U

Urea carboxylase, 141
Urea cycle, 137, 140
Urea, 128
Usher protein, 43

V

Vibrio cholerae, 12
Vinblastine, 57
Vincristine, 57
Visual pigments, 110
Visual signal transaction, 110
Vomiting sickness, 219

W

Wart virus, 23
Wernwe syndrome, 248

X

Xenopus, 226, 229, 233, 287

Y

Yeast, 53

Z

Zinc metalloprotein, 56
Zonation of the primate endometrium, 66

❑❑❑